Photovoltaics: Engineering and Technology for Solar Power

Photovoltaics: Engineering and Technology for Solar Power

Edited by Catherine Waltz

SYRAWOOD
PUBLISHING HOUSE

New York

Published by Syrawood Publishing House,
750 Third Avenue, 9th Floor,
New York, NY 10017, USA
www.syrawoodpublishinghouse.com

Photovoltaics: Engineering and Technology for Solar Power
Edited by Catherine Waltz

International Standard Book Number: 978-1-68286-456-2 (Hardback)

Cataloging-in-publication Data

Photovoltaics : engineering and technology for solar power / edited by Catherine Waltz.
     p. cm.
Includes bibliographical references and index.
ISBN 978-1-68286-456-2
1. Photovoltaic power generation. 2. Solar energy. 3. Solar cells. 4. Direct energy conversion. I. Waltz, Catherine.
TK1087 .P46 2017
621.312 44--dc23

Printed in the United States of America.

TABLE OF CONTENTS

PREFACE

While understanding the long-term perspectives of the topics, the book makes an effort in highlighting their impact as a modern tool for the growth of the discipline. This book is a valuable compilation of topics, ranging from the basic to the most complex advancements in the field of photovoltaics which are deployed for the transformation of light into electricity with the help of semiconducting materials. From theories to research to practical applications, case studies related to all contemporary topics of relevance to this field have been included in this book. It aims to shed light on the various applications of photovoltaics that are used across the world for example; rooftop and building integrated systems, standalone systems and power stations, etc. As this field is emerging at a rapid pace, the contents of this book will help the readers understand the engineering and technology of photovoltaic and solar energy.

This book is the end result of constructive efforts and intensive research done by experts in this field. The aim of this book is to enlighten the readers with recent information in this area of research. The information provided in this profound book would serve as a valuable reference to students and researchers in this field.

At the end, I would like to thank all the authors for devoting their precious time and providing their valuable contributions to this book. I would also like to express my gratitude to my fellow colleagues who encouraged me throughout the process.

Editor

Geometrical optimization and electrical performance comparison of thin-film tandem structures based on pm-Si:H and μc-Si:H using computer simulation[*]

F. Dadouche[1,a], O. Béthoux[2], M.E. Gueunier-Farret[2], E.V. Johnson[3], P. Roca i Cabarrocas[3], C. Marchand[2], and J.P. Kleider[2]

[1] Institut d'Électronique du Solide et des Systèmes (InESS), CNRS, 23 rue du Loess, BP 20 CR, 67037 Strasbourg Cedex 2, France
[2] Laboratoire de Génie Électrique de Paris, CNRS UMR 8507, SUPELEC, Université Paris-Sud, UPMC Univ Paris VI, 11 rue Joliot-Curie, Plateau de Moulon, 91192 Gif-sur-Yvette Cedex, France
[3] Laboratoire de Physique des Interfaces et Couches Minces, École polytechnique, CNRS, 91128 Palaiseau, France

Abstract This article investigates the optimal efficiency of a photovoltaic system based on a silicon thin film tandem cell using polymorphous and microcrystalline silicon for the top and bottom elementary cells, respectively. Two ways of connecting the cells are studied and compared: (1) a classical structure in which the two cells are electrically and optically coupled; and (2) a new structure for which the "current-matching" constraint is released by the electrical decoupling of the two cells. For that purpose, we used a computer simulation to perform geometrical optimization of the studied structures as well as their electrical performance evaluation. The simulation results show that the second structure is more interesting in terms of efficiency.

1 Introduction

The technological progress that has been made in the development of thin film silicon solar cells has led to a significant reduction in the cost per peak watt generated by such devices. Thin film silicon materials such as hydrogenated amorphous (a-Si:H), polymorphous (pm-Si:H) and microcrystalline silicon (μc-Si:H) have become a serious alternative to monocrystalline silicon for the fabrication of solar cells, as the production cost can be drastically reduced through numerous mechanisms: (i) in contrast to the high temperature process (>1400 °C) used in preparing mono or polycrystalline silicon, the plasma enhanced chemical vapor deposition (PECVD) technique, which is a widely used deposition process to fabricate thin film solar cells, needs relatively low temperatures (<300 °C); (ii) the thin film semiconductor can be deposited directly on low-cost large-area substrates; (iii) high deposition rates combined with low defect density silicon thin films have been obtained using PECVD or other deposition techniques [1–4] leading to good efficiency solar cells.

To further decrease the cost per watt for thin film devices, a common design strategy is to increase conversion efficiency through the use of multijunction cells. In tandem devices, two PIN cells made with materials of different bandgap energies are fabricated in series [5–7]. A low bandgap material such as microcrystalline silicon can be used as the bottom cell in conjunction with amorphous silicon (top cell) to extend the spectral range of high collection efficiency [5,8]. Also used as the top cell in such structures is polymorphous silicon, a nanostructured material deposited by PECVD at high pressure and RF power, in a regime where silicon clusters and nanocrystals synthesized in the plasma contribute to the growth along with silicon radicals [9]. It has been reported in previous studies that pm-Si:H has better electronic properties and stability than conventional a-Si:H [10–13]. Moreover, the pm-Si:H optical gap being slightly larger than the a-Si:H one, the use of pm-Si:H in a tandem structure in place of a-Si:H allows one the possibility to increase the open-circuit voltage of the entire device and therefore to increase the electric output of the photovoltaic modules.

We present herein a comparative numerical modeling study of two tandem pm-Si:H/μc-Si:H cell structures: (1) a conventional tandem cell for which the two elementary PIN cells are superimposed by successive layer deposition; and (2) an assembly of two electrically decoupled PIN cells. After having introduced the two structures in detail, including the relevant cell parameters, we report

[*] This article has been previously published in PV Direct, the former name of EPJ Photovoltaics.

[a] e-mail: foudil.dadouche@iness.c-strasbourg.fr

on the simulation procedure used to optimize the power delivered by each structure, and finally discuss our results.

2 Technical details

In a conventional tandem device, the two elementary cells are directly stacked by successive layer deposition, which means that they are both optically and electrically coupled [5–7]. To provide the current to the load, this structure requires only two contact electrodes connected to the top cell P layer and to the bottom cell N layer. In the following, we will call this design the "two-wire structure". Using this interconnection design, the two cells are physically series connected and thus have the same current flowing through them. This introduces an important constraint, because the thickness of each cell has to be precisely chosen in order to share the same short circuit current. Otherwise, the cell with the higher short circuit current will have to work at a shifted operating point due to the lower current of the other cell. This will lead to a degraded performance compared to the optimal, current-matched situation.

This design requirement leads to the idea of an electrical cell decoupling in order to independently target the maximum power of each cell, the current-matching constraint being released. In this configuration, each cell has its own electrodes connected to its own P and N layers. Thus, the two cells in such a combination are optically coupled and electrically decoupled, and we will refer to this design as a "four-wire structure". These two tandem cell structures are presented in Figure 1.

We focus here on tandem pm-Si:H/μc-Si:H cells. The pm-Si:H PIN cell needs a small intrinsic layer (i-layer) thickness (several hundreds of nm) to convert its useful spectrum. This property is linked to the high absorption coefficient of this material due to its direct-like band-gap. On the other hand, the μc-Si:H PIN cell requires a thicker i-layer (a few μms) so as to compensate its lower absorption coefficient. In the tandem configuration, the solar spectrum is more used more efficiently, as the top pm-Si:H cell will absorb the energy of photons with less thermalization loss, whereas the bottom μc-Si:H cell will transfer the infrared energy that would normally go unabsorbed.

In order to quantify the power benefit one can expect from using the four-wire structure instead of the traditional two-wire one, we have used numerical modelling software dedicated to studying heterojunction solar cells, "AFORS-HET" (Automat FOR Simulation of HETerostructures). This software has been developed by the Hahn-Meitner Institut (now Helmholtz Zentrum) in Berlin [14]. Macroscopic characteristics of different layer structures and layer interfaces can be simulated in the dark or under illumination, taking into account optical reflections at any existing interfaces. A different sub-gap defect density spectrum can be introduced for each layer.

In the case of a two-wire structure, the tunnel-recombination effect which occurs at the N(pm-Si:H)/P(μc-Si:H) interface and which allows the passage of current between the two sub-cells is not

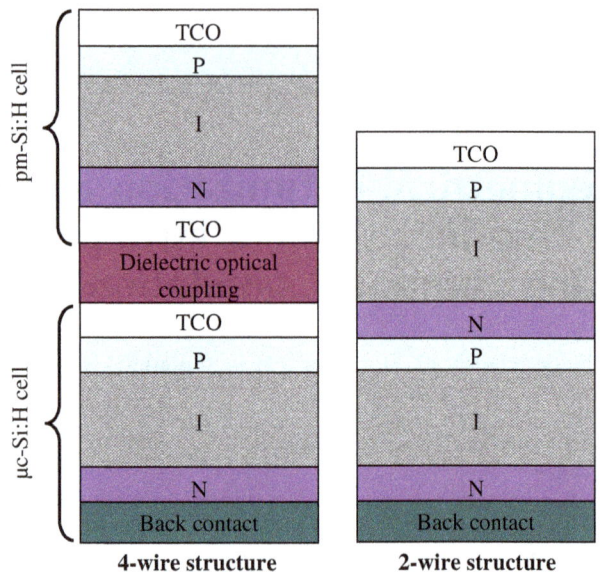

Fig. 1. Two-wire versus four-wire tandem structures.

included in the simulation. Thus, it is well adapted to the simulation of independent individual solar cells (four-wire structure) since the elementary cells are decoupled. For the two-wire structure simulation, we calculated the J-V characteristics of each cell separately, taking into account the optical coupling. The J-V characteristics of the two-wire structure are then reconstructed considering that the current is the same in each sub-cell (fundamental characteristic of a conventional tandem cell). The maximal power can then be calculated and compared to that of the four-wire tandem structure. Note that this procedure neglects the losses that might be due in practice from a non ideal tunnel-recombination between the cells, so the solar cell performance calculated on the two-wire tandem might be somewhat overestimated.

Polymorphous silicon and microcrystalline silicon are both characterized by defects in their energy bandgap. Two kinds of defects can be mainly observed: the deep defects linked to the dangling bonds and the network defects linked to the weak bonds.

The first defect category, even if it is known to be of amphoteric type [15,16], can be modelled by two Gaussian continuous distributions of monovalent states [17]:

$$D(E) = D_{\max} \exp \left[-\frac{(E - E_{\max})^2}{2E_0^2} \right]$$

with a peak value $D_{\max}$, peak position $E_{\max}$ and the standard deviation σ_0 that depend on the quality of the film (that depends itself on the deposition conditions) as well as on the doping.

The second category is represented by an extension of the valence band and the conduction band on either side of the forbidden band. Those extensions are modelled by two exponential bandtails. The description of the valence

Table 1. Main electrical parameters of the pm-Si:H and μc-Si:H intrinsic layers introduced in the simulation.

Parameter	pm-Si:H intrinsic layer	μc-Si:H intrinsic layer
Mobility gap E_G (eV)	1.85	1.23
Donor characteristic energy E_{UD} (eV)	0.047	0.02
Acceptor characteristic energy E_{UV} (eV)	0.03	0.01
Prefactor G_{D0}, G_{V0} (cm^{-3} eV^{-1})	4×10^{21}	4×10^{21}
Donor density of states Gaussian peak $D_{\max}$ (cm^{-3} eV^{-1})	As-deposited state: 5×10^{15} Light soaked state: 1×10^{17}	1.7×10^{16}
Donor position of the Gaussian peak $E_{\max}$ (eV), E_V taken as reference	0.8	0.5
Acceptor density of states Gaussian peak $D_{\max}$ (cm^{-3} eV^{-1})	As-deposited state: 5×10^{15} Light soaked state: 1×10^{17}	1.7×10^{16}
Acceptor position of the Gaussian peak $E_{\max}$ (eV)	1.3	0.8
Standard deviation σ_{0D}, σ_{0V} (eV)	0.2	0.18

bandtail is given by:

$$g_{Vbt}(E) = G_{V0} \exp \left[-\frac{E - E_V}{E_{UV}} \right]$$

where E_{UV} is the characteristic energy width of the tail, and G_{V0} the DOS at the valence band edge. An analogous expression holds for the conduction band tail, with a characteristic energy width E_{UD}.

The material parameters for the μc-Si:H and pm-Si:H cells used in the numerical calculations originate from several references [18–21]. The main parameters of intrinsic layers introduced in our simulation are given in Table 1.

The refractive index of each layer, from which the absorption and the reflection of the incoming photons according to their wavelength can be calculated, has been derived from spectroscopic ellipsometry measurements. It should be noted that no light scattering due to texturing was used in this study, and therefore the absolute values of current-density for a given layer thickness will be lower than typically observed in devices using textured substrates.

3 Simulation procedure

To optimize the two cells so that the global structure (two- or four-wire structures) can produce the maximum power, it is necessary to tune the thickness of each pm-Si:H and μc-Si:H i-layer. The thickness of the top pm-Si:H cell plays the key role, as it additionally determines the part of the incident photon flux that is transmitted to the bottom μc-Si:H cell. Moreover, in the micromorph tandem a-Si:H/μc-Si:H cell approach, the thickness of the μc-Si:H cell is on the order of several micrometers [7,22]. However, in order to reduce production costs, one should reduce the thickness of the μc-Si:H layer as much as possible. Therefore, we have decided to fix the thickness of the intrinsic part of the μc-Si:H cell at a reasonable value of 1.5 μm and to sweep the width of the intrinsic pm-Si:H layer. The P

and N layers are mainly used to create the junctions and the internal electrical field in the I-layers, and should be kept as thin as possible. We also therefore fixed the thickness of these very thin layers at values that are typical for PIN cells. These values are summarized in Table 2. Moreover, both junctions are sandwiched between two SnO$_2$ transparent electrodes. In order to enhance the photon absorption probability in the bottom cell, the microcrystalline cell is designed with an Ag back reflector.

To define the thickness range of the pm-Si:H intrinsic layer, we took into account the ageing process, which occurs during the first months of solar illumination. Exposure to solar illumination causes the creation of new dangling bonds created by breaking weak bonds, observed as the so-called light-soaking (LS) or Staebler-Wronski effect [23]. The DOS of pm-Si:H after light-soaking was modeled by increasing the magnitude of the dangling bond Gaussian distribution ($D_{\max}$). We present in Figure 2 the pm-Si:H cell efficiency as a function of the pm-Si:H layer thickness for different values of $D_{\max}$ introduced in the simulation.

It can be observed that for a constant i-layer thickness, the cell efficiency deteriorates with an increase in defect density, as caused by the light soaking process, as the DOS increase shortens the charged carriers' diffusion length. For a given DOS, the efficiency shows an optimum in i-layer thickness due to recombination growing more quickly with thickness than the number of photogenerated electron-hole pairs. This optimal thickness is lowered by an increase of the DOS. In our simulation, this optimal thickness is located beyond 4 μm for a low peak DOS in the pm-Si:H I layer, representative of a cell in the as-deposited state ($D_{\max} = 5 \times 10^{15}$ cm^{-3} eV^{-1}), then decreases to 2.7 μm after intermediate degradation caused by light-soaking ($D_{\max} = 1 \times 10^{16}$ cm^{-3} eV^{-1}), and finally stabilizes around 0.5 μm for a fully light-soaked cell ($D_{\max} = 1 \times 10^{17}$ cm^{-3} eV^{-1}). We need to take into account this last data which represents the point that will guarantee us the good function of the cell. After several months of utilization, a cell with I-layer more than 500 nm

Table 2. Layer thicknesses for pm-Si:H and μc-Si:H cells as used in simulation.

	pm-Si:H cell		μc-Si:H cell	
	Material	Thickness (nm)	Material	Thickness (nm)
P-Layer	a-SiC:H	15	μc-SiC:H	25
Buffer I-layer	a-SiC:H	2	–	–
I-layer	pm-Si:H	Variable	μc-Si:H	1500
N-layer	a-Si:H	20	μc-Si:H	20

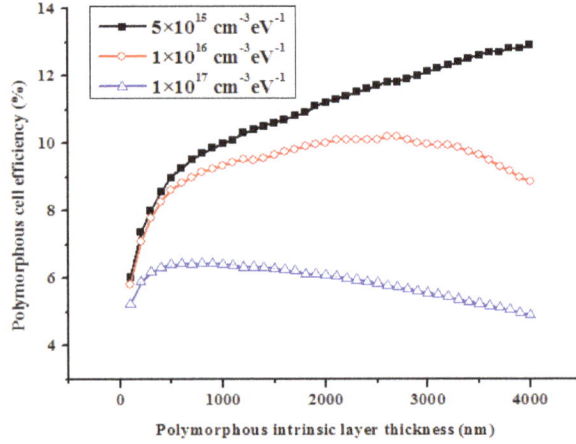

Fig. 2. Pm-Si:H cell efficiency as a function of pm-Si:H intrinsic layer thickness for three values of the Gaussian distribution peak value D_{max} (expressed in cm^{-3} eV^{-1}): 5×10^{15} ($\blacksquare$), 1×10^{16} ($\circ$) and 1×10^{17} ($\triangle$).

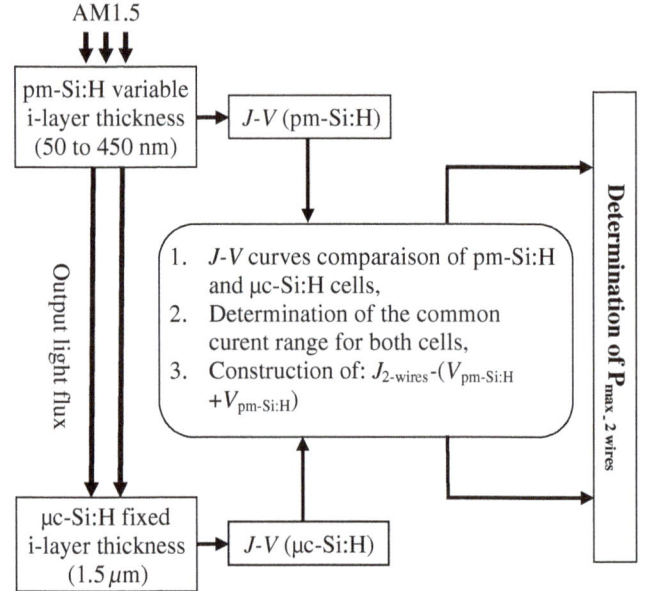

Fig. 3. Simulation steps for a two-wire tandem structure.

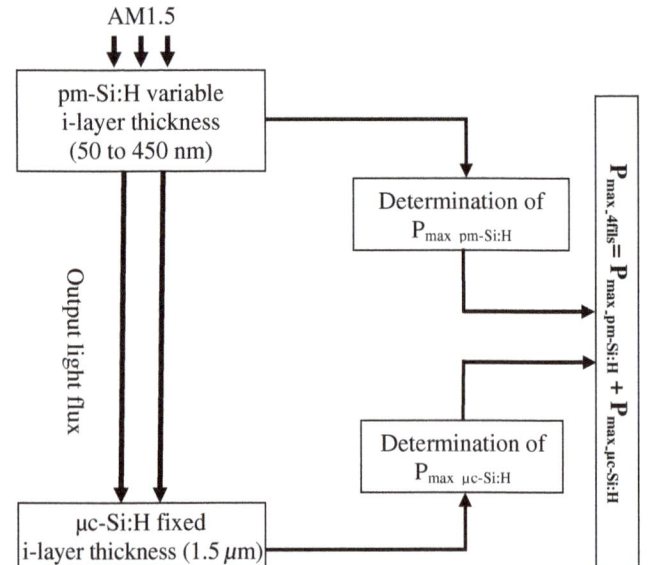

Fig. 4. Simulation steps for a four-wire tandem structure.

thick would work far less efficiently than one with a thinner I-layer.

To conclude, when thinking of a long term use, the i-layer thickness of the pm-Si:H cell must not exceed 500 nm. This maximal thickness value is even thinner for greater values of the DOS. In this study, our light-soaked cell is described by a D_{max} of 1×10^{17} cm^{-3} eV^{-1}, and so the maximal i-layer thickness is chosen as 450 nm. At the opposite side of the sweep range, technological considerations limit the thinnest possible i-layer to 50 nm. Consequently, we have varied the I-layer thickness of the pm-Si:H cell from 50 nm to 450 nm with a step of 50 nm.

Regarding the former results, we have optimized the top cell intrinsic layer thickness in both two- and four-wire structures following the procedure illustrated in Figure 3 for the two-wire cell and in Figure 4 for the four-wire cell.

The different steps can be summarized as follows:

1. Application of standard AM1.5 illumination at the top pm-Si:H cell.
2. Variations of the intrinsic layer thickness from 50 nm to 450 nm with a 50 nm step.
3. For each thickness, the output light flux of the pm-Si:H cell is calculated and is used as an input flux of the μc-Si:H cell. J-V and P-V curves are then computed for both cells.
4. For the four-wire structure, using the P-V curves, the maximum power is determined by adding the maximum power of elementary cells.

5. For the two-wire structure, the output currents have to be matched. So, we first determine the common current range in both cells. We then get for each current density J, the voltage of the global multi-junction structure $V_{pm-Si:H} + V_{\mu c-Si:H}$. We hence plot the $J - (V_{pm-Si:H} +$

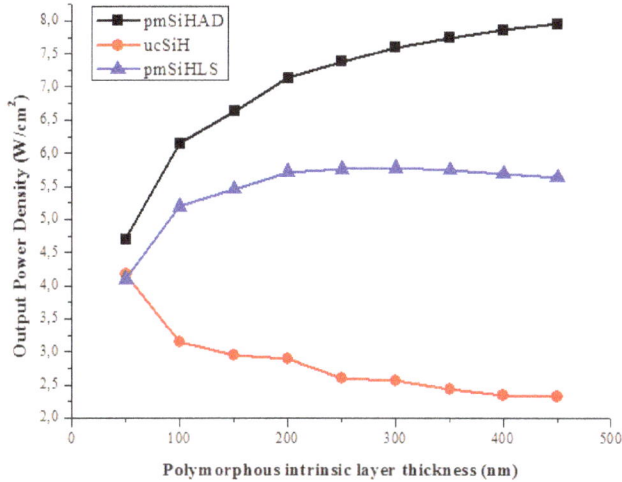

Fig. 5. Maximum output power of elementary cells versus pm-Si:H i-layer thickness.

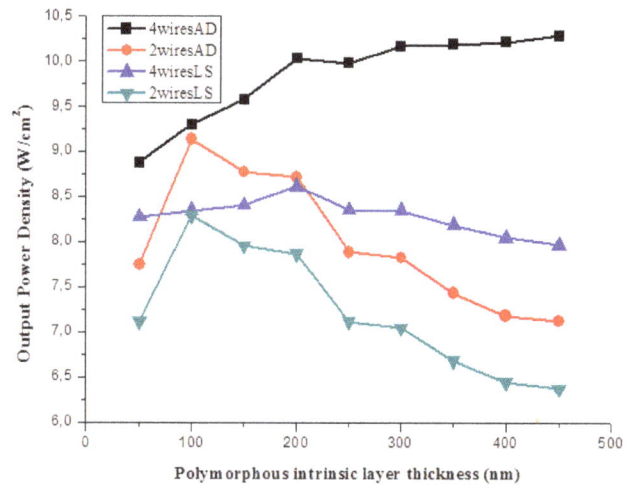

Fig. 6. Maximum output power of both tandem structures as a function of pm-Si:H i-layer thickness (AD and LS state shown).

$V_{\mu c-Si:H}$) and $P - (V_{pm-Si:H} + V_{\mu c-Si:H})$ curves of the two-wire tandem structure. From this data, we can establish the maximum power of the structure.

4 Results and discussion

We present in this section simulation results for both elementary cells and for both two- and four-wire tandem structures.

The pm-Si:H and μc-Si:H PIN cell simulations allow us to compute the variation in maximum device output power with polymorphous cell i-layer thickness. These variations are plotted in Figure 5 for the thickness range under consideration (50 nm to 450 nm) for the pm-Si:H i-layer. Our initial assumptions are confirmed:

- for a pm-Si:H cell in the as-deposited (AD) state, the wider the intrinsic layer, the better the efficiency,
- the LS pm-Si:H cell provides its maximum power for a 300 nm thick intrinsic layer,
- the μc-Si:H cell maximum power is directly linked to the number of photons coming out of the top pm-Si:H cell. By this simple fact, the thinner the top cell, the more efficient the bottom cell.

As described in Figures 3 and 4, we determined the maximal power of two- and four-wire structures as a function of the pm-Si:H i-layer thickness. These results are presented in Figure 6 in both the AD and LS state.

This figure reveals interesting differences between the two structures. In the case of the as-deposited top pm-Si:H cell, we notice that the four-wire structure is always more efficient, regardless of top cell thickness. The maximum power delivered by the four-wire structure monotonically increases with increasing top cell thickness, as all photons absorbed in the top cell are used more efficiently than those in the bottom cell due to less thermalization loss. No offsetting effect is present due to very low defect

density. The two-wire structure must cope with the current matching constraint, so the two-wire structure performance depicts a maximum point around 100 nm as this constraint prevents the pm-Si:H cell from operating at its maximum power point. Under the condition of good pm-Si:H electronic properties, we can draw a 12.6% gain by using the four-wire structure in comparison to the traditional two-wire structure. This gain may be even greater in the case of a thicker top pm-Si:H cell.

In reality, the pm-Si:H cell thickness will be limited by the Staebler-Wronski effect. This is observed for the case of a light-soaked pm-Si:H cell, where the four-wire device output power exhibits a maximum at 200 nm. Again, the absolute value of these numbers will be shifted with respect to actual values, as no light-diffusion by textured substrates is included.

We must also underline that in the four-wire structure, the total device efficiency is more robust with respect to variations of the pm-Si:H thickness. The maximum power fluctuations do not exceed 7.5% for this structure, whereas for the two-wire structure the decrease is more pronounced and reaches 23%. This parametric robustness of the four-wire structure is typical of the fact that even though one cell faces electronic defects, the second one is not modified.

In the light-soaked case, the benefit of using a four-wire structure instead of a two-wire one seems to be small, about 4% for the optimum thickness. But this value may be actually much more important. First, the simulation software does not take into account the tunnel-recombination junction effect which occurs in the N(pmSi:H)/P(μcSi:H) junction and which degrades the two-wire tandem structure performance. Second, all the simulations have been implemented with the standard AM1.5 solar flux whereas the incident solar spectrum is subject to extensive variations due to influences such as incidence angle, cloud cover, etc. These variations influence the power delivered by of each cell and hence make the "current-matching" condition that much more limiting.

As shown by the above simulations, the four-wire structure will be much less sensitive to such variations due to the decoupling of the elementary cells.

5 Summary and conclusions

Through numerical simulation, we have performed a comparative study of thin film pm-Si:H/μc-Si:H tandem cells with two different interconnection designs: a conventional, "two-wire" structure where the two PIN cells are superimposed and electrically coupled, and a "four-wire" structure where the two PIN cells are optically coupled but electrically decoupled. The aim of this study was to quantify the output power benefit one can expect from using the four-wire structure instead of the traditional two-wire one. This benefit was studied both before and after material degradation through light soaking.

The results reveal that the four-wire structure is more efficient in both as-deposited and light-soaked state, although the obtained power benefit of the four-wire structure is only 4% when comparing optimized structures in the light-soaked state. However, this benefit may be underestimated, as variations in the photon flux due to outdoor conditions were not modeled. Moreover, we note the robustness of the four-wire design to i-layer thickness variations; its peak output power fluctuations do not exceed 7.5% for the range studied, whereas the thickness effect on the two-wire structure is more pronounced, and results in an output power decrease up to 23%. This may have important consequences regarding robustness to fluctuations during the cell fabrication process.

The work was carried out under the project *"Association Tandem Optimisée pour le Solaire (ATOS)"* supported by *"Agence Nationale de la Recherche (ANR)"*.

References

1. Y.M. Soro, A. Abramov, M.E. Gueunier-Farret, E.V. Johnson, C. Longeaud, P. Roca i Cabarrocas, J.P. Kleider, J. Non-Cryst. Solids **354**, 2092 (2008)
2. A. Matsuda, M. Takai, T. Nishimoto, M. Kondo, Sol. Energy Mater. Sol. Cells **78**, 3 (2003)
3. H. Kakiuchi, M. Matsumoto, Y. Ebata, H. Ohmi, K. Yasutake, K. Yoshii, Y. Mori, J. Non-Cryst. Solids **351**, 741 (2005)
4. A.H. Mahan, Y. Xu, E. Iwaniczko, D.L. Williamson, B.P. Nelson, Q. Wang, J. Non-Cryst. Solids **299–302**, 2 (2002)
5. J. Meier, E. Vallat-Sauvain, S. Dubail, U. Kroll, J. Dubail, S. Golay, L. Feitknecht, P. Torres, S. Fay, D. Fischer, A. Shah, Sol. Energy Mat. Sol. Cells **66**, 73 (2001)
6. T. Matsui, H. Jia, M. Kondo, Prog. Photovolt.: Res. Appl. **18**, 48 (2010)
7. A.V. Shah, H. Schade, M. Vanecek, J. Meier, E. Vallat-Sauvain, N. Wyrsch, U. Kroll, C. Droz, J. Bailat, Prog. Photovolt.: Res. Appl. **12**, 113 (2004)
8. P.D. Veneri, P. Aliberti, L.V. Mercaldo, I. Usatii, C. Privato, Thin Solid Films **516**, 6979 (2008)
9. P. Roca i Cabarrocas, Th. Nguyen-Tran, Y. Djeridane, A. Abramov, E. Johnson, G. Patriarche, J. Phys. D: Appl. Phys. **40**, 2258 (2007)
10. C. Longeaud, J.P. Kleider, P. Roca i Cabarrocas, S. Hamma, R. Meaudre, M. Meaudre, J. Non-Cryst. Solids **227–230**, 96 (1998)
11. R. Butté, R. Meaudre, M. Meaudre, S. Vignoli, C. Longeaud, J.P. Kleider, P. Roca i Cabarrocas, Philos. Mag. B **79**, 1079 (1999)
12. M. Meaudre, R. Meaudre, R. Butté, S. Vignoli, C. Longeaud, J.P. Kleider, P. Roca i Cabarrocas, J. Appl. Phys. **86**, 946 (1999)
13. J.P. Kleider, C. Longeaud, M. Gauthier, M. Meaudre, R. Meaudre, R. Butté, S. Vignoli, P. Roca i Cabarrocas, Appl. Phys. Lett. **75**, 3351 (1999)
14. R. Stangl, A. Froitzheim, M. Kriegel, T. Brammer, S. Kirste, L. Elstner, H. Stiebig, M. Schmidt, W. Fuhs, in *Proc. of the 19th Eur. Photovoltaic Solar Energy Conf., 2004, Paris, France*, pp. 1497–1500
15. R.A. Street, N.F. Mott, Phys. Rev. Lett. **35**, 1293 (1975)
16. R.A. Street, K. Winer, Phys. Rev. B **40**, 6236 (1989)
17. J. Meier, U. Kroll, E. Vallat-Sauvain, J. Spitznagel, U. Graf, A. Shah, Sol. Energy **77**, 983 (2004)
18. S. Tchakarov, U. Dutta, P. Roca i Cabarrocas, P. Chatterjee, J. Non-Cryst. Solids **338–340**, 766 (2004)
19. E. Klimovsky, A. Sturiale, F.A. Rubinelli, Thin Solid Films **515**, 4826 (2007)
20. E.V. Johnson, M. Nath, P. Roca i Cabarrocas, A. Abramov, P. Chatterjee, J. Non-Cryst. Solids **354**, 2455 (2008)
21. M. Nath, P. Roca i Cabarrocas, E.V. Johnson, A. Abramov, P. Chatterjee, Thin Solid Films **516**, 6974 (2008)
22. S. Lee, M. Gunes, C.R. Wronsky, N. Maley, M. Bennet, Appl. Phys. Lett. **59**, 1578 (1991)
23. D.L. Staebler, C.R. Wronsky, Appl. Phys. Lett. **31**, 292 (1977)

Investigation of silicon heterojunction solar cells by photoluminescence under DC-bias

Guillaume Courtois[1,2,a], Parsathi Chatterjee[3,2], Veinardi Suendo[4], Antoine Salomon[1], and Pere Roca i Cabarrocas[2]

[1] Total Énergies Nouvelles, La Défense, France
[2] LPICM, CNRS – École Polytechnique, Palaiseau, France
[3] Energy Research Unit, Indian Association for the Cultivation of Science, Kolkata, India
[4] Inorganic and Physical Chemistry Research Division, Institut Teknologi Bandung, Indonesia

Abstract Photoluminescence measurements on solar cells are usually carried out under open-circuit conditions. We report here on an innovative approach, in which the samples are simultaneously illuminated and DC-biased, so that the luminescence can be monitored under several operating points, that is to say several injection levels, ranging from short-circuit conditions to the light-emitting regime of the device. The experiments were performed on in-house made c-Si/a-Si:H heterojunction solar cells illuminated by a continuous green laser diode and positively biased. The luminescence spectra obtained this way were compared to those obtained with no light excitation source, which corresponds to usual electroluminescence mode and dark $J(V)$. Firstly, the obtained luminescence spectra have shown the expected exponential dependence on the applied voltage. Furthermore, given that the amplitude of the emitted luminescence is proportional to the radiative recombination rate, this approach enables to indirectly characterise the non-radiative recombination phenomena. In the case of HJ solar cells with intrinsic thin layers processed on high quality FZ-wafers, non-radiative recombination is dominated by the defects at the c-Si/a-Si:H interface. The luminescence measurements presented here therefore give information on the quality of the surface passivation. An estimation of the interface defect density was achieved by comparing our experimental results with modelling.

1 Introduction

Among the different types of solar cells, crystalline silicon (c-Si)/hydrogenated amorphous silicon (a-Si:H) heterojunction (HJ) with intrinsic thin layer solar cells combine the advantages of crystalline cells, for instance high photocurrent, with those of thin film silicon solar cells: strong open-circuit voltage and low temperature coefficient. Moreover, given that the a-Si:H thin layers are deposited at low temperature, namely 200 °C, by plasma processes, the thermal budget of this technology is low. Although HJ solar cells have been commercialised for fifteen years and a record cell efficiency of 24.7% has been reported[1], there is still an issue regarding the properties of the c-Si/a-Si:H interfaces, which condition the solar cell performances [1]. More precisely, since the cells are processed on high quality monocrystalline silicon wafers, the

efficiency of the interface defect passivation is of crucial importance. As a matter of fact, the higher the interface defect density, the higher the surface recombination rate, and thus the poorer collection of photogenerated carriers. Surface recombination has been investigated through several techniques, such as photoconductance [2] or surface photovoltage [3]. In this paper, spectral luminescence will be emphasised.

In recent years, luminescence techniques have become more and more popular for the monitoring of defects in semi-conductor materials and solar cells. Since luminescence provides a fast and non-destructive assessment of radiative recombination, it turns out to be an indirect means for characterising non-radiative phenomena, which are for the cells in question, dominated by interface recombinations. Depending on whether the samples are optically or electrically biased, luminescence is referred to as either photoluminescence (PL) or electroluminescence (EL). Whereas PL constitutes a versatile technique, implementable from wafer to finalised devices, EL requires a finalised cell. Spatially resolved luminescence has been used to get mappings of various characteristics, from carrier lifetime to series resistance inhomogeneities [4]. PL measurements are generally carried out with the illuminated sample in open-circuit conditions, whereas EL is extracted from the light-emitting regime of a diode in the dark. When the optical power (respectively the electrical one) is increased, the amount of photogenerated (injected) carriers increases, hence an enhanced recombination – including radiative recombination – rate, which eventually enhances the luminescence signal.

In the case of a completed device, combination of both excitations is of course possible. This PL under DC-bias allows monitoring several operating modes of the device into question.

We report here on spectrally resolved EL and PL under DC-bias from HJ solar cells. The spectra obtained with and without light excitation have been compared; the experimental results have then been compared to model-generated ones. An estimation of the actual interface defect density (N_{id}) of our samples could be extracted by modelling the experimental light $J(V)$ characteristics.

2 Experimental procedure

The experiments were carried out on 2×2 cm^2 in-house made heterojunction solar cells processed on double-side polished, float zone, n-type, 280 μm-thick monocrystalline silicon wafers whose resistivity was of 3 Ohm cm. Once the wafers had undergone a hydrofluoric acid dip, thin (5 nm) intrinsic layers ensuring the interface passivation as well as 15 nm-thick doped a-Si:H layers constituting the front emitter and the back surface field were deposited by radio-frequency plasma enhanced chemical vapour deposition (RF-PECVD) at 200 °C [5]. Indium tin oxide (ITO) layers were then deposited by magnetron sputtering on both front and rear sides as anti-reflective coating and electrodes. Finally, the front contact grid and the full sheet back reflector were both made of evaporated silver (Fig. 1). In order to check the general behaviour of the cells, $J(V)$ characteristics were measured under 1 sun on a AAA class solar simulator. Efficiencies of more than 15% were assessed, associated with V_{oc}, J_{sc} and fill factor of 690 mV, 31 mA/cm^2 and 73%, respectively.

Coming to the luminescence investigations, the photo-excitation was performed using a continuous laser diode delivering an optical power of 15 mW at 532 nm, wavelength at which the reflectivity of the sample displayed a minimum. The spotlight section was around 5 mm^2, which corresponds to an impinging flux on the sample of around 10^{18} photons cm^{-2} s^{-1}. A SourceMeter® by Keithley was used simultaneously as DC-supply and ammeter, in order to DC-bias the samples and at the same time acquire $J(V)$ curves. All measurements were carried out at room temperature. As for the detection part of the setup, the collected luminescence signal transited via an optical fibre to a grating spectrometer by Horiba and finally reached a

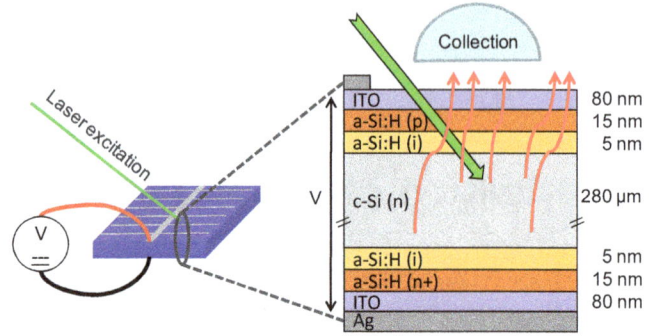

Fig. 1. Schematic of a HJ solar cell in our experimental configuration, showing the DC-supply, as well as the optical excitation supplied by a green laser.

mono-crystal InGaAs detector cooled down by liquid nitrogen. Over a set of measurements, the sample and the optical fibre remained fixed, so that all luminescence spectra were acquired from the same area on the device, coinciding with the illuminated area when the laser diode was turned on (Fig. 1).

3 Simulation model

The one-dimensional electrical-optical model ASDMP (Amorphous Semiconductor Device Modelling Program) [6, 7], that has recently been extended to equally model HJ solar cells [8], solves the Poisson's equation and the two carrier continuity equations under steady state conditions for the given device structure. The complex refractive indices for each layer of the structure are required as input and have been measured in-house by spectroscopic ellipsometry.

The emitted photon flux density from n-type c-Si wafers can be described by the generalized Planck's/Kirchoff's law, which takes into account non-homogeneous carrier distribution and hence a non-constant quasi-Fermi level splitting, as well as a non-homogeneous generation profile and reabsorption of luminescence photons, and reads as follows [9]:

$$d\Phi_{lum}(\hbar\omega) = \frac{(\hbar\omega)^2}{4\pi^2\hbar^3c^2}\exp\left(\frac{-\hbar\omega}{kT}\right)$$

$$\times \left[\int_0^t A(\hbar\omega,x)\exp\left(\frac{E_{Fn}(x)-E_{Fp}(x)}{kT}\right)dx\right]d\hbar\omega, \quad (1)$$

where t is the thickness of the wafer. $A(\omega,x)$, the absorptivity of light of a particular energy ($\hbar\omega$) at a given position inside the wafer, and the quasi-Fermi level splitting $E_{Fn}(x)-E_{Fp}(x)$ inside it can be calculated by model ASDMP. Details of the procedure are given in reference [9], where ASDMP has been implemented to the simulation of luminescence in heterostructures.

For a good luminescent diode (i.e., in which the amorphous/crystalline interface defect density (N_{id}) is less than $\sim 5 \times 10^{11}$ cm^{-2}), $E_{Fn} - E_{Fp}$ is constant and equal to qV,

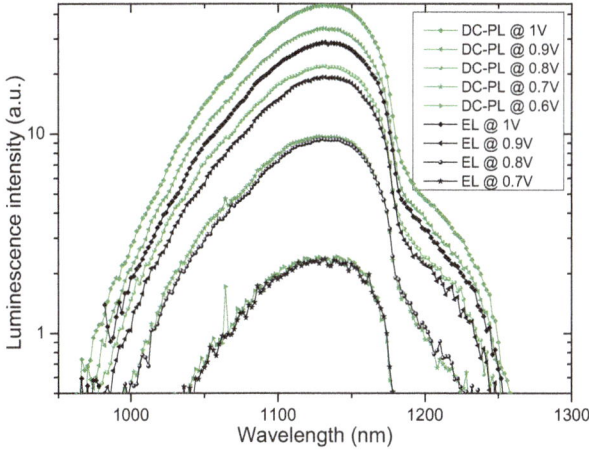

Fig. 2. Experimental EL and PL under DC-bias spectra.

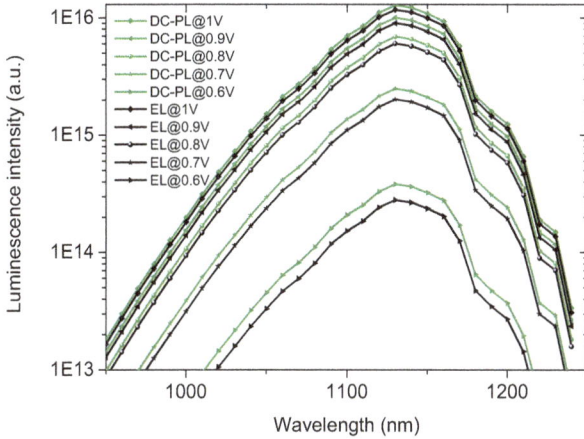

Fig. 3. Model-generated EL and PL under DC-bias spectra.

where V is the applied voltage. Alternatively it is equal to qV_{oc}, when, as is generally the case, the PL is measured under open-circuit conditions. Model-generated EL and PL under DC-bias results are presented in Figures 3 and 4b and will be discussed in Section 5.

4 Experimental results

EL spectra were obtained under applied voltages up to 1 V, at the same time as dark $J(V)$ curves. By turning the laser on, everything else kept constant, both light and DC-bias excitations were combined, hence a luminescence mode referred to as PL under DC-bias.

Experimental luminescence spectra are displayed on Figure 2. On all spectra, the maximal luminescence intensity occurs at 1130 nm, which corresponds to band-to-band radiative recombination in c-Si. If the amorphous silicon were to contribute to the luminescence, it would emit below 950 nm, where no luminescence is detected. As a matter of fact, the a-Si:H layers are too thin to contribute to the luminescence signal. As to the shoulder appearing on the high wavelength side, it has been accounted for by recombination bringing into play the creation of two phonons [10].

Obviously, both EL and PL under DC-bias intensities increase with the applied voltage (see Eq. (1)). For a given applied voltage, PL under DC-bias turns out to be more intense than pure EL, which originates in the superposition of photogenerated carriers to the injected ones. Moreover, when moving from EL to PL under DC-bias, the shape of the spectra remains unaltered; the total integrated flux, corresponding to the luminescence yield, is therefore proportional to the peak intensity. That being stated, we will henceforth go on dealing with the values of the peak intensity when comparing the measurements one to another.

The variation of the peak intensity with the applied voltage is presented on Figure 4a for both EL and PL under DC-bias modes. Two regimes are underlined: at low applied voltages, that is to say below 0.7 V, both curves exhibit a diode behaviour, the luminescence intensity increasing exponentially with the applied voltage, whereas for higher voltages a kind of saturation occurs, associated with a slower increase in the luminescence intensities. A surprising feature is the discrepancy between the ideality factors extracted from the $0.5-0.7$ V range of applied voltage, which turn out to be of 1.3 and 2.2 for EL and DC-PL, respectively, whilst a similar dependence upon the applied voltage was expected. The mismatch observed experimentally probably stems from lateral effects due to the competition between the illuminated and the shaded areas. Even more surprising from our point of view is the third value of the ideality factor obtained from J_{dark}, which turned out to be of 1.7, so between the previous ones. Note that, for applied voltages around 0.6 V, a single-diode model suffices to render the dominating current transport [11].

5 Discussion and comparison with modelling

In order to be sure that our experimental results were not affected by a possible change in temperature induced by either the DC-bias up to 1 V – which corresponds to an electrical power of 350 mW – or the impinging optical power, we had a closer look at the high energy region, namely above 1.2 eV (i.e. below 1000 nm). The expression for absorptance A is given in equation (2) [12]. In equation (1) used in ASDMP, multiple reflections are considered until all the long wavelength luminescent photons are either reabsorbed in the c-Si wafer, or lost by absorption in the rear contact or in the non-luminescent front layers, or by reflection at the front. Equation (2) involves front-side and back-side reflectivities (R_f and R_b, respectively), as well as the absorption coefficient in crystalline silicon (α). All these quantities are spectrally dependent. t still stands for the thickness of the wafers, namely 280 μm.

$$A(\hbar\omega) = \frac{(1-R_f)\left[1-R_b e^{-2\alpha t}-(1-R_b)e^{-\alpha t}\right]}{1-R_f R_b e^{-2\alpha t}}$$
$$\xrightarrow[\hbar\omega \geqslant 1.2\ \text{eV}]{} 1-R_f. \qquad (2)$$

The more energetic the radiation, the higher the value of α. Namely, for photons, α is higher than 10 cm^{-1} [13],

Table 1. Solar cell output parameters and the diode ideality factor of the DC-PL peak tion intensity versus V curve, the EL peak intensity versus V curve and the dark $J(V)$ characteristic of the deposited HJ solar cell on n-type wafer, compared to modelling results.

	J_{sc} (mA cm^{-2})	V_{oc} (V)	FF	Efficiency (%)	Diode ideality factor, n		
					DC-PL peak	EL peak	J_{dark}
Experiment	31.0	0.692	0.73	15.5	2.2	1.3	1.7
Model	31.36	0.691	0.77	16.8	1.06	1.04	1.14

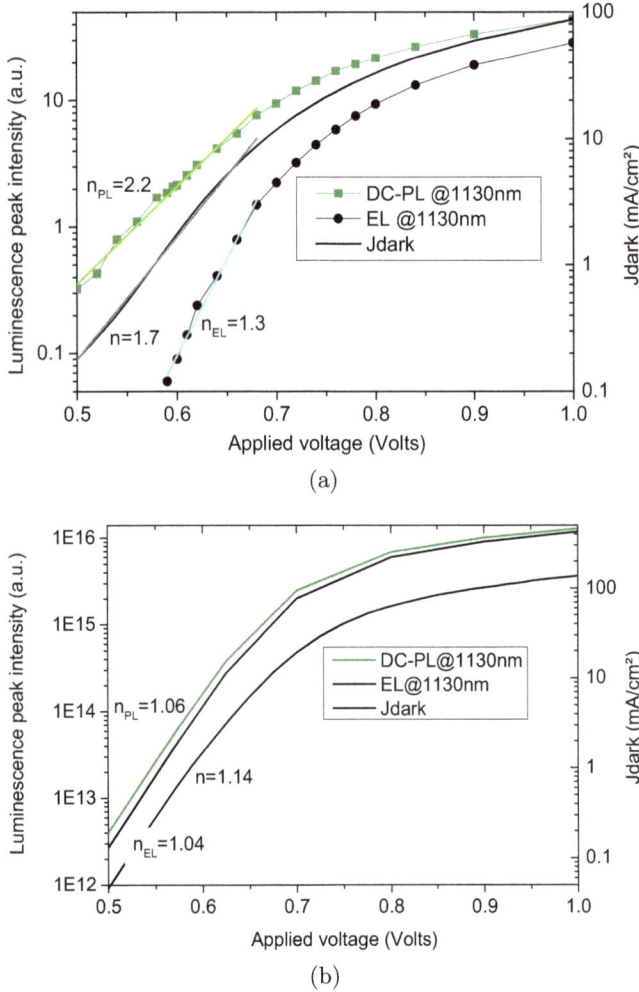

(a)

(b)

Fig. 4. Variations of (a) measured and (b) model-generated EL and PL peak intensities under DC-bias with respect to the applied voltage. Comparison with (a) the measured and (b) the modelled dark $J(V)$ characteristic.

therefore $2\alpha t$ is greater than 5 and the expression of A simplifies to $(1 - R_f)[1 - (1 - R_b)e^{-\alpha t}]$. Now, R_b has been evaluated to 93% over the $900-1000$ nm range, thus A reasonably reduces to $1 - R_f$, which varies very little from 80% for the considered a-Si:H/ITO stack deposited on polished c-Si. Subsequently, above 1.2 eV, the spectral dependence of $d\Phi_{lum}$ becomes restrained in the product $(\hbar\omega)^2 \exp(\frac{-\hbar\omega}{kT})$. Consequently, it reads (Eq. (6)):

$$d\log(\Phi_{lum}(\hbar\omega)) = \left(2\log(\hbar\omega) - \frac{\hbar\omega}{kT} + C\right) d\hbar\omega, \quad (3)$$

where C accounts for the spectrally independent quantities. Considering energies and temperatures of the order of 1 eV and 300 K, respectively, the term $\log(\hbar\omega)$ in the above expression becomes negligible, and thus the slope of the high-energy tail of the luminescence peak, when plotted in logarithmic scale, is simply $-1/kT$. It is clear from Figure 2 that the experimental luminescence spectra all show the same slope below 1000 nm, which attests to the absence of significant temperature change over the experiments.

Modelling of the illuminated $J(V)$ characteristics under AM 1.5 light of the structure shown in Figure 1 was carried out in order to get an idea of the defect density at the a-Si:H/c-Si interfaces. It has been previously found [9] that when N_{id} exceeds 10^{12} cm^{-2} at the emitter/c-Si interface, the PL intensity is too weak for accurate measurement. Regarding the material properties, a doping density of 2×10^{15} cm^{-3} was assumed in the n-doped wafer, corresponding to a resistivity of 3 Ohm cm. The other input parameters are also similar to those given in reference [9]. A comparison to the experimental measurements is shown in Table 1. Luminescence features were also modelled under voltage bias, while for PL excitation, additional generation by a 530 nm light was considered in parallel to the voltage excitation.

The HJ cell V_{oc} is known to be strongly sensitive to the defect density at the emitter/c-Si interface [8,9] and this fact has been used to estimate N_{id} at this junction from modelling and turns out to be 4.5×10^{11} cm^{-2}. Photovoltaic performance is almost insensitive to the c-Si/BSF N_{id} below 5×10^{11} cm^{-2} [8,9,14] and a value of 10^{11} cm^{-2} has been assumed here in the modelling calculations. However, the calculated FF does not agree with the measured value, because of the high series resistance of the contacts, estimated to be $\sim$ 3 Ohm. The calculated value of the FF would correspond to the measured value corrected for the series resistance of the contacts. Also compared in the same table are the diode ideality factors for the dark $J(V)$ characteristics, and the PL under DC-bias and EL peak intensities (at $\lambda = 1130$ nm).

The calculated PL and EL curves under applied voltages are plotted in Figure 3. The first satisfactory feature is that they are of the same shape as the experimental ones (see Fig. 2). In addition, the variations of EL and PL under DC-bias peak intensities with respect to the applied voltage, as well as J_{dark} (Fig. 4), show, like in the experimental curves, a break of slope at 0.7 V. Indeed, above this value, the electrical properties of the device are dominated by the resistive effects. Nevertheless, the ideality factors extracted from the $0.5-0.7$ V region of these curves proved to be of 1.06, 1.04 and 1.14 for PL under DC-bias, EL and J_{dark}, respectively, so very close to each other, which was expected. Indeed, EL yield and J_{dark} are supposed to display the same dependence upon the applied voltage. Moreover, assuming the EL yield is proportional to $\exp(\frac{qV}{nkT})$, one would expect for the PL yield under DC-bias: $\phi_{DCPL} \propto \exp(\frac{q(V+\Delta V)}{nkT})$ where ΔV accounts for the extra junction voltage conveying the enhanced splitting of the quasi-Fermi levels induced by the laser excitation. As a result, when plotted in logarithmic scale, a simple shift is expected between the $EL(V)$ and DC-PL(V) curves, as displayed in Figure 3.

The ideality factors obtained from modelling turn out to differ significantly from the experimental ones. As a matter of fact, a homogeneous illumination on the entire device is taken into account in the one-dimensional model, whereas strong lateral effects occur in our experimental configuration.

6 Conclusion and perspectives

By combining optical and electrical excitations, luminescence spectra from in-house elaborated heterojunction with intrinsic thin layer solar cells were obtained at many operating points. As expected on this kind of devices processed on high quality monocrystalline wafers, whose properties are limited by the interface features, EL and PL under DC-bias spectra proved to be similar in shape, conveying regular carrier distributions within the wafer. In both cases, the luminescence yields exhibited an exponential dependence with the applied voltage. However, the ideality factors calculated from EL, PL under DC-bias and dark $J(V)$ curve significantly differed from each other and were far from unity. This feature may be due to lateral resistive losses which would alter the local diode voltage in some fashion, all the more so as only the investigated fraction of the cell was illuminated. Next set of experiments will be performed on fully illuminated cells. Our experimental results were compared to 1D ASDMP modelling, allowing an estimation of the front interface defect density, which proved to be of $\sim$4.5 $\times$ 10^{11} cm^{-2}. Experiments and modelling agreed on the detrimental effect of series resistance at voltages higher than 0.7 V.

We intend to extend this study to samples processed on textured wafers and we expect, that, in conjunction with photoconductance measurements, this study will help discriminate between field effect and chemical passivation, the latter being less sensitive to the injection level.

This work was performed within the joint solar R&D team between Total and LPICM.

References

1. S. De Wolf, A. Descoeudres, Z.C. Holman, C. Ballif, Green **2**, 7 (2012)
2. M. Garin, U. Rau, W. Brendle, I.A.R. Martin, J. Appl. Phys. **98**, 093711 (2005)
3. L. Korte, A. Laades, K. Lauer, R. Stangl, D. Schaffarzik, M. Schmidt, Thin Solid Films **517**, 6396 (2009)
4. T. Trupke et al., in *Proceeding of the 22nd EPVSC, Milan, 2007*
5. P. Roca i Cabarrocas, J.B. Chévrier, J. Huc, A. Lloret, J.Y. Parey, J.P.M. Schmitt, J. Vac. Sci. Technol. A **9**, 2331 (1991)
6. P. Chatterjee, J. Appl. Phys. **76**, 1301 (1994)
7. P. Chatterjee, J. Appl. Phys. **79**, 7339 (1996)
8. M. Rahmouni, A. Datta, P. Chatterjee, J. Damon-Lacoste, C. Ballif, P. Roca i Cabarrocas, J. Appl. Phys. **107**, 054521 (2010)
9. A. Datta, M.-H. Song, J. Wang, M. Labrune, S. Chakroborty, P. Roca i Cabarrocas, P. Chatterjee, J. Non-Cryst. Solids **358**, 2241 (2012)
10. G. Weiser, S. Kazitsyna-Baranovski, R. Stangl, J Mater. Sci.: Mater. Electron. **18**, 93 (2007)
11. T.F. Schulze, L. Korte, E. Conrad, M. Schmidt, B. Rech, J. Appl. Phys. **107**, 023711 (2010)
12. R. Brüggemann, in *Physics and Technology of Amorphous-Crystalline Heterostructures Silicon Solar Cells*, edited by W.G.J.H.M. van Sark, L. Korte, F. Roca (Springer-Verlag, Berlin, Heidelberg, 2012)
13. M.A. Green, M.J. Keevers, Prog. Photovolt.: Res. Appl. **3**, 189 (1995)
14. S. de Wolf, in *Physics and Technology of Amorphous-Crystalline Heterostructures Silicon Solar Cells*, edited by W.G.J.H.M. van Sark, L. Korte, F. Roca (Springer-Verlag, Berlin, Heidelberg, 2012)

3

Rapid thermal annealing of sputter-deposited ZnO:Al films for microcrystalline Si thin-film solar cells

H. Koshino[1], Z. Tang[1], S. Sato[1], H. Shimizu[2], Y. Fujii[3], T. Hanajiri[3], and H. Shirai[1,a]

[1] Graduate School of Science and Engineering, Saitama University, 255 Shimo-Okubo, Sakura, 338-8570 Saitama, Japan
[2] Saitama Industrial Technology Centre (SAITEC) 3-12-28 Kami-Aoki, Kawaguchi, 333-0844 Saitama, Japan
[3] Bio-Nano Electronics Research Centre, Toyo University, 2100 Kujirai, Kawagoe, 350-8585 Saitama, Japan

Abstract Rapid thermal annealing of sputter-deposited ZnO and Al-doped ZnO (AZO) films with and without an amorphous silicon (a-Si) capping layer was investigated using a radio-frequency (rf) argon thermal plasma jet of argon at atmospheric pressure. The resistivity of bare ZnO films on glass decreased from 10^8 to 10^4–10^5 Ω cm at maximum surface temperatures T_{max}s above 650 °C, whereas the resistivity increased from 10^{-4} to 10^{-3}–10^{-2} Ω cm for bare AZO films. On the other hand, the resistivity of AZO films with a 30-nm-thick a-Si capping layer remained below 10^{-4} Ω cm, even after TPJ annealing at a T_{max} of 825 °C. The film crystallization of both AZO and a-Si layers was promoted without the formation of an intermixing layer. Additionally, the crystallization of phosphorous- and boron-doped a-Si layers at the sample surface was promoted, compared to that of intrinsic a-Si under the identical plasma annealing conditions. The TPJ annealing of n$^+$-a-Si/textured AZO was applied for single junction n-i-p microcrystalline Si thin-film solar cells.

1 Introduction

Silicon thin-film solar cells are a promising candidate for future photovoltaic power generation. The most advanced approach employs hydrogenated amorphous and microcrystalline silicon (a-Si:H, μc-Si:H) as active layers in single and multi-junction cells, which are fabricated using a very-high-frequency (VHF) plasma-enhanced chemical vapor deposition method (PE-CVD) from a mixture of SiH$_4$ and H$_2$ [1, 2]. Silicon thin-film solar cells in a p-i-n (superstrate) configuration require a transparent conductive oxide (TCO) films as a front contact. Such contacts must have a low series resistance and a high transparency in the visible and near-infrared regions. Furthermore, a surface topography is required to ensure light-scattering and subsequent light trapping inside the silicon solar cell structure [3,4]. Among various TCOs, impurity-doped zinc oxide, ZnO, films such as ZnO:Al (AZO) and ZnO:Ga (GZO) are attractive because of their low material cost, non-toxicity, relatively low resistivity, and high visible transmission. In addition, AZO can be easily textured by wet chemical etching with diluted hydrochloric acid (0.5% HCl). Recently, there has been particular interest in improving of the properties of impurity-doped ZnO, i.e., the suppression of the free-carriers in multi-junction Si thin-film solar cells [5]. Deposition of AZO and GZO films as a TCO have been extensively studied using several fabrication techniques, i.e., metal-organic chemical

vapor deposition (CVD), magnetron sputtering, and sol-gel methods [6–10], followed by subsequent thermal annealing in vacuum to improve their optical and electrical properties. Among them, the thermal annealing of AZO with n$^+$-a-Si capping layer on top of AZO is one of interests for further improvement of the performance of TCO films [11, 12]. In addition, for further high efficiencies of single- and multi-junction solar cells, a new Si thin-film technology is needed, i.e., polycrystalline Si thin film and its alloys [13,14].

In this study, we demonstrate the rapid thermal annealing for AZO film with and without an a-Si capping layer using an inductively-coupled radio-frequency (rf) argon thermal plasma jet (TPJ) at atmospheric pressure and applied for the n-i-p microcrystalline Si (μc-Si) thin-film solar cells.

2 Experimental details

The 800-nm-thick AZO films used in this study were fabricated on corning 1737 glass (25×15 mm^2) using a rf magnetron sputtering technique in argon and a 99.99% AZO ceramic targets with 2-wt%-Al$_2$O$_3$ at a substrate temperature T_s of 300 °C and a working pressure of 0.9 Pa. The AZO was textured by wet chemical etching with diluted hydrochloric acid (0.5% HCl$_{aq}$) [15]. The resistivity of the as-deposited AZO films was approximately 3–6×10^{-4} Ω cm. 30-nm-thick n$^+$- and p$^+$-a-Si:H layers were fabricated on flat and textured AZO-coated glass using a capacitively-coupled rf PE-CVD of a dichlorosilane

---

a e-mail: shirai@fms.saitama-u.ac.jp

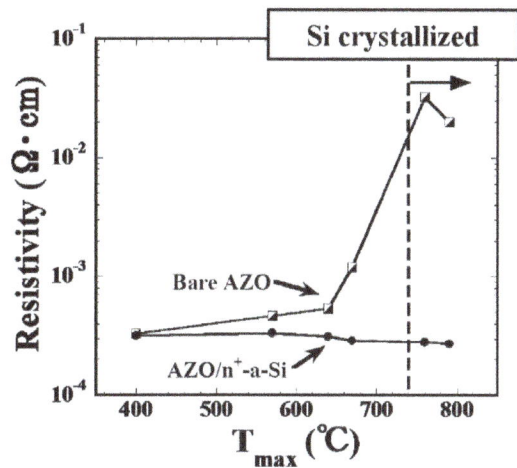

Fig. 1. Resistivity vs. maximum surface temperature T_{max} of AZO films with and without a 30-nm-thick n^+-a-Si capping layer after TPJ annealing.

(SiH_2Cl_2) and H_2 mixture with and without an additional 1-%-PH_3(B_2H_6), respectively, at a T_s of 250 °C. In the subsequent solar cell processing steps, much higher temperatures would be reached during thermal plasma exposure. Therefore, the a-Si:H films were pre-heated to 400 °C for 90 min prior to the thermal plasma annealing to remove the residual hydrogen from the films. The TPJ-annealing of ZnO and AZO films with and without an a-Si:H capping layer was performed with a variable substrate stage velocity v_{sub} and a 2 KW inductively-coupled rf plasma source in a tri-axial quartz tube having an inner diameter of 10 mm. The maximum surface temperature T_{max} was controlled by adjusting v_{sub}. The quartz tube-substrate to substrate spacing was 15 mm. The surface temperature profiles during TPJ annealing was monitored using a non-contact thermometry [16].

TPJ-annealed AZO and crystallized Si films were characterized by grazing incidence X-ray diffraction (GI-XRD), X-ray photoelectron spectroscopy (XPS), photoluminescence (PL), and spectroscopic ellipsometry (SE). The Hall mobility μ and carrier density N_e of the TPJ-annealed ZnO and AZO films were determined by a standard Van der Pauw method. The optical characterization of the textured AZO before and after TPJ annealing was performed using a double-beam UV-visible spectrometer equipped with an integrating sphere (Perkin Elmer, Lamda 35) at wavelengths ranging from 250 to 1800 nm. The Si thin-film solar cells were fabricated with a single-junction n-i-p μc-Si solar cell structure. The 2-μm-thick μc-Si layer was fabricated from a SiH_2Cl_2 and H_2 mixture at 500 °C.

3 Results

3.1 TPJ annealing of ZnO and AZO films with and without an a-Si capping layer

Figure 1 shows the resistivity of AZO films with and without capping 30-nm-thick n^+-a-Si layer upon the TPJ

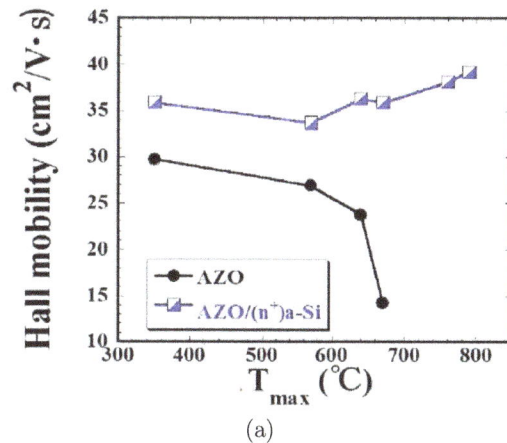

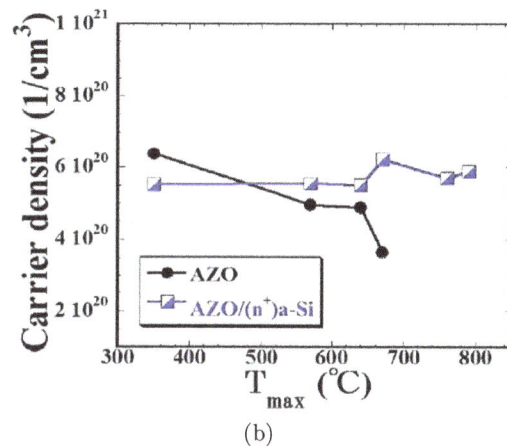

Fig. 2. (a) Hall mobility μ and (b) carrier density N_e of TPJ-annealed AZO films at different T_{max}s. The n^+-Si layer on top of AZO was etched out using reactive ion etching after TPJ annealing. The TPJ annealing was performed by changing v_{sub} from 55 to 22 mm/s.

annealing plotted against T_{max}. Here, the TPJ annealing was performed on the identical region by repeating the plasma exposure at v_{sub}s from 55 down to 10 mm/s to minimize the thermal damage of previous annealing history. The resistivity increased markedly from 10^{-4} to 10^{-2} Ω cm at T_{max}s above 650 °C for bare AZO films on glass. On the other hand, it increased slightly and/or was almost independent of v_{sub} of an order of 10^{-4} Ω cm up to a T_{max} of 825 °C (v_{sub}: 25 mm/s) for AZO films with a 30-nm-thick n^+-a-Si capping layer. The film crystallization of both a-Si and AZO layers was observed at T_{max} above 740 °C. These results agreed qualitatively with those of conventional solid-phase-crystallization (SPC) of n^+-a-Si coated AZO films at 600 °C for 22 h [11,12]. Similar result was also obtained in the AZO films capping p^+-a-Si layer. Similar results were also observed in the boron-doped and intrinsic a-Si layers.

In Figure 2, the μ and N_e of TPJ-annealed ZnO and AZO layers with and without a-Si capping layers are summarized as a function of T_{max}s. The Hall measurement was performed for TPJ-annealed AZO films on flat glass after etching out top n^+-μc-Si layer. The μ and N_e in bare

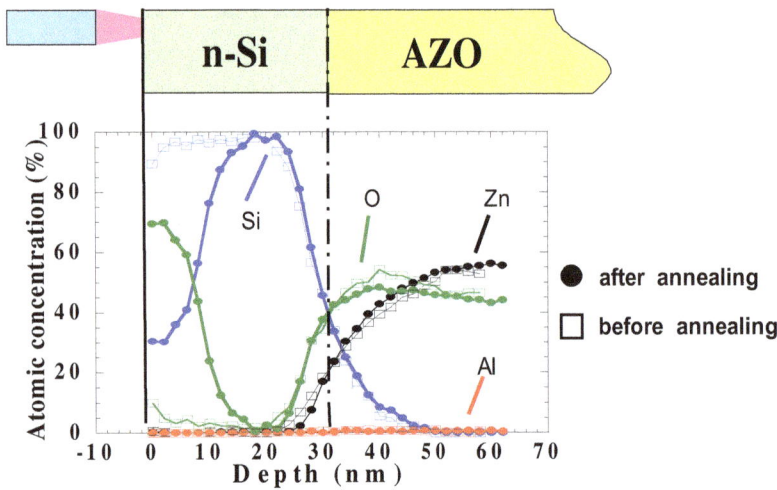

Fig. 3. Depth profiles of atomic concentrations, Si, Zn, Al, and O for n$^+$-a-Si/AZO heterostructures before and after TPJ annealing monitored by XPS. The measurement was performed each after 5-nm-thick etching by Ar plasma.

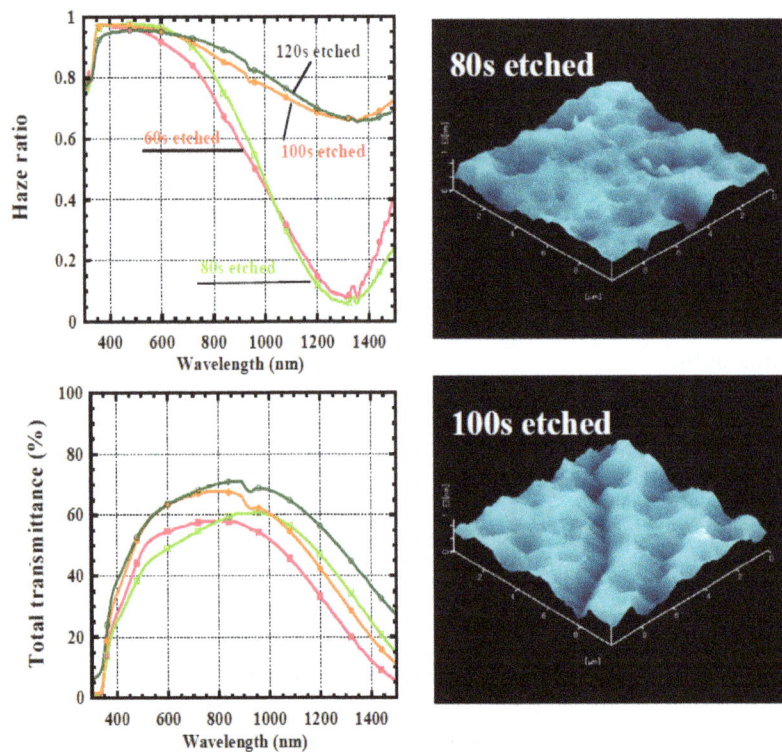

Fig. 4. Total transmittance T_{total} and haze spectra of TPJ-annealed AZO films with different surface morphology. The wet chemical etching was conducted after TPJ annealing by a 0.5% HCl at different dipping times. The TPJ annealing was performed at a v_{sub} of 25 mm/s and a $T_{\max}$ of 650 °C. The inset shows the AFM images of the 80 and 100 s chemically etched AZO film.

ZnO films decreased to 13–18 cm^2/Vs and 3×10^{20} cm^{-3}, respectively, with increasing $T_{\max}$. On the other hand, the μ increased from 30 to 39 cm^2/Vs for AZO with n$^+$-a-Si capping layer. While the N_e was almost independent of v_{sub} and remained on the order of 10^{20} cm^{-3} even after TPJ annealing at a $T_{\max}$ of 825 °C. These results suggest that the film crystallization of AZO layer promoted with any increase in the O and Zn related defects. The PL characterization also revealed that the generation of O and Zn related defects was suppressed by the a-Si capping layer.

Depth profiles of the atomic concentrations of Si, Zn, Al, and O of the AZO a 30-nm-thick n$^+$-a-Si capping layer are shown in Figure 3, as measured by XPS before and after TPJ annealing. No significant changes in the depth profiles of Al, Zn, or Si and thicknesses were observed in the Si/AZO heterostructure except for the formation of an oxidized Si layer at the top surface. In addition, the SE characterization also revealed that the film crystallization of a-Si ansd AZO layers was enhanced and no significant generation of intermixing layer were formed at a-Si and AZO interface. These findings imply that the film crystallization of both AZO and Si layers was enhanced without significant oxygen diffusion from underlying AZO to top a-Si layer.

3.2 Texturing of AZO and its effect of the crystallization of a-Si capping layer

Figure 4 shows the total transmittance T_{total} and haze spectra of TPJ-annealed AZO films with different surface morphology. The TPJ annealing was conducted at a $T_{\max}$

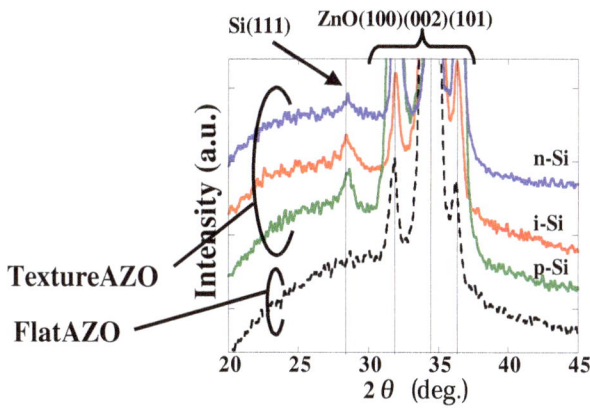

Fig. 5. GI-XRD patterns of TPJ-annealed 30-nm-thick n^+-, p^+-, and intrinsic- crystallized Si films on textured AZO-coated glass substrates after TPJ annealing at a T_{max} of 825 °C. The result of intrinsic- crystallized Si on flat AZO is also shown as a reference.

of 650 °C and subsequently, the wet chemical etching using a 0.5% HCl was carried out at different dipping times for 2-μm-thick AZO films. The AFM image of corresponding textured AZO is also shown on the top. The uniform crater structures were formed by adjusting the HCl rinse time. High haze values were obtained of above 70% in entire spectra range for ~100 s (800 nm thickness) etched AZO films. The T_{total} in the 1000–1200 nm regions corresponding to the free carrier absorption was also suppressed with increasing the HCl etching time. These results originate from less free carrier density. The Hall measurement revealed that the μ and N_e were 45 cm^2/V s and 10^{20} cm^{-3} in the 100 s etched AZO films. On the other hand, the TPJ annealing of HCl-etched AZO films showed relatively high resistivity because the etching speed was higher. Thus, the texturing of AZO was conducted after TPJ annealing to design the optical management without deteriorating electrical properties.

In Figure 5, the GI-XRD patterns are shown for the 30-nm-thick n^+-, p^+-, and intrinsic-crystallized Si films fabricated on textured AZO-coated glass substrates after TPJ annealing. The result of intrinsic-crystallized Si film on flat AZO is also shown as a reference. Apparently, both Si and AZO layers were crystallized by the TPJ annealing for all samples, although the degree of the film crystallization depended on the impurity concentration and film thickness. In addition, the Si (111) diffraction peak intensities in intrinsic-, n^+-, and p^+-a-Si films were enhanced on textured AZO rather than the flat. Similar results were also reported elsewhere for p-i-n μc-Si thin-film solar cells [17]. These results originate from that SPC of a-Si by TPJ annealing is promoted not from the top surface but the bottom. These results suggest that the texturing AZO surface promotes the inhomogeneous temperature profile toward the film depth, which is a possible origin to enhance the nucleation and grain growth rather than the flat. Therefore, the use of texturing AZO is effective to promote the film crystallization of top a-Si layer, although its degree depends on impurity type, P and B, and their concentrations in a-Si precursors. These findings imply that the

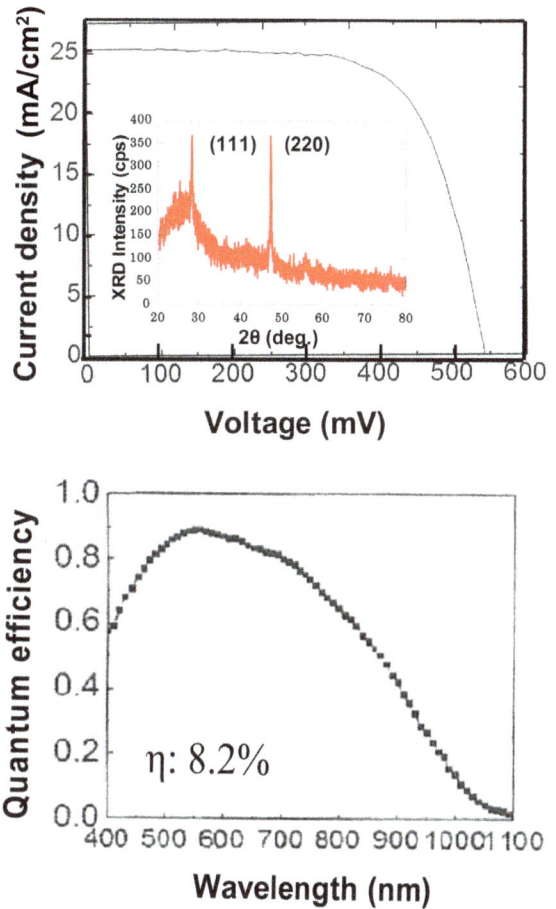

Fig. 6. *I-V* characteristics and QE spectra of n-i-p μc-Si thin-film solar cells with textured AZO. The intrinsic μc-Si:H:Cl layer used was 2 μm thickness and fabricated from a SiH$_2$Cl$_2$ and H$_2$ mixture by rf PE-CVD. Inset shows the XRD pattern of corresponding μc-Si films.

TPJ annealing of n^+-a-Si/textured AZO heterostructure is a possible technique for the optical management for Si thin-film solar cells.

Figure 6 shows the current-voltage, *I-V* curve and quantum efficiency (QE) spectra of μc-Si single-junction solar cells using textured AZO. Inset shows the corresponding μc-Si film. The current density J_{sc} improved markedly with no decrease in open circuit voltage V_{oc} by introducing the textured AZO because of better light scattering. The efficiency of 8.2% (V_{oc}:542 mV, J_{sc}: 25.4 mA/cm^2, FF: 68%) has been obtained for μc-Si:H n-i-p solar cell with 2 μm i-layer and 0.253 cm^2 area at a deposition rate of 5 Å/s. These findings suggest that the TPJ annealing of AZO with n^+-a-Si capping layer is a possible method for further high performance of μc-Si thin-film solar cells at high temperature regime.

4 Conclusions

We demonstrated the effects of a-Si capping layers on ZnO:Al (AZO) during rf thermal plasma jet annealing. Both AZO and a-Si layers were crystallized by the TPJ annealing at a maximum surface temperature T_{max} above 825 °C. In addition, the film crystallization of a-Si layer was promoted efficiently on textured AZO by TPJ annealing with no significant diffusion of oxygen at the interface. The role of a-Si capping layer on AZO during TPJ annealing is the suppression of the fluctuation of the band potential due to the segregation of Al_2O_3 phase in ZnO. Higher haze values above 70% were obtained in the entire spectra range from 400 to 1500 nm maintaining.

This work was supported partially by a Grant-in-Aid for Scientific Research from the Ministry of Education, Culture, Sports, Science and Technology of Japan.

References

1. B. Rech, T. Repmann, S. Wieder, M. Rusuke, U. Stephan, Thin Solid Films **502**, 300 (2006)
2. K. Yamamoto, M. Yoshimi, Y. Tawada, S. Fukuda, S. Sawada, T. Meguro, H. Takata, T. Suezaki, Y. Koi, K. Hayashi, T. Suzuki, M. Ichikawa, A. Nakajima, Sol. Enery Mater. Sol. Cells **74**, 449 (2002)
3. B. Rech, H. Wagner, Appl. Phys. A **69**, 155 (1999)
4. S.J. Tark, M.G. Kang, S. Park, J.H. Jang, J.C. Lee, W.M. Kim, J.S. Lee, D. Kim, Thin Solid Films **9**, 1318 (2009)
5. O. Kluth, B. Rech, L. Houben, S. Wieder, G. Schope, C. Beneking, H. Wagner, A. Loffl, H.W. Schock, Thin Solid Films **351**, 247 (1999)
6. S. Mridha, D. Bask, J. Phys. D **40**, 6902 (2007)
7. M. Suchea, S. Christoulakis, N. Katsarakis, T. Kisopoulos, G. Kiriakidis, Thin Solid Films **51**, 6562 (2007)
8. B.-Z. Dong, H. Hu, G.J. Fang, X.Z. Zhao, D.Y. Zheng, Y.P. Sun, J. Appl. Phys. **103**, 073711 (2007)
9. R. Romero, D. Leinen, E.A. Dalchiele, J.R. Ramos-Burrado, F. Martin, Thin Solid Films **515**, 1942 (2006)
10. M. Volintiru, B. Creatore, J. Kniknie, C.I.M.A. Spee, M.C. Van de Sanden, J. Appl. Phys. **102**, 043709 (2007)
11. C. Becker, E. Conrad, P. Dogan, F. Fenske, B. Gorka, T. Hanel, K.Y. Lee, B. Rau, F. Ruske, T. Weber, M. Berginski, J. Hupkes, S. Gall, B. Rech, Sol. Energy Mater. Sol. Cells **93**, 825(2009)
12. S. Gall, C. Becker, E. Conrad, P. Dogan, F. Fenske, B. Gorka, K.Y. Lee, F. Ruske, B. Rech, Sol. Energy Mater. Sol. Cells **93**, 1004 (2009)
13. B. Rech, T. Repmann, M.N. Van den Donker, M. Berginski, T. Kilper, J. Hupkes, S. Calnan, H. Stiebig, S. Wieder, Thin Solid Films **511-522**, 548 (2006)
14. K.Y. Lee, C. Becker, M. Muske, Appl. Phys. Lett. **91**, 241911 (2007)
15. O. Kluth, B. Rech, L. Houben, S. Wieder, G. Schope, C. Beneking, H. Wagner, A. Loffl, H.W. Schock, Thin Solid Films **251**, 247 (1999)
16. N. Ohta, T. Imamura, H. Shimizu, T. Kobayashi, H. Shirai, Phys. Stat. Sol. A **207**, 574 (2010)
17. H. Shirai, Y. Sakurai, M. Ye, K. Haruta, T. Kobayashi, Proc. Matter. Res. Soc. Symp. Proc. **989**, 0989-A13-04 (2007)

4

Evolutionary process development towards next generation crystalline silicon solar cells: a semiconductor process toolbox application

J. John[1,a], V. Prajapati[1,2], B. Vermang[1,2], A. Lorenz[1,2], C. Allebe[1,2], A. Rothschild[2], L. Tous[2], A. Uruena[2], K. Baert[2], and J. Poortmans[2]

[1] IMEC, Kapeldreef 75, Leuven, Belgium
[2] Katholieke Universiteit Leuven, Leuven, Belgium

Abstract Bulk crystalline Silicon solar cells are covering more than 85% of the world's roof top module installation in 2010. With a growth rate of over 30% in the last 10 years this technology remains the working horse of solar cell industry. The full Aluminum back-side field (Al BSF) technology has been developed in the 90's and provides a production learning curve on module price of constant 20% in average. The main reason for the decrease of module prices with increasing production capacity is due to the effect of up scaling industrial production. For further decreasing of the price per wattpeak silicon consumption has to be reduced and efficiency has to be improved. In this paper we describe a successive efficiency improving process development starting from the existing full Al BSF cell concept. We propose an evolutionary development includes all parts of the solar cell process: optical enhancement (texturing, polishing, anti-reflection coating), junction formation and contacting. Novel processes are benchmarked on industrial like baseline flows using high-efficiency cell concepts like i-PERC (Passivated Emitter and Rear Cell). While the full Al BSF crystalline silicon solar cell technology provides efficiencies of up to 18% (on cz-Si) in production, we are achieving up to 19.4% conversion efficiency for industrial fabricated, large area solar cells with copper based front side metallization and local Al BSF applying the semiconductor toolbox.

1 Introduction

The photovoltaics (PV) sector is a strongly growing industrial sector with a compound annual growth rate of 33% over the last 3 decades (Fig. 1). It is expected that this growth rate could remain up to 40%/year for this decade as a result of the efforts made worldwide to reduce dependence on fossil fuel and the CO_2-emissions related to electricity generation. Exemplary in this respect is the decision of the European Commission to go for a share of 20% renewable energy sources in 2020 in the European energy mix (with a share as high as 30% for electricity generation). As a result of this sustained growth, the photovoltaic sector which measures at this moment $25-30$ billion \$ (the value of the PV-systems market) in financial terms, will become a plus 100 billion \$ sector in 2020.

The cost of Si material constitutes about 1/3 of the solar cell module cost [1]. In order to be less dependent on price fluctuations of polysilicon feedstock and wafers, and to eventually realize cost targets down to 0.5 euro/Wp, an evolution towards a reduction of "grams of pure Si/Wp"

is taking place. As one does not want to sacrifice solar cell efficiency despite the use of thinner wafers, this requires quite drastic changes for crystalline Si solar cell technology. As a basic trend one could state that the objective is to reduce the grams of Silicon per Wp by a factor of 2 with an efficiency increase of roughly 20% relative (from $17-18\% \rightarrow >20\%$ for industrial crystalline Si solar cells) [2].

2 Experimental results

2.1 Efficiency improvement potential: the toolbox application

The output efficiency of mono-crystalline Silicon solar cells in production is $\sim$17.5%. This is $\sim$12.5% less than predicted by theory taking the Auger Recombination limit into account. These losses can be specified in three main parts: electrical losses in the bulk and the surface due to SRH recombination processes, optical losses due to insufficient optical confinement of the cell and resistive losses on the contacts. Figure 3 depicts the distribution of the losses in mono-crystalline Silicon solar

a e-mail: joachim.john@imec.be

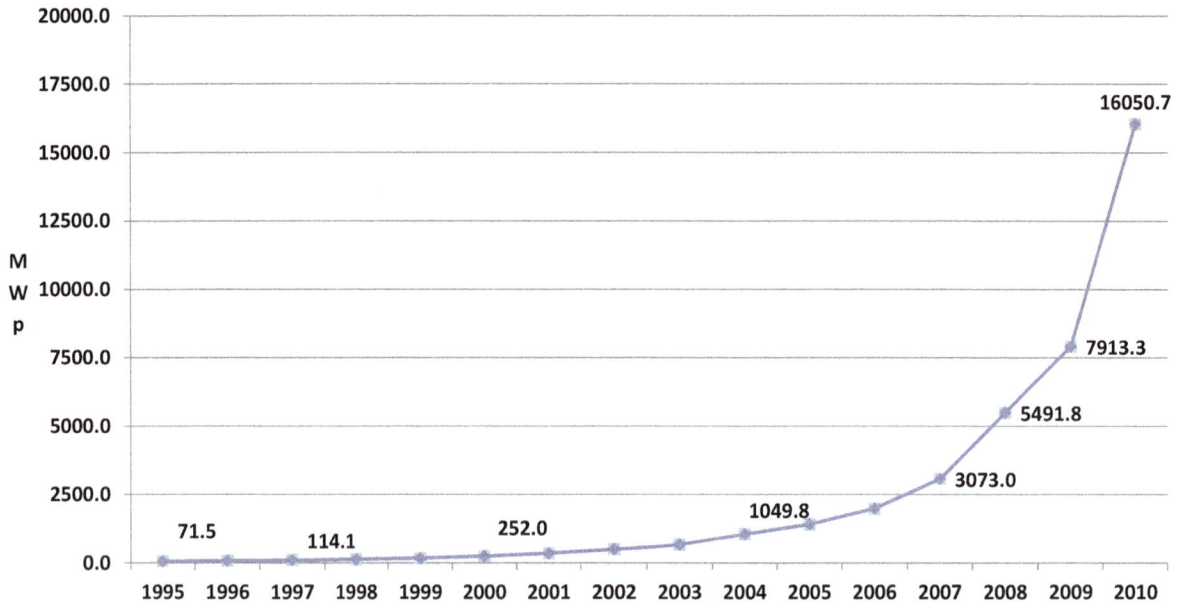

Fig. 1. PV Industry Growth 1995 to 2010 [source: Navigant consulting].

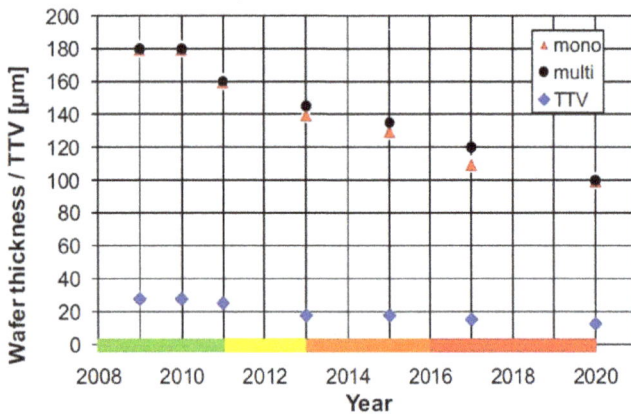

Fig. 2. Thickness development roadmap for crystalline silicon solar cells, wafer thickness and total thickness variation is depicted, the green code indicates that technical solutions are known, yellow means industrial solution is known but not yet in production. Orange means interim solution is known, too expensive or not suitable for production, whereas red means that no solutions for high-volume manufacturing of such thin wafers with high yield are available yet. [source: International Technology Roadmap for Photovoltaics (ITRPV.net). Results 2010] [2].

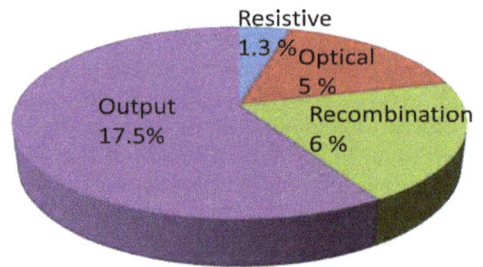

Fig. 3. Losses in monocrystalline cz-Si solar cells. Maximum efficiency $= 29.8\%$ (Auger limit), the recieved efficiency in production is $\sim 17.5\%$, losses due to recombination $\sim 6\%$, optical losses due to insufficient optical confinement $\sim 5\%$ and 1.3% resistive losses at the contacts [3].

cells after MacDonalds [3]. To overcome the losses successive improvement of the different solar cell processes is required. The Passivated Emitter and Rear Locally diffused cell concept depicted in Figure 4 has a number of features added to overcome the losses present in the full Aluminum back-side field solar cell.

The semiconductor process toolbox is benchmarked in an industrial Passivated Emitter and Rear Locally diffused cell concept (i-PERL) on 148 cm^2, 150 um thick, (1−3 Ω cm) cz-Si material, as depicted in Figure 4.

In comparison to the full Al BSF cell, the following features are added:

– Fine line front metallization (reduce shadowing losses).
– Shallow or deep emitter (reduce recombination losses in the emitter, enhance blue responsivity).
– Dielectric rear passivation (reduce surface recombination losses).
– Laser ablated vias in rear passivation (reduce contacting recombination losses).
– Textured front side and polished rear side (enhance optical confinement, enhance infrared responsivity).
– Physical vapor deposition (PVD) of back-side metallization (reduce metal consumption, contactless processing, increase optical confinement, e-beam or sputtering of Al).

The SEMI PV road map predicts a decrease of the emitter saturation current below 100 fA/cm^2 in 2020 (Fig. 5). In order to reach this value the emitter formation process has to be optimized. The challenges are that higher ohmic emitters with lower surface concentration have to

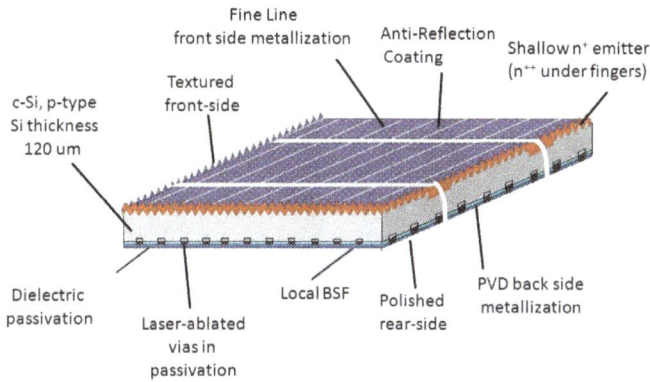

Fig. 4. Industrial Passivated Emitter and Rear Locally diffused cell concept (i-PERL), 120–150 um thick, 1–3 Ω cm, 156 cm^2, cz-Si material.

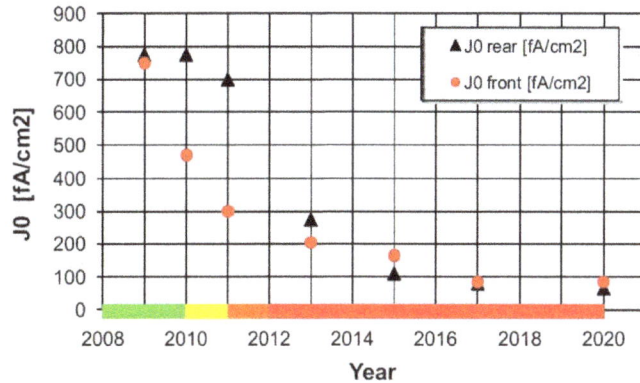

Fig. 5. Front saturation currents (emitter saturation current) and rear saturation current development [2].

be passivated with novel dielectrics that enhance the optical properties of the anti-reflection coating.

2.2 Anti-reflection coating (ARC)

Furthermore, PECVD SiN is one of the most expensive process steps. By further increasing the production capacity as predicted the use of gases like silane will increase. In future solar cell production the circumvention of silane would be preferable. SiN produced with the silane free precursor from Sixtron Applied Materials has been applied in a solar cell production run. Local Al BSF cells has been manufactured and a best cell result of 18.6% has been achieved using the silane free SiN as an ARC. Using it in the rear side passivation stack on top of SiO$_x$, best cell results in local Al BSF solar cells have been measured to be 18.3% [4].

2.3 Pre-passivation cleaning

Cleaning is an underestimated process for the next-generation crystalline Si solar cells. If efficiencies >20% are to be obtained, maintaining high bulk lifetimes is required.

The present cleaning sequences within photovoltaic manufacturing has not been developed for this purpose. High lifetime processing will require very efficient cleaning and handling methods in view of metal contaminants. It is obvious that there is an valuable knowledge base within the microelectronics (development of ultraclean surface processes) to be taken advantage off, although it must be realized that eventually the allowable surface contamination level at a cleaned surface will be lower for crystalline Si solar cells with efficiency potential >21% than for a typical clean in advanced CMOS-processing. For the latter a lower level metallic contamination of 10^{10} cm^{-2} is acceptable but for crystalline Si solar cells metal contamination levels of 10^9 cm^{-2} might be required [5]. This is a serious challenge in terms of cost-effectiveness of the cleaning and drying process as well as on the level of characterization of such low levels of metallic contaminants on non-mirror polished or even textured Si-surfaces.

We have improved the homogeneity of ALD-grown Al$_2$O$_3$-layers for surface passivation by an adapted cleaning and drying using a Marangoni dryer [6]. Also the reduction of interface contamination in case of a-S:H heterojunctions is key to obtain high open-circuit voltages [7]. Figure 6 depicts the influence of surface conditioning cleanings of the minority carrier lifetime. The lifetime measurements have been performed on 4-inch fz-Silicon wafer (1–3 Ω cm) using QSSPC [8].

2.4 Rear-side passivation

In the framework of the investigation for high-k dielectrics which are necessary to achieve low gate leakage currents in scaled CMOS transistors, atomic layer deposition (ALD) of Al$_2$O$_3$ has been investigated intensively in the past [9]. Although the density of interface traps at the Si-Al$_2$O$_3$-interface is low, the high negative charge present in this material is an issue for CMOS transistor because it affects the threshold voltage of the device. This on the other hand is useful for applications in photovoltaic devices where the negative charge gives rise to a highly accumulated surface in p-type substrates or highly inverted surfaces in n-type substrates. As a result very low surface recombination velocities have been measured on both n- and p-type substrates [10] as well as low emitter saturation current densities on B- and P-emitters. The advantages of Al$_2$O$_3$ layers have been demonstrated in fz-Si, small area, high-efficiency crystalline Si solar cells with efficiencies up to 23% [11].

Introducing these layers in industrial solar cell flows (large area, cz-Si, 1–3 Ω cm) efficiencies up to 19% have been reported by Gatz et al. [12] on cz-Si material (2–3 Ω cm) with a thickness of 180 um. A best cell conversion efficiency of 19.1% has been achieved in IMEC [13] on cz-Si material (1–3 Ω cm) with a thickness of 150 um.

2.5 Junction formation

Achieving enhanced cell performance requires optimal dimensional control of doping profiles. Ion implantation

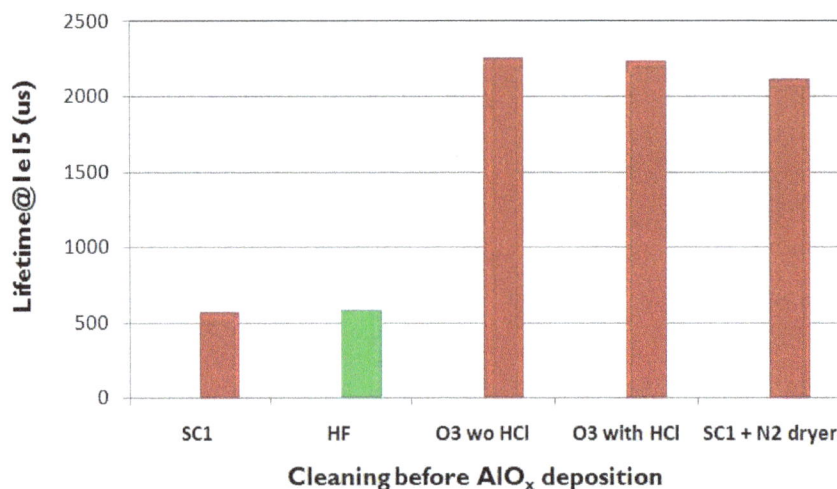

Fig. 6. Influence of surface conditioning provided by different cleanings on minority carrier lifetime. The lifetime is given in microseconds and measured at an injection level of 1e15 cm^{-3}. SC_1: NH_3:H_2O_2:H_2O mixture, HF: hydrofluoric acid, O_3 wo HCl: ozone without hydrochloric acid, O_3 with HCl: ozone with hydrochloric acid, SC_1+N_2 dryer: NH_3:H_2O_2:H_2O mixture + Nitrogen dryer.

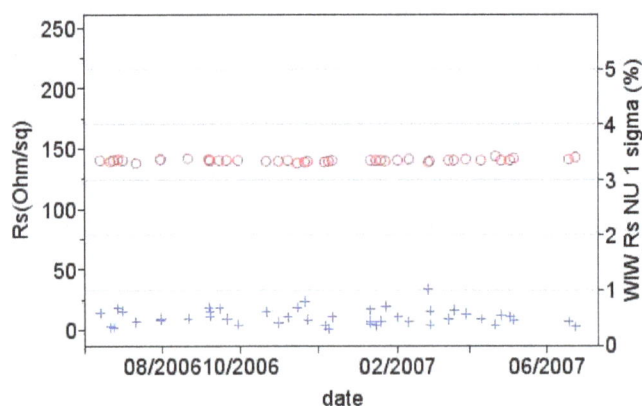

Fig. 7. Wafer to wafer reproducibility recorded over one year in a P-implantation system at IMEC aiming on 120 Ω/sq emitter.

as an emitter passivation dielectric layer. For advanced emitter, suited for the implementation in PERL cell concepts we have achieved an emitter saturation current density of 17 fA/cm^2 on 140 Ω/sq sheet resistance emitter with $POCl_3$ diffusion and TO_x+SiN passivation. We have achieved 55 fA/cm^2 emitter saturation current density on 132 Ω/sq sheet resistance emitter with P-implantation and thermally grown silicon oxide (TO_x) passivation (Fig. 8). The achieved results are compared with literature values of Moschner et al. [14] and Kerr et al. [15].

We have reported earlier [16] conversion efficiencies of over 18.8% for n-type emitter. Now we reached up to 19% conversion efficiency with a shallow 120 Ω/sq implanted emitter (independently confirmed by ISE Cal Lab).

2.6 Contacting

In the roadmap outlined by the SEMI-PV Group the amount of Ag/W_p is to be reduced given the weight of the Ag-cost in the total cost. In addition, this reduction or eventually fully avoiding Ag is required to ensure sustainability of crystalline Si solar production on longer term. The use of Ag would exclude production levels much higher than 100 GW_p/year [17]. Options to replace Ag are Al or Cu with the last one having the advantage of lower resistivity.

In the microelectronics sector the replacement of Al by Cu in advanced CMOS processing took place in the time period around 2000. This replacement was enabled by the use of ALD and barrier technology to avoid direct contact between the Cu contact and Si which would lead to the destruction of the junctions by the rapid indiffusion of Cu already at moderate temperatures. In CMOS technology these barriers are based on elemental metals like Ti or Ta, nitrides (TaN, . . .) or silicides. Other potential issues caused by introducing copper contacts are ghost plating

with its excellent areal uniformity and run-to-run producibility provide a possible alternative to diffusion for shallow emitters or doping profiles difficult to achieve by diffusion processes. Wafer to wafer reproducibility, recorded over one year, for a 120 Ω/square emitter based on P-implantation was found to vary by 1.4% whereas the variation on the within wafer non uniformity was as low as 0.6% (Fig. 7). The combination with hard masks can also lead to substantial reduction in the number of steps to achieve locally doped regions in PERL and IBC cell concepts.

Emitter saturation current density (Joe) has been extracted from lifetime measurements and plotted versus corresponding emitter sheet resistance values. For sheet resistances above 100 Ω/sq, Joe below 100 fA/cm^2 has been reached using conventional PECVD SiN emitter passivation. Saturation current density values lower than 10 fA/cm^2 could be reported on 200−400 Ω/sq emitter using a stack of thermally grown silicon oxide (TO_x) and SiN

Table 1. Large area cell results (cz-Si, $1-3\,\Omega\,cm$) reached with AlO_x based rear-passivation dielectric layer stack. The AlO_x passivation and the SiN_x capping layers have a thickness of 10 and 110 nm, respectively.

Cell type		Size (cm^2)	Jsc (mA/cm^2)	Voc (mV)	FF $(\%)$	Eta $(\%)$
Al_2O_3 pass. PERC	Average (4 cells)	148.25	38.0 $\pm\,0.2$	643 $\pm\,1$	77.6 $\pm\,0.2$	19.0 ±0.1
	Best cell	148.25	38.2	645	77.7	19.1
SiO_x pass. i-PERC	Best cell	148.25	37.8	638	77.7	18.7

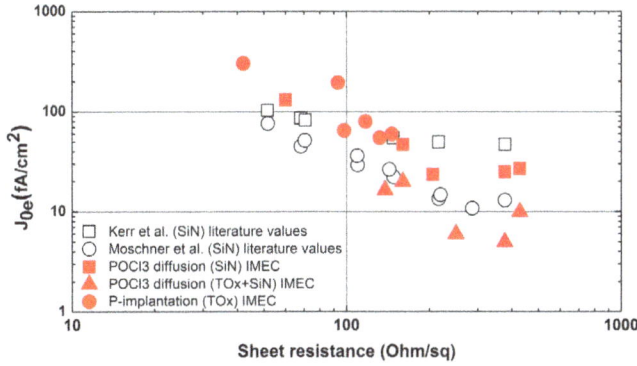

Fig. 8. Emitter saturation current density vs. emitter sheet resistance extracted from lifetime measurements. Full symbols are representing IMEC results, while hollow symbols are representing values published in literature for emitters passivated with SiN_x [14,15]. Full squares are $POCl_3$ diffused emitter passivated with PECVD SiN, Full triangles are $POCl_3$ diffused emitter passivated with a thermally grown Silicon oxide (TO_x) and SiN stack, full circles are P-implanted emitter passivated with thermal oxide. These lifetime measurements have been performed in IMEC on $1-3\,\Omega\,cm$, 4-inch, fz-Silicon wafer and extracted from QSSPC measurements.

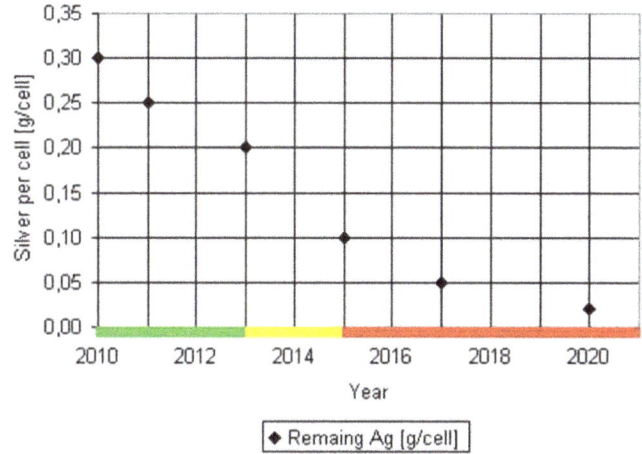

Fig. 9. Predicted development of the weight of silver in gram/cell in silicon solar cell manufacturing [2].

Table 2. Overview of efficiencies obtained with various barrier layer structures on large-area crystalline Si solar cells with local Al-BSF at the rear side and Cu-based contacts at the front side. All barrier layers have been applied by physical vapor deposition (sputtering).

Contact Layer	Emitter	J_{sc} (mA/cm^2)	V_{oc} (mV)	FF $(\%)$	η $(\%)$	R_s $(\Omega\,cm^2)$
Ti	60	38.4	639	79.1	19.4	5
Ti	135	38.8	651	75.6	19.1	1.1
Ta	60	38.2	640	77.9	19.0	0.82
TaN	60	38.3	636	77.8	19.0	0.78
T1/7iN	60	38.1	638	77.7	18.9	0.76
Ni	60	38.3	639	77.7	19.0	0.69

(diffusion through dielectric pinholes/defects during plating), reliability issues (effective barrier during subsequent processing and at operating conditions (25 years)), and corrosion (copper corrosion of the Cu capping) as shown in Figure 10. The Cu-layers in advanced circuits are normally realized by electroplating whereas the barrier layers are grown by sputtering or ALD.

We have applied different barrier layers by means of physical vapor deposition (sputtering) under Cu based contacts on the front side of the solar cell, conversion efficiencies between 19 and 19.5% were obtained on large-area solar cells using layers like Ti, Ta, TaN and NiSi$_2$. Best cell results are summarized in Table 2. By optimizing the metal grid spacing at the front side to the sheet resistance of the emitter for the $120\,\Omega$/square case, simulations and analytic modeling have predicted for this technology efficiencies up to 20% [18].

3 Conclusions

The photovoltaic sector is confronted with the challenge to reduce cost whilst at the same time increasing efficiency to reach grid parity as soon as possible and to be on equal footing with other sources of renewable energy like wind energy. For crystalline Si solar cells the gap between the theoretical limit of 30% (auger limit) and the manufactured cell efficiency of 17.5% for mono-crystalline material has to be bridged. To do so, there is still plenty of room to absorb and adapt technologies up till now limited to micro-electronics. Several examples were given, showing that this is indeed occurring at the moment for techniques like implantation, atomic layer deposition, Cu-plating and barrier layer technology with the obvious requirement that costs should be brought down to make it compatible with PV cost requirements. The interaction between the 2 sectors might not be limited to taking over elements from the technology toolbox but might also extend to the more operational issues dealing with statistical process control,

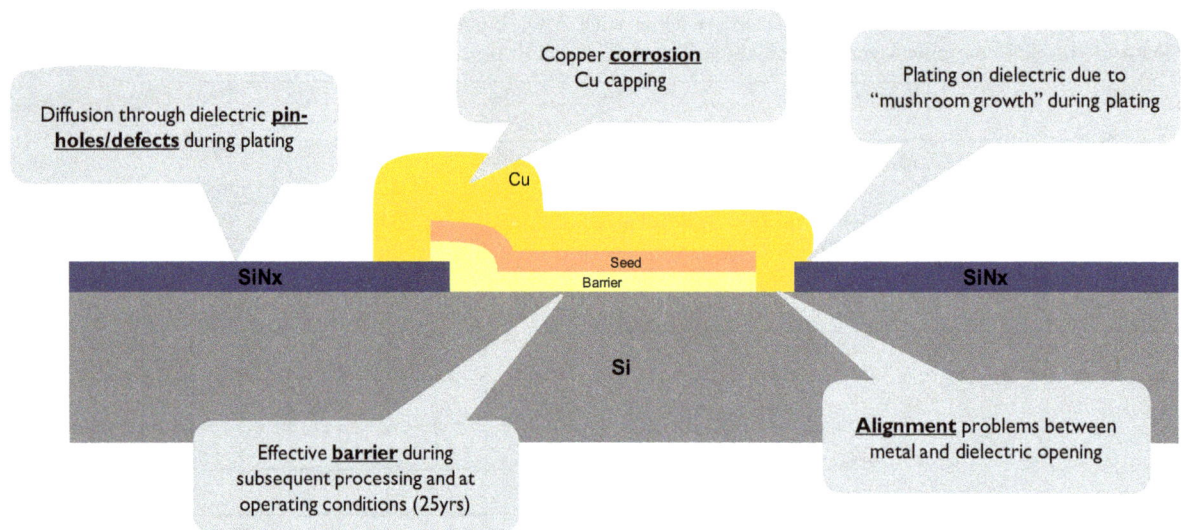

Fig. 10. Issues related to Cu-metallization (schematic).

Table 3. Best cell efficiencies achieved with the semiconductor toolbox technology implemented into an i-PERL process flow (large area cells, cz-Si, $1-3$ Ω cm).

Toolbox process implemented in i-PERL cell (cz-Si, $1-3$ Ω cm, 150 um thick, 148 cm^2)	Best cell efficiency [%]
Silane free SiCN as ARC	18.6
Implanted n+ Emitter	18.9
AlO$_x$/SiN$_x$ rear passivation stack	19.1
Cu plated front contact	19.4

quality insurance and in-line analysis. A crucial role is given to the system suppliers, their willingness to adapt to the requirements of photovoltaic processes will at the end decide over the possible implementation of a developed technology in a silicon solar cell manufacturing line.

We have developed a semiconductor process toolbox for further decreasing the opto-electrical losses in industrial large area crystalline silicon solar cells. The developed processes are finally integrated into an industrial PERL cell concept that acts as technology demonstrator. The process integration has been performed on large area 148 cm^2, $1-3$ Ω/cm resistivity, cz-silicon substrates with a thickness of 150 um. The successful implementation of the following processes in local Al BSF cells has been demonstrated and is depicted in Table 3 (best cell efficiency).

References

1. W. Sinke, C. del Cañizo, G. del Coso Sánchez, 1 € per watt-peak advanced crystalline silicon modules: the crystalclear integrated project, *Proceedings of the 23rd EU PVSEC – PV Conf* (Valencia, 2008)
2. International Technology Roadmap for Photovoltaics (ITRPV.net) Results, http://www.itrpv.net/doc/roadmap_itrpv_2011_brochure_web.pdf
3. D.H. MacDonalds, Ph.D. thesis, ANU, Australia, 2001
4. V. Prajapati, J. John, J. Poortmans, R. Mertens, Silane free high-efficiency industrial silicon solar cells using dielectric passivation and local BSF, *Proceedings of the 25th EU PVSEC – PV conf.* (Valencia, 2010)
5. A. Istranov, T. Buonassisi, M. Picketta, M. Heuer, E. Weber, Mater Sci. Eng. B **134**, 282 (2006)
6. M.M. Heyns, T. Bearda, I. Cornelissen, S. DeGendt, R. Degraeve, G. Groeseneken, C. Kenens, D.M. Knotter, L.M. Loewenstein, P.W. Mertens, S. Mertens, M. Meuris, T. Nigam, M. Schaekers, I. Teerlinck, W. Vandervorst, R. Vos, K. Wolke, IBM J. Res. Devel. **34** (1999)
7. D.A. Buchanan, E.P. Gusev, E. Cartier, H. Okorn-Schmidt, K. Rim, M.A. Gribelyuk, A. Mocuta, A. Ajmera, M. Copel, S. Guha, N. Bojarczuk, A. Callegari, C. D'Emic, P. Kozlowski, K. Chan, R.J. Fleming, P.C. Jamison, I. Brown, R. Arndt, 80 nm poly-silicon gated n-FETs with ultra-thin AI203 gate dielectric for ULSI applications, *IEDM Proceedings* (2000), pp. 223–226
8. R. Sinton, A. Cuevas, Appl. Phys. Lett. **69**, 2510 (1996)
9. G. Agostinelli, A. Delabie, P. Vitanov, Z. Alexieva, H. Dekkers, S. De Wolf, G. Beaucarne, *Solar Energy Mater. Solar Cells* **90**, 3438 (2006)
10. A. Richter, S. Henneck, J. Benick, M. Hörteis, M. Hermle, S.W. Glunz, Firing Stable Al$_2$O$_3$/SiN$_x$ Layer Stack Passivation for the Front Side Boron Emitter of n-type Silicon Solar Cells, *Proceedings of the 25th European Photovoltaic Solar Energy Conference and Exhibition*, Valencia, pp. 1453–1456
11. J. Benick, N. Bateman, M. Hermle, Very Low Emitter Saturation Current Densities on Ion Implanted Boron Emitters, *Proceedings of the 25th European Photovoltaic Solar Energy Conference and Exhibition*, Valencia, pp. 1169–1173
12. S. Gatz, H. Hannebauer, R. Hesse, F. Werner, A. Schmidt, T. Dullweber, J. Schmidt, K. Bothe, R. Brendel, Phys. Status Solidi RRL **5**, 147 (2011)
13. B. Vermang, H. Goverde, A. Lorenz, A. Uruena, G. Vereecke, J. Das, J. Meersschaut, P. Choulat, E. Cornagliotti, A. Rothschild, J. John, J. Poortmans, R. Mertens. On the blistering of

Al2O3 passivation layers for p-type Si PERC, *Proceedings of the 26th European Photovoltaic Solar Energy Conference and Exhibition* (Hamburg, 2011)

14. J. Moschner, J. Henze, J. Schmidt, R. Hezel, *Prog. Photovolt. Res. Appl.* **12**, 21 (2004)

15. M. Kerr, J. Schmidt, A. Cuevas, J. Bultman, J. Appl. Phys. **89**, 71

16. T. Janssens, N.E. Posthuma, B.J. Pawlak, E. Rosseel, J. Poortmans, Implantation for an Excellent Definition of Doping Profiles in Si Solar Cells, *Proceedings of the 25th European Photovoltaic Solar Energy Conference and Exhibition* (Valencia, 2010), pp. 1179−1181

17. A. Feltrin, A. Freundlich, Renew. Energy **33**, 180 (2008)

18. K. Van Wichelen, L. Tous, A. Tiefenauer, C. Allebé, T. Janssens, P. Choulat, J.L. Hernàndez, E. Cornagliotti, M. Debucquoy, A. Ruocco, J. John, P. Verlinden, F. Dross, K. Baert, Towards 20.5% efficiency PERC Cells by improved understanding through simulation, published in *Proceedings of Silicon PV 2011 (Energy Procedia)* (Freiburg, Germany, 2011)

5

Feasibility of using thin crystalline silicon films epitaxially grown at 165 °C in solar cells: A computer simulation study

S. Chakraborty[1], R. Cariou[2], M. Labrune[2,3], P. Roca i Cabarrocas[2], and P. Chatterjee[1,2,a]

[1] Indian Association for the Cultivation of Science, 700032 Kolkata, WB, India
[2] Laboratoire de Physique des Interfaces et des Couches Minces, École Polytechnique, 91128 Palaiseau, France
[3] Total S.A., Gas & Power – R&D Division, 92400 Courbevoie, France

Abstract We have previously reported on the successful deposition of heterojunction solar cells whose thin intrinsic crystalline absorber layer is grown using the standard radio frequency plasma enhanced chemical vapour deposition process at 165 °C on highly doped P-type (100) crystalline silicon substrates. The structure had an N-doped hydrogenated amorphous silicon emitter deposited on top of the intrinsic epitaxial silicon layer. However to form the basis of a solar cell, the epitaxial silicon film must be chiefly responsible for the photo-generated current of the structure and not the underlying crystalline silicon substrate. In this article we use detailed electrical-optical modelling to calculate the minimum thickness of the epitaxial silicon layer for this to happen. We have also investigated by modelling the influence of the a-Si:H/epitaxial-Si and epitaxial-Si/c-Si interface defects, the thickness of the epitaxial silicon layer and its volume defect density on cell performance. Finally by varying the input parameters and considering various light-trapping schemes, we show that it is possible to attain a conversion efficiency in excess of 13% using only a 5 micron thick epitaxial silicon layer.

1 Introduction

The rapid growth of the photovoltaic industry and the resultant shortage of silicon feedstock supply in the last decade, prompted interest in thin crystalline wafers obtained either by thinning down thick ones or by developing epitaxial growth processes. The latter option has gained particular importance [1–3] in view of the fact that the epitaxial silicon film can be "lifted off" from the c-Si substrate (or any other suitable substrate) on which it is grown, and be transferred to a foreign substrate [4], thus allowing for cost-saving via c-Si substrate re-use. In addition, Petermann et al. [4] demonstrated that it is possible to attain ∼19% efficiency in heterojunction solar cells with only a 43 micron thick intrinsic epitaxial silicon layer. However all these approaches involve temperature processes in excess of 600 °C, which limits the range of suitable substrates and often require post-hydrogenation to passivate the defects in the epitaxial silicon layer.

In the wafer equivalent approach used by Cariou et al. [5] the epitaxial silicon (epi-Si) films are deposited in a standard industrial radio frequency plasma enhanced chemical vapour deposition (RF-PECVD) system

on highly doped c-Si (100) substrates at 165 °C, allowing for the additional advantage of a low thermal budget with a standard industrial tool. In these epi-Si films high crystalline quality and low stress levels have been confirmed from HRTEM and Raman measurements. At such low growth temperatures it may seem normal for the epitaxial film to have a high density of defects; however a high fill factor of ∼79% has been achieved in ITO/N-a-Si:H/I-epi-Si (1.7 microns)/P++-c-Si/Aluminium type solar cells fabricated from these epi-Si layers, testifying to the excellent quality of the epitaxial films produced. In fact our low temperature material comes with high hydrogen content, which provides de facto good defect passivation. Moreover since for a satisfactory diffusion length, $L_{Eff} > 3d$, where d is the epitaxial layer thickness, a thinner film demands lower material quality. So far the maximum thickness of the epi-Si layer achieved, retaining good crystalline quality is ∼5 microns.

However to form the basis of a solar cell, the epitaxial silicon (epi-Si) film must be responsible for the photogenerated current of the structure. So far, in this so-called wafer equivalent approach, people made the assumption that the PV response coming from the highly doped wafer is negligible. But, to our knowledge, no quantitative study has been completed to determine whether a part of this

ᵃ e-mail: parsathi_chatterjee@yahoo.co.in

current is coming from the underlying P^{++}-c-Si substrate, nor has the minimum thickness of the epi-Si layer been determined, so that the major current contribution comes from it. This knowledge is of utmost importance, since our final aim is to achieve a lift-off from the c-Si wafer allowing for the re-use of the latter. As far as the epi-Si film itself is concerned, it has already been demonstrated [6,7] that by using a SiF_4 plasma etching step on the wafer substrate before the deposition of the epi-Si film, it is possible to produce a crystalline silicon layer having a porous interface with the c-Si wafer, which allows the epi-Si film to be easily detached from the substrate. It is naturally desirable that after such detachment, the photo-current of the device remains more or less the same. Also as it is likely that the first few layers of the epi-Si film at its interface with the P^{++}-c-Si may be more disordered, we need to know the minimum thickness of epitaxy required to reduce the defect density at the epi-Si/P^{++}-c-Si interface and thereby to minimise recombination at this interface, as well as to obtain a low defect density inside the epi-Si layer itself. Moreover the maximum open-circuit voltage (V_{oc}) and efficiency achieved so far are 0.55 V and 7% respectively in an actual solar cell having a 2.4 μm thick epi-Si layer and all flat interfaces. Therefore we need to optimise the structure to bring out the potential conversion efficiency achievable in epi-Si solar cells, remembering that it may not be possible (or desirable) to attain a thickness of ∼43 μm for the epi-Si films [4] at temperatures below 200 °C, which is the primary advantage of the present method, from the point of view of reduced material cost and low thermal budget.

In order therefore to answer the above questions and in general to study the feasibility of using such thin crystalline silicon films epitaxially grown at 165 °C in solar cells, we have simulated the deposited solar cells using the one dimensional detailed electrical-optical model "amorphous semiconductor device modelling program (ASDMP) [8,9]", to determine the fraction of the total photo-current actually coming from the epi-Si layer. We have also used it to extract the parameters that characterize these devices. Using these parameters to study the sensitivity of the solar cell output to various device and material parameters, we try to pinpoint the parameters to which the solar cell output is most sensitive and whose improvement is essential for the success of epi-Si solar cells. Finally, using the optical model built into ASDMP, we try to predict the maximum efficiency achievable in such structures for a practical thickness of 5 μm for the absorber epi-Si layer.

2 Experimental details

Heavily boron-doped (100)-oriented Si wafers with a resistivity of 0.02−0.05 Ω cm and a thickness of 525 μm were used as substrates for the epitaxial growth, as well as the electrical contact for the solar cell. The native oxide was removed from the surface of the c-Si wafer just before loading it into the multi-plasma monochamber (13.56 MHz) capacitively coupled RF-PECVD

reactor [10]. Undoped epitaxial Si layers of various thicknesses (0.9, 1.7 and 2.4 μm) were deposited from the dissociation of 6% silane in a hydrogen gas mixture under a total pressure of 2000 mTorr and a RF power density 60 mW cm^{-2}, resulting in a deposition rate of 1.5 Å s^{-1}. A standard N^+-a-Si:H emitter layer was deposited on top after passivating the epi-Si surface with a thin (∼3 nm) intrinsic a-Si:H layer without breaking vacuum and holding the temperature constant at 165 °C throughout the deposition process. The area of the cells (2 × 2 cm^2 for the largest ones) was defined by sputtering ITO through a shadow mask and evaporating aluminium grid contacts above. More details are given in reference [5]. All interfaces are flat and no light-trapping schemes have been introduced. External quantum efficiency (EQE) and current density-voltage (J-V) measurements under AM1.5 illumination were carried out to determine the solar cell output.

3 Simulation model

The one-dimensional "Amorphous Semiconductor Device Modeling Program (ASDMP) [8]", later extended to also model crystalline silicon and HIT cells [9], solves the Poisson's equation and the two carrier continuity equations under steady state conditions for a given device structure, and yields the dark and illuminated J-V and QE characteristics. The program is ab-initio in its electrical part, which is described in references [11,12]. The expressions for the free and trapped charges, the recombination term, the boundary conditions and the solution technique in this program are similar to the AMPS computer code [13]. The transport over the potential barriers at the contacts if any is by thermionic emission and across the N-a-Si:H/epi-Si heterojunction by both thermionic emission and electron diffusion. This is because in this structure electrons are collected at the N-a-Si:H end and, as will be seen later in Figure 5a, at this hetero-junction the conduction band edge is closer to the Fermi level on the I-epi-Si side than in N-a-Si:H, due to the conduction band discontinuity. This means that the free electron density is in fact higher on the epi-Si side than in N-a-Si:H, resulting also in electron diffusion across this heterojunction. The phenomenon is similar to the "inversion layer" observed at the N-a-Si:H/P-c-Si and P-a-Si:H/N-c-Si heterointerfaces [14]. The gap state model consists of the tail states and two Gaussian distribution functions to simulate the deep dangling bond states in the case of the amorphous layers, while in the epi-Si layer and c-Si substrate the tails are absent. The defect density on the surfaces of the epi-Si film is modelled by a defective layer 5 nm thick. This means that a volume defect density of ∼2×10^{17} cm^{-3} translates into a surface defect density (N_{ss}) of 10^{11} cm^{-2}.

The generation term has been calculated using a semi-empirical model [15] that has been integrated into ASDMP [8]. Both specular interference effects (for flat surfaces) and diffused reflectance, transmittance and light-trapping effects for structures with textured interfaces, are taken into account. The complex refractive indices of each layer are required as input to the modelling program. Light

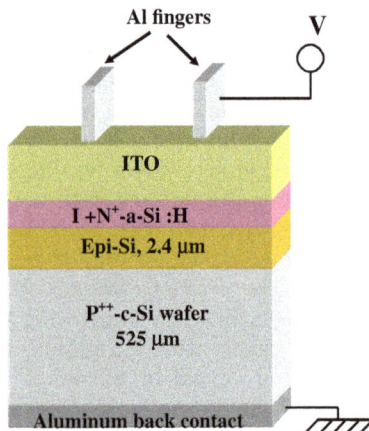

Fig. 1. Schematic diagram of the epitaxial silicon solar cell.

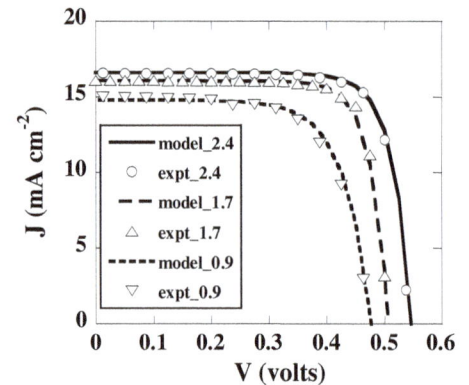

Fig. 2. The experimental (expt) and simulated (model) illuminated J-V characteristics of the epi-Si NIP cells, where the thickness of the epi-Si layer is 2.4 μm, 1.7 μm and 0.9 μm.

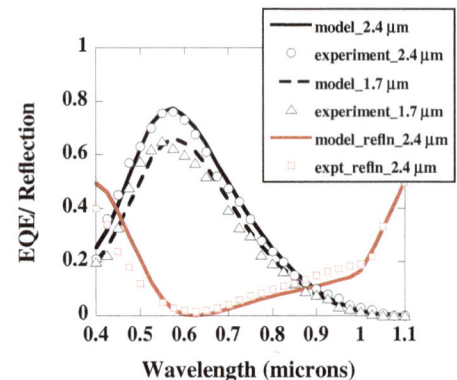

Fig. 3. The experimental (expt) external quantum efficiency (EQE) curves of the 2.4 μm and 1.7 μm epi-Si NIP cells and the reflection from the 2.4 μm device (symbols) under AM1.5 light, 0 V, compared to model results (lines).

enters through the ITO/emitter "front contact" which is taken at $x = 0$. Voltage is also applied here, while the Al back contact is at earth potential. The P$^+$-c-Si/Al "rear contact" at the back of the c-Si wafer is taken at $x = L$, where L is the total thickness of the semiconductor layers. The epitaxial c-Si film is assumed to have the same band gap as the c-Si substrate, so that there is only a single hetero-junction (HJ) at the N-a-Si:H/I-epi-Si interface. The top ITO/N-a-Si:H barrier height is assumed to be 0.2 eV, while that of the rear P$^+$-c-Si/Al is 1.06 eV.

4 Simulation of experiments

The structure modelled is shown in Figure 1. Since the macroscopic structural properties of epi-Si are consistent with c-Si, as deduced from Raman spectroscopy, spectroscopic ellipsometry and transmission electron microscopy measurements [5], the band gap and the complex refractive indices of the epi-Si layer are assumed to be the same as that for c-Si. Therefore this cell has only a single band discontinuity at the N-a-Si:H/I-epi-Si junction. Since it has been shown [16–18] that for an a-Si:H layer deposited via PECVD on c-Si (or, in this case on the epi-Si layer), the major part of the band offset is on the valence band side, we have apportioned two-thirds of the band discontinuity onto the valence band, and the rest onto the conduction band. This large valence band discontinuity (ΔE_v) at the N-a-Si:H/epi-Si junction has a beneficial influence on carrier collection, as it produces a strong field on the photo-generated holes at this junction in the right direction, viz., towards the back of the device, where holes are collected in this structure. Figure 2 compares the experimental illuminated J-V characteristics of epi-Si solar cells of three different thicknesses: 2.4 μm, 1.7 μm and 0.9 μm to model results, while Figure 3 compares the external quantum efficiency (EQE) of the two thicker cells and the reflection from the 2.4 μm cell to our simulations. In Table 1 we compare the measured and calculated solar cell output parameters.

In order to base our model predictions and sensitivity calculations on realistic device parameters, we have modelled the light J-V and EQE characteristics mentioned

Table 1. Comparison of the measured and simulated solar cell output parameters for the deposited epi-Si solar cells.

Case		J_{sc} (mA cm^{-2})	V_{oc} (V)	FF	Efficiency (%)
2.4 μm	experiment	16.6	0.546	0.77	7.0
cell	model	16.6	0.547	0.778	7.07
1.7 μm	experiment	16.1	0.510	0.786	6.4
cell	model	16.1	0.507	0.786	6.4
0.9 μm	experiment	15.0	0.478	0.66	4.7
cell	model	14.8	0.476	0.70	4.95

above. Table 2 and its caption present the input parameters extracted from the above modelling. Of course the parameters marked by superscript "a" in Table 2, such as the thickness of the individual layers, the band gap, the doping and defect densities inside the emitter (deduced from measured activation energies), and the carrier mobilities in the epi-Si layer and the highly doped defective P$^+$-c-Si wafer have been measured. The doping density in the P$^+$-c-Si wafer (marked by superscript "c") has been supplied by the manufacturer, while yet other parameters, marked by superscript "b" in Table 2, have been taken from the literature. The main parameters obtained by fitting the measured illuminated J-V and EQE curves

Table 2. The input parameters used to model the 2.4 μm epi-Si layer cell. Note that the quantities marked with superscript "a" have been measured, those marked with superscript "b" have been taken from the literature, and that marked with superscript "c", supplied by the manufacturer. The other parameters and the interface defect densities have been extracted by simulation. In this case (2.4 μm cell), the defect density at the top N-a-Si:H/I-epi-Si interface is 10^{11} cm^{-2}, while that at the rear I-epi-Si/P$^+$-c-Si interface is 10^{12} cm^{-2}. For the case of the 1.7 μm cell all parameters are the same as for the 2.4 μm cell, only the N-a-Si:H layer appears to be $\sim$16 nm from modelling (as explained in the text) and the rear interface defect density is higher: 1.35×10^{12} cm^{-2}. However, for the 0.9 μm cell the defect density at both interfaces of the epi-Si layer had to be equal to 10^{12} cm^{-2} and the volume defect density 1.2×10^{16} cm^{-3}. The charged and neutral capture cross-sections (σ) of these defect states, have been assumed to be one order of magnitude higher than the corresponding σ's inside the volume of epi-Si for all cases, while a 0.5 μm region adjacent to the top DL is assumed to have σ's intermediate between the two and a defect density twice that in the rest of the volume.

Parameters	N-a-Si:H	I-epitaxial Si	P-c-Si
Thickness (μm)[a]	0.012	2.4	525
Mobility gap (E_μ) (eV)	1.8[a]	1.12[a,b]	1.12[b]
Doping density (cm^{-3})	1.1×10^{19a}	0	3.0×10^{18c}
Electron affinity (eV)[b]	4.0	4.22	4.22
Eff. DOS in bands N_c^b	2×10^{20}	2.8×10^{19}	2.8×10^{19}
(cm^{-3}) N_v^b	2×10^{20}	1.04×10^{19}	1.04×10^{19}
Charac. En. E_D (E_A) (eV)[b]	0.05 (0.03)	–	–
G_{D0}, G_{A0} (cm^{-3} eV^{-1})[b]	4.0×10^{21}	–	–
Elec. (hole) mobility (cm^2/V S)	20 (4)	400 (125)[a]	300(100)[a]
Gaussian defect den. (cm^{-3})	6.0×10^{18a}	10^{15}	6.0×10^{17}
Donor Gaussian peak pos w.r.to VB (eV)	0.7	0.4	0.4
Acceptor Gaussian peak pos w.r.to CB (eV)	0.6	0.4	0.4
Capture cross-sections (cm^2)			
mid-gap: charged (neutral)	10^{-14} (10^{-15})	10^{-16} (10^{-17})	4×10^{-17} (4×10^{-18})
tails: charged (neutral)	10^{-14} (10^{-15})	–	–

therefore are the defect densities at the front and back of the epi-Si layer and the defect density in the volume of this layer. The charged and neutral capture cross-sections (σ) of these defect states, have been assumed to be one order of magnitude higher than the corresponding σ's inside the volume of epi-Si for all cases; while a 0.5 μm region adjacent to the top DL is assumed to have σ's intermediate between the two and a defect density twice that in the remainder of the epi-Si film. Without this latter assumption, we find that the V_{oc} and FF of the cells are overestimated. The defect densities at different points of the structure were extracted in the following way:

In order to fix the defect density at the I-epi-Si/P$^+$-c-Si interface in the 2.4 μm epi-Si layer cell, it was noted that a value of $N_{ss, back} \leqslant 7 \times 10^{11}$ cm^{-2}, leads to a higher V_{oc} but a *lower* FF than the experimental case. A higher V_{oc} for lower $N_{ss, back}$ is to be expected, but the lower FF is not. This fact will be explained in detail in Section 5.3.1. Moreover all aspects of solar cell performance deteriorate for $N_{ss, back}$ higher than 10^{12} cm^{-2} (Tab. 3). This prompted us to fix $N_{ss, back}$ for the 2.4 μm device at 10^{12} cm^{-2}, while the lower V_{oc} of the 1.7 μm device could be matched by assuming a higher $N_{ss, back}$ of 1.35×10^{12} cm^{-2} for this case. On the other hand, there is practically no sensitivity of the solar cell output to $N_{ss, front}$ (Tab. 3) upto $N_{ss, front} = 10^{12}$ cm^{-2}; V_{oc} and FF begins to deteriorate rapidly for higher values of $N_{ss, front}$. Therefore our choice of $N_{ss, front} = 10^{11}$ cm^{-2} at the top N-a-Si:H/I-epi-Si interface for the solar cells with 2.4 μm and 1.7 μm I-epi-Si

layers has been dictated by the fact that for these cases the defects at this hetero-interface have been carefully passivated by depositing a 3 nm thick intrinsic a-Si:H layer on epi-Si before the deposition of the emitter N-a-Si:H. However for the 0.9 μm device, it was found that a front interface defect density less than 10^{12} cm^{-2} leads to a slight over-estimation of V_{oc} and FF and so this latter value was taken as $N_{ss, front}$ for the 0.9 μm epi-Si case. On the other hand there is considerable sensitivity of V_{oc} and FF to the Gaussian defect density inside the epi-Si layer beginning from a low value of 10^{14} cm^{-3} for this parameter. Therefore this parameter could be fixed more uniquely and we extracted values of 10^{15} cm^{-3}, 10^{15} cm^{-3} and 1.2×10^{16} cm^{-3} for the 2.4 μm , 1.7 μm and 0.9 μm devices respectively. However as already stated in all cases, a 0.5 μm region next to the top DL had to be assumed more defective with higher σ's (Tab. 2 caption) than in the rest of the epi-Si film to ensure that the model calculations do not overestimate the V_{oc} and FF. It maybe relevant here to point out that the EQE curves in Figure 2 did not help us much in determining the defect densities at different points of the device. It was in fact found that the short wavelength QE (SWQE) right up to the peak EQE were determined, for a given thickness of the flat ITO window layer, by the thickness of the N-a-Si:H emitter and the capture cross-sections of its defect states. The EQE beyond the peak was determined by the thickness of the epi-Si layer. From these arguments one would expect the same SWQE for both the 2.4 μm and the 1.7 μm cells.

Table 3. Sensitivity to the defect density at the rear epi-Si/P-c-Si interface ($N_{\text{ss, back}}$) and to that at the front N-a-Si:H/epi-Si interface ($N_{\text{ss, front}}$).

$N_{\text{ss, front}}$ (cm^{-2})	$N_{\text{ss, back}}$ (cm^{-2})	2.4 μm cell			
		J_{sc} (mA cm^{-2})	V_{oc} (V)	FF	Efficiency (%)
	10^{10}	16.63	0.592	0.748	7.37
	10^{11}	16.63	0.591	0.749	7.36
10^{11}	5×10^{11}	16.63	0.582	0.758	7.34
	10^{12}	16.63	0.547	0.778	7.07
	5×10^{12}	16.61	0.394	0.663	4.35
	10^{13}	16.60	0.360	0.629	3.76
10^{10}		16.63	0.547	0.778	7.07
10^{11}		16.63	0.547	0.778	7.07
5×10^{11}	10^{12}	16.68	0.546	0.778	7.09
10^{12}		16.75	0.542	0.780	7.09
5×10^{12}		16.77	0.404	0.648	4.39
10^{13}		16.41	0.369	0.596	3.61

As this is not the case (Fig. 2), we had to assume that the N-a-Si:H emitter is thicker – 16 nm instead of 12 nm for the 1.7 μm cell. A 12 nm N-a-Si:H layer with higher capture cross-sections for the 1.7 μm cell was found to yield the same solar cell output, but the EQE is over-estimated.

In summary it has been shown that the defect density at the top N-a-Si:H/I-epi-Si interface is 10^{11} cm^{-2}, while at the rear I-epi-Si/P-c-Si interface it is 10^{12} cm^{-2} for the 2.4 μm cell, as inferred from modelling. In the case of the 1.7 μm cell all parameters, including the defect density inside the epi-Si layer are the same as for the 2.4 μm cell; only the N-a-Si:H layer appears to be ~16 nm and the rear interface defect density is higher: 1.35×10^{12} cm^{-2}. However for the 0.9 μm cell, the interface defect density at both interfaces had to be taken as equal to 10^{12} cm^{-2}; as well the volume defect density inside the epi-Si layer was found from modelling to be more than 1 order of magnitude higher: ~1.2×10^{16} cm^{-3}. Clearly the epi-Si layer has to be at least 1.7 μm thick and preferably more than ~2 μm thick for good photovoltaic (PV) performance. This point will be discussed further in Section 5.2.

5 Results and discussion

In this section we will first try to understand the origin of the photo-generated current – whether mainly from the epi-Si layer or from the c-Si wafer. We will also try to understand at what thickness of the epi-Si layer does the main contribution to the photo-current come from it. We will also try to determine the optimum thickness of the epi-Si layer for the highest cell efficiency. Thereafter, the sensitivity of the solar cell output to the defect states on the top surface of the epi-Si film (facing the emitter layer), the rear surface (contacting with the doped P^{+}-c-Si wafer), and the volume defect density inside the epi-Si film, will be investigated.

It may be noted from Table 1 that the epitaxial solar cells have J_{sc} and V_{oc} lower than those of the standard

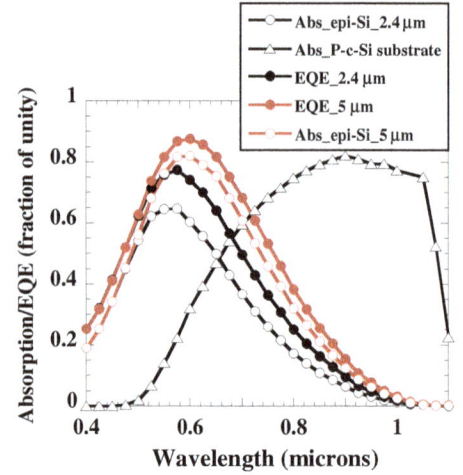

Fig. 4. Model calculations of the absorption (Abs) in the 2.4 μm epitaxial silicon (epi-Si) layer and the P-type c-Si substrate (black lines with open circles and open triangles respectively) and the external quantum efficiency (EQE – black line with closed circles) of the 2.4 μm cell. Also shown are the calculated values of the absorption in the epi-Si layer (red line with open circles) and the EQE (red line with closed circles) if the epi-Si layer were to be 5 μm thick.

diffused P/N junction mono-c-Si solar cell, while the fill factor (FF) is comparable. One obvious reason for the rather low J_{sc} is that in the present calculations all interfaces are assumed to be flat, resulting in current losses in the poor quality P^{+}-c-Si wafer. In other words, no light-trapping schemes are included. We will therefore study by modelling the influence of texturing the top and bottom faces of the epi-Si film on J_{sc}. Regarding the low V_{oc}, one reason is surely the rather high defect density at the epi-Si/P-c-Si interface in our solar cells. Here there is no band discontinuity, and holes photo-generated inside the c-Si substrate can easily back diffuse to give a high recombination at this interface when the defect density here is high as is the case for all the deposited cells ($N_{\text{ss}} \geqslant 10^{12}$ cm^{-2}). Other factors like the volume defect density in the epi-Si layer itself may also play a role. Finally we need to investigate whether an inversion of the structure by growing the epi-Si layer on an N-type instead of on a P-type c-Si substrate can lead to the possibility of a higher V_{oc}. We will investigate or comment on these factors in the following sections.

5.1 Origin of photo-generated current in these structures

In order to understand the origin of the photo-generated current in this structure – whether mainly from the epi-Si layer or from the c-Si wafer – we plot in Figure 4 the absorption profile in the epi-Si and c-Si layers as well as the external quantum efficiency of the 2.4 μm cell. We note that a significant contribution to the peak EQE comes from the heavily doped P^{+}-c-Si wafer, which has however been assumed defective. This latter fact is

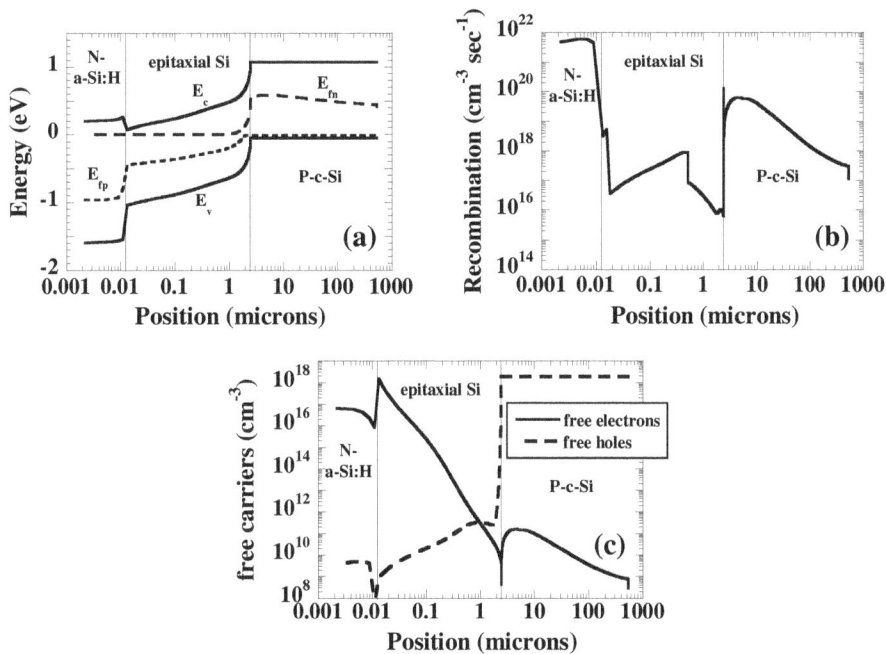

Fig. 5. (a) The energy band diagram, (b) the recombination rate and (c) the free carrier population as a function of position in the device for the solar cell having a 2.4 μm epitaxial silicon intrinsic layer under AM1.5 light and short-circuit conditions.

responsible for the low EQE of the device at long wavelengths. In fact results indicate that electron-hole pairs photo-generated in the front part of the c-Si wafer are collected but not those generated by light deep inside it. To understand this, in Figure 5a we plot the band diagram of the structure, shown schematically in Figure 1. Because of the high electric field at the epi-Si/P-c- Si interface (as evident from Fig. 5a) electrons and holes photo-generated in the front part of the P-c-Si wafer are well-separated. Since the epi-Si layer is of high quality (Tab. 2), the electrons get relatively easy passage through it and are collected at the front contact. The corresponding holes can diffuse through the P-c-Si wafer without recombining, as this wafer is rich in holes. However there is little field in the interior of the c-Si wafer (as maybe inferred from Fig. 5a) to separate the photo-generated carriers, and with the large number of defect states assumed in this heavily P-doped substrate (Tab. 2), considerable recombination occurs (Fig. 5b) resulting in a very low collection from the bulk of the wafer, and consequent low EQE over the longer wavelengths (Figs. 3, 4). It is interesting to note from Figure 5b that the recombination inside the epitaxial silicon layer is lower on either side compared to the centre. This is because of the high field at the interfaces and since there are very few holes available at the front N-a-Si:H/epi-Si interface (because of the strong favourable field here on the holes due to the large valence band discontinuity (ΔE_v) pushing the holes towards the back of the device); and because the free electron population is much lower than that of the free holes at and near the rear epi-Si/P-c-Si interface (Fig. 5c). However, photo-generated hole trapping at the rear interface when the defect density here is high (10^{12} cm^{-2} in this case) produces a large band bending (Fig. 5a) leading to high field very close to the

interface and lowering the field over the interior of the epi-Si layer in this 2.4 μm device. Thus the recombination in the interior of the epi-Si layer is high (Fig. 5b). However the strong change in the recombination rate at ~0.5 μm is due to the fact that over a ~0.5 μm region adjacent to the top DL, higher capture cross-sections and defect density than those in the rest of the epi-Si film had to be assumed as already discussed and mentioned in the caption to Table 2.

For comparison we have also shown in red on Figure 4, the calculated values of the light absorbed in the epitaxial Si layer, if it was 5 μm thick (open circles with line) and the corresponding EQE (closed circles with line). We now find that the contribution to the EQE of the device mainly comes from the light absorbed in the thicker epitaxial layer (5 μm), leading to higher currents and efficiency, as will be discussed in Section 5.2.

5.2 What is the optimum thickness of the epi-Si layer for the highest cell efficiency?

The sensitivity of the solar cell output to the thickness of the epi-Si layer is shown in Figure 6, which has been drawn using the parameters of the 2.4 micron cell (Tab. 2), except for the 1.7 μm cell. This is because the thickest (2.4 μm) cell has the best parameters of all the cells so far deposited (Tab. 2 and discussion in Sect. 4). In the case of the 1.7 μm cell, the actual parameters extracted by modeling this case (Sect. 4 and caption of Tab. 2) have been utilised. The abrupt fall in the V_{oc} when the epi-Si layer thickness is 1.7 μm, is due to a higher defect density at the rear epi-Si/P-c-Si interface for this case (Sect. 4). Moreover we have not included the case of 0.9 μm in Figure 6,

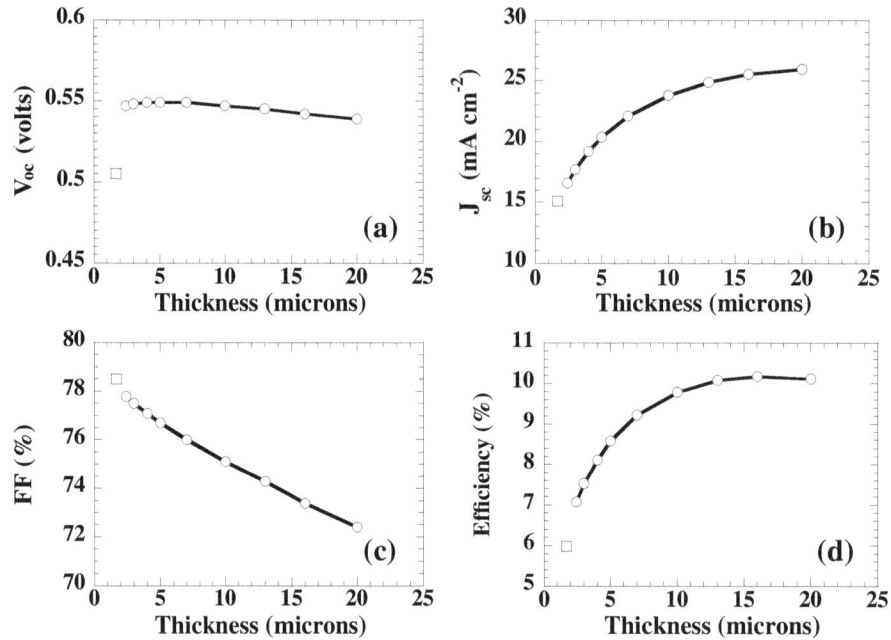

Fig. 6. The solar cell output as a function of the thickness of the epitaxial silicon layer. The sharp fall in V_{oc} of the 1.7 micron cell is due to the fact that modelling of this cell's J-V and EQE characteristics indicate that it has a more disordered interface layer at the epi-Si/P-c-Si interface than is present in the 2.4 micron cell. Therefore a different symbol is used for this case. The thicker cells have been assumed to have the same parameters as the 2.4 micron cell.

as the results of Table 1 indicate an all-round deterioration of cell performance due to a general increase of defect densities for the 0.9 μm cell (Tab. 2 and discussion in Sect. 4). We note from Figure 6d that the efficiency continues to increase up to an epitaxial layer thickness of 17 μm; however, the increase is more prominent for up to 10 microns.

It may be noted that in these results no light-trapping schemes have been considered. The optimum thickness of 17 μm is likely to be radically reduced when one or both surfaces of the epi-Si layer is textured.

5.3 Sensitivity of the solar cell output to the defect states at the rear epi-Si/P-c-Si interface and to those at the front N-a-Si:H/epi-Si interface

In the first part of Table 3 we show the sensitivity of the solar cell output to the defect states at the rear epi-Si/P-c-Si interface and in the second part, the sensitivity to the front N-a-Si:H/epi-Si interface.

5.3.1 Sensitivity of the solar cell output to the defect states at the rear epi-Si/P-c-Si interface

From Table 3 we note that up to $N_{\mathrm{ss,\,back}} = 5 \times 10^{11}$ cm^{-2}, the efficiency remains constant, a slight fall in V_{oc} being compensated by a corresponding increase in FF. Even up to $N_{\mathrm{ss,\,back}} = 10^{12}$ cm^{-2}, FF continues to increase, although now it is not fully able to annul the fall in V_{oc} and the efficiency begins to decrease. However beyond a defect density of 10^{12} cm^{-2}, V_{oc}, FF

and efficiency fall drastically, only the current remaining unchanged. To understand the sharp fall in V_{oc} beyond $N_{\mathrm{ss,\,back}} = 5 \times 10^{11}$ cm^{-2} and especially after 10^{12} cm^{-2}, we plot in Figure 7 the electric field at and near the rear DL of the 2.4 μm epi-Si layer cell under AM 1.5 light and open-circuit condition (Fig. 7a) and in Figure 7b the field in an expanded scale inside the epi-Si layer close to this DL. They indicate that for high values of N_{ss} in this DL on the rear surface of epi-Si, the field is localised, resulting in less field penetration into the epi-Si layer and lowering the V_{oc}.

In Figure 8a we trace the band diagram and in Figure 8b the recombination over the 2.4 μm epi-Si layer under AM1.5 light and *short*-circuit conditions. They indicate a sharp increase in band bending at the epi-Si/P-c-Si interface for the highest $N_{\mathrm{ss,\,back}}$, resulting in less field penetration into the bulk of the epi-Si layer. However, it is interesting to note the increase in the FF up to $N_{\mathrm{ss,\,back}} = 10^{12}$ cm^{-2}, that is particularly sharp at the latter value of $N_{\mathrm{ss,\,back}}$. In order to understand we repeat the recombination plot in Figure 8c, this time focusing on the back defective layer on epi-Si and plot the electric field with the same focus in Figure 8d. From this, it appears that the increased FF at $N_{\mathrm{ss,\,back}} = 10^{12}$ cm^{-2} is the effect of competition between recombination in the volume of the epi-Si layer (Fig. 8b) and increased field in the front part of the P-type c-Si wafer (Fig. 8d). We have demonstrated in Section 5.1 that carriers from the front part of the c-Si wafer do contribute to the photo-current in the 2.4 μm epi-Si device. This is also the reason for the constancy of the photo-current even for high values of $N_{\mathrm{ss,\,back}}$ (e.g. 10^{13} cm^{-2}), the increased recombination

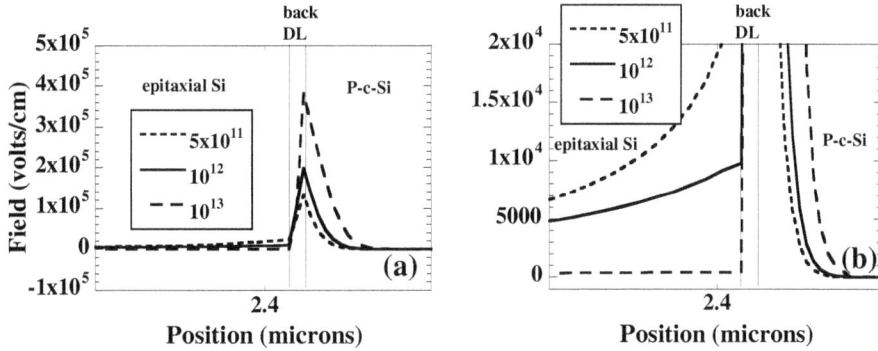

Fig. 7. The electric field (a) inside and near the rear defective layer (DL) on the epi-Si surface and (b) this field shown in an expanded scale inside the epi-Si layer close to this DL, for the 2.4 μm epi-Si layer cell under AM1.5 light and open-circuit conditions with the defect density at the epi-Si/P-c-Si interface as a parameter.

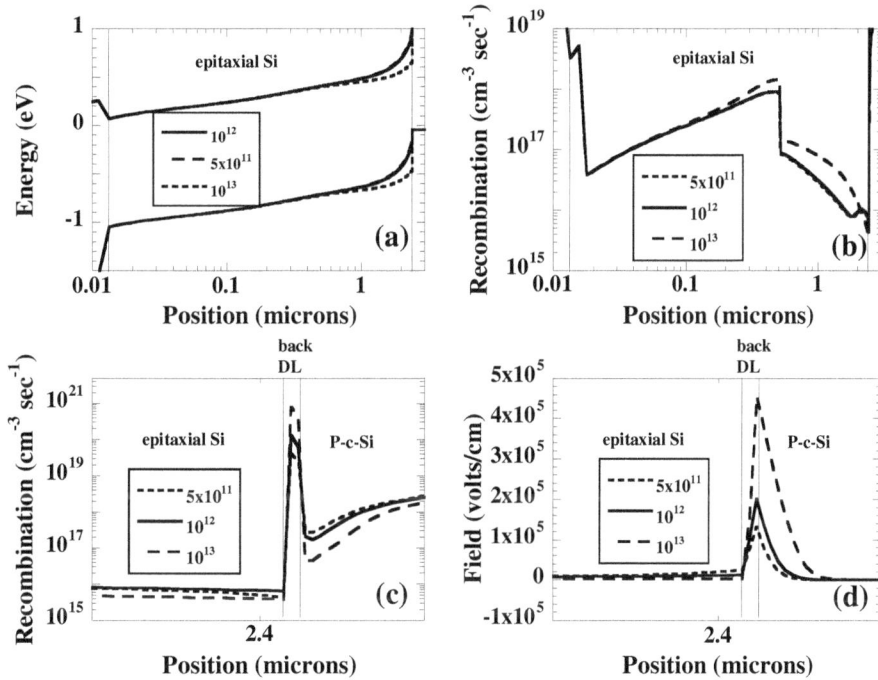

Fig. 8. (a) The energy band diagram, (b) the recombination over the epi-Si layer and (c) around the back interface region; and (d) the electric field over the latter region for the 2.4 μm epi-Si layer cell under AM1.5 light and short-circuit conditions with the defect density at the epi-Si/P-c-Si interface as a parameter.

over the epi-Si layer (Fig. 8b) in this case being compensated for by the higher field (Fig. 8d) and therefore shows decreased recombination (Fig. 8c) over the front part of the c-Si wafer.

To come back to the reasons for the abrupt increase of the FF at $N_{ss, back} = 10^{12}$ cm^{-2}, we note from Figure 8d that with increasing $N_{ss, back}$ the electric field improves appreciably over the front part of the c-Si wafer leading to reduced recombination from this region (Fig. 8c). However for values of $N_{ss, back}$ higher than 10^{12} cm^{-2}, we find from Figure 8b that recombination increases sharply over the volume of the epi-Si layer because of the flattening of the bands (Fig. 8a), so that the extra electrons saved from recombination in the c-Si wafer recombine with back diffusing holes from c-Si inside the epi-Si layer. Thus the collection of carriers and therefore the FF decrease

for $N_{ss, back} > 10^{12}$ cm^{-2}. However, up to $N_{ss, back} = 10^{12}$ cm^{-2}, the improved field and hence reduced recombination in the front part of the P-c-Si wafer (Figs. 8d and 8c) is not matched by a corresponding increase of recombination in the epi-Si layer (Fig. 8b), so that the collection of carriers and hence the FF improve.

5.3.2 Sensitivity of the solar cell output to the defect states at the front N-a-Si:H/epi-Si interface

This is shown in the latter part of Table 3. We note that while the solar cell output begins to deteriorate for $N_{ss, back} > 5 \times 10^{11}$ cm^{-2}, in the present case the output is practically unchanged up to $N_{ss, front} = 10^{12}$ cm^{-2}. The reason for this is as follows: We have already stated

Table 4. Sensitivity to the Gaussian defect density inside the epitaxial silicon layer.

DOS in epi-Si (cm^{-3})	2.4 μm cell			
	J_{sc} (mA cm^{-2})	V_{oc} (V)	FF	Efficiency (%)
10^{13}	16.62	0.558	0.815	7.56
10^{14}	16.62	0.556	0.812	7.51
10^{15}	16.62	0.547	0.778	7.07
5×10^{15}	16.60	0.511	0.721	6.12
10^{16}	16.54	0.484	0.7	5.6
5×10^{16}	15.92	0.410	0.646	4.22
10^{17}	15.19	0.379	0.622	3.58

in Section 4 that the major part of the band discontinuity lies on the valence band side for such PECVD deposited samples. It is this large valence band discontinuity (Fig. 5a) that results in a strong N-a-Si:H/epi-Si beneficial interface field that sweeps the photo-generated holes in the right direction, in other words towards the back P-c-Si/metal contact. Thus there are very few holes near the N-a-Si:H/epi-Si junction (Fig. 5c), resulting in negligible recombination that is responsible for the total lack of sensitivity of J_{sc} to $N_{ss, front}$ and of even the V_{oc} and FF up to $N_{ss, front} = 10^{12}$ cm^{-2}. It is only when $N_{ss, front}$ exceeds 10^{12} cm^{-2} that V_{oc}, FF and hence the efficiency fall sharply.

5.4 Sensitivity to the Gaussian defect density in the epitaxial silicon layer

Since, the macroscopic structural properties of these epi-Si layers are consistent with c-Si [5] we have assumed that as in c-Si, there are no tails in the band gap of this epitaxial silicon layer. The average Gaussian defect density in it as extracted by modelling is 10^{15} cm^{-3} (Tab. 2). We have also shown in Figure 4 that the major contribution to the quantum efficiency even for the 2.4 μm epitaxial silicon cell comes from the epitaxial layer itself. It is therefore only to be expected that there would be a strong sensitivity to this parameter. This is shown in Table 4 and in fact all the solar cell output parameters deteriorate when the defect density inside the epitaxial silicon layer increases. In addition the deposited 2.4 μm cell, whose dangling bond defect density, as extracted by modelling (Tab. 2) is 10^{15} cm^{-3} may in fact attain a FF of 0.812, equal to c-Si or "Heterojunction with Intrinsic Thin layer (HIT)" solar cells, if its dangling bond defect density can be reduced to 10^{14} cm^{-3}.

5.5 Cumulative improvement of the epitaxial silicon solar cell output as predicted by modelling

In this sub-section we begin with the solar cell output parameters already attained in the deposited solar cells and point out the improvements suggested by modelling. In the process we also briefly describe the effect of randomly texturing one or both faces of the epitaxial silicon

layer on device performance, using the optical model integrated into ASDMP [8]. In the model a ray impinging on a rough interface is divided into 4 components – specular reflection and transmission obeying the Fresnel's law and diffused reflection and transmission that in general have both angular and wavelength dependence. For the diffused components, coherence is assumed to be lost, so that the point where the light impinges is like a new source emitting in all directions. However no wavelength dependence is assumed for the rear textured interface (epi-Si/P-c-Si) at present in this model. Cumulative improvement of the cell output is given in Table 5.

We find that increasing the epi-Si layer thickness to 5 μm increases the current output of the device, and although the FF falls slightly, there is a net gain in efficiency of 1.5%. Our experimental group has recently been successful in depositing epi-Si layers 5 μm thick (not yet reported). Therefore we hold this thickness constant for the subsequent improvement steps. Diminution of the epi-Si/P-c- Si interface defect density ($N_{ss, back}$) from 10^{12} cm^{-2} to a moderate 5×10^{11} cm^{-2} mainly brings up the V_{oc}. Thereafter we initially consider the effect of randomly texturing only the top N-a-Si:H/epi-Si interface and finally both the top and bottom interfaces of the epi-Si layer. The resultant increase in the short-circuit current density leads to a final J_{sc} of 29.21 mA cm^{-2} and an efficiency of 13.02% for a 5 μm thick epi-Si layer solar cell. In fact the final current level achievable compares favourably with that of mono-c-Si solar cells.

Table 5 indicates that with light trapping effects in place, we can now achieve satisfactory values of both J_{sc} and FF. Only the V_{oc} is still far inferior to HIT solar cells. In fact the structure studied is a NIP solar cell, the emitter being N-a-Si:H. It is known that due to a more favourable band diagram, the reverse structure PIN (emitter P-a-Si:H), that is, when the epi-Si layer is deposited on a highly N-doped c-Si substrate, should have a higher V_{oc} (higher built-in potential) although at the cost of a lower FF, because in this case the valence band discontinuity would act to hamper hole extraction from the device. It may however be pointed out that the final aim of depositing such epi-Si solar cells is to lift the epi-Si film off the c-Si wafer, thus leading to cost saving from wafer reuse. When this is done both the epi-Si film deposited on the P-type wafer and that deposited on the N-type wafer would have valence band discontinuities from amorphous contacts deposited at both ends. In fact the resultant epi-Si cells would have double hetero-junctions, one at either end of the epi-Si layer. In such cases, the valence band discontinuity hampers hole collection either at the emitter or the BSF ends. Thus the FF advantage of the NIP structure would be lost. However the superior V_{oc} potential of the PIN structure would remain, so that a PIN structure epi-Si solar cell is expected to also show an improved V_{oc}.

6 Conclusions

We have simulated a thickness series of epitaxial silicon solar cells grown on a highly doped P-type c-Si substrate,

Table 5. Cumulative improvement of the solar cell output parameters by improving various device and material parameters. The parameter that is improved at each stage leading to a higher efficiency, is given in bold italics.

Different solar cell input parameters	J_{sc} (mA cm^{-2})	V_{oc} (V)	FF	Efficiency (%)
Cell deposited experimentally (2.4 μm epi-Si layer) – all flat interfaces	16.60	0.546	0.77	7.00
Model results for the above cell	16.63	0.547	0.778	7.07
Epitaxial layer thickness = 5 μm	*20.35*	0.549	0.767	8.57
$N_{ss, back}$ reduced to 5×10^{11} cm^{-2}	20.36	*0.580*	0.747	8.83
N-a-Si:H/epi-Si interface textured	*25.53*	0.589	0.750	11.27
Both top & bottom of epi-Si textured	*29.21*	0.594	0.750	13.02

with a flat ITO/N-a-Si:H window design. We have assumed that the macroscopic structural properties of the epi-Si film, such as the band gap, no band tails and the complex refractive indices are the same as that of c-Si, based on experimental findings on the epi-Si films [5]. Modelling indicates that optimised parameters are not attained in this type of solar cells for thickness $\leqslant 1.7$ μm of the epi-Si film, while the 2.4 μm deposited cell has low interface and bulk defect densities. However for the current in this N-a-Si:H/I-epi-Si/P$^+$-c-Si structure to come almost entirely from the epi-Si film, we need to increase the epitaxial silicon layer thickness further, to at least 5 μm. Simulations also reveal that solar cell output is more sensitive to the defect states at the rear epi-Si/P-c-Si interface than to those at the top N-a-Si:H/epi-Si junction, mainly because the valence band discontinuity at the latter interface yields a strong favourable field that aids hole collection. Using model ASDMP to calculate the sensitivity of the solar cell output to various device parameters, we show that an efficiency of 8.83% is practically attainable with a 5 μm epi-Si layer, which maybe increased further to 13.02% by introducing suitable light-trapping effects, thus increasing J_{sc} to 29.21 mA cm^{-2}. Final optimisation results presented in Table 5 indicate that V_{oc} still remains inferior to HIT solar cells. The inverse structure, with the epi-Si layer grown on a heavily doped N-type c-Si wafer needs to be studied experimentally and by modelling and may hold out promise of improved V_{oc} and a further gain in conversion efficiency.

This study was funded by the Ministry of New and Renewable Energy, Government of India, and by the *Centre National de la Recherche Scientifique* (CNRS), France. The computer modelling program was initially developed (electrical part) by P. Chatterjee during the course of a project funded by MNRE and DST, Government of India, and the optical part was added during her tenure as Marie Curie fellow *École Polytechnique*, Palaiseau, France. It was extended to model also HIT cells under a CSIR project.

References

1. R.B. Bergmann, C. Berge, T.J. Rinke, J. Schmidt, J.H. Werner, Sol. Energy Mater. Sol. Cells **74**, 213 (2002)
2. I. Kuzma-Filipek, K.V. Nieuwenhuysen, J.V. Hoeymissen, M.R. Payo, E.V. Kerschaver, J. Poortmans, R. Mertens, G. Beaucarne, E. Schmich, S. Lindekugel, S. Reber, Prog. Photovolt.: Res. Appl. **18**, 137 (2010)
3. K. Alberi, I.T. Martin, M. Shub, C.W. Teplin, M.J. Robero, R.C. Reedy, E. Iwaniczko, A. Duda, P. Stradins, H.M. Branz, D.L. Young, Appl. Phys. Lett. **96**, 073502 (2010)
4. H. Petermann et al., Prog. Photovolt.: Res. Appl. **20**, 1 (2012)
5. R. Cariou, M. Labrune, P. Roca i Cabarrocas, Sol. Energy Mater. Sol. Cells **95**, 2260 (2011)
6. M. Moreno, D. Daineka, P. Roca i Cabarrocas, Phys. Stat. Sol. C **7**, 1112 (2010)
7. M. Moreno, P. Roca i Cabarrocas, EPJ Photovoltaics **1**, 10301 (2010)
8. P. Chatterjee, M. Favre, F. Leblanc, J. Perrin, Mat. Res. Soc. Symp. Proc. **426**, 593 (1996)
9. M. Nath, P. Chatterjee, J. Damon-Lacoste, P. Roca i Cabarrocas, J. Appl. Phys. **103**, 034506 (2008)
10. P. Roca i Cabarrocas, J.B. Chevrier, J. Huc, A. Lloret, J.Y. Parey, J.P.M. Schmitt, J. Vac. Sci. Technol. A **9**, 2331 (1991)
11. P. Chatterjee, J. Appl. Phys. **76**, 1301 (1994)
12. P. Chatterjee, J. Appl. Phys. **79**, 7339 (1996)
13. P.J. McElheny, J.K. Arch, H.-S. Lin, S.J. Fonash, J. Appl. Phys. **64**, 1254 (1988)
14. J.P. Kleider, A.S. Gudovskikh, P. Roca i Cabarrocas, Appl. Phys. Lett. **92**, 162101 (2008)
15. F. Leblanc, J. Perrin, J. Schmitt, J. Appl. Phys. **75**, 1074 (1994)
16. J.M. Essick, J. David Cohen, Appl. Phys. Lett. **55**, 1232 (1989)
17. H. Matsuura, T. Okuno, H. Okushi, K. Tanaka, J. Appl. Phys. **55**, 1012 (1984)
18. H. Mimura, Y. Hatanaka, Appl. Phys. Lett. **50**, 326 (1987)

6

Novel texturing method for sputtered zinc oxide films prepared at high deposition rate from ceramic tube targets

E. Bunte, H. Zhu, J. Hüpkes[a], and J. Owen

Institut für Energie- und Klimaforschung, IEK5-Photovoltaik, Forschungszentrum Jülich GmbH, 52425 Jülich, Germany

Abstract Sputtered and wet-chemically texture etched zinc oxide (ZnO) films on glass substrates are regularly applied as transparent front contact in silicon based thin film solar cells. In this study, chemical wet etching in diluted hydrofluoric acid (HF) and subsequently in diluted hydrochloric acid (HCl) on aluminum doped zinc oxide (ZnO:Al) films deposited by magnetron sputtering from ceramic tube targets at high discharge power ($\sim$10 kW/m target length) is investigated. Films with thickness of around 800 nm were etched in diluted HCl acid and HF acid to achieve rough surface textures. It is found that the etching of the films in both etchants leads to different surface textures. A two steps etching process, which is especially favorable for films prepared at high deposition rate, was systematically studied. By etching first in diluted hydrofluoric acid (HF) and subsequently in diluted hydrochloric acid (HCl) these films are furnished with a surface texture which is characterized by craters with typical diameter of around $500-1000$ nm. The resulting surface structure is comparable to etched films sputtered at low deposition rate, which had been demonstrated to be able to achieve high efficiencies in silicon thin film solar cells.

1 Introduction

Transparent conducting oxide (TCO) films are the base for a wide range of sophisticated applications such as organic light-emitting diodes [1–3], flat panel displays [4, 5], and thin-film solar cells [6, 7] etc. Nowadays aluminium doped zinc oxide (ZnO:Al) films are becoming increasingly popular as contact and window layer, especially in the field of thin-film photovoltaics. Rough and light scattering surfaces are beneficial for the so-called light-trapping effect in thin-film silicon solar cells and lead to an increase of the conversion efficiency [8–13]. Commonly, the sputtering conditions for sputtered ZnO:Al films are optimized to tune the etching behaviours and to obtain adapted light scattering properties [10, 14, 15]. These surfaces exhibit regularly distributed craters with typical diameter of $1-2$ μm and depth of several 100 nm. The typical root mean square (RMS) roughness of these films is $110-150$ nm. Applying these films as substrate in microcrystalline silicon (μc-Si:H) thin film solar cells leads to an increase in photocurrent compared to the solar cells with smooth ZnO:Al as front contact. However, the deposition rate of the best lab-type radio frequency (RF) sputtering process is too low for industrial production of ZnO:Al films. Mid-frequency (MF) sputtering from rotatable dual magnetrons (RDM) may produce ZnO:Al films at deposition rates of more than 100 nm·m/min [16, 17]. However, the high rate sputtered ZnO:Al films usually exhibit sporadic craters on the surfaces after etching in

diluted hydrochloric acid (HCl) and the light scattering is inferior [16, 17]. In this contribution, we systematically study the influence of different etching methods, which include etching in diluted hydrofluoric acid (HF) and diluted HCl as well as combined two-step etching method in both etchants, on ZnO:Al films sputtered at high discharge power and different substrate temperatures. It was found that ZnO:Al films etched in HF show rather sharp features [18]. This may reduce the fill factor and thus, the efficiency of silicon based thin-film solar cells. In order to smoothen the sharp features and widen the craters, after the initial HF etching step a second etching step in HCl was carried out. Electrical and optical properties as well as the resulting surface topographies are investigated and discussed in detail for high rate films in particular deposited at 350 °C. Due to their low sheet resistance ZnO:Al films deposited at high temperature ($>$300 °C) are relevant for solar cells.

2 Experimental details

ZnO:Al films were prepared on Corning glass substrates (Corning, Eagle XG) in an in-line sputtering system (VISS 300 by Von Ardenne Anlagentechnik, Dresden, Germany). Rotatable dual magnetrons (RDM) with ZnO:Al$_2$O$_3$ (99.5:0.5 wt%) ceramic tube targets (750 mm length) were mounted in the deposition chamber. The sputtering process was operated under mid-frequency excitation at a frequency of 40 kHz. The substrate temperature was varied between 250 °C and 350 °C as measured

[a] e-mail: `j.huepkes@fz-juelich.de`

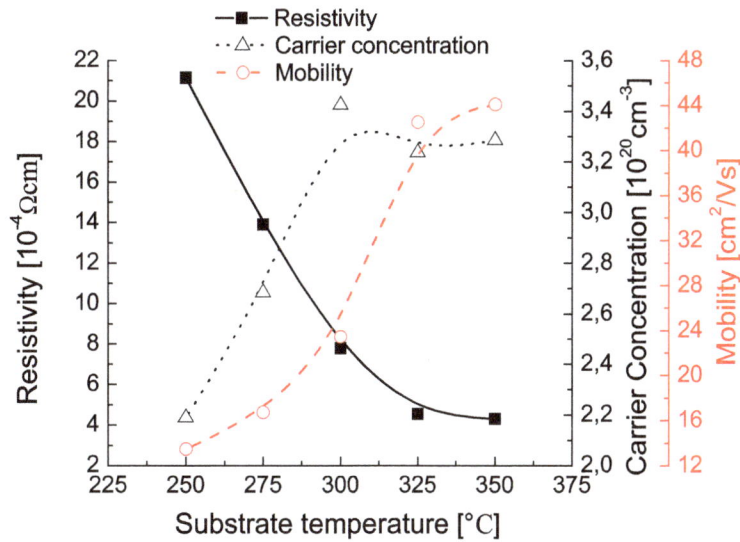

Fig. 1. Electrical properties of high rate ZnO:Al films deposited at different substrate temperatures.

by pyrometers prior to deposition. The applied discharge power was 14 kW and the working pressure was 2.0 Pa. The texturing of the as-deposited flat ZnO films (RMS roughness <10 nm) was carried out by wet chemical etching in diluted hydrochloric (HCl) and hydrofluoric (HF) acid. The concentrations were 0.5 wt% and 1 wt%, respectively. The etching time is 50 s for all films etched in HCl or HF solutions, if no other duration is given. Thickness before and after etching was measured by surface profiler (Dektak 3030, by Veeco Instruments).

The electrical properties of etched ZnO:Al thin films were characterized by four point probe as well as Hall effect measurements. Diffuse and total transmissions of surface textured films were measured by a double beam spectrometer with an integrating sphere (Perkin Elmer Lambda 19). The topographies of the samples were investigated by scanning electron microscopy (SEM, Supra 55VP SmartSEMTM, Carl Zeiss,Germany) and atomic force microscopy (AFM, Nanoscope system from Veeco as well as Surface Imaging System (SIS), Nano Station 300).

3 Results and discussion

3.1 Electrical properties

ZnO:Al films were sputtered at a high discharge power of 14 kW and the substrate temperature was varied between 250 °C and 350 °C as temperature is an important parameter to control material properties and etching behavior of ZnO:Al films [10]. Figure 1 shows the specific resistivity (ρ), carrier concentration (n) and carrier mobility (μ) as a function of the substrate temperature. As found in previous studies [7], with increasing substrate temperature the resistivity of the ZnO:Al films decreases. Here it drops from 2.2×10^{-3} Ω cm to 4×10^{-4} Ω cm and the carrier mobility increases from 14 cm^2/Vs to about 44 cm^2/Vs. The carrier concentration first increases with temperature to a plateau value of about of 3.3×10^{20} cm^{-3} for films

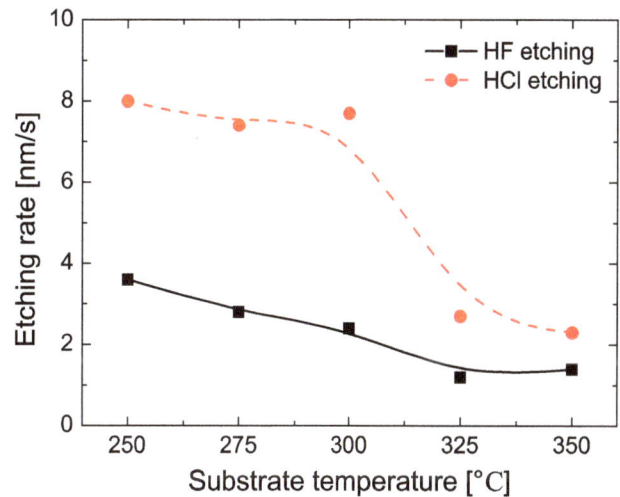

Fig. 2. Dependence of etching rate of high rate ZnO:Al films in HF etchant and HCl etchant on deposited substrate temperatures.

deposited at 300 °C or above. The increase of carrier concentration with substrate temperature is explained by an improved integration of aluminum into the ZnO crystal. At very high substrate temperatures this effect might be compensated by desorption of zinc atoms [19,20]. The carrier mobility in sputtered ZnO:Al films is affected mainly by scattering at the grain boundaries [21–24] and impurity atoms [19,25–27]. Here we consider mainly a decrease in the first scattering mechanism, as we observe an improved mobility in coincidence with higher carrier density [28,29].

3.2 Etching behaviours of ZnO:Al in acids

A. Comparison of HCl etching and HF etching

Figure 2 shows etching rates of high rate ZnO:Al films deposited at different substrate temperatures and then

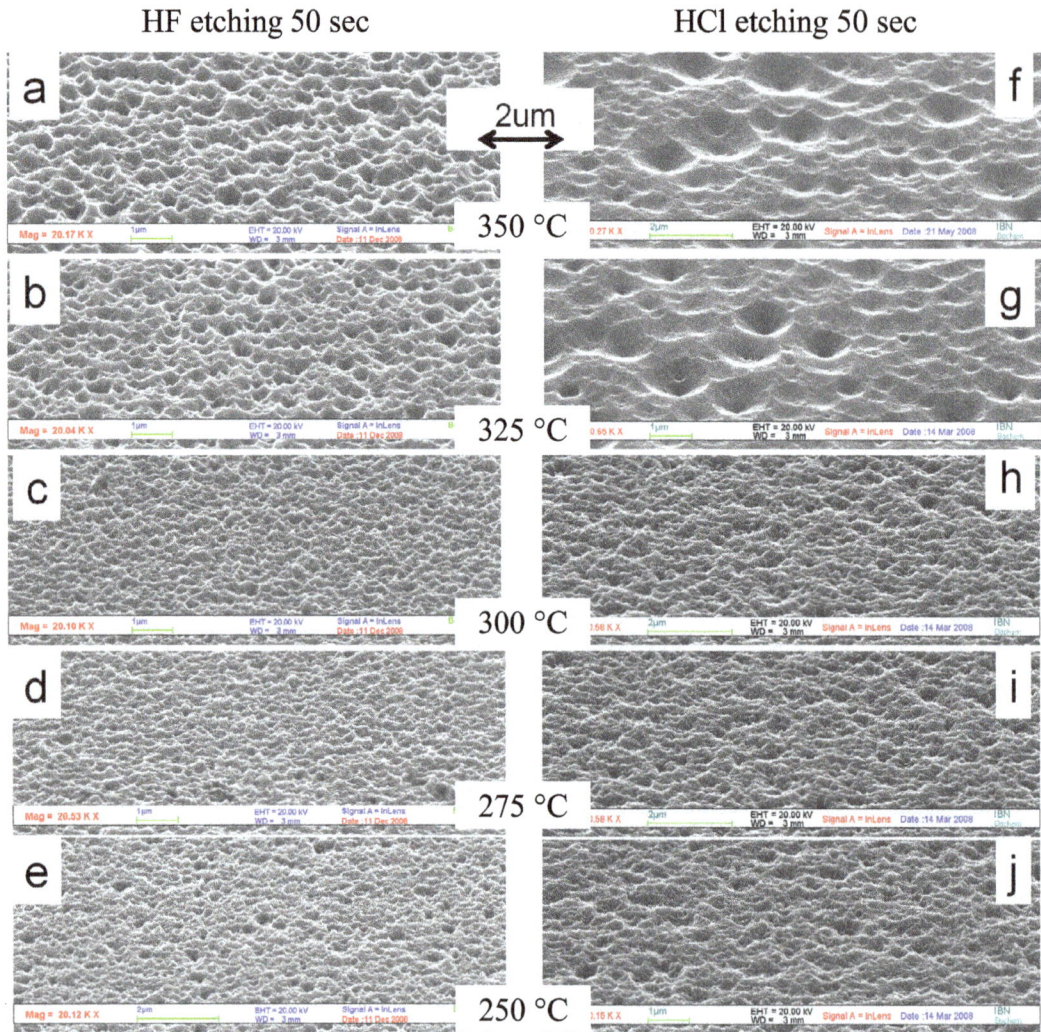

Fig. 3. The surface structures of films deposited at different substrate temperatures after etching in diluted HF solution (1%): (a) 350 °C, (b) 325 °C, (c) 300 °C, (d) 275 °C, (e) 250 °C as well as etching in diluted HCl solution (0.5%), (f) 350 °C, (g) 325 °C, (h) 300 °C, (i) 275 °C, (j) 250 °C.

etched in diluted HF solution and HCl solution, respectively. The etching rate of films decreases with rising substrate temperature for both etching methods. For HF etching method, the etching rate decreases from 3.6 nm/s to 1.5 nm/s with the increasing substrate temperature from 250 °C to 350 °C. For HCl etching method, the etching rate decreases in the same way from 8 nm/s to 2.3 nm/s. The reduction in etch rate at high substrate temperatures may be due to the fact that ZnO:Al films become much more compact by atoms with large diffusion energy at high substrate temperatures [15]. Etching rate in HF is almost a factor of two lower than that in HCl even though the concentration of HF is a factor of two higher than that of HCl. HCl is a strong acid while HF is a weak acid. In general, HCl would be completely dissociated into H^+ and Cl^-. Only a very small fraction of HF molecules are dissociated. The dissociation constant K_α of HF at room temperature is only in the range of $6-7.5 \times 10^{-4}$ while HCl dissociates nearly completely [30]. This means, that

1% HF etches slower than 0.5% HCl due to the lower pH value of the HF solution.

Figure 3 shows the SEM micrographs of the corresponding surface topographies: (a)−(e) after etching in HF (1 wt%), (f)−(j) after etching in HCl (0.5 wt%). The latter topographies of films prepared at high temperatures (350 °C and 325 °C) show a few wide but shallow craters with diameter of about 1−2 μm sporadically distributed on the surfaces as shown in Figure 3(f) and 3(g), respectively. Both surface topographies are similar to type C that Kluth et al. reported [15]. Microcrystalline silicon solar cells on this kind of substrate typically have good fill factors (>70%) but low photocurrents (<20 mA/cm^2 for 1.2 μm absorber layer thickness and a ZnO:Al/silver back contact) [31]. When a substrate temperature below 325 °C is applied during the sputtering process, the surface topographies of the etched films are characterized by small features with diameter below 100 nm (see Fig. 3(h)−3(j)) The height distribution function shows a more or less

Gaussian shape indicating a statistical rough surface. This topography is similar to type A as presented by Kluth et al. [15]. On type A substrates microcrystalline silicon solar cells usually lead to poor electrical performance [31]. Up to now the best surface topography of ZnO:Al films realized by sputtering and etching is characterized by regular, wide and deep craters with $1-2$ μm diameter (denoted as type B by Kluth et al. [15]). Unfortunately, a type B surface has not been achieved only by HCl etching step on the surfaces of these high deposition rate ZnO:Al films. In other words, type B surface cannot be realized or at least is very difficult to achieve through changing the main sputtering parameters i.e. substrate temperature and working pressure. According to the Kluth model [10, 15, 32] we would expect to find type B in a narrow process window with a substrate temperature between 300 °C and 325 °C. However, stop-and-go deposition as well as multi-layer formation during dynamic sputtering may complicate or even limit optimization [33–36].

Upon etching in diluted HF (1 wt %) the ZnO:Al film deposited at 350 °C (see Fig. 3(a)) develops a surface texture with small craters characterized by steep edges and the lateral feature size is typically between 200 nm and 500 nm. With the decrease of substrate temperature, the diameter and depth of these structures gradually decrease (see Fig. 3(b)–3(e)). For substrate temperatures between 300 °C and 250 °C, the films become relatively flat on the micrometer scale, but the nano-roughness looks similar to those obtained by etching in HCl as shown in Fig. 3(h)–3(j), even though the features are still smaller. More characteristic properties of the HF etch process and resulting ZnO:Al films are discussed elsewhere [18].

B. Optical properties

All films after etching in HF or HCl solutions show high average transmission of more than 80% in the visible spectrum region (not show here). The haze is defined as the ratio of diffuse transmission to total transmission and is used to distinguish the light scattering properties of the various surface topographies. Figure 4 shows haze of the ZnO:Al films after etching in HF (see Fig. 4(a)) and in HCl (see Fig. 4(b)).

A trend of increasing haze with higher substrate temperatures was found. The films etched in HF show in principle the same trend; however, the increase of haze with substrate temperature is more pronounced. The increased haze of high rate sputtered films deposited at high substrate temperatures is especially pronounced for wavelengths <600 nm. For ZnO:Al films deposited at 350 °C etched in HF a haze of 37% at 600 nm is found. All films etched in HCl and the ZnO:Al films etched in HF after preparation at low substrate temperature below 300 °C exhibit only low haze.

C. High temperature films etched in HF

To study the evolution of the HF etching process, the ZnO:Al films sputtered at high discharge power of 14 kW

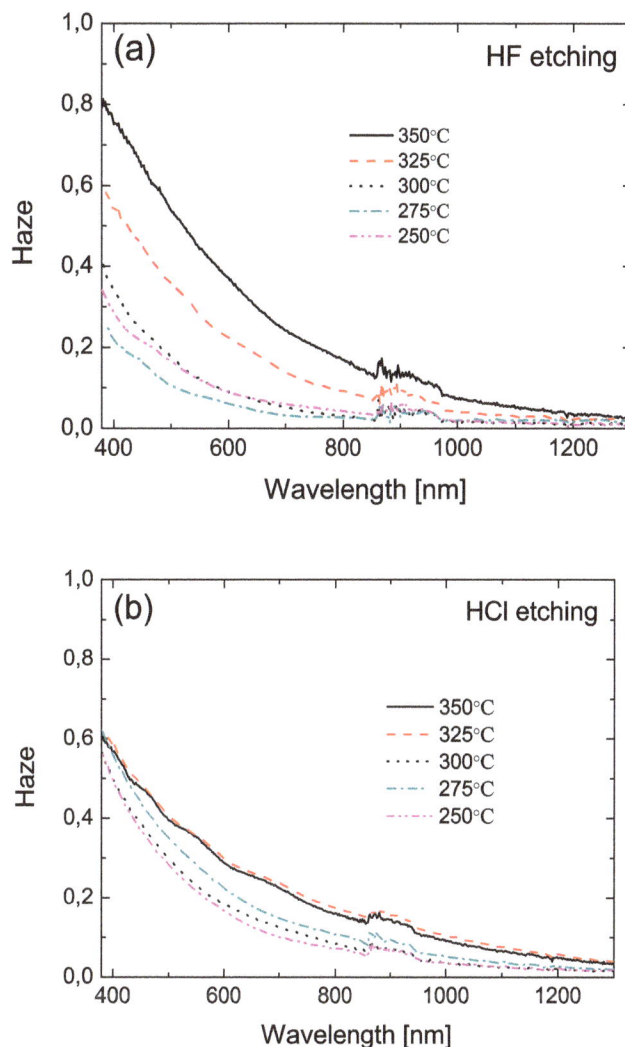

Fig. 4. Spectral haze of high rate ZnO:Al films deposited at various temperature and etched in HF etchant (a) and in HCl etchant (b).

and a substrate temperature of 350 °C were etched in 1% HF solution for different durations between 0 s and 180 s. Figures 5(a)–5(f) exhibit the surface structures after etching for 0 s, 60 s, 90 s, 120 s and 180 s, respectively, which were obtained by AFM measurements. By prolonging etching duration the feature size of the structures increases from typical diameters of around 100 nm for 60 s etching time (compare Fig. 5(b)) to 500 nm for 180 s (Fig. 5(f)). In parallel to the increase of the lateral feature size the vertical feature size increases. The RMS roughness derived from the AFM measurements increases almost linearly from ~10 nm for the unetched sample to 80 nm for the longest etching time (not show here). As already seen in Figure 3(a), films deposited at high rate and high temperature show a surface texture with rather steep edges after etching in HF. This observation is also valid for longer etching duration. Regardless of the increase of the lateral and vertical feature size with etching duration all investigated samples exhibit surfaces with sharp features.

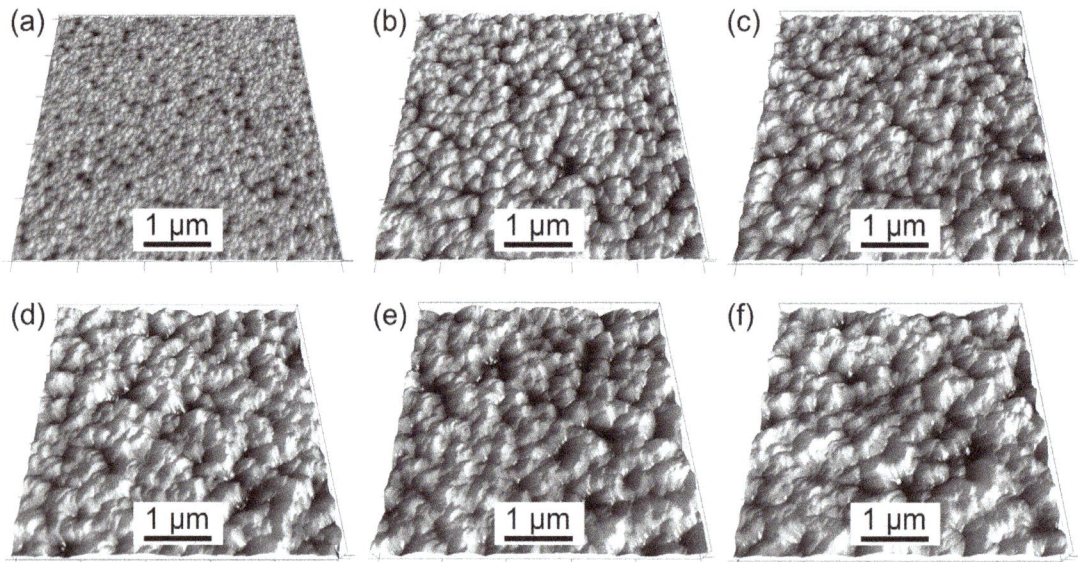

Fig. 5. AFM images of etch evolution upon HF etching for ZnO:Al film deposited at 350 °C. AFM images correspond to different etching time in diluted HF solution: (a) 0 s, (b) 60 s, (c) 90 s, (d) 120 s, (e) 150 s, (f) 180 s.

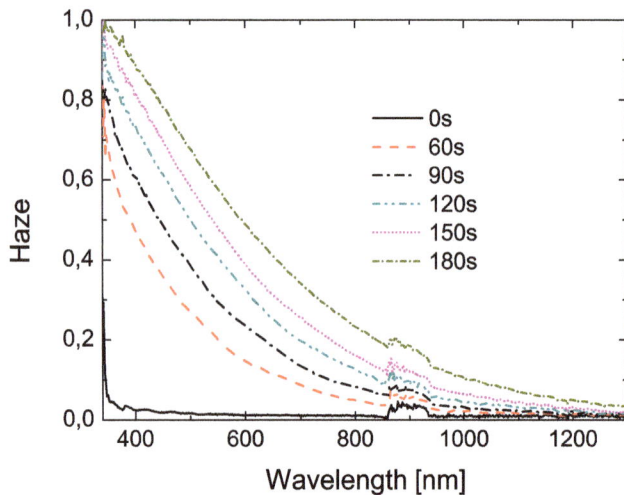

Fig. 6. Haze of high rate ZnO:Al films deposited at 350 °C after etching in diluted HF (1%) solution for various etching times.

Figure 6 shows haze as function of wavelength of ZnO:Al films shown in Figure 5. With increasing RMS roughness typically the haze increases. This also holds for the presented series as can be seen in Figure 6. The un-etched sample exhibits nearly no haze of close to "0" in the whole investigated wavelength spectrum (330 nm to 1300 nm). For the rough surfaces the typical curve progression can be seen: For short wavelengths at around 330 nm the haze is close to 1 and then it decreases towards longer wavelengths. With longer etching duration the haze increases in the observed wavelength range, e.g. at 600 nm, from 15% for the 60 s etched sample to 50% for the 180 s etched sample.

3.3 Two-step etching first in HF solution and then in HCl solution

As seen in the previous section, etching in HF results in a ZnO topography characterized by rather sharp features. This may lead to a deterioration of the fill factor of silicon thin-film solar cells [31]. Zhu et al. [37] developed a two-step etching method for high rate reactively sputtered ZnO:Al films. The idea is to utilize the HF etching to create the high density of craters and widen the craters by HCl etching. Therefore, we also employed such an etching method to our high rate ZnO:Al films prepared at 350 °C. The as-grown high rate ZnO:Al films were first etched in diluted HF solution (1%) for 120 s and then in diluted HCl solution (0.5%) for different etching time from 0 s to 16 s. Moreover, the ZnO:Al only etched in HCl etching for 50 s is also added to this study to make a comparison for different properties.

Figure 7 shows the variations of thickness and sheet resistance as well as RMS roughness of the ZnO:Al films after two-step etching. As-deposited values are given at negative x-values, 0 s corresponds to the film properties after the first HF etching step. The film thickness (solid squares) is reduced by 200 nm by the first HF etching step (see sample "0 s"). Consequently the sheet resistance (solid circuits) increases to 9 Ω. The RMS roughness (solid triangles) of that sample is ∼80 nm. By applying the second HCl etching step, the thickness is further reduced from 550 nm for the "2 s" sample to 450 nm for the "16 s" sample. The sheet resistance increases from 12 Ω to 18 Ω, respectively. The RMS roughness slightly increases up to 100 nm by the HCl etching for 2 s. For prolonged etching in HCl solution (from 4 s to 16 s) the RMS roughness keeps relatively constant. This behavior is expected for the widening effect on the craters by the HCl, because the depth distribution would not be modified. In all HF based

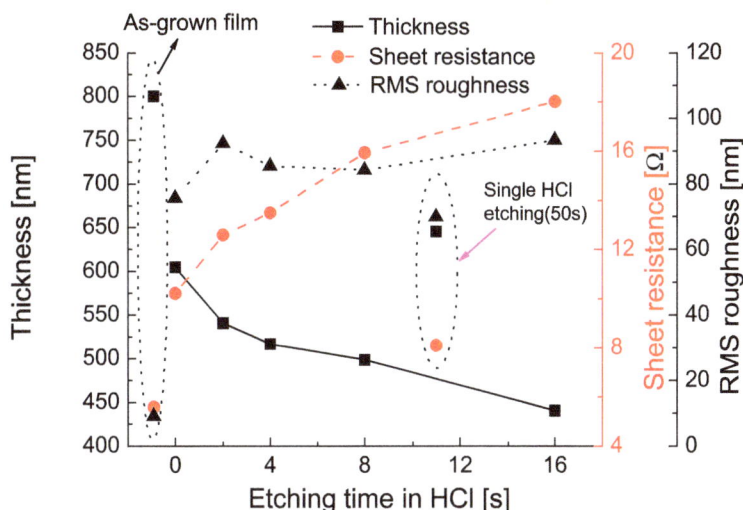

Fig. 7. The variations of thickness and sheet resistance as well as RMS roughness of high rate as-grown and after-etched ZnO:Al films from tube ceramic targets after etching with single step or two steps in HF etchant and HCl etchant.

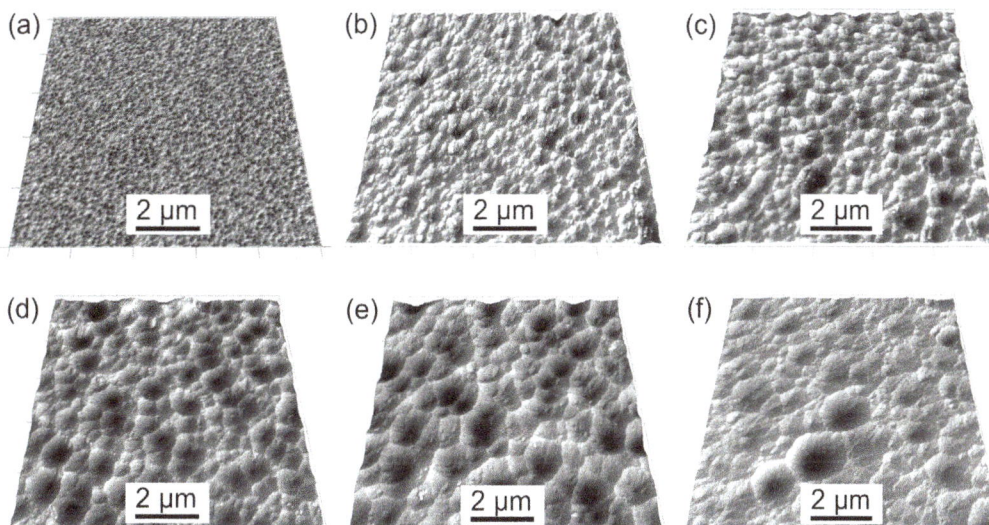

Fig. 8. The morphologies of high rate films etched with one step in HF solution or HCl solution or with two steps first in HF solution and then in HCl solution. The HF etching time / HCl etching time are: (a) 0 s/0 s, (b) 120 s/0 s, (c) 120 s/2 s, (d) 120 s/8 s, (e) 120 s/16 s, (f) 0 s/50 s.

etch processes, the RMS roughness exceeds the value of the ZnO:Al film etched solely in HCl.

Figure 8 shows the topographies as given by AFM measurements of different high rate ZnO:Al films. Figure 8(a) shows the quite flat, unetched film surface. In Figure 8(b) shows the surface structure of ZnO:Al film solely etched in HF (120 s). Its surface exhibits small but steep craters with typical diameters between $100-200$ nm similar to the ones observed in Figure 5 for the surface evolution upon HF etching. By applying a short dip in HCl solution, e.g. 2 s, the crater-like features become wider (see Fig. 8(c)). With the increase of etching time in HCl (8 s, see Fig. 8(d) and 16 s, see Fig. 8(e)) the craters are further enlarged in diameter ending up with diameters around to $1-2$ μm

for the longest applied etching time. This topography, as shown in Figure 8(e) looks similar to type B [15] that is favored for solar cell application. For comparison the rather flat surface of a film etched in HCl for 50 s is shown in Figure 8(f), which typically leads to reduced short circuit currents and consequently low efficiencies [17]. For reactively sputtered ZnO:Al films a similar etching behavior was observed as well [37].

All ZnO:Al films of this two-step etch series show high transmission of more than 85% and low absorption of less than 5% in visible spectrum region (see Fig. 9(a)). Even in the infrared spectrum region, they all show high transmission and low absorption due to low carrier concentration shown in Figure 1. The slight variations of transmission

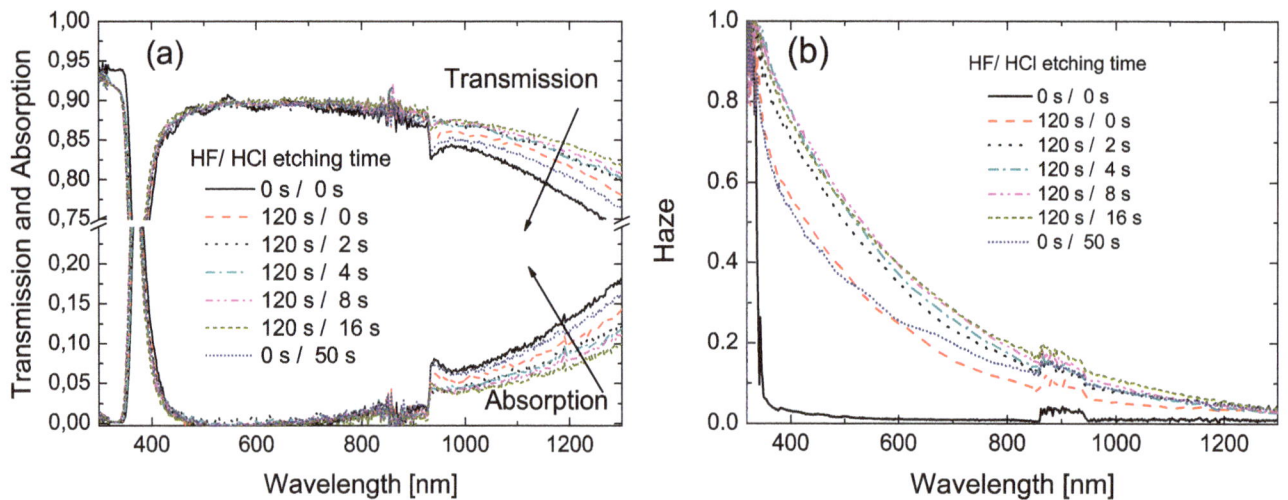

Fig. 9. The transmission and absorption (a) as well as the haze (b) of high rate films etched with one step in HF solution or HCl solution or with two steps first in HF solution and then in HCl solution. etch times are given in the legend.

and absorption are mainly due to the change in thicknesses of films etched for different etching times. The haze is illustrated in Figure 9(b). For comparison the curve of a film etched in HCl is also plotted. Even though the surfaces of the solely HCl and HF etched samples (compare Figs. 8(b) and 8(f), respectively) are significantly different in number, size and shape of their features, the haze of both samples is similar. At 600 nm the haze of both samples is about 0,25. By dipping the HF etched sample into HCl acid solution for short time (2 s), the haze of the sample increases in the whole observed wavelength range, e.g. it is 0,4 at 600 nm. By prolonging the etching time in the HCl solution to 4 s the haze only slightly increases and stays almost constant for longer etching times (8 or 16 s).

4 Conclusions

In this study, the etching behaviour of high rate ZnO:Al films deposited at substrate temperatures in the range of 250−350 °C in HCl and HF was compared. For films deposited at lower substrate temperatures all films show a rather flat surface topography after etching in both etchants. ZnO:Al films deposited at 350 °C and 325 °C display characteristic surface structures with either regular small or a few large craters after etching in HF or HCl, respectively. In contrast to findings for optimized low deposition rate ZnO:Al films [8], prolonged etching does not create regularly distributed large craters by single step etching for our high rate sputtered ZnO:Al films. Combining both etching first in HF and then in HCl leads to regular large crater distribution on the surface of ZnO:Al films with strong and efficient light scattering that are in favor of light trapping in solar cells. Still the deposition of ZnO:Al films at high substrate temperatures (>300 °C) is required to result in highly conductive and highly transparent TCO films with compact structure at the same time to form craters upon etching.

ZnO:Al films with regular distribution of large craters on their surfaces have been achieved previously on reactively sputtered high rate ZnO:Al films etched with such a two-step etching method and proven to result in excellent silicon thin film solar cell performance [37]. However, the etching rate of films here is higher than reactively sputtered films, which could be explained by higher working pressure in our case. Thus the resistance of the films is a bit higher than in the other study, but others reported that sheet resistance well above 10 Ohm is still sufficient and might even be favorable for solar cells, since the free carrier absorption is reduced in these films [38,39].

The authors would like to thank H. Siekmann, J. Worbs, and H.-P. Bochem for extensive technical support. Financial support by the German BMU (contract No. 0327693A) and by the European Commission (contract No. 019670) is gratefully acknowledged. The authors also thank the company W. C. Heraeus for providing the targets.

References

1. S. Besbes, H. Ben Ouada, J. Davenas, L. Ponsonnet, N. Jaffrezic, P. Alcouffe, Materials Science & Engineering C-Biomimetic and Supramolecular Systems **26**, 505 (2006)

2. H. Kim, A. Pique, J.S. Horwitz, H. Mattoussi, H. Murata, Z.H. Kafafi, D.B. Chrisey, Appl. Phys. Lett. **74**, 3444 (1999)

3. C.A. Wang, S.A. VanSlyke, Appl. Phys. Lett. **51**, 913 (1987)

4. Y. Akao, T. Haranoh, Transparent conductive coatings for flat panel displays (Elsevier Science Bv, Amsterdam, 1999)

5. W.J. Lee, Y.K. Fang, J.J. Ho, C.Y. Chen, R.Y. Tsai, D.Y. Huang, F.C. Ho, H.W. Chou, C.C. Chen, J. Elect. Mat. **31**, 129 (2002)

6. B. Rech, O. Kluth, T. Repmann, T. Roschek, J. Springer, J. Müller, F. Finger, H. Stiebig, H. Wagner, Sol. Energy Mat. Sol. Cells **74**, 439 (2002)

7. E. Fortunato, D. Ginley, H. Hosono, D.C. Paine, Mrs Bulletin **32**, 242 (2007)

8. B. Rech, T. Repmann, M.N. van den Donker, M. Berginski, T. Kilper, J. Hupkes, S. Calnan, H. Stiebig, S. Wieder, Thin Solid Films **511**, 548 (2006)

9. S. Calnan, J. Hupkes, B. Rech, H. Siekmann, A. Tiwari, Thin Solid Films **516**, 1242 (2008)

10. M. Berginski, J. Hüpkes, M. Schulte, G. Schöpe, H. Stiebig, B. Rech, M. Wuttig, J. Appl. Phys. **101**, 11 (2007)

11. M. Kondo, A. Matsuda, Thin Solid Films **457**, 97 (2004)

12. M. Kondo, T. Matsui, Y. Nasuno, H. Sonobe, S. Shimizu, Thin Solid Films **501**, 243 (2006)

13. S. Faÿ, L. Feitknecht, R. Schluchter, U. Kroll, E. Vallat-Sauvain, A. Shah, Sol. Energy Mat. Sol. Cells **90**, 2960 (2006)

14. J. Hüpkes, B. Rech, O. Kluth, T. Repmann, B. Zwaygardt, J. Müller, R. Drese, M. Wuttig, Solar Energy Mat. Sol. Cells **90**, 3054 (2006)

15. O. Kluth, G. Schöpe, J. Hüpkes, C. Agashe, J. Müller, B. Rech, Thin Solid Films **442**, 80 (2003)

16. H. Zhu, E. Bunte, J. Hüpkes, H. Siekmann, S.M. Huang, Thin Solid Films **517**, 3161 (2009)

17. E. Bunte, H. Zhu, J. Hüpkes, 23rd European Photovoltaic Solar Energy Conference, Valencia, Spain, 2008, p. 2105

18. J.I. Owen, J. Hüpkes, H. Zhu, E. Bunte, S.E. Pust, Physica Status Solidi (A) **208**, 109 (2010)

19. T. Minami, H. Nanto, S. Takata, Japanese J. Appl. Phys. Part 2-Lett. **23**, L280 (1984)

20. T. Minami, H. Nanto, S. Takata, Appl. Phys. Lett. **41**, 958 (1982)

21. R.L. Petritz, Phys. Rev. **104**, 1508 (1956)

22. K.L. Chopra, S. Major, D.K. Pandya, Thin Solid Films **102**, 1 (1983)

23. J.W. Orton, M.J. Powell, Rep. Progress Phys. **43**, 1263 (1980)

24. K. Ellmer, R. Mientus, Thin Solid Films **516**, 4620 (2008)

25. T. Minami, H. Sato, H. Nanto, S. Takata, Japanese J. Appl. Phys. Part 2-Lett. **24**, L781 (1985)

26. K. Tominaga, K. Kuroda, O. Tada, Japanese J. Appl. Phys. Part 1-Regular Papers Short Notes & Review Papers **27**, 1176 (1988)

27. T. Minami, H. Sato, T. Sonoda, H. Nanto, S. Takata, Thin Solid Films **171**, 307 (1989)

28. K. Ellmer, J. Phys. D-Appl. Phys. **34**, 3097 (2001)

29. F. Ruske, A. Pflug, V. Sittinger, B. Szyszka, D. Greiner, B. Rech, Thin Solid Films **518**, 1289 (2009)

30. D. D. Perrin, Ionization Constants of Inorganic Acids and Bases in Aqueous Solution, 2nd. (Pergamon, Oxford, 1982)

31. H. Zhu, J. Hüpkes, E. Bunte, A. Gerber, S.M. Huang, Thin Solid Films **518**, 4997 (2010)

32. J. Hüpkes, B. Rech, S. Calnan, O. Kluth, U. Zastrow, H. Siekmann, M. Wuttig, Thin Solid Films **502**, 286 (2006)

33. J. Hüpkes, B. Rech, O. Kluth, J. Müller, H. Siekmann, C. Agashe, H.P. Bochem, M. Wuttig, in Amorphous and Nanocrystalline Silicon-Based Films-2003, edited by J.R. Abelson, G. Ganguly, H. Matsumura, J. Robertson, E.A. Schiff, **762**, 405 (2003)

34. V. Sittinger, F. Ruske, W. Werner, B. Szyszka, B. Rech, J. Hupkes, G. Schope, H. Stiebig, Thin Solid Films **496**, 16 (2006)

35. H. Zhu, E. Bunte, J. Hüpkes, S.M. Huang, Thin Solid Films **519**, 2366 (2011)

36. R.J. Hong, X. Jiang, B. Szyszka, V. Sittinger, S.H. Xu, W. Werner, G. Heide, J. Crystal Growth **253**, 117 (2003)

37. H. Zhu, J. Hüpkes, E. Bunte, J. Owen, S.M. Huang, Sol. Energy Mat. Sol. Cells **95**, 964 (2011)

38. S. Faÿ, J. Steinhauser, N. Oliveira, E. Vallat-Sauvain, C. Ballif, Thin Solid Films **515**, 8558 (2007)

39. M. Berginski, J. Hüpkes, W. Reetz, B. Rech, M. Wuttig, Thin Solid Films **516**, 5836 (2008)

7

A modelling study of the performance of conventional diffused P/N junction and heterojunction solar cells at different temperatures

S. Chakraborty[1], A. Datta[1,2], M. Labrune[3,4], P. Roca i Cabarrocas[3], and P. Chatterjee[1,3,a]

[1] Energy Research Unit, Indian Association for the Cultivation of Science, Jadavpur, 700032 Kolkata, India
[2] Haltu High School for Girls (H.S.), Neli Nagar, Haltu, 700 078 Kolkata, India
[3] Laboratoire de Physique des Interfaces et Couches Minces, CNRS, École Polytechnique, 91128 Palaiseau, France
[4] Total S.A., Gas & Power – R&D Division, 92400 Courbevoie, France

Abstract Conventional crystalline silicon (c-Si) diffused P/N junction solar cells remain the largest contributor to solar electricity. In order to retain a high efficiency and as well, reduce the cost of solar electricity, Sanyo has proposed the "heterojunction with intrinsic thin layer (HIT)" solar cells where the emitter and the back surface field layers are deposited using low temperature (<200 °C) plasma processes, thus reducing the thermal budget and allowing for thinner wafers. Since solar cells are used in extremes of climate, we felt that it would be interesting to study the behaviour of c-Si and HIT cells, based on both P- and N-type wafers at different temperatures. Our results indicate that in HIT cells the amorphous doped layers form a heterojunction on the c-Si substrate, with a large valence band discontinuity that acts as a barrier for hole collection, specially at low temperatures. It is the aim of this article to investigate the effect of this valence band offset on solar cell performance at different ambient temperatures.

1 Introduction

Amorphous-crystalline (a/c) silicon "heterojunction with intrinsic thin layer (HIT)" solar cells, proposed by Sanyo [1,2] offer a low cost alternative to standard crystalline silicon (c-Si) solar cells with diffused P/N junction and back surface field (BSF) layers. This is because in HIT cells the P/N junction and BSF layer formation steps take place at a relatively low temperature (∼200 °C) using hydrogenated amorphous (a-Si:H) or polymorphous (pm-Si:H) silicon deposition technology, whereas in conventional c-Si cells the wafer has to be raised to ∼800 °C for junction and BSF layer formation by diffusion. This means not only a lower thermal budget, but also cost reduction from thinner wafers in "HIT" cells, since at ∼200 °C, there is little danger of a thin (∼100 μm) c-Si wafer becoming brittle. At the same time Sanyo [3] has demonstrated that in HIT cells, the high stable conversion efficiencies, characteristic of conventional diffused P/N junction solar cells, can be retained (since the absorber layer remains c-Si) by passivating the defects on the surface of the c-Si wafer, using a thin intrinsic amorphous silicon layer. Moreover, HIT cells have a lower temperature coefficient of conversion efficiency than standard c-Si cells, making them all the more suitable for practical applications [4,5].

Solar cells are used in extremes of climates – in hot regions, where there is plenty of sunshine, as well as in remote mountainous and polar regions, or in space, where a stand-alone source of energy is indispensable. It would, therefore, be interesting to analyze the relative performances of standard c-Si and HIT cells, in particular their illuminated current density-voltage (J-V) characteristics at different ambient temperatures. In this article we have first used detailed electrical-optical modelling of the dark and illuminated J-V characteristics of HIT cells on both P- and N-type wafers, where the hydrogenated amorphous silicon (a-Si:H) layers are deposited by the plasma-enhanced chemical vapour deposition (PECVD) technique [6]; to extract parameters that characterize the individual high efficiency cells. The latter fact guarantees that the defect states on the surface of the c-Si wafer have been well-passivated resulting in state-of-the-art high efficiency HIT cells. These parameters are then used with suitable modifications to take into account temperature effects, to predict the behaviour of these cells between around 175 K and 330 K, vis-à-vis that of standard diffused P/N junction solar cells.

[a] e-mail: parsathi_chatterjee@yahoo.co.in

2 Experiments

To fabricate the HIT solar cells, high quality CZ P-type (doping level $\sim 9 \times 10^{14}$ cm^{-3}) or N-type (doping level $\sim 2 \times 10^{15}$ cm^{-3}) c-Si substrates were used. Their main properties are as follows: $\langle 100 \rangle$ orientation, ~ 14 Ω-cm resistivity, 300 μm thickness and 4-inch diameter. The wafers have flat (non-textured) surfaces. The emitter and the BSF layers were deposited in a capacitively coupled PECVD reactor operating at an RF of 13.56 MHz [6] and resulting in a heterojunction (HJ) at both ends of the c-Si wafer (double HIT structure). The crystalline silicon wafers were dipped into hydrofluoric acid and immediately put into the reactor, which was pumped down to a pressure below 10^{-6} mbar before deposition. On both sides of the N-type c-Si and on the emitter side of the P-type c-Si wafers, ~ 3 nm thick intrinsic amorphous silicon layers were first deposited using conditions that result in hydrogenated polymorphous silicon (pm-Si:H), i.e., at a relatively high pressure in the range of 1200 mTorr and at 200 °C. This material provides excellent passivation of defects on the surface of the c-Si wafer [7]. The highly phosphorous-doped N-a-Si:H layer, that acts as the emitter (~ 8 nm thick) on a P-type c-Si wafer; or as the BSF layer (thickness ~ 20 nm) on N-type c-Si wafer was deposited at 200 °C. The material has a band gap of 1.8 eV and an activation energy of 0.20 eV. On the other face of the wafers, in the same RF-PECVD reactor and at 150 °C, highly boron doped, ~ 20 nm thick P$^+$-a-Si:H layers were deposited. On the emitter layer of these structures an indium tin oxide (ITO) layer between 85 nm and 110 nm thick was deposited in a DC-magnetron sputtering setup. Lastly front contact metallization (with silver) was performed using serigraphy followed by annealing at a temperature of ~ 200 °C under high pressure. A 15 nm aluminium layer was evaporated onto the BSF layer and annealed at a temperature of 150 °C to complete the back contact metallization. The structure of the cells was: for double HIT's on N-type substrate: ITO/ P-a-Si:H/ I-pm-Si:H/ N-c-Si/ I-pm-Si:H/ N^{++}-a-Si:H/ ITO/ Al; and for the double HIT's on P-type substrate: ITO/ N-a-Si:H/ I-pm-Si:H/ P-c-Si/ P^{++}-a-Si:H/ Al.

Basic solar cell parameters were obtained through current density-voltage (J-V) measurements under AM1.5 illumination. The complex refractive indices for each layer of the structure (that are required as input to the modelling program) have been measured by spectroscopic ellipsometry.

3 Simulation model

The "amorphous semiconductor device modelling program (ASDMP) [8,9]", later extended to also model crystalline silicon and HIT cells [10], solves the Poisson's equation and the two carrier continuity equations under steady state conditions for a given device structure, and yields the dark and illuminated J-V and QE characteristics. The gap state model consists of two monovalent donor-like and acceptor-like tail states and two monovalent Gaussian distribution functions (one being of donor type and the other of acceptor type) to simulate the deep dangling bond (DB) states, in the case of the amorphous layers (e.g. [11] (AMPS), [12] (AFORS-HET), [13]); while in the c-Si substrate, the tail states are absent. A more realistic gap state distribution model in a-Si:H would consist of, besides the monovalent band tail states, a deep defect distribution of DBs determined from the defect pool model (DPM) [14, 15]. However, in ASDMP, as in AMPS and AFORS-HET, since only monovalent states can be introduced, the deep DB distribution determined from DPM is replaced by two Gaussian distributions of monovalent states, donor like and acceptor like, separated by a correlation energy. Such a replacement has proved to be quite accurate in a-Si:H [16,17].

The defect density on the surfaces of the c-Si wafer is modelled by a defective layer 3 nm thick. This means for e.g., that a surface defect density of 10^{11} cm^{-2} translates into a volume density of $\sim 3.3 \times 10^{17}$ cm^{-3}. Also the transport over the potential barriers at the contacts if any, is by thermionic emission and across the amorphous/ crystalline heterojunction by both thermionic emission and carrier diffusion. For example in P-a-Si:H/ N-c-Si HIT cells, holes are collected at the P-a-Si:H emitter end and, as will be seen later in Figure 3b, at this hetero-junction the valence band edge is closer to the Fermi level on the N-c-Si side than in P-a-Si:H, due to the valence band discontinuity, giving rise to a thin region on the N-c-Si side where the majority carriers are holes rather than electrons. This layer has been named the "inversion layer" [18], and, unless the defect density on the surface of the c-Si wafer is extremely high, the free hole density here is in fact higher than in P-a-Si:H, resulting also in hole diffusion across this heterojunction.

The generation term has been calculated using a semi-empirical model [19] that has been integrated into AS-DMP [20]. Light enters through the ITO/emitter "front contact" which is taken at $x = 0$. Voltage is also applied here. The amorphous BSF / Al "rear contact" at the back of the c-Si wafer is taken at $x = L$, where L is the total thickness of the semiconductor layers; and is assumed to be at 0 V. For HIT cells on N-type wafers, the P-a-Si:H (emitter) / ITO front contact barrier height is 1.24 eV and the rear N-a-Si:H / Al contact barrier height =0.2 eV; while for HIT cells on P-type wafers, the rear P-a-Si:H (BSF) / Al contact barrier height is 1.20 eV and the front ITO / N-a-Si:H barrier height is 0.2 eV.

In our calculations we have taken into account the changes in the band gap, the effective densities of states at the conduction and valence band edges (N_c, N_v, respectively), carrier mobilities and dopant ionization as a function of temperature. We know that the band gap of c-Si decreases with increasing temperature approximately as [21]:

$$E_g(T) = E_g(0) - \frac{\alpha T^2}{T + \beta}, \qquad (1)$$

where, $E_g(0)$ is the band gap of c-Si at 0 K, T is the absolute temperature, $\alpha = 4.9 \times 10^{-4}$ eV K^{-1} and $\beta = 655$ K.

Also the effective density of states (DOS) at the conduction and valence bands edges (N_c, N_v, respectively) decrease with decreasing temperature, following the 3/2 power law [22]. On the other hand, at a given doping density, the mobilities increase with decreasing temperature [22] as $T^{-2.42}$ in N-type c-Si and as $T^{-2.20}$ in P-type c-Si wafer.

For the case of a-Si:H, the variation of the band gap with temperature ia taken as:

$$E_g(T) = E_g(0) - 2.25 \times 10^{-4}T, \qquad (2)$$

where $E_g(T)$ and $E_g(0)$ are the band gaps of a-Si:H at the given ambient temperature and at 0 K, respectively [13,23]. Obviously therefore the band offsets would be temperature dependent via the above variations of band gaps of c-Si and and a-Si:H with temperature; as well as of their electron affinities. However, we were unable to find any reference in the literature to determine whether the conduction band or the valence band moves with temperature, in other words the variation of the electron affinity with temperature is not known. We arbitrarily chose to hold the electron affinity of both c-Si and a-Si:H constant with temperature. This appears to be a crucial point, since as will be shown later, the deterioration of the performance of N-type HIT cells at low temperatures is primarily linked to the ΔE_v at the amorphous emitter/c-Si interface. In order to understand how important an exact determination of the a/c band offset at a given temperature is for determining HIT cell performance, we also performed some other simulations where we held not only the electron affinities of c-Si and a-Si:H constant with temperature, but also the band gap of a-Si:H. We found that the trend of deteriorating performance of HIT cells at low temperatures remains unchanged in both the above cases, leading us to believe that it is rather the transport mechanism of carriers across this barrier at low temperatures that plays a more vital role in determining HIT cell performance, rather than the exact value of ΔE_v as a function of temperature. In our calculations, we have assumed the same variation of the mobility in a-Si:H with temperature as has been assumed for c-Si above, while the effective DOS at the band edges in a-Si:H have been held constant with temperature. We suppose that these latter parameters, particularly in the doped a-Si:H layers would be less sensitive to temperature and have a small influence on HIT cell performance as a function of temperature. Also a more detailed modelling would require to consider the variation of TCO properties (e.g. barrier heights) on temperature. However, this would affect the simulation results of *both* standard P/N diffused junction and HIT cells and we suspect that the *relative* performance of these two types of cells (the primary aim of this study) would not be appreciably affected.

At room temperature and above we assume full ionization of the donor and acceptor impurity levels. However, we know that at lower temperatures, these levels are not fully ionized, and we calculate the density of the ionized donors and acceptors according to the formulae [22,24]:

$$\frac{N_D^+}{N_D} = \frac{\left\{\sqrt{8\frac{N_D}{N_C}\exp\left(\frac{E_D}{kT}\right)+1}\right\}-1}{4\frac{N_D}{N_C}\exp\left(\frac{E_D}{kT}\right)} \qquad (3)$$

and

$$\frac{N_A^-}{N_A} = \frac{\left\{\sqrt{16\frac{N_A}{N_V}\exp\left(\frac{E_A}{kT}\right)+1}\right\}-1}{8\frac{N_A}{N_V}\exp\left(\frac{E_A}{kT}\right)}, \qquad (4)$$

where k is the Boltzmann constant, T the ambient temperature, N_D^+ and N_A^- are the density of the ionized donors and acceptors, respectively; and N_D, N_A the total density of the donor and acceptor impurity atoms respectively; E_D, E_A the ionization energies of the donor (phosphorous) and acceptor (boron) impurity levels, respectively, taken equal to 0.044 eV and 0.045 eV [22].

4 Simulation of experimental results

We had in previous articles [25,26] simulated the illuminated *J-V* and quantum efficiency (QE) characteristics of HIT solar cells on P-type wafers, with amorphous emitter and BSF layers (double HIT structure) deposited by PECVD [6]. From these simulations we had extracted the parameters characterizing these solar cells, which we will utilize in the course of this article to predict the low temperature behaviour of these cells.

In this article, we simulate the dark and illuminated *J-V* characteristics of double HIT solar cells on N-type c-Si wafer, whose amorphous layers have been deposited under identical conditions (described in Sect. 2) and in the same PEVCD reactor [6]. We have simulated a series of HIT cells having different thickness of the intrinsic pm-Si:H layer covering that surface of the N-c-Si wafer that faces the P-a-Si:H emitter and the impinging light. Table 1 compares the measured solar cell output parameters under AM1.5 light, as well as the dark diode parameters, to the calculated values obtained using ASDMP. We find that in order to match experiments, lower values for the defect density (N_{ss}) on the surface of the c-Si wafer had to be assumed when this intrinsic layer was thicker. In a previous article, while modelling the extensive experimental results of Taguchi et al. [27], where the thickness of this intrinsic layer was varied, we had to make the same assumption, and had come to the conclusion that the main function of this intrinsic layer is to passivate the defects on the surface of the c-Si wafer on which it is deposited. This modelling result is consistent with the fact that in our deposition system [6], without load-lock, the defect density on the wafer surface decreases with increasing deposition time (film thickness) of the I-pm-Si:H buffer layer. It has been shown (e.g. [28]) that reduction of N_{ss} on that side of the N-type c-Si wafer facing the emitter layer, has a beneficial effect on both the open-circuit voltage (V_{oc}) and *FF* of the HIT cell – as is also observed in Table 1 (cases A to C). The surface defective layer (DL) is modelled by a ~3 nm slice on the wafer surface. A similar

Table 1. Modelling of the experimental output parameters of HIT structure solar cells on N-type c-Si wafer at 300 K, having different thickness of the I-pm-Si:H layer on the emitter side (denoted by A, B, C, D). N_{ss} (DL) is the defect density on that surface of the c-Si wafer that faces the emitter. J_0 is the reverse saturation current density and n_0 the diode ideality factor calculated from the experimental and model generated dark J-V characteristics.

Cell name	I-pmSi:H thickness (nm)	N_{ss} (DL) (cm^{-2})		J_{sc} (mA cm^{-2})	V_{oc} (mV)	FF	Efficiency (%)	J_0 (mA cm^{-2})	n_0
A	2.5	9.2×10^{11}	Expt.	30.3	679	0.755	15.5	2.70×10^{-8}	1.30
			Model	31.2	673	0.758	15.9	2.41×10^{-8}	1.57
B	3.3	6.6×10^{11}	Expt.	30.8	683	0.767	16.1	1.60×10^{-7}	1.43
			Model	31.0	684	0.765	16.2	1.78×10^{-7}	1.77
C	4.0	4.5×10^{11}	Expt.	30.8	701	0.796	17.2	1.90×10^{-7}	1.49
			Model	31.1	701	0.807	17.6	1.10×10^{-7}	1.42
D	4.5	1.6×10^{11}	Expt.	30.6	706	0.732	15.8	5.90×10^{-8}	1.40
			Model	30.3	706	0.738	15.8	4.95×10^{-8}	1.67

Table 2. Measured parameters and those extracted by modelling deposited HIT cells on P-type [13–15] and N-type (present study) c-Si wafers. Note that the quantities marked with superscript "a" have been measured, those marked with superscript "b" have been taken from the literature, and those marked with superscript "c", supplied by the manufacturer. The defect density in the defective layer on the surface of the P-type c-Si wafer is 10^{11} cm^{-2} on the emitter side and 10^{12} cm^{-2} on the BSF side, while in the case of the N-type HIT cells these values are 4.5×10^{11} cm^{-2} and 10^{11} cm^{-2} at the emitter and BSF ends respectively, as extracted by modelling.

Parameters	P a-Si:H	I-pmSi:H (buffer)	N c-Si wafer	P c-Si wafer	N-a-Si:H
Layer thickness (μm)[a]	0.02	0.004	280	300	0.008–0.02
Mobility gap (eV)	1.75[a]	1.96[a]	1.12[b]	1.12[b]	1.8[a]
ΔE_v with respect to c-Si (eV)	−0.41	−0.46	0	0	−0.46
Donor (Accep) doping (cm^{-3})	(1.41×10^{19})[a]	0	2×10^{15}[c]	(9×10^{14})[c]	1.45×10^{19a}
Effective DOS in CB (cm^{-3})[b]	2×10^{20}	2×10^{20}	2.80×10^{19}	2.80×10^{19}	2×10^{20}
Effective DOS in VB (cm^{-3})[b]	2×10^{20}	2×10^{20}	1.04×10^{19}	1.04×10^{19}	2×10^{20}
Charac. energy (VB tail) (eV)[b]	0.05	0.05	–	–	0.05
Charac. energy (CB tail) (eV)[b]	0.03	0.03	–	–	0.03
Expon. tail prefact.(cm^{-3} eV^{-1})[b]	$4. \times 10^{21}$	$4. \times 10^{21}$	–	–	4×10^{21}
Elec. (hole) mobility (cm^2/V s)	25 (5)	25 (5)	1500 (500)[b]	1000(450)[b]	20 (4)
Gaussian defect density (cm^{-3})	8×10^{18a}	10^{15}	10^{12}	10^{12}	9×10^{18a}
Neutral σ (tails, midgap cm^2)	10^{-17}	10^{-17}	4×10^{-18}	4×10^{-19}	10^{-17}
Charged σ (tails, midgap cm^2)	10^{-16}	10^{-15}	4×10^{-17}	4×10^{-18}	10^{-16}

DL is assumed to exist at the rear side of c-Si facing the amorphous BSF; however, in previous simulations of HIT cells on N-type wafers [28], we had noted that the defect density here has less influence on device performance and had also explained why. We note that for case D with the thickest I-pm-Si:H layer, the fill factor (FF) and conversion efficiency deteriorate, in spite of a lower defect density (N_{ss}) on the N-c-Si surface. This is due to the fact that in these structures it is the I-pm-Si:H layer that is most resistive; therefore, when this layer is too thick, the FF falls. The short-circuit current density (J_{sc}) is relatively low in all cases as these cells have flat interfaces and no light-trapping schemes have been attempted.

The parameters for double HIT cells on P-type wafers, simulated by modelling their illuminated J-V and QE characteristics and given in references [25, 26], as well as those extracted by modelling the double N-HIT cells above, are summarized in Table 2. Of course the parameters marked by superscript "a" in Table 2, such as the thickness of the individual layers, the band gap, the doping and defect densities inside the emitter and BSF (deduced from measured activation energies), have been measured. Parameters, marked by superscript "b" in Table 2, have been taken from the literature, while the doping density in the N- and P-type c-Si wafers (marked by superscript "c") have been supplied by the manufacturer. The main parameters extracted by fitting the measured dark and illuminated J-V characteristics, therefore, are the defect densities on the surfaces of the c-Si wafers, their charged and neutral capture cross-sections and carrier mobilities inside the amorphous layers. It is relevant here to point out that in this article we are interested in evaluating the relative performances of various types of HIT cells at different ambient temperatures and comparing the results to the performance of standard P/N diffused homo-junction c-Si cells. Therefore, we chose for comparison, HIT cells on N-type wafers whose amorphous layers were deposited in the same PECVD reactor [6] and under identical conditions as for HIT cells on P-type wafers simulated in references [25, 26]. Unfortunately relatively little experimental results are available for this group of HIT cells on N-type wafers. Therefore, in our present simulations we had to depend heavily on our experience of simulating the extensive experimental results of HIT cells on N-type wafers produced by Taguchi et al. [27] and modelled in reference [28], and the interested reader is referred to these detailed simulations to get an insight and explanation of how the above parameters are extracted.

It maybe noted that we were unable to simulate all the above experimental results using the same values of the electron and hole mobilities in the emitter and the I-pm-Si:H layer sandwiched between it and the N-type c-Si wafer, as was also our experience in simulating the experimental results [27] in reference [28]. The electron and hole mobilities (μ_n, μ_p) in cases A, B were 7 cm^2/V s and 1.4 cm^2/V s, respectively, while for case D it was marginally lower (6 cm^2/V s and 1.2 cm^2/V s, respectively). However, a much higher value of these mobilities had to be assumed to simulate case C (25 cm^2/V s and 5 cm^2/V s, respectively). Increase of hole mobility over this region improves hole collection at the ITO/ P-a-Si:H interface and brings up the FF. In fact the high value of the FF and the conversion efficiency for case C may be assigned to higher carrier mobilities in the front amorphous layers, as well as reduced N_{ss}. The outstanding defect passivation properties of pm-Si:H has already been documented [7]. We also know that it has a higher hole mobility than a-Si:H [29]. Therefore, its use as the thin intrinsic layer on the wafer surface facing the emitter should also have a beneficial influence on the FF of the cell. However, up to this stage, such a high FF could only be achieved for the cell with a 4 nm thick intrinsic pm-Si:H layer (Tab. 1).

5 Results and discussion

The parameters given in Table 2 will be used in this section to study the performance of HIT cells at different temperatures using ASDMP. Again since the aim of this article is to compare the performance of standard P/N diffused junction c-Si solar cells and HIT cells, we have for the former case considered an N-type c-Si wafer, with the same properties and doping density as those given in Table 2 for the N-type wafer. In such a homo-junction structure, where the P/N junction and BSF layers are formed by diffusion in a furnace raised to ~800 °C, the P-c-Si emitter layer is assumed to have a thickness of 0.25 μm, with a doping density of 5×10^{18} cm^{-3}, and the heavily N-doped 1 μm thick BSF layer of a similar doping density. Moreover, the surface recombination speeds at the contacts (which form the boundary conditions for the continuity equations) have been assumed to be equal to the thermal velocity (10^7 cm/s) in c-Si P/N diffused junction as well as in HIT cells. However, we know that whereas low defect densities and low recombination speeds

at the a/c interfaces are crucial for attaining a high V_{oc} in HIT cells [10, 28] it is rather the low surface recombination speeds of the minority carriers at the contacts that help improve V_{oc} in standard P/N diffused junction c-Si cells. Since in this study these have been assumed equal to 10^7 cm/s, the V_{oc} of our P/N diffused junction cells are relatively low. In all these calculations, the variations of the band gap of c-Si and the a-Si:H layers, the effective DOS at the band edges of c-Si, the carrier mobilities of both c-Si and a-Si:H, as also the fact that at low temperatures not all dopant atoms are ionized, have been taken into account (see Sect. 3).

Figure 1 compares the illuminated J-V characteristics of (a) a double HIT cell on N-type c-Si, (b) a double HIT cell on P-type c-Si, (c) the standard diffused P/N junction c-Si cell and (d) a front HIT cell on P-type wafer (where only the emitter layer is a-Si:H) at different temperatures. Figure 2 gives the solar cell output parameters – (a) the short-circuit current density, (b) the open-circuit voltage, (c) the fill factor and (d) the conversion efficiency of these cases as a function of temperature. We have in a previous article [28] simulated the solar cell output parameters of HIT cells on N-type wafers developed by Sanyo [27] with varying thickness of the intrinsic a-Si:H layer deposited on the wafer and facing the P-a-Si:H emitter. That study carried out between 343 K and 283 K, shows similar variation with temperature as Figure 2. Comparing Figure 1a to Figure 1b, we find that the HIT cell on N-c-Si has a better low temperature performance than the HIT cell on P-c-Si, the latter already showing an S-shape at 300 K. From Figures 1 and 2, it is obvious that the performance of HIT cells, in particular J_{sc}, FF and the conversion efficiency deteriorate sharply at low temperatures, the front HIT cell on P-c-Si being the only exception to this rule. The reason for this will be explained subsequently. Only the V_{oc} continues to increase. As temperature decreases, carrier densities in the bands decrease. This, therefore, means lower carrier recombination, a lower dark reverse saturation current density (J_{D0}) and hence via the formula:

$$V_{oc} = (nkT/q) \ln \left[(J_{sc}/J_{D0}) + 1 \right], \qquad (5)$$

a higher open-circuit voltage at lower temperatures. This approximate formula ignores series and shunt resistances, but in the case of a c-Si absorber, with its surfaces well passivated to ensure low surface defect density, this is more or less true, even for HIT cells, where the absorber layer remains c-Si. A similar temperature dependence of V_{oc} has been obtained experimentally by Taguchi et al. [27] and while simulating these results [28].

To understand the difference in behaviour of the rest of the output parameters, we must refer to the fundamental difference between the standard diffused P/N junction and the HIT cell. The former is a homo-junction, while in the HIT device (in this article when referring simply to "HIT" cells, we mean double "HIT" cells with heterojunctions at both emitter and BSF ends), the emitter, the BSF and the intrinsic passivating layers are a-Si:H or pm-Si:H, with a band gap considerably higher than c-Si. In PECVD deposited samples, as in the present study,

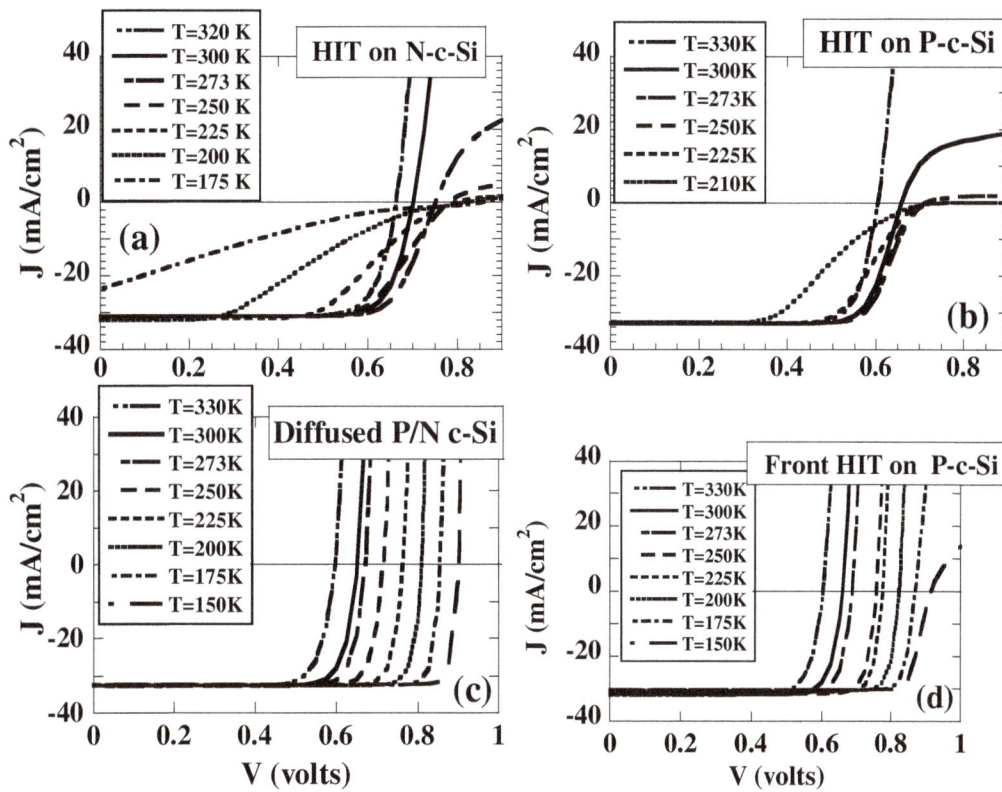

Fig. 1. Calculated illuminated J-V characteristics at various temperatures for different types of solar cells, all having c-Si as the absorber layer: (a) HIT cell on N-type wafer, (b) HIT cell on P-type wafer, (c) the conventional P/N diffused homo-junction solar cell and (d) front HIT cell on P-type wafer.

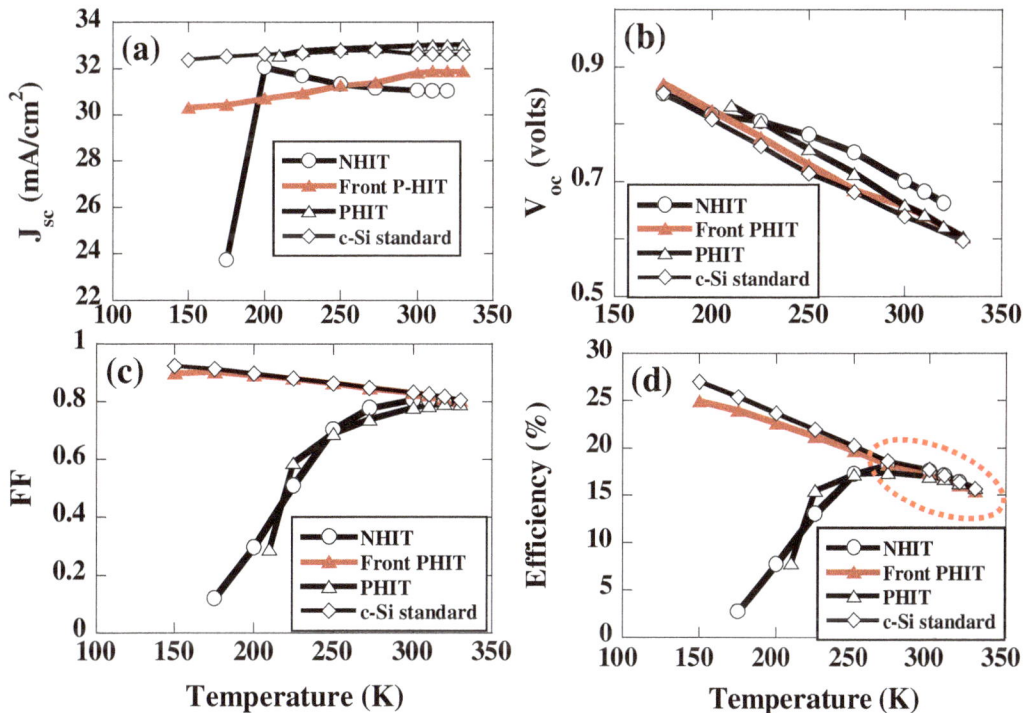

Fig. 2. Variation of the calculated values of (a) J_{sc}, (b) V_{oc}, (c) FF and (d) efficiency as a function of temperature for different types of solar cells having c-Si as the absorber. The lines are guides for the eye. The part encircled in red in (d) is expanded in Figure 5.

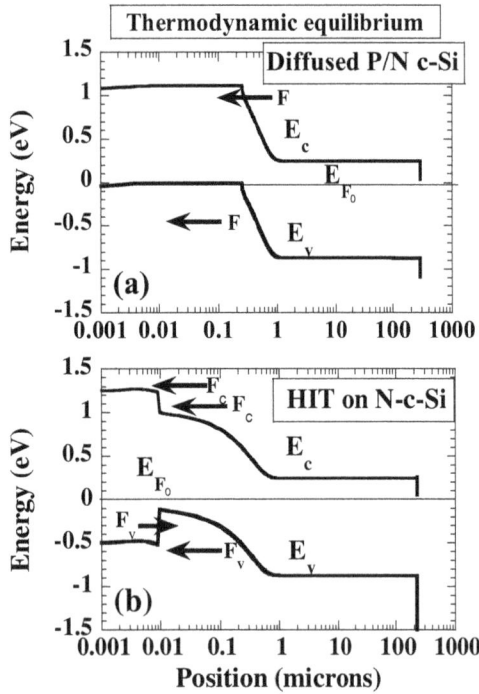

Fig. 3. Band diagrams in thermodynamic equilibrium at 300 K for (a) a conventional P emitter / N absorber homo-junction c-Si cell and (b) a HIT cell on N-type c-Si wafer. The directions of the electric field "F" on charge carriers in (a) and "F_c" on electrons and "F_v" on holes in (b) are indicated.

two-thirds of the band gap discontinuity falls on the valence band edge. In Figure 3, we draw the band diagrams of (a) the conventional P/N homo-junction cell, and (b) a HIT cell on N-type wafer at thermodynamic equilibrium, the main absorber layer in both cases being N-type c-Si with a P-type emitter. The directions of the electric field "F" on charge carriers in Figure 3a and "F_c" on electrons and "F_v" on holes in Figure 3b are indicated.

In these structures under illumination holes are collected at the top emitter side contact. But in the HIT structure, holes are faced with a barrier equal to the valence band discontinuity (ΔE_v) at the amorphous / crystalline HJ. At normal temperatures and provided ΔE_v is not too high [28], photo-generated holes can overcome this barrier by thermionic emission. There is also appreciable hole diffusion across this barrier unless the defect density at this a/c interface is extremely high, on account of the presence of an "inversion layer" on the N-c-Si side, as explained in Section 3. However, it has been shown [27,28] for P-a-Si:H/ N-c-Si HIT cells, that over the voltage range of interest in solar cell performance, tunnelling as means of hole transport is negligible and it is rather in the voltage range 0.1 V < V < 0.4 V that this latter transport mechanism becomes important [27]. At low temperatures thermionic emission decreases, making it difficult for the photo-generated holes to surmount the ΔE_v potential barrier and resulting in S-shaped J-V characteristics (Fig. 1a) and a fall in FF, J_{sc} and efficiency (Figs. 2c, 2a, 2d) for the HIT cell. Similar results have

been obtained by van Cleef et al. [30] in their study of P-a-SiC:H/ N-c-Si hetero-junction solar cells, where some tunnelling, probably multi-step tunnelling via the high defect density broad valence band tail of P-a-SiC:H has been reported. When the HIT cell is deposited on a P-type wafer, holes are collected on the rear BSF end, and then ΔE_v plays a similar role and hampers hole collection at the rear crystalline/ amorphous junction at low temperatures with similar effect on the illuminated J-V characteristics (Fig. 1b). The electrons, collected at the rear BSF end in HIT cells on N-type wafers and at the front emitter end in P-type HIT cells, face a much smaller barrier (ΔE_c), and moreover due to their higher mobility are easily collected and do not contribute to the S-shaped J-V characteristics of the HIT cells at low temperature.

For the case of the conventional diffused P/N homo-junction c-Si solar cell, whose band diagram is given in Figure 3a, there is no band discontinuity and, therefore, no potential barrier for carrier collection under illumination at any point of the device. Therefore, it retains good J-V characteristics (Fig. 1c) and a high value of the FF and efficiency, even at low temperature (Figs. 2c and 2d).

It is interesting to note that a front P-HIT cell (i.e., with a HJ at the emitter / P-c-Si junction only) retains good J-V characteristics for $T > 150$ K (Fig. 1d). This is only to be expected as in P-HIT cells, holes are collected at the rear contact, where for this case there is no ΔE_v potential barrier. The N-a-Si:H / P-c-Si valence band discontinuity at the emitter end in all types of P-HIT cells is actually beneficial, pushing holes in the right direction, viz., towards the rear of the device. Thus, a front P-HIT cell would perform better at low temperatures compared to double HIT cells; however, this case is only of academic interest, since here the BSF layer needs to be formed by diffusion at high temperatures, thus compromising the advantages of the HIT configuration (low temperature process, therefore use of thinner wafers possible).

In Figure 4a, we plot the free hole density as a function of position inside the N-type HIT cell under light and short-circuit conditions at 300 K and 175 K. We find that at 175 K due to a reduction of the thermionic emission component, few holes can surmount the ΔE_v potential barrier, resulting in a pile-up of holes on the c-Si side of the device. Hole pile-up leads to a very strong electric field over a narrow region on either side of the amorphous / crystalline interface and a collapse of the electric field and flat bands over the rest of the depletion region inside c-Si (Fig. 4b), resulting in a sharp fall in the FF and the conversion efficiency (Figs. 2c and 2d) and ultimately also of J_{sc} (Fig. 2a). In fact, at 175 K, the strong accumulation of charge in c-Si can partially deplete even the highly defective P-layer, resulting in a shift of the depletion region from the c-Si side to the amorphous emitter layer (Fig. 4b). Again our results are similar to that of reference [30]. It is also interesting to note from Figure 4b that the ΔE_v potential barrier at 175 K has become extremely narrow, a condition that might favour tunnelling across this barrier. However, to the best knowledge of the authors, no report of such tunnelling across a

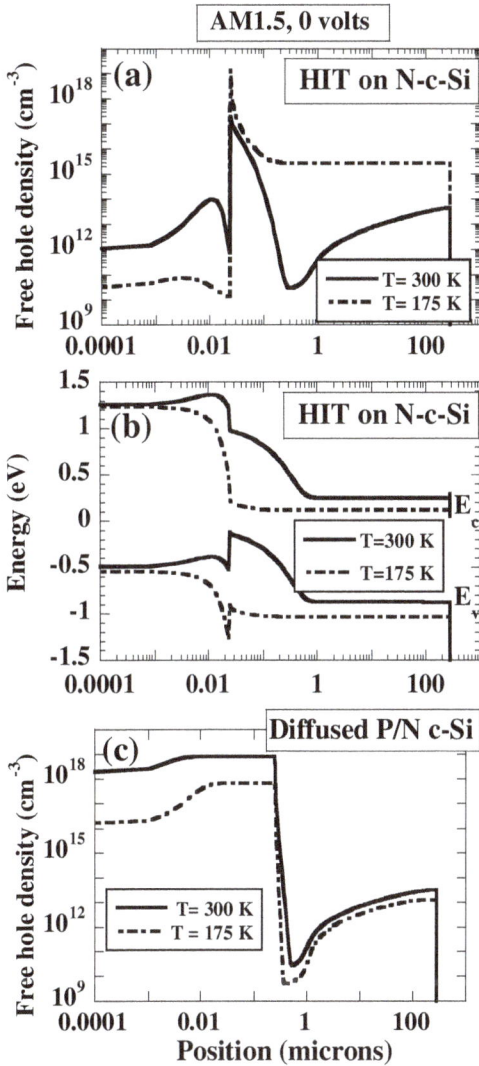

Fig. 4. (a) The free hole density as a function of position in the device and (b) the band diagram, under AM1.5 light and short-circuit conditions, for a HIT cell on N-type c-Si wafer at 300 K and 175 K, showing hole accumulation and virtual disappearance of the depletion region on the c-Si side of the amorphous-crystalline interface at 175 K; (c) the free hole density as a function of position under the same conditions and at the same temperatures for a standard P/N homo-junction c-Si solar cell.

P-a-Si:H / N-c-Si barrier under AM1.5 light has been reported. For comparison we plot the free hole density under illumination and short-circuit conditions for the standard c-Si homo-junction at 300 K and 175 K in Figure 4c and note that there is no charge accumulation in this case, leading to high values of FF (Fig. 2c) and efficiency (Fig. 2d) up to very low temperatures. In fact these parameters would have been even higher, but for the fact that we have taken account of partial donor and acceptor dopant ionization at low temperatures (Sect. 3). Only the current decreases slightly due to a fall in carrier concentration in the bands at low temperatures.

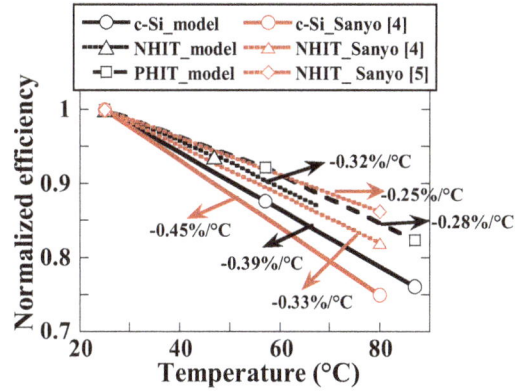

Fig. 5. Comparison of the normalized efficiency of solar cells of different structures obtained from our model with the published results of Sanyo [4,5]. The latter are shown in red, while our results are in black. The temperature coefficient of the normalized conversion efficiency for the conventional P/N diffused junction solar cell is –0.45% / °C (Sanyo [4]), while it is –0.39% / °C in our case. For HIT cells on N-type wafers, Sanyo results vary from –0.33% / °C [4] to –0.25% / °C [5], while it is –0.32% / °C in our case. For HIT cells on P-type wafers we obtain a temperature coefficient –0.28% / °C.

In Figure 5, we have zoomed in on that part of the conversion efficiency versus temperature curve, that is encircled in red (temperature range between 298 K and 330 K) and plotted the normalized efficiency versus temperature. Our curves are compared to the experimental results of Sanyo [4,5]. The temperature coefficient of the normalized conversion efficiency for the conventional P/N diffused junction solar cell is –0.45% / °C [4], while it is –0.39% / °C in our case. For HIT cells on N-type wafers, Sanyo results vary from –0.33% / °C [4] to –0.25% / °C [5], while it is –0.32% / °C in our case. For HIT cells on P-type wafers, we obtain a temperature coefficient –0.28% / °C. Thus both experiments and our modelling results indicate that HIT cells have a lower temperature coefficient of conversion efficiency than the conventional diffused P/N junction solar cell making them more suitable for use at high temperatures.

6 Conclusions

We have compared the performance of standard P/N diffused junction and HIT solar cells at different ambient temperatures. This has been done since, as mentioned in the introduction, solar cells are used in extremes of climate and therefore we need to know which type of solar cell is particularly suited for use under specific conditions. Our study clearly indicates that HIT cells are disadvantageous for use below 250 K, as their performance deteriorates sharply on account of the large amorphous-crystalline valence band discontinuity and reduced thermionic emission across this barrier at low temperatures. However, they are well-adapted for use in hot and sunny regions of the planet, not only because of their lower production cost, but also because they have a lower temperature coefficient of conversion efficiency compared to the standard diffused

P/N junction solar cell. However, as also demonstrated in this study, the latter class of cells is clearly better-suited for low temperature and space applications.

This study was funded by the Ministry of New and Renewable Energy (MNRE), Government of India, and by the *Centre National de la Recherche Scientifique* (CNRS), France. The computer modelling program was developed (electrical part) by P. Chatterjee during a project funded by MNRE and DST, Government of India, and the optical part was added during her tenure as Marie Curie fellow at *École Polytechnique*, Palaiseau, France. It was extended to model also HIT cells under a CSIR project.

References

1. M. Tanaka, M. Taguchi, T. Matsuyama, T. Sawada, S. Tsuda, S. Nakano, H. Hanafusa, Y. Kuwano, Jpn J. Appl. Phys. **31**, 3518 (1992)

2. T. Takahama, M. Taguchi, S. Kuroda, T. Matsuyama, M. Tanaka, S. Tsuda, S. Nakano, Y. Kuwano, in *Proceedings of the 11th European Photovoltaic Solar Energy Conference, Montreux, Switzerland, 1992*, p. 1057

3. T. Kinoshita, D. Fujishima, A. Yano, A. Ogane, S. Tohoda, K. Matsuyama, Y. Nakamura, N. Tokuoka, H. Kanno, H. Sakata, M. Taguchi, E. Maruyama, *Proceedings of the 26th European Photovoltaic Solar Energy Conference, 2011*, p. 871. http://www.pv-tech.org/news/sanyo_a_hit_with_23_solar_cell_efficiency_record

4. H. Sakata, T. Nakai, T. Baba, M. Taguchi, S. Tsuge, K. Uchihashi, S. Kiyama, Proc. IEEE Photovoltaic Spec. Conf. 7 (2000)

5. E. Maruyama, A. Terakawa, M. Taguchi, Y. Yoshimine, D. Ide, T. Baba, M. Shima, H. Sakata, M. Tanaka, in *Proceedings 4th World Conference on Photovoltais Solar Energy Conversion, Hawaii, USA, 2006*, pp. 1455-1460

6. P. Roca i Cabarrocas, J.B. Chevrier, J. Huc, A. Lloret, J.Y. Parey, J.P.M. Schmitt, J. Vacuum Sci. Technol. A **9**, 2331 (1991)

7. I. Martin, M. Vetter, A. Orpella, J. Puigdollers, C. Voz, R. Alcubilla, J. Damon-Lacoste, P. Roca i Cabarrocas, in *Proceedings of the 19th European Photovoltaic Solar Energy Conference, Paris, France, 2004*, p. 1185

8. P. Chatterjee, J. Appl. Phys. **76**, 1301 (1994)

9. P. Chatterjee, J. Appl. Phys. **79**, 7339 (1996)

10. M. Nath, P. Chatterjee, J. Damon-Lacoste, P. Roca i Cabarrocas, J. Appl. Phys. **103**, 034506 (2008)

11. F.A. Rubinelli, S.J. Fonash, J.K. Arch, in *Proceedings of the 6th International Photovoltaic Science & Engineering Conference*, edited by B.K. Das, S.N. Singh (Oxford & IBH publishing Co.Pvt.Ltd., New Delhi, Bombay, Calcutta, 1992), p. 811

12. R. Stangl, M. Kriegel, K.V. Maydell, L. Korte, M. Schmidt, W. Fuhs, *Conference record 31st IEEE Photovoltaic Specialists Conference, 2005*, p. 1556

13. R. Varache, J.P. Kleider, W. Favre, L. Korte, J. Appl. Phys. **112**, 123717 (2012)

14. M.J. Powell, S.C. Deane, Phys. Rev. B **48**, 10815 (1993)

15. M.J. Powell, S.C. Deane, Phys. Rev. B **53**, 10121 (1996)

16. V. Halpern, Philos. Mag. B **54**, 473 (1986)

17. C. Longeaud, J.P. Kleider, Phys. Rev. B **48**, 8715 (1993)

18. J.P. Kleider, A.S. Gudovskikh, P. Roca i Cabarrocas, Appl. Phys. Lett. **92**, 162101 (2008)

19. F. Leblanc, J. Perrin, J. Schmitt, J. Appl. Phys. **75**, 1074 (1994)

20. P. Chatterjee, M. Favre, F. Leblanc, J. Perrin, Mat. Res. Soc. Symp. Proc. **426**, 593 (1996)

21. V. Alex, S. Finkbeiner, J. Weber, J. Appl. Phys. **79**, 6943 (1996)

22. S.M. Sze, *Physics of Semiconductor Devices* (John Wiley & Sons, Inc., New York, London, Sydney, Toronto, 1969)

23. C.R. Wronski, J. Non-Cryst. Solids **141**, 16b (1992)

24. N. Dasgupta, A. Dasgupta, *Semiconductor Devices: Modelling, Technology* (Prentice-Hall of India Private Limited, New Delhi, 2004)

25. A. Datta, J. Damon-Lacoste, P. Roca i Cabarrocas, P. Chatterjee, Sol. Energy Mater. Sol. Cells **92**, 1500 (2008)

26. A. Datta, J. Damon-Lacoste, M. Nath, P. Roca i Cabarrocas, P. Chatterjee, Mater. Sci. Eng. B **159-160**, 10 (2009)

27. M. Taguchi, E. Maruyama, M. Tanaka, Jpn J. Appl. Phys. **47**, 814 (2008)

28. M. Rahmouni, A. Datta, P. Chatterjee, J. Damon-Lacoste, C. Ballif, P. Roca i Cabarrocas, J. Appl. Phys. **107**, 054521 (2010)

29. M. Brinza, G. Adriaenssens, P. Roca i Cabarrocas, Thin Solid Films **427**, 123 (2003)

30. M.W.M. van Cleef, F.A. Rubinelli, R. Rizzoli, R. Pinghini, R.E.I. Schropp, W.F. van der Weg, Jpn J. Appl. Phys. **37**, 3926 (1998)

8

Zn(O, S) layers for chalcoyprite solar cells sputtered from a single target

A. Grimm[1], D. Kieven[1], I. Lauermann[1], M.Ch. Lux-Steiner[1], F. Hergert[2], R. Schwieger[2], and R. Klenk[1,a]

[1] Helmholtz-Zentrum Berlin für Materialien und Energie, Hahn-Meitner-Platz 1, 14109 Berlin, Germany
[2] Bosch Solar CISTech, Münstersche Str. 24, 14772 Brandenburg an der Havel, Germany

Abstract A simplified $Cu(In, Ga)(S, Se)_2/Zn(O, S)/ZnO{:}Al$ stack for chalcopyrite thin-film solar cells is proposed. In this stack the $Zn(O, S)$ layer combines the roles of the traditional CdS buffer and undoped ZnO layers. It will be shown that $Zn(O, S)$ films can be sputtered in argon atmosphere from a single mixed target without substrate heating. The photovoltaic performance of the simplified stack matches that of the conventional approach. Replacing the ZnO target with a ZnO/ZnS target may therefore be sufficient to omit the CdS buffer layer and avoid the associated complexity, safety and recycling issues, and to lower production cost.

1 Introduction

$Zn(O, S)$ is emerging as one of the most promising materials to replace CdS in the buffer layer of chalcopyrite-based thin-film solar cells [1]. Successful preparation technologies include chemical bath deposition and atomic layer deposition. Sputtering, already established in mass production for other layers of the cell, may be another attractive deposition technology. Chalcopyrite cells and modules are prepared with an undoped sputtered ZnO layer (i-ZnO) on top of the CdS. Therefore, we may consider a sputtered $Zn(O, S)$ layer a modification of the standard ZnO layer, eliminating the need for a dedicated buffer layer. Reasonable cell performance for this approach has been reported previously. The $Zn(O, S)$ was prepared by co-sputtering from ZnO and ZnS targets [2] or by reactive sputtering from a ZnS target in an Ar/O_2 gas mixture [3]. These methods are well suited for fundamental investigations because the $S/(O + S)$ ratio can be freely adjusted for an optimal conduction band alignment. On the other hand, in order to develop a true drop-in replacement for the standard ZnO process, a non-reactive process without substrate heating and using a single target is much more appropriate. In this contribution we report selected properties of thin $Zn(O, S)$ films sputtered from single mixed targets and demonstrate their successful application in solar cells.

2 Film preparation and properties

Targets with a nominal composition of $S/(S + O) = 0.4$ (atomic ratio) were procured from a commercial supplier.

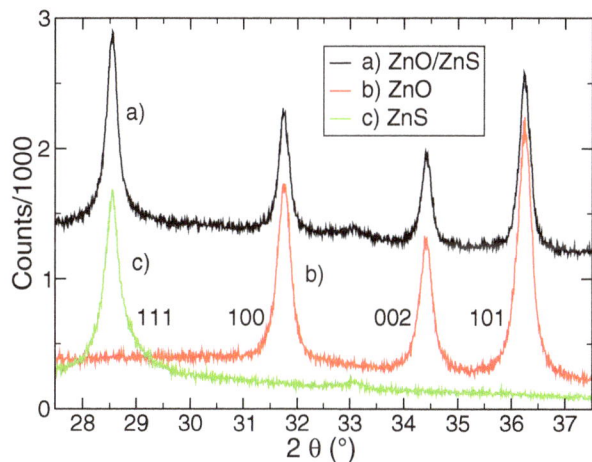

Fig. 1. XRD patterns of powder samples prepared from ZnO, ZnS and ZnO/ZnS sputtering targets. Miller indices are shown assuming hexagonal symmetry (wurtzite) for ZnO and cubic symmetry (zinc blende) for ZnS.

According to our energy dispersive X-ray spectroscopy (EDX) the ratio was 0.35. X-ray diffraction (XRD) patterns recorded with Cu K_α radiation (Fig. 1) revealed that the target material is a two phase mixture of ZnS and ZnO. Films were sputtered in pure Ar using 13.56 MHz (RF) plasma excitation in two different systems with target diameters of 75 (system A) and 125 mm (system B), respectively. The substrate was not moving during deposition. Typical parameters were working gas pressures in the range of $3-9$ μbar and power densities of $1.3-1.7$ W/cm^2 which resulted in deposition rates of about 50 nm/min. The film composition was measured by EDX (film thickness ≈ 0.5 μm on glass/Mo substrates).

[a] e-mail: klenk@helmholtz-berlin.de

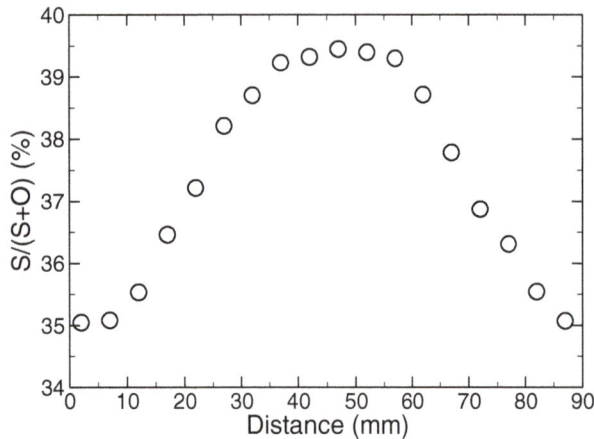

Fig. 2. S/(S + O) ratio across the substrate as measured by EDX.

Table 1. Composition of Zn(O, S) films as measured by EDX.

Pressure (μbar)	Substrate heating	Zn (%)	O (%)	S (%)	S/(S+O) (%)
3	–	50.1	33.7	16.2	0.32
3	200 °C	50.4	37.0	12.6	0.25
9	–	50.2	27.7	22.1	0.44
9	200 °C	50.6	31.4	18.0	0.36

The S/(S + O) ratio in the deposited films roughly reflected the composition of the target, however, it also depended on process parameters (Tab. 1). Films prepared at higher pressures were generally richer in sulphur. Similarly, substrate heating (if any) influences the sulphur content. We also found a radially symmetric inhomogeneity (Fig. 2) in the larger of the two deposition systems (system B without substrate heating). Furthermore, the sulphur content of the films increased slightly with target erosion. In contrast to our previous results with reactive sputtering, the sulphate ($SO4^{2-}$) and hydroxide contaminations as estimated from photo electron spectroscopy (XPS) were minimal even without deliberate substrate heating. XRD patterns of films (Fig. 3) sputtered onto heated substrates indicated a crystalline (wurtzite) structure with lattice constants approximately as expected from the S/(S + O) ratios. Without substrate heating, films did typically show only very weak XRD patterns. Optical transmission and reflection spectra (Fig. 4) were very similar to the ones measured for the reactively sputtered films, with the absorption being very low at longer wavelength and increasing rather slowly when approaching the band gap.

3 Device properties

Devices were prepared by using sequentially prepared glass/Mo/Cu(In, Ga)(S, Se)$_2$ substrates from the Bosch Solar CISTech production line [4]. The full size substrates were cut into smaller pieces (2.5 × 2.5 or 5 × 5 cm^2,

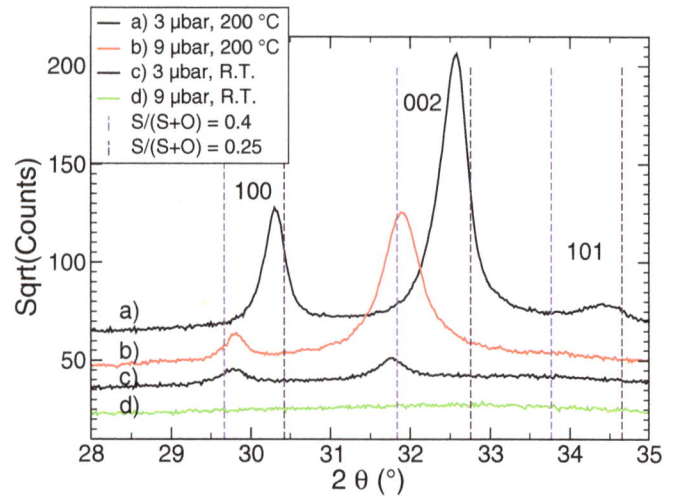

Fig. 3. XRD patterns of Zn(O, S) thin films on glass/Mo substrates prepared with and without substrate heating at two different Ar pressures. Vertical lines indicate calculated peak positions assuming hexagonal symmetry and a linear shift in lattice constants between ZnO and ZnS.

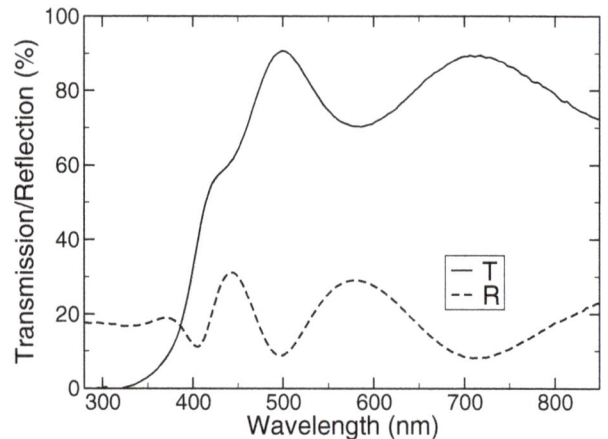

Fig. 4. Optical transmission (T) and reflection (R) of a Zn(O, S) film (thickness 400 nm) sputtered from a mixed target without substrate heating onto a soda lime float glass substrate.

depending on the sputtering system used) and sealed in dry atmosphere for shipping. Some samples were etched in aqueous KCN solution before depositing the Zn(O, S) layer. The thickness of the latter was in the range of 20 to 60 nm. Cells were completed with a sputtered ZnO:Al layer and Ni/Al grids evaporated through shadow masks. 32 cells with an area of 0.5 cm^2 were defined on the 5 × 5 cm^2 substrates by mechanical scribing. No antireflective coating was applied. Current-voltage (jV) characteristics were measured in-house with simulated AM 1.5 illumination without deliberate light soaking or postannealing. The best efficiencies that could be achieved in a completely dry process (without etching) were reasonable but there was a distinct inhomogeneity across the substrate. Higher efficiencies, together with very good homogeneity, were achieved with the etched samples. Figure 5

Table 2. Parameters of the best cells (in-house total area measurements under simulated AM 1.5 illumination, without AR coating).

Window	Open circuit voltage (mV)	Short circuit current density (mA/cm^2)	Fill factor (%)	Efficiency (%)
Zn(O, S)/ZnO:Al	561	37.9	68.1	14.5
CdS/ZnO/ZnO:Al	574	37.5	69.8	14.9

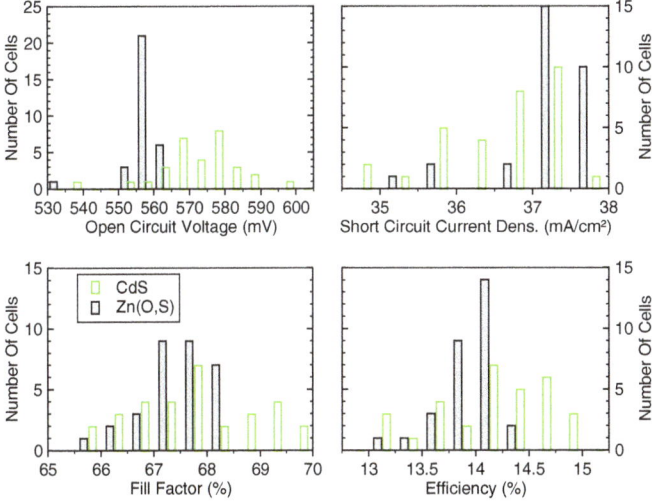

Fig. 5. Distribution of the parameters of small (0.5 cm^2) cells with Zn(O, S)/ZnO:Al and CdS/ZnO/ZnO:Al windows on 5×5 cm^2 substrates (see text) as measured under simulated AM 1.5 illumination.

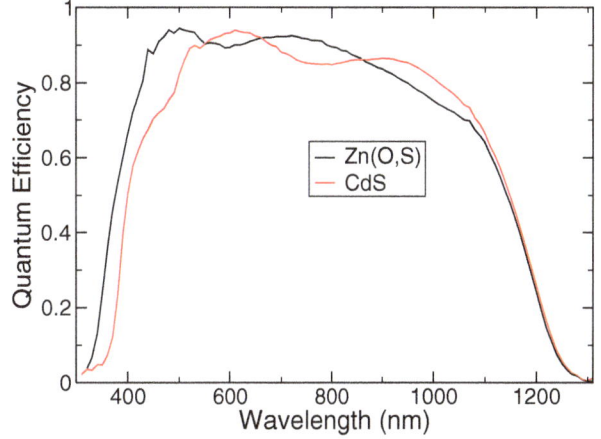

Fig. 6. Quantum efficiencies of CIGSSe/Zn(O, S)/ZnO:Al and CIGSSe/CdS/ZnO/ZnO:Al solar cells.

shows a distribution of cell parameters measured on a 5×5 cm^2 substrate. The performance in this case was comparable to reference cells prepared with the standard CdS/ZnO/ZnO:Al window on two 5×5 cm^2 substrates cut from the same full size plate (absorbers not etched, only every other cell measured). Parameters of the best cells are given in Table 2. Considering these data, the new process results in better device performance than reactive sputtering [5]. The quantum efficiency (Fig. 6) shows better blue response for the cells with Zn(O, S)/ZnO:Al window. Using a tabulated AM 1.5 reference spectrum the calculated active area current densities are 37.1 mA/cm^2 (CdS) and 38.1 mA/cm^2, respectively, in good agreement with the total area short circuit current densities from jV measurements.

4 Discussion

Due to the fact that the films did not always exhibit clear XRD patterns, it is difficult to unambiguously determine which phases are present in the films. Comparison of the optical band gap and cell parameters as a function of the overall S/(O + S) ratio with crystalline films prepared with substrate heating nevertheless suggest that the films are essentially compound ZnO$_{1-x}$S$_x$. Presumably, single phase Zn(O, S) films can be prepared by non reactive RF sputtering from a ZnO/ZnS mixed target without additional substrate heating. The optical band gap seems to

be slightly higher than that of reactively sputtered films (with the same sulphur content) [6]. Lower sputtering pressure leads to slightly better crystallinity which may indicate that the growth is ion assisted. Inhomogeneity of the S/(O + S) ratio and drift of the latter with target erosion may be challenges in scaling-up of the process. Our previous studies conducted with varied S/(O + S) ratios (reactive sputtering) show a very rapid decline in fill factor and photo current density when there is too much sulphur in the Zn(O, S) film, presumably due to current blocking by a too high conduction band spike [6,7]. The process is more tolerant on the oxygen-rich side of the optimum composition where the losses in open circuit voltage are not immediately critical. Compared to the previous results, the optimum sulphur content seems to be somewhat lower for the films sputtered from the mixed target (which may be connected to the different band gap mentioned above). In view of this, the composition of our targets was probably a little bit too sulphur rich. This is reflected in the cell results measured on as-grown substrates where the higher sulphur content beneath the center of the target already leads to partial current blocking and poor fill factor. It is interesting to note that etched absorbers appear to tolerate a higher S/(S + O) ratio. XPS shows (in agreement with literature data) that etching removes sodium containing compounds from the absorber surface. We may speculate that the presence of these sodium compounds induces a dipole at the absorber/Zn(O, S) interface which increases the tendency for too high a conduction band spike. The requirement for wet chemical surface conditioning is of course incompatible with the original goal of completely dry processing. However, we hope that a slightly more oxygen-rich target will remedy this problem. In any case, the device results reported here are clearly superior to the ones achieved previously by sputtering Cd-free materials directly onto a chalcopyrite absorber.

5 Summary and conclusions

RF sputtering from a ZnO/ZnS mixed target produces $ZnO_{1-x}S_x$ films with different degrees of long range ordering. The $S/(S+O)$ ratio in the film reflects the target composition but is also slightly influenced by process conditions. By transitioning from the previous approach (reactive sputtering onto heated substrates) to the one described here, we were able to almost close the efficiency gap between devices with sputtered $Zn(O,S)$ layers and those with standard CdS/ZnO buffer. We have thus shown, that with a simple modification of the standard ZnO target, the Cd-containing buffer layer and its costly wet chemical preparation process may no longer be needed for efficient chalcopyrite solar cells.

We thank the Federal Ministry for the Environment, Nature Conservation and Nuclear Safety (BMU) for financial support of the NeuMaS project. Technical support in preparation and measurements by C. Kelch, M. Kirsch, C. Klimm, and L. Chikhaoui is gratefully acknowledged.

References

1. N. Naghavi, D. Abou-Ras, N. Allsop, N. Barreau, S. Bücheler, A. Ennaoui, C.-H. Fischer, C. Guillen, D. Hariskos, J. Herrero, R. Klenk, K. Kushiya, D. Lincot, R. Menner, T. Nakada, C. Platzer-Björkman, S. Spiering, A.N. Tiwari, T. Törndahl, Prog. Photovolt. Res. Appl. **18**, 411 (2010)
2. A. Okamoto, T. Minemoto, H. Takakura, Jpn J. Appl. Phys. **50**, 04DP10 (2011)
3. A. Grimm, J. Just, D. Kieven, I. Lauermann, J. Palm, A. Neisser, T. Rissom, R. Klenk, Phys. Stat. Sol. RRL **4**, 109 (2010)
4. V. Probst, F. Hergert, B. Walther, R. Thyen, G. Batereau-Neumann, B. Neumann, A. Windeck, T. Letzig, A. Gerlach, in *Proceedings of the 24th European Photovoltaic Solar Energy Conference and Exhibition Munich, 2009*, edited by W.C. Sinke, H.A. Ossenbrink, P. Helm (WIP, Munich, 2009), p. 2455
5. P. Pistor, A. Grimm, D. Kieven, F. Hergert, A. Jasenek, R. Klenk, in *Proceedings of the 37th IEEE Photovoltaic Specialists Conference, Seattle, 2011* (IEEE, New York, 2011), p. 2812
6. A. Grimm, D. Kieven, R. Klenk, I. Lauermann, A. Neisser, T. Niesen, J. Palm, Thin Solid Films **520**, 1330 (2011)
7. D. Kieven, A. Grimm, I. Lauermann, M.Ch. Lux-Steiner, J. Palm, T. Niesen, R. Klenk, Phys. Stat. Sol. RRL **6**, 294 (2012)

9

Passivation effects of atomic-layer-deposited aluminum oxide

R. Kotipalli[a], R. Delamare, O. Poncelet, X. Tang, L.A. Francis, and D. Flandre

ICTEAM, Université catholique de Louvain, Place du Levant 3, 1348 Louvain-la-Neuve, Belgium

Abstract Atomic-layer-deposited (ALD) aluminum oxide (Al_2O_3) has recently demonstrated an excellent surface passivation for both n- and p-type c-Si solar cells thanks to the presence of high negative fixed charges ($Q_f \sim 10^{12}-10^{13}$ cm^{-2}) in combination with a low density of interface states (D_{it}). This paper investigates the passivation quality of thin (15 nm) Al_2O_3 films deposited by two different techniques: plasma-enhanced atomic layer deposition (PE-ALD) and Thermal atomic layer deposition (T-ALD). Other dielectric materials taken into account for comparison include: thermally-grown silicon dioxide (SiO_2) (20 nm), SiO_2 (20 nm) deposited by plasma-enhanced chemical vapour deposition (PECVD) and hydrogenated amorphous silicon nitride (a-SiNx:H) (20 nm) also deposited by PECVD. With the above-mentioned dielectric layers, Metal Insulator Semiconductor (MIS) capacitors were fabricated for Q_f and D_{it} extraction through Capacitance-Voltage-Conductance (C-V-G) measurements. In addition, lifetime measurements were carried out to evaluate the effective surface recombination velocity (SRV). The influence of extracted C-V-G parameters (Q_f,D_{it}) on the injection dependent lifetime measurements $\tau(\Delta n)$, and the dominant passivation mechanism involved have been discussed. Furthermore we have also studied the influence of the SiO_2 interfacial layer thickness between the Al_2O_3 and silicon surface on the field-effect passivation mechanism. It is shown that the field effect passivation in accumulation mode is more predominant when compared to surface defect passivation.

1 Introduction

It is well known that the thermal silicon oxide (SiO_2) is a very good surface passivation material for crystalline Si (c-Si). However, the formation of thermal SiO_2 requires a high-temperature process (>1000 °C) which does not only increase the processing cost, but may also degrade the quality of the silicon wafer. Therefore, passivation materials that can be deposited at low temperatures are required. Hydrogenated amorphous silicon nitride (a-SiNx:H) obtained by plasma-enhanced chemical vapour deposition (PECVD) is commonly used as front surface emitter passivation and anti-reflection coating (ARC) for both n- and p-type low-resistivity c-Si solar cells. This material can be deposited by PECVD at low temperatures i.e. less than 300 °C. Surface passivation of c-Si solar cell with atomic layer deposited (ALD) aluminum oxide (Al_2O_3) is a good candidate for both p- and n-type surfaces as well as highly-doped p-type emitters, due to its very high built-in negative fixed charge density ($Q_f \sim 10^{12}-10^{13}$ cm^{-2}), low interface state density ($D_{it} \leqslant 10^{11}$ eV^{-1} cm^{-2}) and low surface recombination velocity (SRV $\leqslant$ 5 cm s^{-1}). Best reported conditions were obtained by plasma-ALD process

with film thickness ranging from 5 to 30 nm at deposition temperatures between 150−250 °C followed by an annealing step in nitrogen or forming gas atmosphere at 400 °C on low-resistivity Si wafers [1–6].

To understand the involved surface passivation mechanism, Girisch et al. [7] introduced an extended Shockley-Read-Hall (SRH) formalism to model the surface recombination mechanism, which included the effects of band bending due to fixed insulator charges (Q_f) and charged interface states ($D_{it} = qN_{it}$) over the entire band gap. By assuming the case of single defect level at mid-gap, the surface recombination rate (U_s) can be written as an energy independent quantity [3, 7–10].

$$U_s \cong \frac{n_s p_s}{\frac{n_s}{S_p} + \frac{p_s}{S_n}} \qquad (1)$$

where:

- n_s and p_s are the surface concentrations of electrons and holes respectively.
- $S_n S_p$ are the surface recombination velocity parameters given by $S_n = \sigma_n v_{th} N_{it}$, $S_p = \sigma_p v_{th} N_{it}$.
- σ_n, σ_p cross-section of electrons and holes respectively.
- v_{th} being the thermal velocity.

---
[a] e-mail: `Raja.Kotipalli@uclouvain.be`

From the above equation (1) one can estimate the surface recombination velocity (SRV) at a particular injection level given by the excess carrier density(Δn).

$$\text{SRV} = \frac{U_s}{\Delta n}. \qquad (2)$$

Surface recombination rate (U_s), can be reduced by altering two fundamental mechanisms:

(i) *Reducing the interface state densities D_{it} (i.e. Chemical passivation)*

The D_{it} is dependent on material and chemical processes used in the fabrication of the solar cell. For example its reduction can be realized by diffusing hydrogen into the silicon/dielectric interface to replace the dangling bond defects.

(ii) *Reducing the surface concentration of minority carriers (i.e. Field-effect passivation)*

The surface recombination rate (U_s) can be reduced by decreasing one of two carrier concentrations at the silicon surface typically, the minority carrier concentration. This can be achieved by the fixed charges existing in the dielectric layer. Specifically, these charges creates a built-in electric field which shields the minority carrier to be recombined at the surface by driving the device into accumulation or inversion modes depending on the charge sign (positive or negative) and the chosen substrate type [3,8].

In this article we study the surface passivation mechanism induced by negative fixed charges in Al_2O_3 films along with other fixed positive charge dielectrics. After a general introduction about the surface passivation mechanism, we describe the metal insulator semiconductor (MIS) capacitor device fabrication, lifetime sample preparations and electrical characterization techniques we used to extract the main parameters involved in the quality of the passivation, i.e. Q_f, D_{it} and SRV in Section 2. Section 3 reports the experimental results of all the considered dielectrics. A discussion on the influence of films characteristics on the nature of interface passivation will be held. Subsequently, the impact of interfacial SiO_2 layer on the field-effect passivation will be addressed in Section 4. Finally, in Section 5, conclusions are drawn from our Al_2O_3 passivation study of Si surfaces.

2 Samples fabrication and characterization techniques

It clearly appears from equations (1)−(2) that insight knowledge about the density of interface states (D_{it}) and surface concentrations of carrier n_s, p_s (depends on Q_f) gives in-depth information about the passivation quality of dielectric films. To extract these values we have considered MIS capacitors as test vehicle.

2.1 Dielectric film deposition

Thermal SiO_2 was grown up to a thickness of 20 nm at 1050 °C in an ultra-dry oxygen atmosphere using vertical

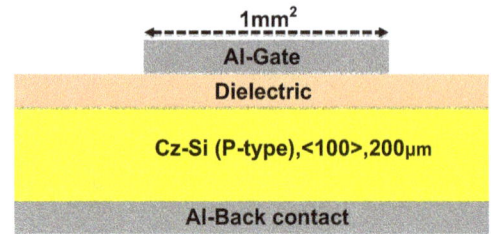

Fig. 1. Schematic of the metal insulator semiconductor (MIS) capacitor.

furnace from Koyo Thermo Systems for a duration of 10 min.

For PECVD SiO_2 20 nm-thick layers were deposited in a parallel plate reactor from Oxford Plasmalab system 100. The deposition parameters used during the film growth were: chamber pressure 0.8 Torr, deposition temperature 300 °C, gas flow: SiH_4-500 sccm, N_2O-20 sccm and O_2-5 sccm. PECVD a-$SiNx$:H 20 nm-thick layers were also deposited using the same Plasmalab system 100. In this case, Silane (SiH_4) and ammonia (NH_3) were used as reactive gases in the chamber. During the deposition, SiH_4 was diluted to 5% in pure nitrogen. The deposition parameters used for the film growth were: chamber pressure 0.8 Torr, deposition temperature 300 °C, radio frequency (RF) power 20 W, plasma frequency 13.56 MHz, gas flow: NH_3-1.8 sccm, SiH_4-10 sccm, and N_2-700 sccm.

In the case of Al_2O_3, 15 nm-thick layers were deposited in a Fiji F200 ALD system from Cambridge NanoTech by thermal atomic layer deposition T-ALD and plasma enhanced atomic layer deposition PE-ALD. In both cases, trimethylaluminum precursor (TMA) from Sigma-Aldrich was used as aluminum source. Depositions were performed at 250 °C for both T-ALD and PE-ALD with argon as a carrier gas. Purge after precursor pulse is mandatory to avoid chemical vapor deposition. Before the deposition, all the samples were kept in the deposition chamber for 1800 s for pumping away H_2O or O_2. Each precursor flows separately through the deposition chamber. TMA pulse duration and purge time were 0.06 s and 10 s respectively for T-ALD and PE-ALD depositions. For T-ALD the precursor was de-ionized water while for PE-ALD, oxygen flow is used instead of water pulse. The flow was 30 sccm and plasma power was 300 W. The pulse duration and purge time of the plasma were 20 s and 5 s, respectively. The growth rates were observed to be 1 Å per cycle for both PE-ALD and T-ALD [11–15].

2.2 Device fabrication and C-V-G measurements

MIS capacitors were fabricated on p-type, Cz $\langle 100 \rangle$ silicon wafers with a resistivity of $(1-3)$ Ω cm as illustrated in (Fig. 1).

Before the deposition of the dielectric layers, all the wafers were cleaned using Piranha solution ($3:1:H_2SO_4:H_2O_2$, for a duration of 20 min at 120 °C). After the Piranha cleaning the wafers were dipped in dilute HF solution (1:50:HF:DI water) at room temperature to remove the native oxide. The etching of native oxide layer is confirmed by the appearance of hydrophobic Si surface. Next, dielectric layers described in the above Section 2.1 were grown or deposited. Then, gate electrodes with an active area of 1 mm^2 were patterned on the front side of the samples using an image reversal lithography step (i.e. resist coating, pre-bake, image exposure with mask, post- exposure bake, flood exposure and development). A 300 nm Al layer was evaporated on the front side of the samples followed by a lift-off in acetone. After front side device fabrication, full-area aluminum back contact (300 nm) is evaporated on the backside of the wafers. Finally, all the samples were annealed in forming gas (N$_2$/H$_2$: 90/10%) ambient at 432 °C for 30 min.

Capacitance-voltage-conductance (C-V-G) measurements were performed with Agilent B1500A Semiconductor device analyzer at different frequencies ranging from 1 kHz to 1 MHz. To confidently extract the interface trap charge densities (D_{it}) at the silicon/dielectric interface we used different available extraction methods namely: High-Low frequency method, Terman method and Conductance method. The fixed charge density (Q_f) in the dielectric was estimated from the flat-band voltage of the low-frequency C-V curve [16–21].

2.3 Sample preparation for lifetime measurements

In our experiments, to emphasize the electronic properties of the dielectrics, we have chosen p-type, Boron doped, 200 μm thick double-side polished, $\langle 111 \rangle$, Float zone (FZ) silicon wafers in order to neglect the bulk lifetime in SRV extractions. A choose of high resistivity >5000 Ω cm wafers in particular is to avoid the bandgap defects induced by impurity dopants which acts like an effective recombination centers for SRH. Since the dopant impurity concentration sets the Fermi level, low-resistivity materials are more sensitive to these defects than high-resistivity material. So the dominant recombination in these samples is only due to Auger and radiative mechanisms. Lifetime measurements were performed using Sinton WCT-120 lifetime tester in both quasi steady state and transient modes. The different dielectric layers under consideration were symmetrically deposited on both sides of the wafers, followed by a forming gas annealing at 432 °C for 30 min to activate the passivation mechanism [22, 23].

3 Passivation analysis using different dielectrics

The C-V characteristics of MOS capacitors measured at 10 kHz for the different considered dielectrics are shown in Figure 2. The flat-band voltage (V_{fb}) of the low-frequency C-V curve allows calculation of the density and the polarity of charges present in the dielectric film from the following equation:

$$Q_f = (\Phi_{ms} - V_{fb})C_{oxide} \qquad (3)$$

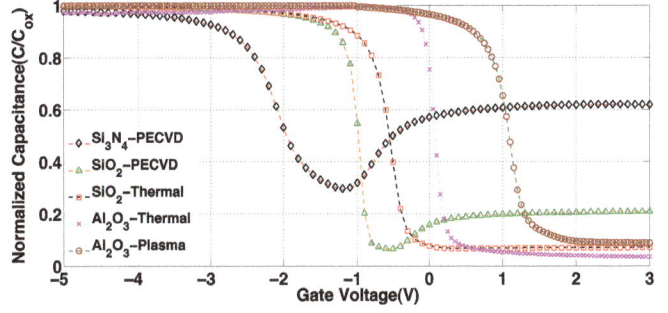

Fig. 2. Normalized C-V Characteristics of different dielectrics at 10 kHz to extract the fixed charge density (Q_f).

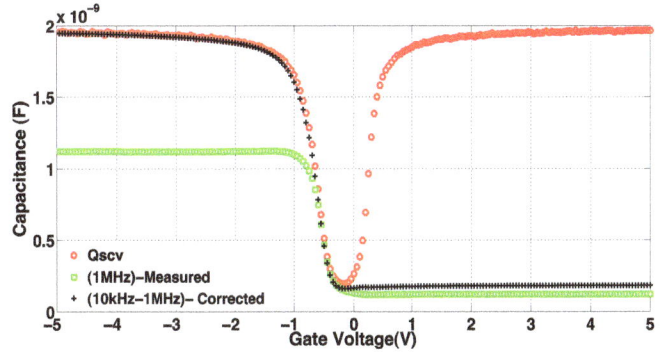

Fig. 3. Typical quasi-static (QSCV) and high-frequency (HF) C-V curves for a MIS capacitor with SiO$_2$ dielectric. The HF curve is also represented after parasitic free (frequency-dispersion related problems in accumulation gate voltages) correction using "dual-frequency five-element small-signal circuit model".

where $\Phi_{ms} = -0.96$ V is the difference between the aluminum and silicon work functions.

The flat-band voltage (V_{fb}) of the MOS capacitors with the PECVD Si$_3$N$_4$, PECVD SiO$_2$ and thermal SiO$_2$ films is negative (i.e. lower than Φ_{ms}), meaning these films contain fixed positive charges (Q_f). For the T-ALD and PE-ALD Al$_2$O$_3$ films, the V_{fb} is positive and Q_f is negative [16–18].

3.1 Interface states densities (D$_{it}$) extraction methodology

The extraction of interface state density (D_{it}) using only one method may suffer from parasitic effects such as high leakage currents through the dielectrics (thin) films, series resistance from the measurement setup and frequency-dispersion related problems in accumulation region. These parasitic effects will definitely alter the measured capacitance-conductance (C-G) values, which will in turn affect the interpreted D_{it} by up to an order of magnitude. To minimize the influence of these effects on the extracted interface states densities (D_{it}), all the measured C-V-G curves were first corrected for parasitic free C-V-G curves using "dual-frequency five-element small-signal circuit model" as shown in (Fig. 3) [19–21, 24].

To assure that the extracted D_{it} values are correctly estimated over the entire band gap and not affected by parasitic effects, we extract the D_{it} using three different methods described below.

Figure 4 exemplifies the measurement results for the MIS capacitor with PECVD SiO_2 dielectric as an example using the different extraction methods as follows [16–21].

– Figure 4b shows the D_{it} extraction using high-low frequency method, which compares the quasi-static C-V (QSCV) curve with a high frequency (1 MHz) C-V curve. In the QSCV measurement the interface traps are assumed to follow the slowly varying dc bias, contributing to interface trap charge capacitance (C_{it}). On the contrary, in the high-frequency C-V measurement the interface traps cannot follow the applied high-frequency ac signal, making the interface trap charge capacitance be zero ($C_{it} = 0$). The value of C_{it} can then be estimated by comparing the difference in capacitance between quasi-static and high frequency C-V curves from depletion – inversion regions (i.e. yielding D_{it} for mid-gap potentials) [20,21].

– Figure 4c illustrates the D_{it} extraction using Terman method, which is based on the stretch-out phenomenon in the experimental high-frequency C-V curve compared to theoretically simulated high-frequency C-V curve (i.e. ideal) with no interface traps. From the ideal C-V curve finding the surface potential (φ_s) for a given capacitance value in the depletion regime, and interpolating it on the experimental gate voltage (V_G) curve gives us the (φ_s–V_G) relation. Repeating this for other points from accumulation to inversion regimes results in a (φ_s–V_G) curve. This φ_s versus V_G curve is stretched-out when compared to theoretical curve without D_{it}, this stretch-out yields the information about the interface state densities [17,19].

– Figure 4d represents the D_{it} extraction using conductance method. This method is based on measuring the equivalent parallel conductance per unit area (Gp) as a function of bias voltage and frequency (ω). This equivalent parallel conductance represents the energy loss caused by capture and emission of carriers from the interface traps when gate bias is swept from accumulation to inversion regimes. Plotting ($\frac{Gp}{\omega}$) with respect to frequencies in the depletion range of gate voltages yields the maximum of energy loss mechanism due to interface states. This peak (maximum energy loss) value of ($\frac{Gp}{\omega}$)$_{max}$ gives direct information on D_{it} [16–18].

Tables 1 and 2 summarize extracted results from C-V measurements on MIS capacitors: oxide capacitance in accumulation (C_{ox}), flat band voltage (V_{fb}), fixed charge density (Q_f) and interface trap charge density (D_{it}) using different methods. The range of Q_f and D_{it} values are estimated considering variations of oxide thickness $\pm(0.1-2)$ nm and substrate resistivity $(1-3)$ Ω cm.

Table 1 highlights that the PECVD Si_3N_4 layer has a high density of positive charge $\sim 4.2 \times 10^{12}$ cm^{-2} compared to other dielectrics. Al_2O_3 dielectric films exhibit

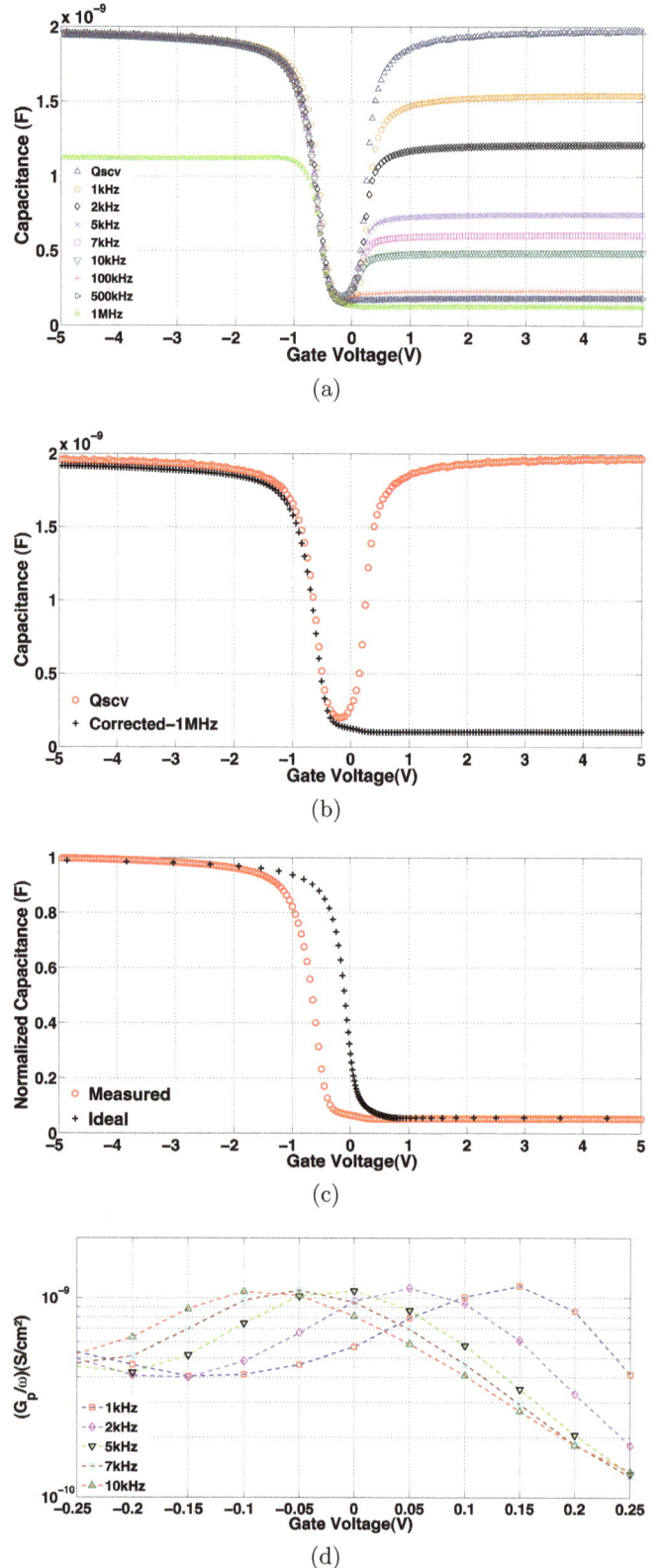

Fig. 4. PECVD-deposited SiO_2 MOS Capacitor (C-V-G) curves after parasitic effects correction using dual-frequency five-element circuit model: (a) C-V curves at different frequencies, (b) D_{it} extraction using high-low frequency method, (c) D_{it} extraction using Terman method, (d) D_{it} extraction using conductance method.

Table 1. Summary of extracted values for fixed charge density (Q_f).

Dielectric layer	C_{ox} (F/cm^2)	V_{fb} (V)	Q_f (cm^{-2})	Literature reported $\sim Q_f$ (cm^{-2}) [3–5, 7–33, 35–42]
SiO$_2$ – Thermal (20 $\pm$ 1 nm)	1.7×10^{-7}	-1.0	$+(3.3{-}4.7) \times 10^{10}$	$+(1{-}20) \times 10^{10}$
SiO$_2$ – PECVD (20 nm $\pm$ 2 nm)	1.7×10^{-7}	-1.3	$+(2.3{-}3.9) \times 10^{11}$	$+(1{-}10) \times 10^{11}$
Si$_3$N$_4$ – PECVD (20 nm $\pm$ 2 nm)	3.0×10^{-7}	-3.0	$+(3.4{-}4.2) \times 10^{12}$	$+(4{-}80) \times 10^{11}$
Al$_2$O$_3$ – Thermal (15 nm $\pm$ 0.1 nm)	5.3×10^{-7}	-0.3	$-(2.1{-}2.3) \times 10^{12}$	$-(3{-}50) \times 10^{11}$
Al$_2$O$_3$ – Plasma (15 nm $\pm$ 0.1 nm)	5.3×10^{-7}	$+0.6$	$-(5.1{-}5.3) \times 10^{12}$	$-(2{-}13) \times 10^{12}$

Table 2. Extracted interface state density (D_{it}) from C-V measurements.

Dielectric layer	D_{it} (cm^{-2} eV^{-1}) (HF-LF method)	D_{it} (cm^{-2} eV^{-1}) (Terman method)	D_{it} (cm^{-2} eV^{-1}) (Conductance method)	Literature reported values [3–5, 7–33, 35–42]
SiO$_2$ – Thermal (20 $\pm$ 1 nm)	$(1.2{-}1.5) \times 10^{10}$	$(1.1{-}2.2) \times 10^{10}$	$(1.0{-}1.5) \times 10^{10}$	$(1{-}10) \times 10^{10}$
SiO$_2$ – PECVD (20 nm $\pm$ 2 nm)	$(2.5{-}2.9) \times 10^{10}$	$(2.3{-}3.5) \times 10^{10}$	$(2.2{-}3.2) \times 10^{10}$	$(5{-}30) \times 10^{10}$
Si$_3$N$_4$ – PECVD (20 nm $\pm$ 2 nm)	$(1.3{-}1.7) \times 10^{11}$	$(1.6{-}2.7) \times 10^{11}$	$(1.3{-}2.4) \times 10^{11}$	$(5{-}50) \times 10^{10}$
Al$_2$O$_3$ – Thermal (15 nm $\pm$ 0.1 nm)	$(1.1{-}1.2) \times 10^{11}$	$(1.0{-}1.6) \times 10^{11}$	$(1.9{-}2.3) \times 10^{11}$	$(6{-}10) \times 10^{10}$
Al$_2$O$_3$ – Plasma (15 nm $\pm$ 0.1 nm)	$(1.6{-}1.8) \times 10^{11}$	$(1.7{-}2.1) \times 10^{11}$	$(2.9{-}3.3) \times 10^{11}$	$(8{-}20) \times 10^{10}$

negative fixed charge densities as high as -5.3×10^{12} cm^{-2} and -2.3×10^{12} cm^{-2} when deposited by PE-ALD and T-ALD respectively. In our experiments, Al$_2$O$_3$ deposited by PE-ALD exhibits more negative charges than T-ALD process. Table 2 summarizes the extracted D_{it} values using different methods. The relative differences can be related to the various specificities and sensitivity ranges of the different extraction methodologies, but also to the fact that the D_{it} values extracted from each method may not be extracted at the same gate voltage (i.e. depletion point). In addition, other errors could be due to non-uniform doping of the substrate, failure to obtain accurate 1 MHz high-frequency curves or inaccurate band bending during low-frequency measurements, which may affect the estimations. The D_{it} orders of magnitudes are correctly estimated and can be used as figures of merit to compare the different dielectrics. However, the D_{it} values extracted using conductance method on Al$_2$O$_3$ PE-ALD, T-ALD samples exhibit slightly higher values due to the asymmetry of the capture cross-sections σ_n/σ_p in these dielectrics, influencing the extractions using this method [16, 25, 26].

3.2 Carrier lifetime measurements

To enable carrier lifetime testing, the respective dielectric layers were deposited on both sides of a wafer to maintain symmetrical structures. Supporting the above extracted C-V-G parameters (Q_f, D_{it}), minority carrier lifetime measurements (Fig. 5) show that thermally grown SiO$_2$ leads to a good passivation quality independent of injection level, mainly due to very low D_{it} [8,9]. The quality of surface passivation is indeed also examined by effective surface recombination velocity (S_{eff}) calculated from the lifetime measurements. A common typical injection point of $\Delta n = 5 \times 10^{15}$ cm^{-3} has been chosen for S_{eff} extraction for the comparison of the different dielectrics [8, 11, 12].

$$\frac{1}{\tau_{eff}} = \frac{1}{\tau_{bulk}} + 2 \left(\frac{S_{eff}}{W} \right) \tag{4}$$

where W is the thickness of the substrate.

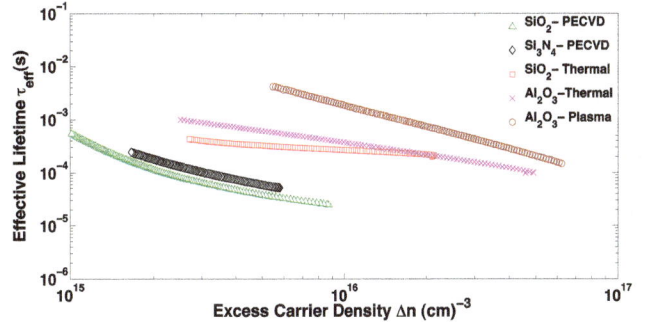

Fig. 5. Injection-level-dependent effective lifetime measurements on different dielectrics layers.

Assuming a very high lifetime, thanks to the use of FZ wafers equation (4) can be simplified and the maximum $S_{eff_{max}}$ can be calculated by

$$S_{eff_{max}} \leqslant \left(\frac{W}{2\tau_{eff}} \right) \tag{5}$$

while the lower limit is the case where no recombination occurs. In reality the value of S lies in-between $(0 < S < S_{eff_{max}})$ depending on the chosen injection level (Δn) [8, 11, 12, 27, 28].

Table 3 presents the surface recombination velocities (SRV) extracted from the lifetime measurements using equation (5) for 200 μm thick wafers covered with the different dielectrics. Samples with PE-ALD show the lowest SRV among all other dielectrics considered in this experiment with a value of less than 5 cm s^{-1} though it presented the highest D_{it}.

Our experimental results have shown that a good silicon surface passivation is achieved with ALD Al$_2$O$_3$ as dielectric film. This is attributed to a high density of negative fixed charges presented in the film, compensating the detrimental role of higher D_{it}.

The source of these negative fixed charges is attributed to the trapped hydroxyl groups in the film due to the deposition process [10, 14, 15, 29]. Another possible reason could be the presence of Al vacancies in the Al$_2$O$_3$ film

Table 3. Extracted τ_{eff}, $S_{eff,max}$ values from lifetime measurements at an injection point $\Delta n = 5 \times 10^{15}$ cm^{-3}.

Dielectric layer	τ_{eff} (μs) at $\Delta n = 5 \times 10^{15}$	$S_{eff,max}$ (cm s^{-1})	Literature reported $\sim S_{eff,max}$ (cm s^{-1}) [3–5, 7–33, 35–42]
SiO$_2$ – Thermal (20 nm)	324	31	10–70
SiO$_2$ – PECVD (20 nm)	36	277	80–400
Si$_3$N$_4$ – PECVD (20 nm)	56	178	30–1000
Al$_2$O$_3$ – Thermal (15 nm)	613	17	5–30
Al$_2$O$_3$ – Plasma (15 nm)	3790	3	2–20

or oxygen interstitials located at the Si/SiO$_X$/Al$_2$O$_3$ interfaces [3, 8, 9, 11, 12, 30]. Al$_2$O$_3$ surface passivation can be described as a combination of both field-effect passivation and chemical passivation. Similar to other dielectrics, Al$_2$O$_3$ also chemically passivates the surface by releasing hydrogen atoms, which diffuse to the Si/dielectric interface to passivate the dangling bond defects. Apart from this, Al$_2$O$_3$ also serves the field-effect passivation mechanism due to its high fixed negative charge density, which is one-two orders of magnitude higher when compared to SiO$_2$ deposited by PECVD and two orders magnitude higher than thermally-grown SiO$_2$ (Tab. 1). This high amount of negative fixed charges in the overlying dielectric film drives the silicon surface into accumulation mode in case of p-type substrate creating a built-in E-field (electric field) at the surface shielding the minority carriers (here electrons) to recombine at the surface This difference in field passivation significantly relaxes the requirements on the interface defect density (D_{it}) at the c-Si/Al$_2$O$_3$ interface. Figure 5 clearly shows two separate groups of curves, two lower curves corresponding to SiO$_2$-PECVD and Si$_3$N$_4$-PECVD and three others. The field effect is more effective in the low injection regime, whereas at high injection, while photo-generated excess charges compensate the fixed charges that induced the field effect, and mainly the "chemical passivation" is dominant [3, 4, 8, 9, 11–13, 24, 30–33].

The shape of the $\tau(\Delta n)$ curves (Fig. 5) can reveal information regarding the involved interface passivation mechanism [34].

- From Tables 1 and 2, PE-ALD and T-ALD Al$_2$O$_3$ layers have almost same level of chemical passivation i.e. interface defect density $D_{it} \sim (1-3) \times 10^{11}$ cm^{-2} eV^{-1}. Thus the difference in lifetimes behavior especially at lower injection range can be solely attributed due to the field passivation. This is mainly induced by fixed charges in dielectrics. Comparing Q_f for the two Al$_2$O$_3$ deposition processes, PE-ALD ($Q_f \sim -5 \times 10^{12}$ cm^{-2}) is more efficient compared to T-ALD ($Q_f \sim -2 \times 10^{12}$ cm^{-2}).
- PECVD-Si$_3$N$_4$ layers exhibit poor chemical passivation ($D_{it} \sim 2 \times 10^{11}$ cm^{-2} eV^{-1}) in combination with a high density of positive $Q_f \sim 4 \times 10^{12}$ cm^{-2}. Even if the net charges is at the same level as Al$_2$O$_3$, the field effect passivation due to positively charged layers is less effective on p-type substrates resulting in a lower passivation quality. Accumulation mode caused by negative charges is more efficient than inversion caused by positive charges on p-type surfaces. This

difference is probably due to the depletion layer beneath the inversion layer where n_s and p_s concentrations are at the same level.
- Low fixed positive charges in PECVD-SiO$_2$ and thermal dielectric layers leads to weak inversion or depletion mode for which the resulting field effect is not efficient. SiO$_2$-thermal lifetime curves exhibit almost injection independent behavior meaning that the dominant passivation mechanism involved at the interface is the chemical passivation, i.e. lowest $D_{it} \sim 2 \times 10^{10}$ cm^{-2} eV^{-1}.

4 Field effect passivation dependency on interfacial (SiO$_2$) layer thickness

Kessels et al. [13] reported that a thin interfacial SiO$_2$ ($\sim$1–2 nm) layer is formed naturally between Si surface and Al$_2$O$_3$ layer. This thermal SiO$_2$ ($\sim$1–2 nm) is the only means of chemical passivation at the interface. The thickness of this SiO$_2$ is too thin to completely passivate the interface states. Another disadvantage of this interfacial oxide is that it may not have the same quality as SiO$_2$ produced by thermal oxidation of Si. In some experiments we have introduced a stack of SiO$_2$/Al$_2$O$_3$ on the Si surface. The main goal of this experiment is to chemically passivate the Si surface by thermally growing SiO$_2$ and maintains field-effect passivation with negative charges present in Al$_2$O$_3$. To perform this and investigate trade-offs between concurrent Q_f and D_{it} reductions, we have thermally-grown SiO$_2$ layers with two different thicknesses: 8 and 20 nm [11, 13].

An Al$_2$O$_3$ film of 15 nm was deposited on SiO$_2$ samples using PE-ALD. The deposition was also performed directly on the Si as a reference sample [3, 11–13, 24, 30–33]. All these samples were treated in the same way as in the earlier experiments for both C-V-G and lifetime measurements except that the Al$_2$O$_3$ layer deposition was performed at 200 °C.

From SRV results, we observe that the SiO$_2$ (8 nm)/Al$_2$O$_3$ (15 nm) stack exhibits the lowest SRV values compared to other samples. This can be correlated to negative Q_f and low D_{it} extracted (Tab. 4) from C-V-G curves (Fig. 6).

As reported by other authors [4, 13], the chemical passivation at the interface of Al$_2$O$_3$/Si occurs during the annealing step, when a very thin interfacial Al$_x$SiO$_y$ layer is created in between the two materials. The formation of this layer is not well understood at this time as oxygen and hydrogen seem to play an important role,

Table 4. Parameters extraction on SiO_2/Al_2O_3 stacks.

Stack type	V_{fb} (V)	Q_{fixed} (cm^{-2})	D_{it} (cm^{-2} eV^{-1}) (Terman method)	τ_{eff} (μs) @ $\Delta n = 2 \times 10^{15}$	$S_{eff,\,max}$ (cm s^{-1})
Al_2O_3(15 nm $\pm$ 0.1 nm)	1.6	$-(8.3-8.7) \times 10^{12}$	$(2.4-3.1) \times 10^{11}$	1110	9
SiO_2(8 nm $\pm$ 1 nm) + Al_2O_3 (15 nm $\pm$ 0.1 nm)	-0.65	$-(3.9-4.4) \times 10^{11}$	$(6.4-8.2) \times 10^{10}$	2320	4
SiO_2(20 nm $\pm$ 1 nm) + Al_2O_3 (15 nm $\pm$ 0.1 nm)	-0.1	$-(1.4-2.2) \times 10^{11}$	$(1.6-2.3) \times 10^{10}$	316	32

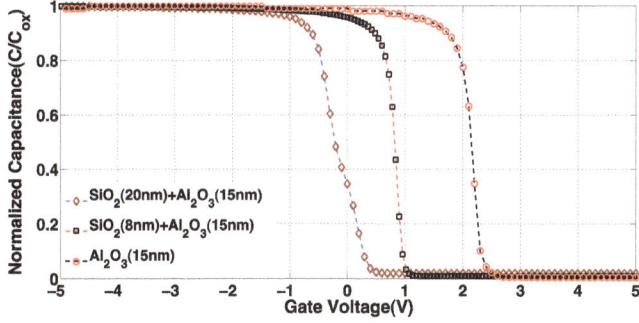

Fig. 6. C-V characteristics of SiO_2/Al_2O_3 stacks.

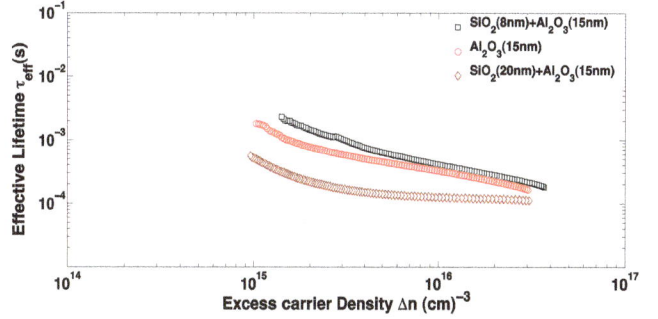

Fig. 7. Lifetime measurements on SiO_2/Al_2O_3 stacks.

but consequences have been demonstrated and shown that electrically active interface traps were reduced. For the reference Al_2O_3 sample, the flat-band (V_{fb}) is positive and the density of negative fixed charges ($Q_f \sim 8.5 \times 10^{12}$ cm^{-3}) and the interface defect density ($D_{it} \sim 2.7 \times 10^{11}$ eV^{-1} cm^{-2}) are higher than any other sample in this experiment. However, an SRV of 9 cm s^{-1} is obtained from lifetime measurements meaning that the field effect passivation is predominant in this sample and slightly relaxes the requirement for lower D_{it} values.

For the sample with SiO_2 (8 nm)/Al_2O_3 (15 nm) stack, the V_{fb} is negative but still exhibits negative fixed charges, however, twenty times lower than in the reference sample (only Al_2O_3). It is important to notice that the D_{it} also reduces by four times due to the presence of thermal-SiO_2 layer, which reduces the defects at the interface and also the field induced by Al_2O_3 layer. This reduction of the effective field when using an SiO_2 layer can be explained by the fact that the charge centroid is driven away from the silicon surface with increasing SiO_2 thickness, as well as by the contribution of fixed positive charges in SiO_2 layer resulting in overall reduction of net effective negative charge density. However, the trade-off between Q_f and D_{it} obtained in these conditions leads to a much better effective lifetime at all injection levels. The difference is larger at low injection level where the field effect is the more efficient showing that 8 nm-SiO_2 does not shield too much the Al_2O_3 charges [3,4,8,9,11–13,24,27,28,30–33].

In the case of thicker SiO_2 layer (20 nm), the "chemical interface" between Si and SiO_2 is the same as thinner SiO_2 (8 nm) layer. We can observe from Figure 7 that the effective lifetime is affected over the complete range of injection level. C-V measurements on this sample confirm that the silicon oxide layer has reduced the interface trap charge density ($D_{it} \sim 2.1 \times 10^{10}$) and also the fixed charges den-

sity ($Q_f \sim -1.9 \times 10^{11}$). The effective lifetime is even lower than with Al_2O_3 layer alone and has returned to the values previously obtained for SiO_2 only, meaning that field-effect passivation has been completely lost and the only means of passivation is due to chemical-passivation [10].

5 Conclusion

Electronic properties (Q_f, D_{it}) of different dielectrics were extracted. In addition, parasitic C-V-G corrections were applied to accurately estimate the interface trap charge density. Extracted parameters were discussed and compared with lifetime measurements to understand the passivation mechanisms involved at the interface.

In case of the PE-ALD Al_2O_3 layer, the extracted fixed charge density is negative and about -5.2×10^{12} cm^{-2}, which provides an effective field-effect passivation for impeding the surface recombination of minority carriers. Interface trap charge density D_{it} has been calculated and is found to be about 3×10^{11} cm^{-2} eV^{-1} as a mean value in the depletion gate voltage range. Such high negative fixed charge density resulted in surface recombination velocity less than 3 cm/s due to formation of accumulation regime at the silicon surface.

We have demonstrated the dependency of field-effect passivation on the thickness of SiO_2 interfacial layer. From the C-V-G parameter extractions and lifetime measurements we concluded that an optimal thickness of SiO_2 (here 8 nm) reduces the interface state densities while still maintaining field-effect passivation. Thick SiO_2 layer reduces the net negative charge effect in the overall dielectric and may lead to a loss of the field-effect passivation.

We have observed that accumulation mode leads to better passivation than inversion mode. In our experiments, high negative fixed charges dielectrics were less sensitive to interface trap defects and exhibits better passivation behavior over all injection range. More generally, for all dielectrics, field-effect passivation mainly drives surface recombination at low injection, as chemical passivation is more predominant at higher level.

The authors thank the teams of the nanofabrication shared facility WINFAB and electrical characterization WELCOME platforms at UCL for their technical support. This work is supported by FRS-FNRS Belgium.

References

1. B. Hoex, Ph.D. thesis, Technische Universiteit Eindhoven, 2008
2. S. Dauwe, Ph.D. thesis, University of Hannover, 2004
3. G. Dingemans, W.M.M. Kessels, J. Vac. Sci. Technol. A **30** (2012)
4. G. Dingemans, W.M.M. Kessels, ECS Trans. **41**, 293 (2011)
5. B. Hoex, J. Schmidt, P. Pohl, M.C.M. van de Sanden, W.M.M. Kessels, J. Appl. Phys. **104**, 044903 (2008)
6. B. Hoex, M.C.M. van de Sanden, J. Schmidt, R. Brendel, W.M.M. Kessels, Phys. Stat. Sol. RRL **6**, 4 (2012)
7. R.B.M. Girisch, R.P. Mertens, R.F. De Keersmaecker, IEEE Trans. Electron Devices **35**, 203 (1988)
8. S. Rein, *Lifetime Spectroscopy: A Method of Defect Characterization in Silicon for Photovoltaic Applications* (Springer Series in Materials Science, 2008)
9. G. Dingemans et al., in *35th IEEE PVSC, Honolulu, Hawaii, 2010*
10. A.G. Aberle, S. Glunz, W. Warta, J. Appl. Phys. **71**, 4422 (1992)
11. B. Hoex, F.J.J. Peeters, M. Creatore, M.A. Blauw, W.M.M. Kessels, M.C.M. van de Sanden, J. Vac. Sci. Technol. A **24**, 1823 (2006)
12. B. Hoex, S.B.S. Heil, E. Langereis, M.C.M. van de Sanden, W.M.M. Kessels, Appl. Phys. Lett. **89**, 042112 (2006)
13. W.M.M. Kessels, J.A. van Delft, G. Dingemans, M.M. Mandoc, *Review on the Prospects for the Use of Al_2O_3 for High-efficiency Solar cells, Workshop* (Dept. of Applied Physics, Eindhoven University of Technology (TU/e), The Netherlands)
14. J.E. Crowell, J. Vac. Sci. Technol. A **21**, S88 (2003)
15. R.S. Johnson, G. Luckovsky, I. Bauvmol, J. Vac. Sci. Technol. A **19**, 1353 (2001)
16. D.K. Schroder, *Semiconductor material and device characterization*, 3rd edn. (John Wiley & Sons Inc., Hoboken, 2006)
17. E.H. Nicollian, J.R. Brews, *MOS (Metal Oxide Semiconductor) physics and technology* (John Wiley & Sons, New York, 1982)
18. R.F. Pierret, *Semiconductor Device Fundamentals* (Addison Wesley Publication, 1996)
19. L.M. Terman, Solid-State Electron. **5**, 285 (1962)
20. C.N. Berglund, IEEE Trans. Electron Devices **13**, 701 (1966)
21. R. Castagné, A. Vapaille, Surf. Sci. **28**, 157 (1971)
22. D.L. Meier, J.M. Hwang, R.B. Campbell, IEEE Trans. Electron Devices **35** (1988)
23. M.J. Kerr, A. Cuevas, J. Appl. Phys. **91**, 2473 (2002)
24. W.H. Wu, B.Y. Tsui, Electron Device Lett. IEEE **27** (2006)
25. F. Werner, A. Cosceev, J. Schmidt, Energy Procedia **319**, 27 (2012)
26. R. Engel-Herbert, Y. Hwang, S. Stemmer, J. Appl. Phys. **108**, 124101 (2010)
27. J. Schmidt, A. Merkle, R. Brendel, B. Hoex, M.C.C. van der Sanden, W.M.M. Kessels, Prog. Photovolt.: Res. Appl. **16**, 461 (2008)
28. B. Hoex, J. Schmidt, R. Bock, P.P. Altermatt, M.C.M. van de Sanden, W.M.M. Kessels, Appl. Phys. Lett. **91**, 112107 (2007)
29. D. Hoogeland, K.B. Jinesh, F. Roozeboom, W.F.A. Besling, M.C.M. van de Sanden, W.M.M. Kessels, J. Appl. Phys. **106**, 114107 (2009)
30. S.W. Glunz, D. Biro, S. Rein, W. Warta, J. Appl. Phys. **86**, 683 (1999)
31. B. Hoex, J.J.H. Gielis, M.C.M. van de Sanden, W.M.M. Kessels, J. Appl. Phys. **104**, 113703 (2008)
32. S. Steingrube, *Recombination models for defects in silicon solar cells*, Ph.D. thesis, Leibniz Universität Hannover, 2011
33. X. Tang et al., J. Vac. Sci. Technol. A **30** (2012)
34. C. Leendertz, N. Mingirulli, T.F. Schulze, J.P. Kleider, B. Rech, L. Korte, Appl. Phys. Lett. **98**, 202108 (2011)
35. J. Schmidt, B. Veith, F. Werner, D. Zielke, V. Tiba, P. Poodt, F. Roozeboom, A. Li, A. Cuevas, R. Brendel, *Industrially Relevant Al2O3 Deposition Techniques For The Surface Passivation Of Si Solar Cells*, in *Proceedings of the 25th European Photovoltaic Solar Energy Conference, Valencia, Spain, 2010*
36. A.G. Aberle, *Crystalline Silicon Solar Cells: Advanced Surface Passivation and Analysis of Crystalline Silicon Solar Cells* (Centre for Photovoltaic Engineering, University of New South Wales Sydney, Australia, 1999)
37. P. Saint-Cast et al., Appl. Phys. Lett. **95**, 151502 (2009)
38. T.-T. Li, A. Cuevas, Phys. Stat. Sol. RRL **3**, 160 (2009)
39. G. Dingemans et al., *Firing Stability of Atomic Layer Deposited Al2O3 for C-Si Surface Passivation*, in *Proceedings of the 34th IEEE Photovoltaics Specialists Conference, Philadelphia, Pennsylvania, USA, 2009*
40. J. Schmidt, B. Veith, R. Brendel, Phys. Stat. Sol. RRL **3**, 287 (2009)
41. J. Benick, A. Richter, M. Hermle, S.W. Glunz, Phys. Stat. Sol. RRL **3**, 233 (2009)
42. R. Hezel, K. Jaeger, J. Electrochem. Soc. **136**, 518 (1989)
43. A.G. Aberle, Prog. Photovolt.: Res. Appl. **8**, 473 (2000)
44. G. Agostinelli, P. Vitanov, Z. Alexieva, A. Harizanova, H.F.W. Dekkers, S. De Wolf, G. Beaucarne, *Surface Passivation of Silicon by Means of Negative Charge Dielectrics*, in *Proceedings of the 19th European Photovoltaic Solar Energy Conference, Paris, France, 2004*, pp. 132–134
45. S.M. George, Chem. Rev. **110**, 111 (2010)

10

Effect of annealing on properties of sputtered and nitrogen-implanted ZnO:Ga thin films

K.S. Shtereva[1,2,a], V. Tvarozek[1], I. Novotny[1], P. Sutta[3], M. Milosavljevic[4], A. Vincze[5], M. Vojs[1], and S. Flickyngerova[1]

[1] Institute of Electronics and Photonics, Slovak University of Technology, Ilkovicova 381219 Bratislava, Slovakia
[2] Department of Electronics, University of Rousse, Studentska 8, 7017 Ruse, Bulgaria
[3] New technologies-Research Center, University of West Bohemia, Plzen Czech Republic
[4] VINČA Institute of Nuclear Sciences, Laboratory for Atomic Physics, Belgrade, Serbia
[5] International Laser Centre, Bratislava, Slovakia

Abstract Thin films of gallium-doped zinc oxide (ZnO:Ga) were deposited on Corning glass substrates by rf diode sputtering and then implanted with 180 keV nitrogen ions in the dose range of $1 \times 10^{15} \div 2 \times 10^{16}$ cm^{-2}. After the ion implantation, the films were annealed under oxygen and nitrogen ambient, at different temperatures and time, and the effect on their microstructure, type and range of conductivity, and optical properties was investigated. Post-implantation annealing at 550 °C resulted in n-type conductivity films with the highest electron concentration of 1.4×10^{20} cm^{-3}. It was found that the annealing parameters had a profound impact on the film's properties. A p-type conductivity (a hole concentration of 2.8×10^{19} cm^{-3}, mobility of 0.6 cm^2/V s) was observed in a sample implanted with 1×10^{16} cm^{-2} after a rapid thermal annealing (RTA) in N$_2$ at 400 °C. Optical transmittance of all films was >84% in the wavelength range of 390–1100 nm. The SIMS depth profile of the complex ^{30}NO$^-$ ions reproduces well a Gaussian profile of ion implantation. XRD patterns reveal a polycrystalline structure of N-implanted ZnO:Ga films with a c-axis preferred orientation of the crystallites. Depending on the annealing conditions, the estimated crystallite size increased $25 \div 42$ nm and average micro-strains decreased $1.19 \times 10^{-2} \div 6.5 \times 10^{-3}$ respectively.

1 Introduction

At present, impurity-doped ZnO thin films with high conductivity and transparency are used as an electrode material for amorphous silicon (a-Si) and Cu(In, Ga)Se$_2$ (CIGS) photovoltaic devices (PV), and have been investigated for electrodes for organic PV and organic light-emitting diodes (OLEDs) [1, 2]. A ZnO semiconductor (a wide direct band gap of 3.37 eV at room temperature and a large exciton binding energy of 60 meV) has the potential to create new transparent electronics products, such as transparent thin film transistors (TTFT) and light-emitting diodes (LED) [3]. The major obstacle to ZnO device applications is the impossibility for controllable and reproducible p-type doping. Main reasons of failure to dope ZnO p-type are: (1) the low or limited solubility of the acceptor dopants; (2) the high activation energies of these dopants; (3) the tendency to form spontaneously compensating defects; and (4) hydrogen or other impurities, acting as unintentional extrinsic donors in ZnO. Among others, nitrogen has been investigated as a promising acceptor dopant and a number of groups,

including ours, reported p-type conductivity in nitrogen-doped ZnO films (ZnO:N) [4,5]. The use of ion implantation to dope semiconductors has become increasingly important. In comparison to diffusion, ion implantation offers the precise process control (doping levels and doping uniformity, extreme purity of the dopant) and therefore, has been used for device fabrication. Moreover, ion implantation enables the introduction of dopants that are not soluble or diffusible, and hence, provides better opportunities for p-type doping in ZnO. Although there were few reports on p-type doping in ZnO via nitrogen implantation [6,7], more studies are needed to understand phenomena and optimize parameters of the ion implantation processing (energy, dose). Implantation causes damage and disorder that have to be annealed to improve the material quality. Therefore, the right choice of post-implantation annealing parameters (ambient, temperature, time), is an important step to prepare a good quality material.

In this work we investigated influence of the post-implantation annealing parameters (ambient, temperature, time), on the physical properties of sputtered N-implanted ZnO:Ga thin films, and the possibility for p-type doping.

[a] e-mail: KShtereva@ecs.uni-ruse.bg

2 Experimental methods

In this study were used ZnO:Ga thin films deposited on Corning 7059 (a baria alumina borosilicate composition, an alkali level of <0.3%) glass substrates by rf diode sputtering from a ceramic $ZnO:Ga_2O_3$ (98 wt%:2 wt%) target, a mixture of ZnO (99.99% purity) and Ga_2O_3 (99.99% purity). The film thickness of ~500 nm was determined by Dektak 150 instrument. The ZnO:Ga films were n-type conductivity with the highest electron concentration of 7.2×10^{19} cm^{-3} and mobility of ~1.2 cm^2/V s measured from a Hall-effect system with a magnetic field of 0.15 T at a room temperature (RT). Nitrogen ions were implanted at 180 keV under normal incidence into the sputtered ZnO:Ga films with doses of 1×10^{15}, 5×10^{15}, 1×10^{16} and 2×10^{16} cm^{-2}. Prior to implantation, the theoretical nitrogen profiles were calculated by means of Monte Carlo simulations using the SRIM code for the various doses [8]. The projected range (R_P) at 180 keV was determined to be 340 nm. Post-implantation annealing was done under an oxygen (O_2) (at 550 °C for 30 min) and nitrogen (N_2) (at 400 and 550 °C for 10 s) ambient. The structure and preferred orientation of crystallites were evaluated by X-ray diffraction (XRD) on X'pert Pro powder diffractometer (symmetric ϑ-ϑ geometry) equipped with an ultra-fast linear semiconductor detector PIXcel using CuKα radiation ($\lambda = 0.154$ nm). Secondary ion mass spectrometry (SIMS) depth profiles of the various ionic species were acquired with a TOF-SIMS IV analyzer, product of ION TOF GmbH, Muenster, using a Cs$^+$ primary ion beam with energy of 2 keV. Optical spectrophotometry measurements were carried out from the UV region to the near IR region by an Ava Spec-2048 Fiber Optic Spectrometer.

3 Results and discussions

All non-implanted ZnO:Ga films showed a c-axis preferred orientation of crystallites with a dominant (002) diffraction line at $2\theta = 34.29°$, and weak lines at 2θ of $31.53°$ and $35.76°$ that corresponds to (100) and (101) crystal planes of ZnO. They were n-type conductivity with quite uniform parameters, an electron concentration in the range of $5 \times 10^{19} \div 7.2 \times 10^{19}$ cm^{-3}, mobility of $1.2 \div 1.6$ cm^2/V s and resistivity of $7.6 \times 10^{-2} \div 7.8 \times 10^{-2}$ Ω cm.

The resistivity of the films increased and the carrier concentration decreased after nitrogen implantation. The higher resistivity in implanted samples can be explained with: (1) nitrogen introduction and formation of N_O acceptors that partly compensate donor defects in the film, and (2) the damage caused by ion implantation. Some support for this first assumption is the unstable p-type conductivity in some samples implanted with low nitrogen doses before annealing. It indicates that the implanted nitrogen produces holes which compensate electrons, and hence the carrier concentration is reduced. Calculations based on the first-principles pseudo potential method within the local-density functional approximation (LDA) show that the type of the donors that compensate nitrogen acceptors depends on the nitrogen doping level and the

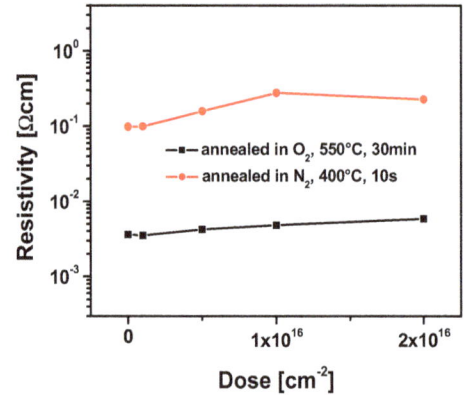

Fig. 1. Resistivity of N-implanted ZnO:Ga films, annealed in O_2 at 550 °C for 30 min, and in N_2 at 400 °C for 10 s, as a function of the dose.

type of the nitrogen source [9]. At low nitrogen doping levels, nitrogen acceptors are compensated mostly by oxygen vacancies, whereas at high doping levels, the major compensating donors are nitrogen composed donors. Hence, n-type conductivity in samples implanted with high nitrogen doses can be explained with self-compensation of the N_O acceptors by nitrogen composed donors, such as nitrogen molecules and donor defect complexes. Furthermore, implantation of high nitrogen doses can introduce a considerable amount of nitrogen interstitials in the film that may become a source of electrons as well as holes. During implantation, N^+ ions penetrate the sample with high energy. Their collisions with the native atoms can displace some of these atoms and thus provide additional donor defects.

The electrical parameters of N-implanted and annealed ZnO:Ga thin films are listed in Table 1. It was found that the annealing parameters had a profound impact on the electrical properties of the film. The resistivity fell and the carrier mobility increased after annealing in O_2 at 550 °C for 30 min, but all samples were n-type as determined by room temperature Hall-effect measurements. These outcome can arise from the activation of gallium atoms that move from interstitial to Zn sites and become effective surplus donors to the native donors in ZnO [10, 11]. The conversion of the conduction type from n- to p-type was not observed in N_2 annealed samples either, when the annealing was performed at 550 °C for a reduced time of 10 s. As observed from other groups [12, 13], the low temperatures are more favorable for nitrogen incorporation in the ZnO lattice. Therefore, next annealing experiments have been carried out in N_2 atmosphere at 400 for 10 s. The resistivity of the N-implanted ZnO:Ga films increased after annealing in N_2 at 400 °C for 10 s compared to the resistivity of the films annealed in O_2 at 550 °C (Fig. 1). The electron concentration did not change significantly while their mobility decreased more than one order of magnitude (Tab. 1), most likely due to the structural defects that could not be completely removed at the lower annealing temperature. The carrier mobility depends on the scattering events that take place inside the semiconductor. In case of polycrystalline and other high defect materials,

Table 1. Dependence of the properties of N-implanted ZnO:Ga films on the annealing treatment.

Dose (cm^{-2})	Annealing conditions	Resistivity (Ω cm)	Concentration (cm^{-3})	Mobility (cm^2/V s)	Type	Transmittance (%)
1×10^{15}	N$_2$, 400 °C, 10 s	0.1	1×10^{20}	0,5	n	83
	O$_2$, 550 °C, 30 min	3.5×10^{-3}	7.6×10^{19}	23	n	87
5×10^{15}	N$_2$, 400 °C, 10 s	0.16	5×10^{19}	0.7	n	85
	O$_2$, 550 °C, 30 min	4×10^{-3}	6.4×10^{19}	23	n	87
1×10^{16}	N$_2$, 400 °C, 10 s	0.46	2.8×10^{19}	0.6	p	90
	O$_2$, 550 °C, 30 min	4.8×10^{-3}	6×10^{19}	21	n	87
2×10^{16}	N$_2$, 400 °C, 10 s	0.23	9×10^{19}	0.3	n	84
	N$_2$, 550 °C, 10 s	4.3×10^{-3}	1.4×10^{20}	11	n	85
	O$_2$, 550 °C, 30 min	5.8×10^{-3}	5.1×10^{19}	21	n	87

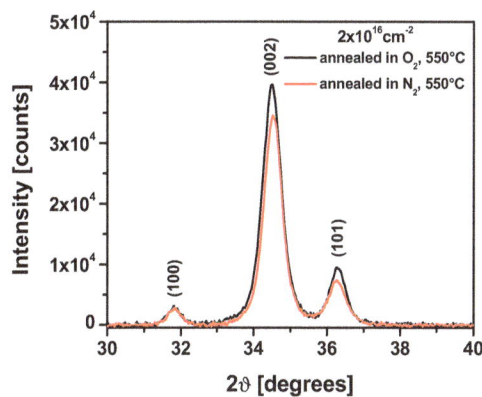

Fig. 2. X-ray diffraction patterns of ZnO:Ga films implanted with 2×10^{16} cm^{-2}, annealed in O$_2$ at 550 °C for 30 min and N$_2$ at 550 °C for 10 s.

ionized impurity scattering and scattering by neutral impurity atoms and defects tend to dominate. Impurities and defects can alter the physical parameters of semiconductors and, in some case, can dominate these parameters. At low annealing temperatures, some impurity atoms remain on interstitial positions. Elevating the annealing temperature can activate some of the impurity gallium atoms to move from interstitial to Zn sites, making them effective donors, and/or some of the interstitial impurity nitrogen atoms to substitute for an O atom, thus, becoming acceptors. Hence, with increasing annealing temperature, scattering by neutral impurity atoms and defects decreases, which causes the increase in mobility. The simultaneous formation of donors and acceptors, and prevalence of the first ones, can explain n-type conductivity of the film. The conductivity type was converted to p-type (hole concentration of 2.8×10^{19} cm^{-3}, mobility of 0.6 cm^2/V s, resistivity of 0.46 Ω cm) in the sample implanted with a 1×10^{16} cm^{-2} dose. This result suggests that low temperature treatment can be more favorable for p-type doping in ZnO:Ga films [6], but it is not sufficient to claim the preparation of p-type ZnO.

The dominant (002) diffraction line in the XRD patterns of N-implanted ZnO:Ga films annealed in O$_2$ and N$_2$ atmosphere at 550 °C (Fig. 2), shows that these films

are preferentially c-axis oriented. The post-implantation annealing reduced the structure damage. The higher integrated intensity of a (002) diffraction line suggests more textured structure after annealing in O$_2$ for 30 min than after annealing in N$_2$ for 10 s. The peaks' positions are shifted toward higher 2θ angles and their values for the corresponding diffraction lines are as follow: 31.79°/(100), 34.47°/(002), and 36.27°/(101). A shift to higher 2θ angles corresponds to the reduction of the interplanar spacing d most likely due to Ga atoms that occupy the Zn lattice site. This assumption is supported from the rise in the carrier concentrations, 4×10^{17} cm$^{-3} \div 5.1 \times 10^{19}$ cm^{-3} after a post-implantation annealing in O$_2$ (550 °C, 30 min), and 8×10^{19} cm$^{-3} \div 1.4 \times 10^{20}$ cm^{-3} after a N$_2$ anneal (550 °C, 10 s). These results demonstrate the relationship between structural and electrical properties. The reduction of the interplanar spacing will introduce tensile strains (stress) into the film. The biaxial lattice stress is given by the formula [14]

$$\sigma_1 + \sigma_2 = -\frac{E}{\mu} \frac{d - d_0}{d_0} \qquad (1)$$

where E is Young's modulus, μ is Poisson's ratio, d_0 is the reference strain-free interplanar spacing and d is the interplanar spacing obtained from the experiment. In Figure 3 are compared stresses from a (002) line for annealing in O$_2$ and N$_2$ at 550 °C. It can be seen that annealing in O$_2$ is more favorable with regard to the film crystalline structure quality. Other authors reported the similar observations [15]. Depending on the annealing conditions, the estimated crystallite size in the films doped with a 2×10^{16} cm^{-2} dose increased, 25 nm (not-annealed), 31 nm (annealed in O$_2$ at 550 °C for 30 min) and 42 nm (annealed in N$_2$ at 550 °C for 10 s). The crystallite size was estimated using the Scherrer's formula:

$$\langle D \rangle = \frac{K\lambda}{\beta_C^f \cos\Theta} \qquad (2)$$

where $K = 2\sqrt{\ln 2/\pi} \approx 0.94$ is the Scherrer's constant, λ is the wavelength of the X-rays used, β_C^f is the pure (physical) Cauchy component of integral breadth of the line taken in radians and Θ is the Bragg's angle [16].

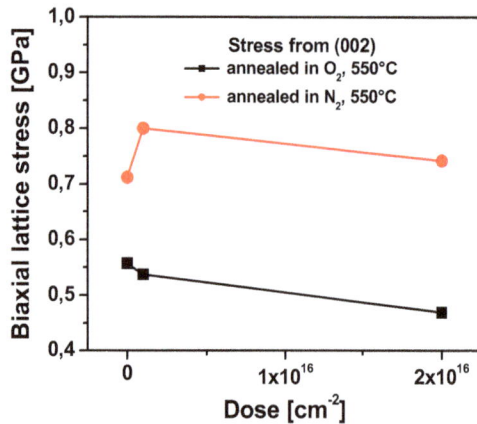

Fig. 3. A biaxial lattice stress from a (002) diffraction line as a function of the dose for N-implanted ZnO:Ga films, annealed in O_2 at 550 °C for 30 min, and annealed in N_2 at 550 °C for 10 s.

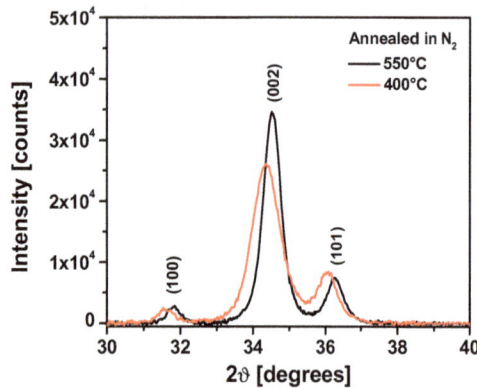

Fig. 4. X-ray diffraction patterns of ZnO:Ga films implanted with 1×10^{16} cm^{-2}, annealed in N_2 at 400 and 550 °C.

The integrated intensity decreased and the diffraction lines position shifts toward lower 2θ diffraction angles after annealing in N_2 at 400 °C (Fig. 4). According to the Bragg's law, the reduction of the diffraction angle corresponds to a lattice expansion and an increase of the interplanar spacing d that introduces compressive stress into the film. These structure changes can arise from N atoms that occupy O sites and create N_O acceptors, and/or other nitrogen related defects and defect complexes in the ZnO lattice. Since, the incorporation of nitrogen atoms into ZnO is more effective at low temperatures [12], the low temperature range may provide better conditions for p-type doping in ZnO. However, it caused the lattice deformation due to the formation of nitrogen related defects in addition to other interstitial and substitution defects that remain after a low temperature anneal.

Implantation and diffusion behavior of nitrogen in the sputtered ZnO:Ga thin films were studied through secondary ion mass spectrometry measurements of non-implanted and N-implanted ZnO:Ga films, before and after annealing treatment. Nitrogen in ZnO is evaluated by monitoring of complex $^{30}NO^-$ ions. The depth profiles of

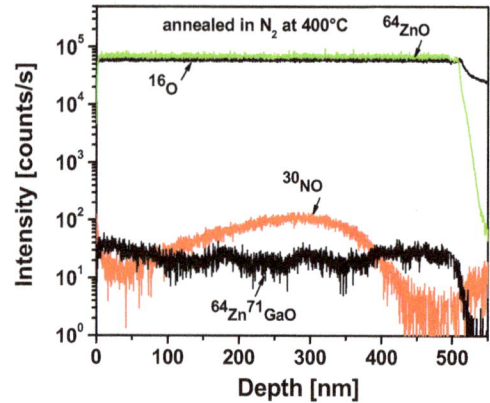

Fig. 5. SIMS depth profiles of N-implanted ZnO:Ga films annealed in N_2 at 400 °C for 10 s.

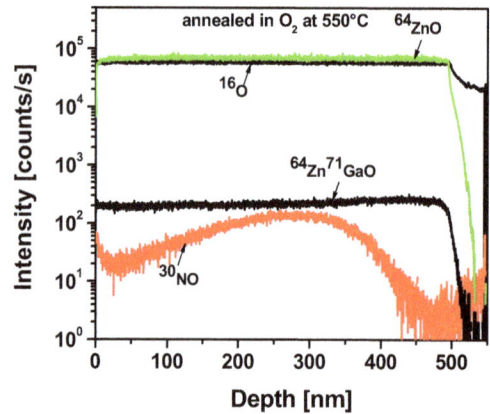

Fig. 6. SIMS depth profiles of N-implanted ZnO:Ga films annealed in O_2 at 550 °C for 30 min.

$^{30}NO^-$ before and after annealing are roughly Gaussian and the depth of the concentration peak is close to the calculated value of the projected range ($\sim$340 nm) obtained by SRIM. The depth profile of the complex $^{30}NO^-$ ions in a p-type film obtained after annealing under N_2 at 400 °C for 10 s is plotted in Figure 5. Figure 6 shows the depth profiles of the negative ions in an n-type film after annealing under O_2 at 550 °C for 30 min. In addition to ^{64}ZnO, ^{16}O and $^{30}NO^-$, all samples contain considerable amounts of $^{64}Zn^{71}GaO$ that can be a source of Ga impurities and n-type conductivity, since Ga is a donor in ZnO. It can be seen that annealing in N_2 at 400 °C caused $^{64}Zn^{71}GaO$ redistribution and its reduction that may be the reason for p-type conductivity in this film. The SIMS results make it clear that the performed annealing treatments under oxygen or nitrogen, at chosen temperatures and times, do not lead to the uniform distribution of the nitrogen impurity in the film. It is in agreement with other authors' observations, which have denoted nitrogen as a less diffusible element in ZnO [17,18]. Hence, the implanted film can be considered a double-layer structure, which can influence the determination of the sign of the Hall coefficient, and the conduction type [19].

In Figure 7 are compared the optical transmittance spectra (including the glass substrate) of the as deposited

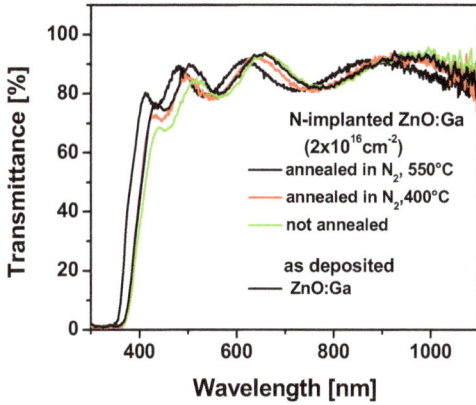

Fig. 7. Optical transmittance spectra of: (a) as deposited ZnO:Ga films (blue line), and N-implanted ZnO:Ga films (implant dose of 2×10^{16} cm^{-2}) (b) not annealed (green line); (c) annealed in N_2 at 400 for 10 s (red line); (c) annealed in N_2 at 550 for 10 s (black line).

Fig. 8. Plot of the absorption squared $(\alpha h\nu)^2$ versus phonon energy for (a) as deposited ZnO:Ga films (blue line), and N-implanted ZnO:Ga films (implant dose of 2×10^{16} cm^{-2}) (b) not annealed (green line); (c) annealed in N_2 at 400 for 10 s (red line); (c) annealed in N_2 at 550 for 10 s (black line).

ZnO:Ga films and N-implanted ZnO:Ga films (implant dose of 2×10^{16} cm^{-2}), not annealed and annealed in N_2 at 400 and 550 °C for 10 s. The transmittance data were acquired over the wavelength range of $200 < \lambda < 1100$ nm. The optical transmittance depends on both, the implant dose and the annealing conditions (an ambient, time and temperature). All N-implanted films are highly transparent with an average transmittance >84% in the wavelength range of 390–1100 nm. Average transmittance of 87% that was obtained after annealing in O_2 at 550 °C resulted from a better film crystallinity. For the samples annealed in N_2, average transmittance increased from 84 to 85% as annealing temperature was increased from 400 to 550 °C.

The transmittance data were used to calculate the absorption coefficients and to determine the optical band gap of the N-implanted ZnO:Ga films. ZnO is known as a direct band gap material therefore, the direct electron transition from valence to conduction bands was assumed for these calculations. The absorption coefficient was calculated by the following equation:

$$\alpha = \frac{1}{t} \ln \frac{1}{T} \qquad (3)$$

where t is the thickness of the film and T is the transmittance. The optical band gap, for allowed direct transitions, can be expressed by the equation [20]:

$$\alpha h\nu = A \left(h\nu - E_g \right)^{1/2} \qquad (4)$$

where $h\nu$ is a photon energy, E_g is an optical band gap and A is a constant. The square of the absorption coefficient against photon energy is plotted in Figure 8 for as deposited ZnO:Ga films, and N-implanted ZnO:Ga films (implant dose of 2×10^{16} cm^{-2}), not annealed and annealed in N_2 at 400 and 550 °C for 10 s. The optical band gap of the films was estimated by extrapolating the linear portion of each curve. The optical band gap of the not annealed N-implanted ZnO:Ga film is ~3.19 eV. Annealing caused widening of the optical band gap. The films

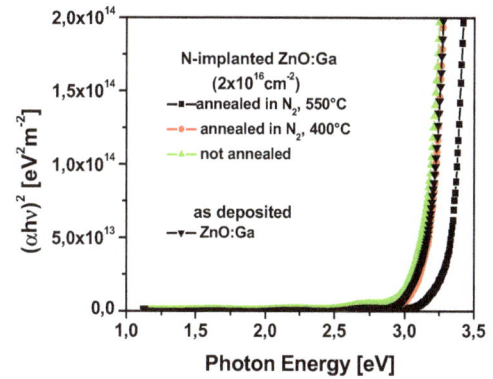

annealed in N_2 atmosphere at 400 °C exhibited an optical band gap of 3.23 eV, which widen to 3.3 eV after annealing at 550 °C.

4 Conclusion

The sputtered ZnO:Ga thin films were implanted with 180 keV N ions with the doses ranging from 1×10^{15} to 2×10^{16} cm^{-2}. The SIMS depth profiles of the complex $^{30}NO^-$ ions before and after annealing are roughly Gaussian and the depth of the concentration peak is close to the calculated value of the projected range obtained by SRIM. The non-implanted ZnO:Ga films were n-type conductivity with the electron concentration in the range of 5×10^{19}–7.2×10^{19} cm^{-3} and a c-axis preferred orientation of the crystallites. It was found that the annealing conditions had a profound impact on the film properties. The resistivity of the films increased after nitrogen implantation and it decreased significantly after post-implantation annealing under both O_2 and N_2 ambient at 550 °C. Lower temperature annealing in N_2 at 400 °C caused the rise in resistivity mainly as a result from the reduction in mobility. The conductivity type was converted to p-type (hole concentration of 2.8×10^{19} cm^{-3}, mobility of 0.6 cm^2/V s, resistivity of 0.46 Ω cm) in the sample implanted with a 1×10^{16} cm^{-2} dose, suggesting that annealing at low temperatures can be more favorable for obtaining p-type ZnO:Ga films via nitrogen implantation. The dominant (002) diffraction line in the XRD patterns of all N-implanted and annealed ZnO:Ga films shows that these films have a c-axis preferred orientation of the crystallites. The correlation was found between the annealing conditions and the microstructure, electrical and optical properties of these films. Post-implantation annealing reduces the structure damages and biaxial lattice stresses. The transition from compressive to tensile stress was observed after annealing in O_2 and N_2 at 550 °C. Hence, the carrier concentration increased (4×10^{17} cm^{-3} ÷ 5.1×10^{19} cm^{-3}), resistivity decreased and average transmittance of 87%

was obtained after post-implantation annealing in O_2 at 550 °C. These results are consistent with the shift of the band edge to lower wavelengths and the band gap widening. The films were under compressive stress and had higher resistivity due to nitrogen incorporation and formation of N_O acceptors after annealing in N_2 at 400 °C.

Presented work was supported by the the CENTEM project, reg. No. CZ.1.05/2.1.00/03.0088 cofunded from the ERDF the OP RDI programe of the MEYS CR within the project No. 1M06031, the APVV project LPP-0094-09 and SK-SRB-0012-09 and the SK Grant Agency VEGA project 1/0220/09. Support from the Serbian Ministry of Science and Technological Development (Project 171023) and the EU FP7 Project SPIRIT are also acknowledged.

References

1. H. Liu, V. Avrutin, N. Izyumskaya, Ü. Özgür, H. Morkoç, Superlatt. Microstruct. **48**, 458 (2010)

2. S.-H.K. Park, J.-I. Lee, C.-S. Hwang, H.Y. Chuj, Jpn J. Appl. Phys. **44**, L242 (2005)

3. R.L. Hoffman, B.J. Norris, J.F. Wager, Appl. Phys. Lett. **82**, 733 (2003)

4. X. Li, S.E. Asher, S. Limpijumnong, B.M. Keyes, C.L. Perkins, T.M. Barnes, H.R. Moutinho, J.M. Luther, S.B. Zhang, S.-H. Wei, T.J. Coutts, J. Cryst. Growth **287**, 94 (2006)

5. V. Tvarozek, K. Shtereva, I. Novotny, J. Kovac, P. Sutta, R. Srnanek, A. Vincze, Vacuum **82**, 166 (2008)

6. A.N. Georgobiani, A.N. Gruzintsev, V.T. Volkov, M.O. Vorobiev, V.I. Demin, V.A. Dravin, Nucl. Instrum. Meth. Phys. Res. A **514**, 117 (2003)

7. Ch.-Ch. Lin, S.-Y. Chen, S.-Y. Cheng, H.-Y. Lee, Appl. Phys. Lett. **84**, 5040 (2004)

8. M. Milosavljević, D. Peruško, V.Milinović, P. Gašpierik, I. Novotný, V. Tvarožek, in *Proceedings of 8th International Conference on Advanced Semiconductor Devices & Microsystems, Smolenice, Slovak Republic*, edited by J. Breza, D. Donoval, E. Vavrinsky (2010), p. 183

9. E.Ch. Lee, Y.S. Kim, Y.G. Jin, K.J. Chang, Physica B **912**, 308 (2001)

10. V. Khranovskyy, U. Grossner, V. Lazorenko, G. Lashkarev, B.G. Svensson, R. Yakimova, Superlatt. Microstruct. **42**, 379 (2007)

11. X. Yu, J. Ma, F. Ji, Y. Wang, X. Zhang, H. Ma, Thin Solid Films **483**, 296 (2005)

12. S.H. Park, J.H. Chang, H.J. Ko, T. Minegishi, J.S. Park, I.H. Im, M. Ito, D.C. Oh, M.W. Cho, T. Yao, Appl. Surf. Sci. **254**, 7972 (2008)

13. K. Nakahara, H. Takasu, P. Fons, A. Yamada, K. Iwata, K. Matsubara, R. Hunger, S. Niki, J. Cryst. Growth **237–239**, 503 (2002)

14. P. Šutta, Q. Jackuliak, in *Proceedings of 2nd International Conference on Advanced Semiconductor Devices & Microsystems, Smolenice, Slovak Republic*, edited by J. Breza (1998), p. 227

15. Ch.-Ch. Yang, Ch.-Ch. Lin, Ch.-H. Peng, S.-Y. Chen, J. Cryst. Growth **285**, 96 (2005)

16. R. Delhez, Th.H. de Keijser, E.J. Mittemeijer, Z. Fresenius, Anal. Chem. **312**, 1 (1982)

17. D.-Ch. Park, I. Sakaguchi, N. Ohashi, Sh. Hishita, H. Haneda, Appl. Surf. Sci. **203-204**, 359 (2003)

18. J. Lee, J. Metson, P.J. Evans, U. Pal D. Bhattacharyya, Appl. Surf. Sci. **256**, 2143 (2010)

19. V. Vaithianathan, Sh. Hishita, J.H. Moon, S.S. Kim, Thin Solid Films **515**, 6927 (2007)

20. S.M. Sze, in *Physics of Semiconductor Devices*, 2nd edn. (John Wiley & Sons, Inc., New York, 1981), p. 72

11

Experimental verification of optically optimized $CuGaSe_2$ top cell for improving chalcopyrite tandems*

M. Schmid[1,a], R. Caballero[1], R. Klenk[1], J. Krč[2], T. Rissom[1], M. Topič[2], and M.Ch. Lux-Steiner[1]

[1] Helmholtz Zentrum Berlin für Materialien und Energie, Hahn-Meitner-Platz 1, 14109 Berlin, Germany
[2] University of Ljubljana, Faculty of Electrical Engineering, Tržaška 25, 1000 Ljubljana, Slovenia

Abstract An efficient tandem solar cell requires a top cell which is highly transparent below the energy gap of its absorber. Previously we had reported on a theoretically optimized $CuGaSe_2$ top cell stack based on realistic material properties. It promised a significant increase in optical transparency and, consequently, enhanced $CuGaSe_2/Cu(In,Ga)Se_2$ tandem efficiency. Here we present the first steps taken towards the experimental realization of this optimized tandem. We started with a mechanically stacked device which achieved 8.5% efficiency. Optical measurements of the improved top cells and corresponding photo current densities of the filtered bottom cell are reported. The experimental findings are in agreement with the optical modeling. These data are used to assess the level of tandem performance that could be accomplished in the near future and to discuss the priorities of further research.

1 Introduction

Up to now, tandem and multi junction solar cells are the only concept exceeding the Shockley-Queisser efficiency limit of 30% under solar illumination without concentration [1]. High transparency of the top cell below its energy gap is – apart from efficient absorption above E_g – a crucial requirement for an efficient tandem cell. The chalcopyrites constitute a system of absorber materials with energy gaps suitable for tandem cells, e.g. $CuGaSe_2$ with $E_g = 1.68$ eV and $Cu(In,Ga)Se_2$ with $E_g = 1.1$ eV. However, the best $CuGaSe_2/Cu(In,Ga)Se_2$ tandem efficiency published so far is 7.4% [2]. This relatively low value is fundamentally related to the low efficiency of the top cell together with a top cell transparency of only 60% in its sub-gap range.

Previously we had developed an optical model of the n-ZnO/i-ZnO/CdS/CuGaSe$_2$/SnO$_2$:F/glass solar cell that allowed for the description of the optical properties of this top cell [3]. Based on this model in [4] an optimized top cell stack had been derived that showed significant improvement in transparency. In this paper we will give the experimental proof of enhanced top cell transmission and resulting gain in bottom cell performance in the tandem. An improved efficiency of the mechanically stacked $CuGaSe_2/Cu(In,Ga)Se_2$ tandem will be reported which is however still far from surpassing the $Cu(In,Ga)Se_2$ single cell efficiency. Our discussion will point out options and requirements for building an efficient chalcopyrite tandem.

2 Experimental and results

Chalcopyrite absorbers were prepared by physical vapor deposition in a three stage process [5]. For the bottom cell the approx. 2 μm thick $Cu(In,Ga)Se_2$ was deposited onto a molybdenum back contact, whereas for the $CuGaSe_2$ top cell a transparent back contact is required. SnO$_2$:F with a thickness of approximately 850 nm was used in the initial configuration. The junction of each cell was formed by chemical bath deposition of CdS and sputtering of intrinsic and Al-doped ZnO. The standard configuration of a $Cu(In,Ga)Se_2$ bottom cell is ZnO:Al(200 nm)/i-ZnO(125 nm)/CdS(50 nm)/Cu(In,Ga)Se$_2$(2000 nm)/Mo(800 nm)/glass substrate.

The starting structure and the layer thicknesses of the $CuGaSe_2$ top cell are indicated as initial stack (J) in Table 1. The theoretically optimized top cell stack derived in [4] is set beside. It is characterized by 1) reduced layer thicknesses; 2) careful adaptation of layer thicknesses to anti-reflection behavior; 3) reduced reflection by an anti-reflection coating on top and 4) a substrate simulating monolithic integration. Furthermore, the experimentally realized stacks are given: first the absorber was grown with a thickness approaching 1 μm (opt. stack exp. (A) in Table 1), then the thicknesses of the front ZnO layers were reduced (stack (B)). Note; for the thinner transparent back contact, the SnO$_2$:F was replaced by ZnO:Al,

---

* This article has been previously published in PV Direct, the former name of EPJ Photovoltaics.

a e-mail: martina.schmid@helmholtz-berlin.de

Table 1. Layer structure and thicknesses of the CuGaSe$_2$ top cells: standard structure of the initial stack (J) compared to theoretically optimized structure and stepwise adaptation of the latter one in the experiment by stacks (A) to (C).

layer thickness (nm)	initial stack (J)	optim. stack theory	optim. stack exp. (A)	optim. stack exp. (B)	optim. stack exp. (C)
MgF$_2$	0	120	0	0	0
ZnO:Al	455	90	455	105	105
i-ZnO	95	50	95	50	50
CdS	50	65	50	50	50
CuGaSe$_2$	1600	1050	1190	1190	1190
SnO$_2$:F or	835	90	835	835	
ZnO:Al					105
substrate	glass	CdS	glass	glass	glass

which has comparable optical properties but was in contrast to SnO$_2$:F available with arbitrary thickness (stack (C)). Hence, stack (C) implements the theoretically derived modifications 1) and 2). Steps 3) and 4) were not implemented because monolithic integration is not yet feasible and the MgF$_2$ anti-reflection coating lacks long-term stability. In addition, the exact tuning of the layers to the optimal (anti-reflective) thicknesses is difficult to achieve in the experiment.

Figure 1 shows the transmission spectra measured for the various CuGaSe$_2$ top cell stacks. The lowest curve gives the measurement of the initial stack (J) which reaches a maximum transparency of 60%. A reduction of the absorber layer thickness for approx. 1/3 leads to an increase in top cell transparency of 8% in the wavelength range from 700 to 1200 nm (stack (A)). Further overall enchancement of the top cell transparency (including the long wavelengths) is obtained by reduction of the thicknesses of the front ZnO layers (stack (B)). The optimized structure (C) finally features an average transparency of 80% and stays at a constant level in the wavelength region defined by the energy gaps of the top CuGaSe$_2$ ($\lambda_g \approx 700$ nm) and the bottom Cu(In,Ga)Se$_2$ ($\lambda_g \approx 1200$ nm) absorber. The 80% transmission of the optimized stack means an increase of 20% absolute compared to the initial stack. The observed improved transparency fitted well to the optical model established in Diplot [6], see the dashed curve in Figure 1. The parameters were derived from the modeling of the initial stack (compare [3]) but corrected to include reduced layer thicknesses and related small changes in material properties. The increase in transparency in the long-wavelength regime is due to reduced free charge carrier absorption in the front and back transparent conducting oxide of the top cell. Close to the energy gap, additional reduced defect absorption of the thinner CuGaSe$_2$ absorber contributes to the enhanced transmission.

The increased transparency of the top cell above its E_g is crucial for improving the short circuit cur-

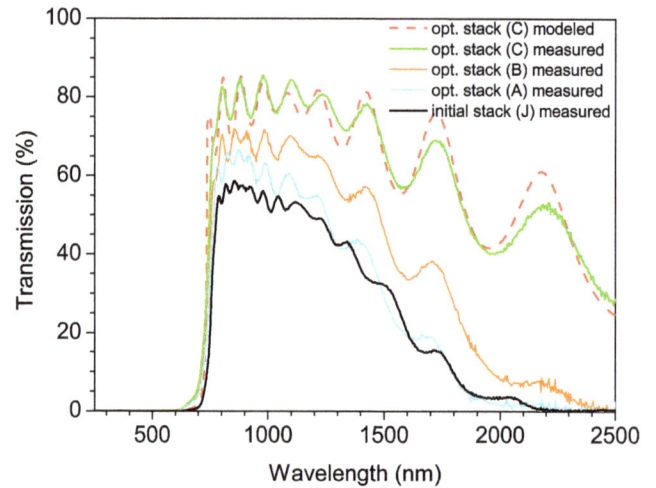

Fig. 1. Measured transparency of the CuGaSe$_2$ top cell in the initial configuration (J) and of the experimentally optimized structures (A), (B) and (C) (see Table 1); comparison to modeling results for the optimized stack.

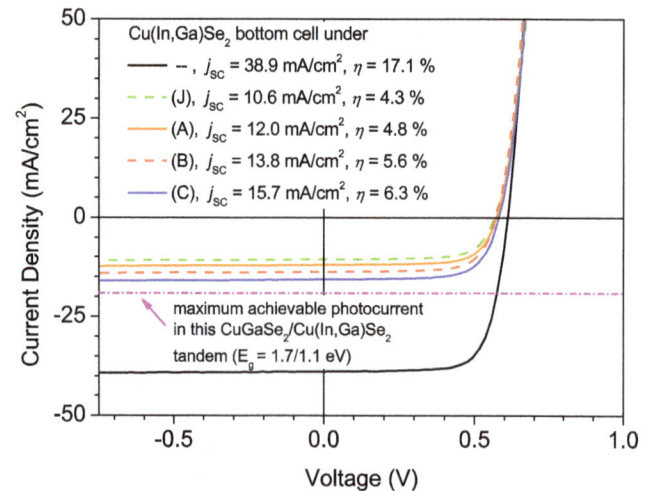

Fig. 2. Measured j-V characteristics of a Cu(In,Ga)Se$_2$ bottom cell filtered by the CuGaSe$_2$ top cell stacks as specified in Table 1. The curve of the unfiltered device is given as a reference. Current densities and efficiencies of the bottom cell are indicated.

rent density of the bottom cell. The j-V characteristics of a Cu(In,Ga)Se$_2$ bottom cell filtered by the various CuGaSe$_2$ top cell stacks are shown in Figure 2. Corresponding short circuit current densities j_{SC} and efficiencies η of the bottom cell are indicated. The current density of 38.9 mA/cm^2 for the unfiltered bottom cell decreased to 10.6 mA/cm^2 under the initial stack (J). When filtering with the optimized top cell stack (C), however, the remaining current density increases to $j_{SC} = 15.7$ mA/cm^2. This corresponds to 82% of the maximum achievable value which is given by the split of the solar spectrum according to the 1.7 / 1.1 eV energy gap pair. Figure 2 shows how the current density improves in step with the top cell improvements and how it approaches the theoretical limit. The measured values are in agreement with theoretical

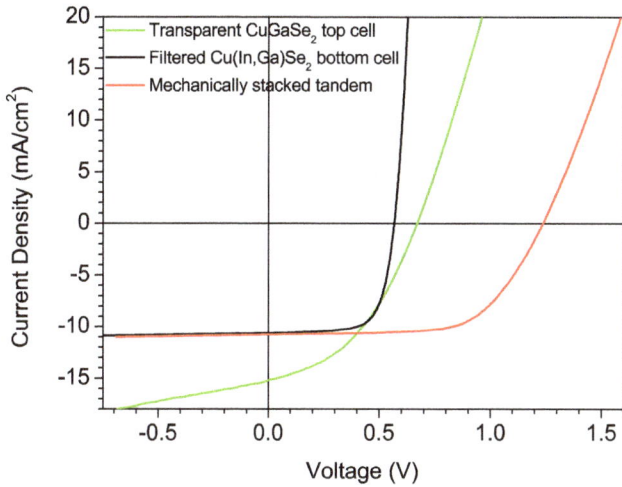

Fig. 3. Solar cell characteristics of a $CuGaSe_2$/$Cu(In,Ga)Se_2$ tandem solar cell and the related $CuGaSe_2$ top and filtered $Cu(In,Ga)Se_2$ bottom cell; for the electrical parameters see Table 2.

	j_{SC} (mA/cm^2)	V_{OC} (V)	FF (%)	η (%)
Top	15.2	0.67	41.8	4.3
Bottom	10.6	0.57	70.4	4.3
Tandem	10.7	1.24	64.1	8.5
Bottom single	38.9	0.61	72.8	17.1

Table 2. Solar cell parameters of a $CuGaSe_2$/$Cu(In,Ga)Se_2$ tandem solar cell and the related $CuGaSe_2$ top and filtered $Cu(In,Ga)Se_2$ bottom cell; for the j-V characteristics see Figure 3.

calculations for the cases (A), (B) and (C) within an error of 5% (not shown here). In the experiment, the efficiency of the $Cu(In,Ga)Se_2$ bottom cell shaded by the $CuGaSe_2$ top cell increases from 4.3 to 6.3%.

3 Discussion

Our latest results of a mechanically stacked $CuGaSe_2$/-$Cu(In,Ga)Se_2$ tandem in the initial configuration are presented in Figure 3. The j-V characteristics are shown for the $CuGaSe_2$ top cell (J), the $Cu(In,Ga)Se_2$ bottom cell filtered by it and the mechanical stack calculated from the top and bottom cell curves by considering series connection of the two single devices. A tandem efficiency of 8.5% was determined. This value surpasses previously reported efficiencies [2]. It is based on a top cell efficiency of 4.3% and a bottom cell efficiency of 4.3%, for detailed electrical data see Table 2. As these data show, the device is bottom cell limited regarding the photo current. The top cell photo

current is still higher in the initial configuartion but will be lowered to better match when using the optimized top cell structures.

Under the improved transparency of the experimentally optimized $CuGaSe_2$ top cell stack (C), the $Cu(In,Ga)Se_2$ bottom cell efficiency reached 6.3%, see Figure 2. Assuming an unchanged top cell performance of 4.3%, the tandem device might reach over 10% efficiency, which is however still lower than the efficiency of the single bottom cell. From the optical point of view, the improvement of the top cell performance has been successfully performed. The improvement of the electrical properties of the $CuGaSe_2$ top cell presents the major task of tandem optimization in the future. The present record efficiency of a $CuGaSe_2$ solar cell is 9.7% on molybdenum [7] and 4.3% on transparent back contact [2]. If the electrical performance of the top cell becomes comparable to the one of the bottom cell – thus also reaching 20% as a single junction device – a tandem efficiency of 26% can be expected. This is the value predicted from our theoretical calculations, compare [4].

4 Summary

In conclusion, the optical model of the chalcopyrite tandem derived before [3, 4] has found experimental verification in this paper. The accuracy of the model as well as predicted design improvements have been shown. The data will be useful to guide further research concerning top cells with superiour optical and electrical properties, to quantify any progress made in the experiment, and to extrapolate feasible tandem efficiencies.

This work was supported by the EC-Project ATHLET, No. 019670. The authors thank C. Kelch, M. Kirsch, K. Kraft and T. Münchenberg for technical assistance and J. Hüpkes from the FZ Jülich for providing substrates.

References

1. W. Shockley, H.J. Queisser, J. Appl. Phys. **32**, 510 (1961)
2. S. Nishiwaki, S. Siebentritt, P. Walk, M.Ch. Lux-Steiner, Progr. Photovolt.: Res. Appl. **11**, 243 (2003)
3. M. Schmid, R. Klenk, M.Ch. Lux-Steiner, Sol. Energy Mater. Sol. Cells **93**, 874 (2009)
4. M. Schmid, J. Krč, R. Klenk, M. Topic, M.Ch. Lux-Steiner, Appl. Phys. Lett. **94**, 053507 (2009)
5. R. Caballero, S. Siebentritt, K. Sakurai, C.A. Kaufmann, H.W. Schock, M.Ch. Lux-Steiner, Thin Solid Films **515**, 5862 (2007)
6. E. Lotter, http://www.diplot.de
7. M. Saad, H. Riazi, E. Bucher, M.Ch. Lux-Steiner, Appl. Phys. A **62**, 181 (1996)

12

2D modelling of polycrystalline silicon thin film solar cells

Ana-Maria Teodoreanu[1,a], Felice Friedrich[1], Rainer Leihkauf[1], Christian Boit[1], Caspar Leendertz[2], and Lars Korte[2]

[1] Technische Universität Berlin, Semiconductor Devices Division, PVcomB, Einsteinufer 19, Sekr. E2, 10587 Berlin, Germany
[2] Helmholtz-Zentrum Berlin, Institute for Silicon Photovoltaics, Kekuléstrasse 5, 12489 Berlin, Germany

Abstract The influence of grain boundary (GB) properties on device parameters of polycrystalline silicon (poly-Si) thin film solar cells is investigated by two-dimensional device simulation. A realistic poly-Si thin film model cell composed of antireflection layer, (n^+)-type emitter, 1.5 μm thick p-type absorber, and (p^+)-type back surface field was created. The absorber consists of a low-defect crystalline Si grain with an adjacent highly defective grain boundary layer. The performances of a reference cell without GB, one with n-type and one with p-type GB, respectively, are compared. The doping concentration and defect density at the GB are varied. It is shown that the impact of the grain boundary on the poly-Si cell is twofold: a local potential barrier is created at the GB, and a part of the photogenerated current flows within the GB. Regarding the cell performance, a highly doped n-type GB is less critical in terms of the cell's short circuit current than a highly doped p-type GB, but more detrimental in terms of the cell's open circuit voltage and fill factor.

1 Introduction

Polycrystalline silicon (poly-Si) is an attractive absorber material for thin film solar cells. Ideally, the high stability against degradation of crystalline silicon can be combined with low-cost production. The reduced optical thickness of thin-film cells leading to incomplete absorption of the solar spectrum, and thus to low short circuit currents J_{SC}, can be quite successfully remedied by different light trapping approaches [1, 2]. Current research on poly-Si focuses on minimizing the critical influence of grain boundaries (GBs) as centers of recombination in the material, which act on the cell's open circuit voltage V_{OC}. Indeed, high efficiencies of 20.4% and corresponding high V_{OC}s of 664 mV were already achieved with multicrystalline silicon wafer solar cells [3]. However, the best poly-Si thin film solar cells today show significantly lower efficiencies of 10.4% [4] and record V_{OC}s of up to 582 mV [5–7], depending on the poly-Si material manufacturing method and contacting scheme. This demonstrates that there is a need but also a potential of improvement of the poly-Si material.

In contrast to their multicrystalline counterpart (i.e. wafer-based cell with diffused junction) poly-Si thin-film solar cells feature a number of layers with different functionality in very close proximity, rendering the local cell properties highly non-uniform. As the standard solar cell characterization methods like current-voltage (J-V) characteristics in the dark and under illumination yield only global properties, the results are usually interpreted in terms of an effective medium approach for the absorber. However, the application of this approximation is not always appropriate and relevant information can be gained by separating the material properties of grain and grain boundary. A straightforward way to investigate the individual effect of grain boundaries on the solar cell performance (e.g. on J-V characteristics) is device simulation. A number of studies on the influence of grain boundaries in silicon-based devices can be found in references [8–13]. In general, the GB is modeled as an interface layer with a specific trap density and interface recombination velocity. The literature results show that the cell efficiency deteriorates, especially when the GB is horizontal and/or located in the space charge region (SCR) [10]. However, an accumulation of impurities or dopant atoms as well as charge carrier transport within the GB, as observed experimentally in reference [14], cannot be adequately investigated with this approach.

In the present study, poly-Si thin film solar cells are investigated by 2D modelling and simulations with the numerical device simulator Sentaurus TCAD [15]. A basic 2D model of the poly-Si thin film solar cell was developed consisting of a low-defect crystalline grain and a highly defective grain boundary layer. The performance of poly-Si

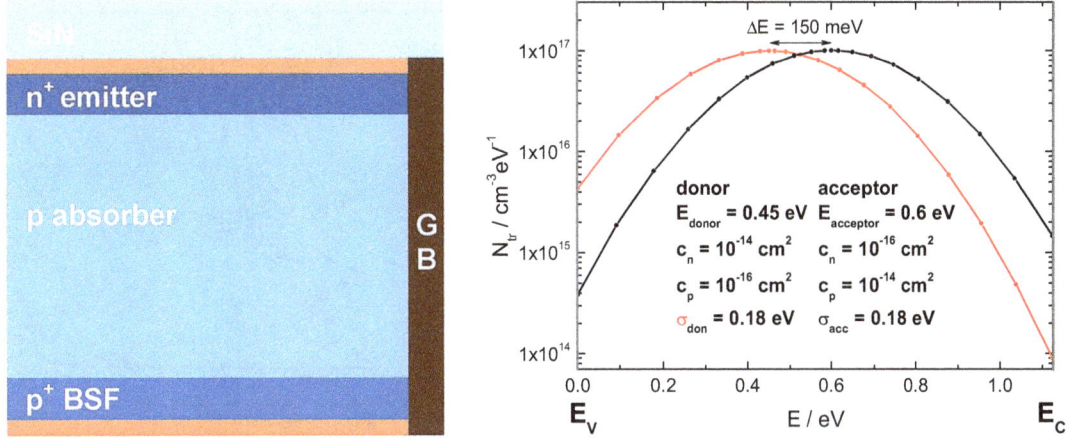

Fig. 1. Left: Structural model of the poly-Si solar cell unit composed of a p-type crystalline Si absorber grain (2 μm width), p^+ back surface field and n^+ emitter with an adjacent vertical grain boundary layer (5 nm width). The contacts, defined as ohmic, are depicted in orange. The silicon nitride (SiN) top layer represents the antireflection coating. Right: Assumed defect distribution in the GB layer over the energy in the band gap for the particular case of a GB defect density of 10^{17} cm^{-3} eV^{-1}. For details see Tables 1 and 2.

Table 1. Parameters of the cell's layers: emitter, absorber, BSF and GB.

	Emitter	Absorber	BSF	GB
type	n^+	p	p^+	variable
doping density/cm^{-3}	1.2×10^{20}	1.5×10^{16}	1.5×10^{19}	variable
defect density/cm^{-3}	10^{19}	10^{10}	10^{19}	variable
thickness	35 nm	1.5 μm	65 nm	1.6 μm
width	2 μm	2 μm	2 μm	5 nm

solar cells with an n-type and a p-type grain boundary, respectively, is compared to the performance of a reference cell without grain boundary. The variation in GB doping type is intended to reflect segregation of doping atoms or impurities at the GB or emitter diffusion through the GB. Within our study, only two parameters of the GB layer are varied: the doping concentration and the defect density. While the influence of the GB doping type is ambivalent dependent on the parameter range, the cell's V_{OC} in general deteriorates in the presence of a GB.

2 Modelling approach

For the implementation of non-horizontal (in the present case vertical) GBs in an optoelectronic solar cell model it is essential to use a 2D/3D numerical device simulator. The results presented in this study were obtained with Sentaurus TCAD from Synopsys [15]. Basic silicon parameters were taken from AFORS-HET [16]. For the optical modelling, we used the transfer matrix method (TMM) implemented in Sentaurus TCAD [17]. The simulations were performed at standard testing conditions (AM1.5 global spectrum, 100 mW/cm^2 radiant power density and 25 °C operating temperature).

The basic structure of the simulated poly-Si thin-film solar cells is shown in Figure 1 together with the assumed defect distribution in the GB over the energy in the band gap. Table 1 lists the parameters of the solar cell layers.

The poly-Si growth is assumed to be columnar with a lateral Si grain size of 2 μm and a film thickness of 1.6 μm. The vertical grain boundary is assumed to be 5 nm wide, which is in agreement with experimental observations [18]. In addition to the p-type absorber an n^+ emitter as well as a p^+ back surface field (BSF) were considered. The adjacent electrical contacts are assumed to be ohmic and transparent and at boundaries of the device that are not contacts Neumann boundary conditions were applied. Finally, a 100 nm thick SiN layer was implemented as antireflection coating for the optical TMM calculation of the generation rate.

The absorber grain is p-type crystalline silicon (c-Si) doped 1.5×10^{16} cm^{-3} and having a typical low defect concentration of 10^{10} cm^{-3} with capture cross sections for electrons and holes of 10^{-14} cm^2, represented by a single defect in the middle of the band gap. The emitter and BSF layers are highly doped with 1.2×10^{20} cm^{-3} and 1.5×10^{19} cm^{-3}, respectively, and have a single defect of 10^{19} cm^{-3} concentration in the middle of the bandgap, with capture cross sections for electrons and holes of 10^{-14} cm^2. The band gap, the mobilities for electrons and holes, and the densities of states of the valence and conduction bands are standard doping-dependent parameters of c-Si [19].

The GB layer is modeled as a highly defective silicon layer with a continuous dangling bond-like density of states distribution in the band gap (cf. Fig. 1, right) [20]. The electronic properties and defect specifications for the

Table 2. Parameters of the GB layer. For the density of states in the band gap are specified: the energetic position of the maximum defect densities for the donor-type and the acceptor-type defect distribution $E_{\text{donor, acceptor}}$, the maximum defect density of the distributions N_{tr}, the capture cross sections for electrons and holes $c_{\text{n,p}}$ and the standard deviation σ.

Layer properties		
bandgap, E_{G}	1.059 eV	
density of states of the conduction band,	8.020×10^{18} cm^{-3}	
N_{C} density of states of the valence band, N_{V}	7.566×10^{18} cm^{-3}	
electron mobility, μ_{n}	193.60 cm^2/V s	
hole mobility, μ_{p}	68.93 cm^2/V s	
Density of states in the band gap		
defect type	donor	acceptor
$E_{\text{donor, acceptor}}$	0.40 eV	0.65 eV
N_{tr}	10^{16}–10^{22} cm^{-3} eV^{-1}	
c_{n}	10^{-14} cm^2	10^{-16} cm^2
c_{p}	10^{-16} cm^2	10^{-14} cm^2
σ	0.18 eV	

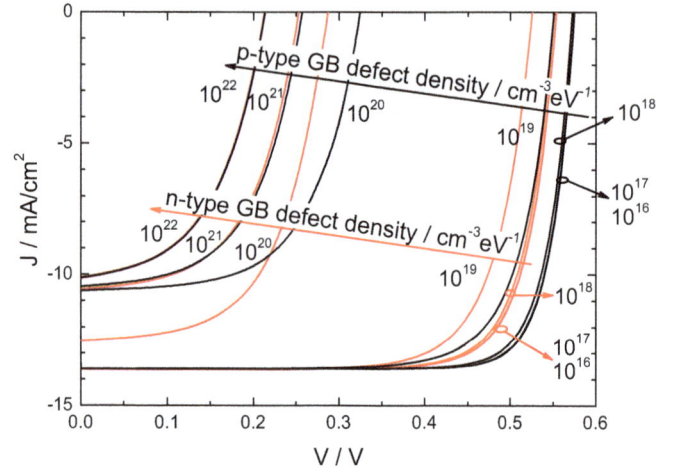

Fig. 2. Calculated J-V characteristics under illumination of the structures A (black line), B (blue lines) and C (red lines) for a highly doped GB with $N_{\text{A,D}} = 10^{19}$ cm^{-3} at varying GB defect densities.

GB layer are summarized in Table 2. For this layer the doping type (p or n), the doping density N_{A} or N_{D}, respectively, and the defect density N_{tr} (corresponding to the maximum defect density of the Gaussian distribution in Fig. 1) were varied, the former ranging from 10^{15}–10^{20} cm^{-3} and the latter from 10^{16}–10^{22} cm^{-3} eV^{-1}. All other parameters such as the band gap, density of states of the valence and conduction band, and mobilities were kept constant. The values are based on reference [19] for heavily doped c-Si with $N_{\text{A}} = 1.5 \times 10^{19}$ cm^{-3}. The dangling bond-like defects are represented by two Gaussian distributions within the band gap, shifted relative to each other by the correlation energy ΔE, which was determined to be in the range of 100 meV to 200 meV [20]. We chose for our simulations an average $\Delta E = 150$ meV.

In the following, the reference cell without GB layer will be denoted A, the cell with p-type GB layer B and the cell with n-type GB layer C.

3 Simulation results

Current-voltage characteristics under illumination were calculated for the reference cell A (without GB) as well as cell B (with p-type GB) and cell C (with n-type GB) for varying GB doping concentration $N_{\text{A,D}}$ and GB defect concentration N_{tr}.

Figure 2 shows the J-V characteristics for a highly doped GB layer with $N_{\text{A,D}} = 10^{19}$ cm^{-3} and varying N_{tr}. The highest V_{OC} is observed for the reference cell as well as cell B in the low GB defect range of 10^{16} and 10^{17} cm^{-3} eV^{-1}. A deterioration of the J_{SC} is generally not observed until the GB defect density exceeds 10^{19} cm^{-3} eV^{-1}. For higher defect densities in the GB of 10^{21} and 10^{22} cm^{-3} eV^{-1}, the cell's V_{OC} as well as J_{SC} are significantly reduced for both cell structures B and C in an equal way. A remarkable difference of the solar cell

characteristics between the structure B and C is found at the intermediate GB defect density of 10^{20} cm^{-3} eV^{-1}: we observe higher J_{SC} for cell C and higher V_{OC} for cell B.

The solar cell parameters J_{SC}, V_{OC}, fill factor FF and efficiency η extracted from the illuminated J-V curves are shown in Figure 3 for the whole range of GB doping concentration $N_{\text{A,D}}$ and GB defect density N_{tr}. For better comparability, the solar cell parameters of the cells B and C were normalized to the values calculated for reference cell A.

We can distinguish three regimes: **(1)** the high defect density regime $N_{\text{tr}} \gg N_{\text{A,D}}$, **(2)** the low defect density regime $N_{\text{tr}} \ll N_{\text{A,D}}$ and **(3)** the intermediate regime, where the GB defect density is in the range of GB doping density.

For defect densities higher than the doping level **(1)**, corresponding to the lower left corner of Figure 3, the solar cell performance is almost independent on the doping level or type. An increased defect density in the GB leads to an overall decrease of the solar cell efficiency of up to 84%. Most affected is the cell's V_{OC} with up to 64% followed by the FF with up to 39% and the J_{SC} with up to 26%.

The regime **(2)** of $N_{\text{tr}} \ll N_{\text{A,D}}$, corresponding to the upper right corner in Figure 3, is defined by equal J_{SC} values for cell types B and cell C, that are also close to the reference cell value. In contrast, V_{OC}, FF and η are higher for cell B.

In the intermediate regime **(3)** the J_{SC} of cell C is higher than that of cell B whereas the V_{OC}, FF and η of cell B are higher than those of cell C.

4 Discussion

The interplay between GB doping concentration and GB defect density determines the Fermi level in the GB layer, which is in general different from the Fermi level

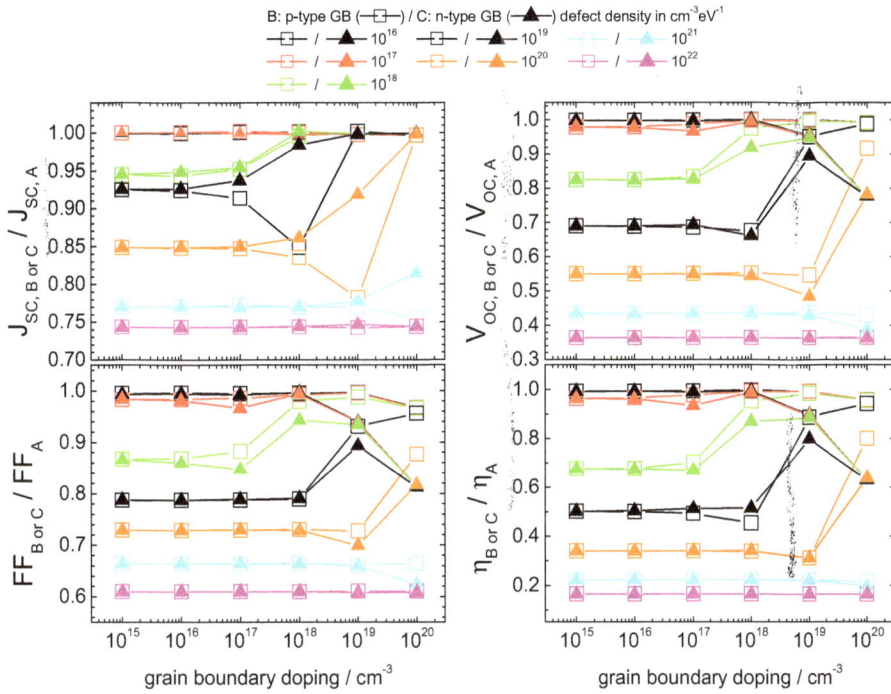

Fig. 3. Simulations of the solar cell parameters of the structure B (with p-type GB) and C (with n-type GB) normalized to the corresponding solar cell parameters of structure A (without any GB, reference cell) $J_{\mathrm{SC,B\ or\ C}}/J_{\mathrm{SC,A}}$, $V_{\mathrm{OC,B\ or\ C}}/V_{\mathrm{OC,A}}$, $FF_{\mathrm{B\ or\ C}}/FF_{\mathrm{A}}$ and $\eta_{\mathrm{B\ or\ C}}/\eta_{\mathrm{A}}$ for varying GB doping concentration (abscissae) and for different GB defect densities (symbol slopes).

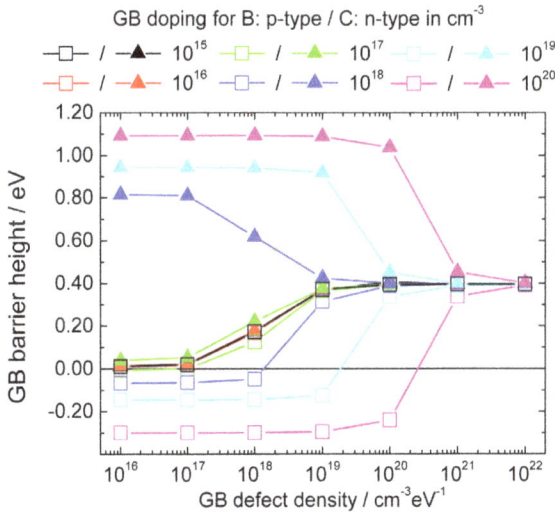

Fig. 4. Calculated potential barrier height at the interface grain-GB, in the bulk of the solar cell. The results are shown over the GB defect density (abscissae) for different GB doping concentrations (symbol slopes) and for the structures B and C.

inside the grain. Thus, a potential barrier forms in the structure at the interface grain-GB. The height of this barrier relative to the conduction band maximum was calculated for a position in the field-free bulk far from the BSF and emitter. Figure 4 shows the potential barrier height over the GB defect density for the specified doping range for cells B and C.

If we consider the three regimes defined above: the regime **(1)** of higher defect density $N_{\mathrm{tr}} \gg N_{\mathrm{A,D}}$ is governed by a Fermi level pinning effect, leading to a potential barrier height of $\sim$0.4 eV, which is independent of doping. This explains the similarity of the J-V characteristics for cases B and C in this regime in Figure 2 and of the GB doping-independent solar cell parameters at high N_{tr} in Figure 3. This corresponds to the case of for example a non-passivated poly-Si absorber.

Only in regime **(2)** for $N_{\mathrm{tr}} \ll N_{\mathrm{A,D}}$, the barrier height is determined by the respective doping, leading to a negative potential barrier for p-type GB and a pronounced positive potential barrier in the cell with n-type GB. Due to these potential barriers, the p-type GB layer acts electron-repulsive – like an additional "back surface field" and the n-type GB acts hole repulsive – forming an additional pn junction at the interface grain-GB.

Figure 5 shows the electron and hole current densities for the three structures at short circuit conditions in the case of $N_{\mathrm{A,D}} = 10^{20}$ cm^{-3} and $N_{\mathrm{tr}} = 10^{16}$ cm^{-3} eV^{-1}. Indeed, for p-type doping, the simulation shows a local quenching of the space charge region (SCR) in the vicinity of the GB, and for n-type doping an extension of the pn junction along the GB. However, the additional pn junction which forms at the interface grain-GB proves to be detrimental for the cell efficiency, which decreases by over 30% mainly due to the decrease in V_{OC}. In literature, beneficial effects of extended pn junctions are discussed [21]. We also observe such effects in the intermediate regime **(3)**. Here, a larger J_{SC} is determined for the n-type GB in cell C compared to cell B. However, the V_{OC}

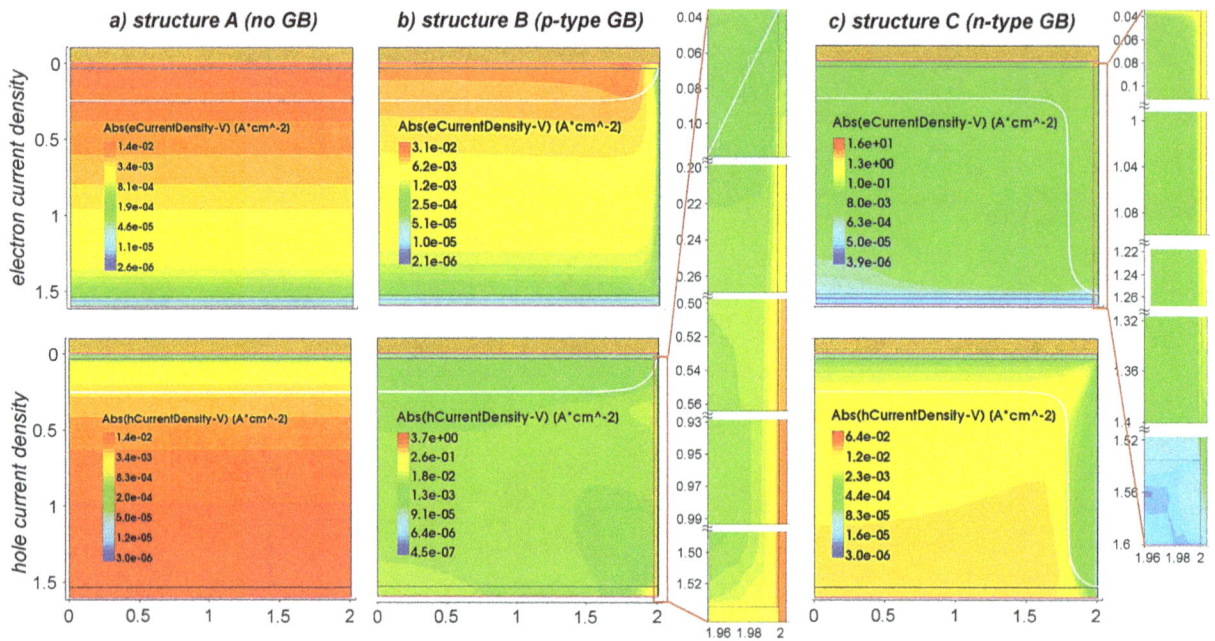

Fig. 5. Exemplary 2D simulations of the electron and hole current density distributions in the cell at short circuit conditions for (a) structure A, (b) structure B and (c) structure C. The GB doping concentration is 10^{20} cm^{-3} and the GB defect density 10^{16} cm^{-3} eV^{-1}. The boundary of the space charge region is marked with a white line. The zoom-in into the GB layer shows the extremely high majority-carrier current in the GB increasing towards the respective majority carrier contacts.

and FF of the p-type GB cell exceed those of the n-type GB cell.

Further, for the higher doping regime (**2**), the simulation results in Figures 5b and 5c indicate the formation of a conductive channel extending along the GB and in its vicinity, from emitter to BSF. This corresponds to the case of an enhanced emitter diffusivity within the GB or an accumulation of dopant atoms in the GB, respectively.

The respective electron and hole current densities within the GB layer are depicted in Figure 6 for the whole parameter range. For n- as well as p-type GB there is a high majority-carrier current density for GB doping concentrations of 10^{18}–10^{20} cm^{-3} (regime (**2**)). This current density is about two orders of magnitude higher than the GB current density for lower doping concentrations, forming in the high-doping regime the conductive channel. The presence of such a conductive channel is of course detrimental for the solar cell, as it effectively corresponds to a shunt of the cell. This explains the decrease in fill factor and efficiency, respectively – observed in Figure 3 in this regime. This effect is even more detrimental for a GB directly connected to the ohmic contact region (not shown here).

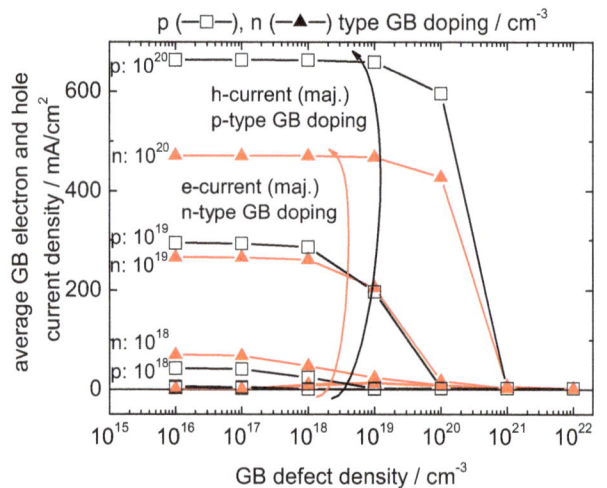

Fig. 6. Average GB electron (red symbols) and hole (blue symbols) current density shown over the GB defect density for different GB doping concentrations for p-type GB (open squares) and n-type GB (full triangles). The majority-carrier (maj.) current increase is highlighted.

5 Conclusions and outlook

The present simulation study shows, that despite the positive effects like the extension of pn junction or the formation of a BSF, that doped GB layers could bring along, both n- and p-type grain boundaries deteriorate the performance of a polycrystalline thin film solar cell. The most important factor for cell performance deterioration is the GB defect density, notably for the regime where the GB defect concentration is higher than the GB doping concentration, which features Fermi level pinning. Another important factor of the cell's characteristics is the formation conductive channel along the GB and in its vicinity, which characterizes the regime of high GB doping concentration and low GB defect density.

The simulation study can further be extended by the implementation of a transparent conductive oxide layer to refine the contacting of the grain and GB layer as well as a detailed analysis of the dark J-V characteristics.

This work was supported by the Federal Ministry of Education and Research (BMBF) and the state government of Berlin (SENBWF) in the framework of the program "Spitzenforschung und Innovation in den Neuen Ländern" (Grant No. 03IS2151B).

References

1. B. Rech, H. Wagner, Appl. Phys. A: Mater. Sci. Proc. **69**, 155 (1999)

2. R. Brendel, *Thin-Film Crystalline Silicon Solar Cells*, 1st edn. (Wiley-VCH, Weinheim, 2003)

3. O. Schultz, S.W. Glunz, G.P. Willeke, Prog. Photovolt. Res. Appl. **12**, 553 (2004)

4. M.J. Keevers, T.L. Young, U. Schubert, M.A. Green, in *Proceedings of the 22nd European Photovoltaic Solar Energy Conference, 3-7 September 2007, Milan, Italy, 2007*

5. D. Amkreutz, J. Müller, M. Schmidt, T. Hänel, T.F. Schulze, Prog. Photovolt. Res. Appl. **19**, 937 (2011)

6. J. Dore et al., EPJ Photovoltaics **4**, 40301 (2012)

7. J. Haschke, L. Jogschies, D. Amkreutz, L. Korte, B. Rech, Sol. Energy Mater. Sol. Cells **115**, 7 (2013)

8. A.K. Ghosh, C. Fishman, T. Feng, J. Appl. Phys. **51**, 446 (1980)

9. J.G. Fossum, F.A. Lindholm, IEEE Trans. Electron Devices **27**, 692 (1980)

10. M.A. Green, J. Appl. Phys. **80**, 1515 (1996)

11. S.A. Edmiston, G. Heiser, A.B. Sproul, M.A. Green, J. Appl. Phys. **80**, 6783 (1996)

12. P.P. Altermatt, G. Heiser, J. Appl. Phys. **92**, 2561 (2002)

13. K.R. Taretto, Ph.D. thesis, Institut für Physikalische Elektronik-Universität Stuttgart, 2003

14. M. Kittler, M. Reiche, Adv. Eng. Mater. **11**, 249 (2009)

15. www.synopsys.com/Tools/TCAD/

16. R. Stangl, C. Leendertz, in *Physics and Technology of Amorphous-Crystalline Heterostructure Silicon Solar Cells*, edited by W.G.J.H.M. Sark, L. Korte, F. Roca, volume 0 of *Engineering Materials* (Springer, Berlin, Heidelberg, 2011), pp. 445–458

17. Sentaurus Device User Guide Version E-2010.12, December 2010, Synopsys

18. M. Klingsporn et al., private communication and presented at EMRS, Strasbourg, France, Spring 2012

19. G. Masetti, R. Severi, S. Solmi, IEEE Trans. Electron Devices **ED-30**, 764 (1983)

20. R. Schropp, M. Zeman, *Amorphous and Microcristalline Silicon Solar Cells* (Kluwer Academic Publishers, Boston, 1998), Chap. 6

21. A. Zerga, E. Christoffel, A. Slaoui, in *3rd World Conference on Photovoltaic Energy Conversion (WCPEC-3), Osaka, Japan, May 12-16 2003*, edited by K. Kurokawa, L. Kazmerski, B.M. Nelis, M. Yamaguchi, C. Wronski, W. Sinke (2003), Vol. 2, pp. 1053–1056

13

Light induced electrical and macroscopic changes in hydrogenated polymorphous silicon solar cells

K.H. Kim[1,2,a], E.V. Johnson[2], A. Abramov[2], and P. Roca i Cabarrocas[2]

[1] TOTAL S.A., Gas & Power – R&D Division, Courbevoie, France
[2] Laboratoire de Physique des Interfaces et des Couches Minces (UMR 7647 CNRS), École Polytechnique, 91128 Palaiseau, France

Abstract We report on light-induced electrical and macroscopic changes in hydrogenated polymorphous silicon (pm-Si:H) PIN solar cells. To explain the particular light-soaking behavior of such cells – namely an increase of the open circuit voltage (V_{oc}) and a rapid drop of the short circuit current density (J_{sc}) – we correlate these effects to changes in hydrogen incorporation and structural properties in the layers of the cells. Numerous techniques such as current-voltage characteristics, infrared spectroscopy, hydrogen exodiffusion, Raman spectroscopy, atomic force microscopy, scanning electron microscopy and spectroscopic ellipsometry are used to study the light-induced changes from microscopic to macroscopic scales (up to tens of microns). Such comprehensive use of complementary techniques lead us to suggest that light-soaking produces the diffusion of molecular hydrogen, hydrogen accumulation at p-layer/substrate interface and localized delamination of the interface. Based on these results we propose that light-induced degradation of PIN solar cells has to be addressed from not only as a material issue, but also a device point of view. In particular we bring experimental evidence that localized delamination at the interface between the p-layer and SnO$_2$ substrate by light-induced hydrogen motion causes the rapid drop of J_{sc}.

1 Introduction

Hydrogenated amorphous silicon (a-Si:H) is widely used as an absorber layer for thin film solar cells. It has generated tremendous scientific and technical interest for two reasons: firstly, a-Si:H has several interesting material properties that have opened up many opportunities for semiconductor device applications, such as a high absorption coefficient and a high optical bandgap. Secondly, the glow discharge deposition technique, also referred to as plasma enhanced chemical vapor deposition (PECVD), has enabled the production of a-Si:H films at a low temperature ($\sim$200 °C) over large areas ($\sim$several m^2). This low process temperature allows the use of a wide range of low cost substrates such as glass, metal and polymer foils. In addition, a-Si:H can be easily doped and alloyed by adding the appropriate gases to a source gas, usually silane. However, it also displays a serious technical drawback, the so-called Staebler-Wronski effect (SWE) [1]. Indeed, were it not for their light-induced degradation, numerical simulations have predicted that a-Si:H based single junction solar cells could realistically display up to 12% conversion efficiency [2], thus competing with multicrystalline silicon solar cells. Generally, SWE is synonymous of light-induced creation of metastable defect states,

which work as recombination centers, and it mostly affects the intrinsic layer of PIN solar cells. Therefore, in order to overcome the light-induced degradation, the efforts to improve the stability of a-Si:H involved the development of various types of intrinsic materials using hydrogen diluted silane gas mixtures such as microcrystalline silicon (μc-Si:H), protocrystalline silicon (pc-Si:H) [3] and polymorphous silicon (pm-Si:H) [4]. This last material is characterized by the formation of silicon nanocrystals in the plasma, which contribute to deposition along with SiH$_x$ radicals. This results in a nanostructured material whose amorphous phase has a medium range order and is more relaxed than that found in standard a-Si:H [5,6]. The volume fraction of small crystallites in this material is less than 10% [7]. Contrary to pc-Si:H, pm-Si:H films can be made thick enough to be used as an absorber layer of thin film solar cells because the microstructure of pm-Si:H does not depend on the thickness of the film or on the nature of the substrate, as its deposition mechanism mainly relies on the silicon nanocrystals synthesized in the plasma, so the nanocrystals distribute to all over the film thickness. Moreover, such deposition conditions coincide with those which result in a higher deposition rate compared to a-Si:H [8]. Through spectroscopic ellipsometry, Rutherford Backscattering and ERDA measurements, it has been shown that pm-Si:H films are denser than a-Si:H films,

[a] e-mail: ka-hyun.kim@polytechnique.edu

in spite of their high hydrogen content, in the range of 15–20% [7]. The peculiar structure of pm-Si:H results in a low defect density (of the order of 10^{14} cm^{-3}eV^{-1} at Fermi level as measured by SCLC and modulated photocurrent) and higher resistance to light-soaking than a-Si:H [8–10]. In particular improved hole transport appears to be a key point for the application of this material in solar cells [11].

In this work, we focus on the metastability of pm-Si:H PIN solar cells compared to a-Si:H ones. Our results show that pm-Si:H PIN solar cells have reduced SWE, but show strong changes in their structure, which take place at the early stage (∼two hours) of light-soaking. In particular, we aim at explaining the increase in open-circuit voltage (V_{oc}) and fast decrease of short-circuit current density (J_{sc}) in pm-Si:H PIN solar cells and establish a relationship between these changes in the electrical properties and structural changes related to hydrogen diffusion to the substrate/p-layer interface and local delamination of the interface during light-soaking.

2 Experiments

The PIN solar cells were deposited at 175 °C by the radio-frequency (RF, 13.56 MHz) glow discharge PECVD method in a multiplasma-monochamber reactor [12]. Standard a-Si:H was obtained by the dissociation of pure silane at low pressure (50 mTorr) and low RF power density (5 mW/cm^2). Intrinsic pm-Si:H films were deposited under carefully controlled plasma conditions using hydrogen-diluted silane gas mixture. In this work, our pm-Si:H layers were deposited at high pressure of 2 Torr and RF power density of 30 mW/cm^2. The solar cells had the following structure: glass/textured SnO$_2$:F/p-type hydrogenated amorphous silicon carbon (a-SiC:H)/a-SiC:H buffer/intrinsic layer/n-type a-Si:H/Al contact. The area of the cells was 0.126 cm^2. a-Si:H and pm-Si:H PIN layer stacks were also deposited on Corning Eagle glass, flat TCO and highly resistive (>10^4 Ω cm) FZ c-Si substrates for more comprehensive studies. The thickness of p and n doped layers were about 170 and 130 Å respectively, and the intrinsic layer thickness varied from 2000 to 5000 Å.

To study the macroscopic and microscopic changes in the layers and devices, a number of characterization techniques were employed. For the light-induced degradation studies, sets of current-density-voltage ($J(V)$) measurements at various stages of the light-soaking were performed. The $J(V)$ measurements during the early stages of light-soaking were taken more frequently than later, in order to accurately monitor the dynamics during this critical phase. Between each $J(V)$ measurement, the cells were light-soaked in the open-circuit condition using an Oriel-Apex Xe lamp. The cells were both light-soaked and had their $J(V)$ curves measured under an intense illumination of 200 mW/cm^2 (equivalent to 2 suns), to accelerate light-induced degradation. During light-soaking, the PIN solar cells were fan cooled to limit illumination induced heating. The temperature of the PIN solar cells, measured by PT100 thermometer during light-soaking, stayed under 50 °C.

Structural changes during light-soaking were characterized through in-situ micro Raman scattering measurements. A He–Ne laser (632 nm) and a back-scattering collection configuration were used for the Raman measurements. For the light-soaking, a white light source with an intensity of 80 mW/cm^2 was applied from the glass side. The test was done for both a-Si:H and pm-Si:H PIN layer stacks on Corning Eagle glass, and the study on the pm-Si:H PIN was done twice to check its reproducibility. During light-soaking, Raman spectra were measured every five minutes, and for the second pm-Si:H PIN test, the measurement was done every two minutes.

In-situ surface morphology characterization during light-soaking was obtained from AFM measurements. Tapping mode was used to prevent the cantilever from dragging across the surface and resulting in surface damage, as well as providing higher resolution. The samples were exposed to light from the n layer side, with an intensity of 150 mW/cm^2. The scan rate was chosen to be slow because the surface displays low roughness and small features. Scanning area size was varied from 500 × 500 nm^2 to 2 × 2 μm^2. Sets of AFM images were analyzed by surface grain extraction, from which the surface grain size and distribution were obtained. In addition, a set of pm-Si:H PIN layer stacks co-deposited on various substrates such as Corning Eagle glass, textured SnO$_2$:F (Asahi-U), and flat ZnO:Al, was light-soaked for longer periods (up to 100 h under 100 mW/cm^2) and the changes in their topology were characterized by scanning electron microscopy (SEM) and atomic force microscopy (AFM). Veeco's Dimension 5000 was used for AFM, and Hitachi 4800 was used for SEM.

Another important feature in the evolution of film and device properties under light-soaking is the behavior of the hydrogen they contain. To study the characteristics of Si-H bond breaking and subsequent dangling bond recombination as a consequence of light-soaking, infrared absorption was measured for both a-Si:H and pm-Si:H solar cells. Infrared spectra in transmission mode were measured with a Nicolet 6700 Fourier transform infrared spectrometer on samples grown on ⟨100⟩ highly resistive (>10^4 Ω cm) FZ c-Si substrates. Its resolution was set to 4 cm^{-1}. The transmission spectra, resulting from the average over 32 scans, were normalized to the transmission of the c-Si substrates. As a complementary study for infrared absorption, hydrogen exodiffusion experiments were performed on pm-Si:H PIN stacks. During these experiments, the base vacuum was 10^{-7} mbar and the heating rate was 10 °C/min. The effused hydrogen was detected by a quadruple mass spectrometer (QMS), and recorded in a continuous manner with the increase in temperature, to obtain the hydrogen exodiffusion spectrum. For this study, a pm-Si:H PIN solar cell was prepared on Corning Eagle glass substrate and then cut into three pieces: one as a control, a second one light-soaked for one hour, and a third one light-soaked for 20 h.

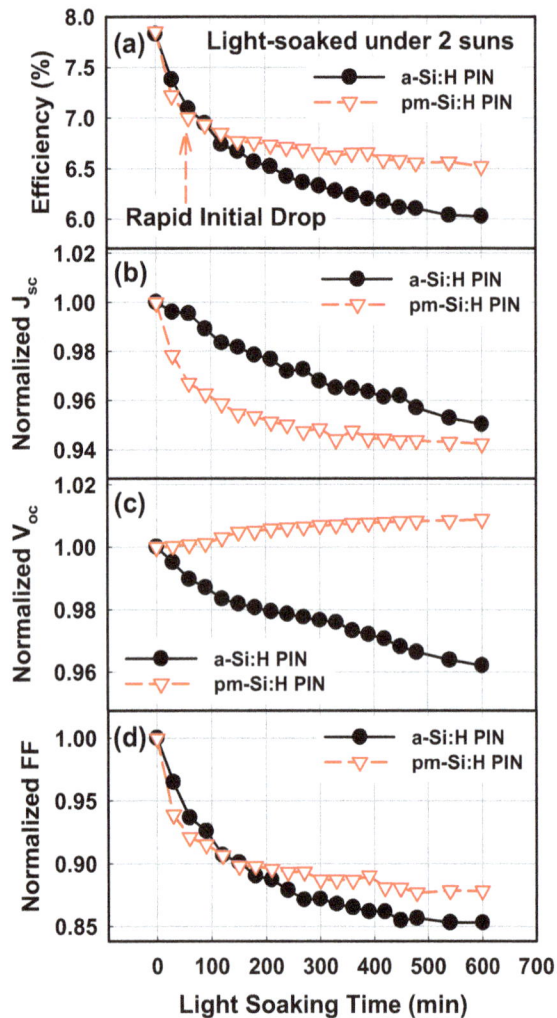

Fig. 1. Evolution of a-Si:H and pm-Si:H PIN solar cell (a) power conversion efficiency (b) normalized J_{sc} (c) normalized V_{oc} and (d) normalized Fill Factor during light-soaking under an illumination of 200 mW/cm^2.

To characterize the variations in the material properties during light-soaking, spectroscopic ellipsometry was performed on a pm-Si:H PIN solar cell deposited on flat glass. The cell was light-soaked for 200 min and five SE spectra were taken during this period. The resulting spectra were modeled using the Tauc-Lorentz dispersion model and the material parameters were obtained from the modeling [7].

3 Results

3.1 Evolution of the solar cell parameters as a function of light-soaking time

Figure 1 shows the power conversion efficiency, normalized J_{sc}, normalized V_{oc} and normalized *Fill Factor* of PIN solar cells as a function of light-soaking time. One can see a clear difference in the light-induced degradation behavior

of the pm-Si:H and the a-Si:H PIN solar cells. The conversion efficiency behavior shows that the pm-Si:H solar cell is more stable than that having a-Si:H as intrinsic layer. However, the pm-Si:H solar cell shows a rapid efficiency drop at the early stages (about first two hours) of light-soaking. The J_{sc} of the pm-Si:H solar cell (Fig. 1b) shows a faster initial drop followed by a phase of more moderate degradation at longer light-soaking times, while the a-Si:H solar cell demonstrates a monotonic degradation and no clear distinction into such "phases". The faster initial decrease of J_{sc} is the major contribution to the initial efficiency degradation of pm-Si:H solar cell, and it differs in character from the classic stretched exponential behavior observed in a-Si:H. The difference in behavior between the two types of cells is also striking for V_{oc} (Fig. 1c), which increases for the pm-Si:H cell, whereas it drops for the a-Si:H solar cell during light-soaking. This behavior cannot be explained by heating of the cells. Indeed if this were the case, then we should observe a decrease in V_{oc} and an increase in J_{sc}. Therefore, the increase of V_{oc} provides a strong indication that heating effects are negligible during light-soaking, in agreement with the Pt-100 thermo-resister probe measurement indicating that the temperature of the PIN solar cells stayed under 50 °C during light-soaking. It is important to note that the data in Figure 1 were continuously recorded during the light-soaking, without turning off the lamp. Further results on the irreversibility of pm-Si:H solar cells characteristics will be presented in a forthcoming publication.

Various hypothesis have been proposed to explain the increase in V_{oc}: (i) the activation of boron in the p-layer of the solar cells during light-soaking [4,13–18], (ii) light-induced changes in the intrinsic mixed-phase material [19–21] and (iii) decrease in valence band tail and state re-distribution at p/i interface [22]. Most of those studies deal with solar cells based on a-Si:H materials deposited by dissociating silane-hydrogen gas mixtures, while our standard a-Si:H was deposited by dissociating pure silane. Therefore, the physical origin of the V_{oc} kinetics of the a-Si:H solar cells in the literature can differ from that of our standard a-Si:H PIN solar cells. Besides, common physical origin in the discussion of the V_{oc} increase is hydrogen motion, and it is rather related to the highly hydrogen diluted silane gas mixture of pm-Si:H deposition. Another interesting point on the V_{oc} increase in the literature is that none of those studies addressed comprehensive study with kinetics of J_{sc}. It has also been reported that a-Si:H shows very fast creation of charged gap states during light-soaking which could be related to the fast degradation kinetics of the pm-Si:H cell [23], but midgap state creation should lead to a decrease in V_{oc} in addition to that of the overall efficiency. Therefore, we consider that fast state creation is insufficient for the explanation behind the particular degradation of pm-Si:H solar cells and that other phenomena must be involved. To better understand the light-induced changes reported in Figure 1, a set of complementary techniques was used to characterize the changes in the structure of the films during or after light-soaking.

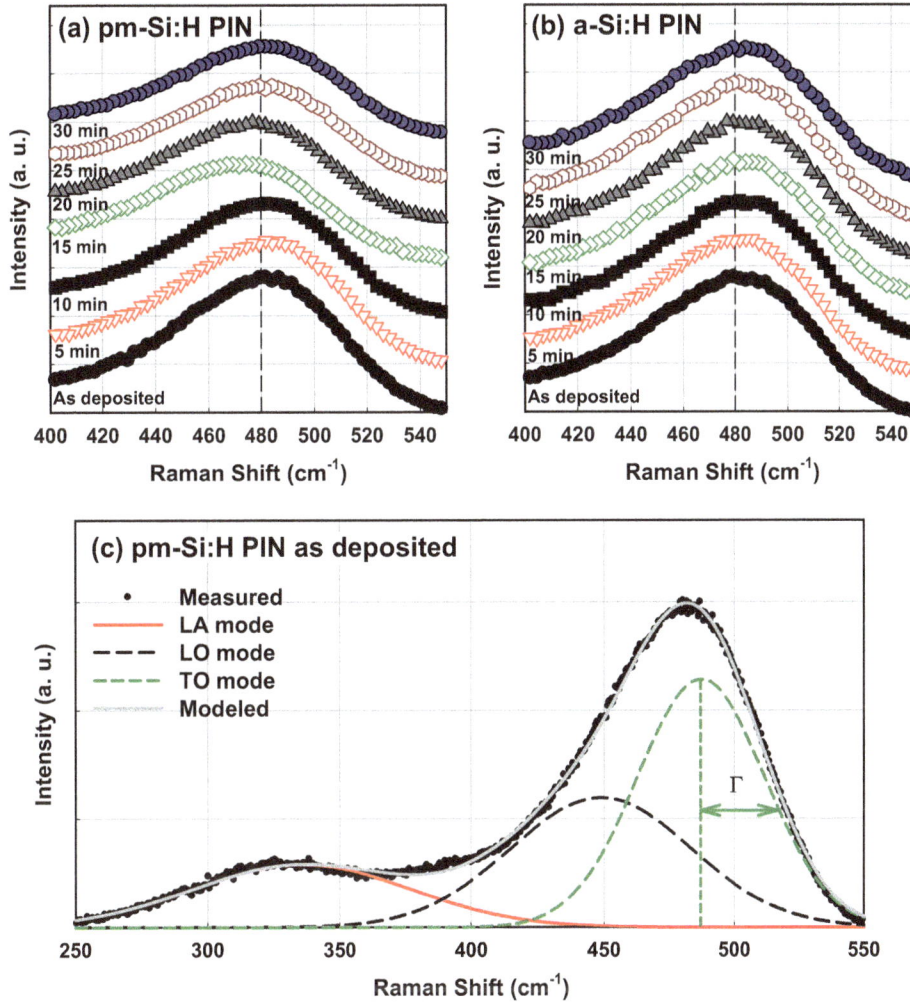

Fig. 2. Raman spectra evolution during light-soaking (LS) of (a) pm-Si:H, (b) a-Si:H PIN solar cells and (c) Raman spectrum of as-deposited pm-Si:H PIN solar cell and its deconvolution. Note that spectra evolution data are presented for LS times up to 30 min.

3.2 In-situ Raman studies

Figures 2a and 2b show the evolution during light-soaking of the Raman spectra of the a-Si:H and pm-Si:H PIN stacks on glass, for which the spectra have been normalized to the intensity of the TO mode. In contrast with a-Si:H PIN stack, the Raman spectra of the pm-Si:H PIN displays a peak shift during light-soaking, as well as a small broadening, and these two changes were observed in multiple samples. To quantify these changes, TO peak position and RMS bond angle deviation, $\delta\theta$, were determined through deconvolution of the measured spectra. Figure 2c shows the Raman spectrum of as-deposited pm-Si:H PIN stack and its deconvolution. The Raman spectra of both pm-Si:H and a-Si:H consist of a broad peak located at around 480 cm^{-1}, which corresponds to the TO mode in the amorphous phase, and the LO and LA modes are additionally observed at lower wave numbers around 330 cm^{-1} and 440 cm^{-1}, respectively [24–26].

The half width of the TO band (Γ) is a sensitive measure of local disorder, and correlates with rms deviations

in the bond angle ($\delta\theta$) from the ideal tetrahedral bond angle of 109.4°. The shift of the peak position can be treated as a measure of the average strain resulting from built in stress. To minimize the influence of the weak vibrational modes, LA and LO, one can perform a best fit on each spectrum from 250 to 550 cm^{-1} by three Gaussian curves and a straight base line [24]. The RMS bond angle deviation was deduced using the following formula, proposed by Beeman et al. [27, 28].

$$\Gamma = 15 + 6\delta\theta.$$

Figure 3 shows the TO peak position and $\delta\theta$ change during light-soaking. One can clearly see that the TO peak of pm-Si:H PIN shifts towards lower wavenumbers during light-soaking, whereas the two a-Si:H PINs show no change. Since the shift of the Raman peak towards lower wavenumbers could be due to local heating from laser excitation, the pm-Si:H PIN was allowed to cool in the dark for 20 min after light-soaking. However, the peak position remained same, as shown in Figure 3a. Physically, a phonon is a quantized mode of vibration occurring in a

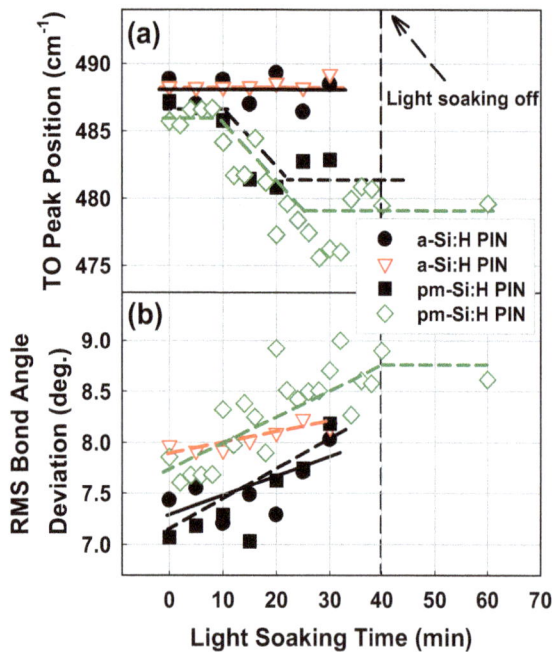

Fig. 3. Evolution of (a) TO peak position and (b) RMS bond angle deviation as deduced from Raman scattering measurements during light-soaking. On some samples we have checked that these material properties did not change after turning off the light.

rigid crystal lattice, such as the atomic lattice of a solid. The Raman spectra peak shift towards lower k can be regarded as a consequence of increasing the inter-atomic distance, that is, tensile stress. On the other hand, it should be noted that for all the samples, the TO band position in the as-deposited state is above at 480 cm^{-1} which can be related to compressive stress in the as-deposited films. However, while the peak position does not change for the a-Si:H PIN stack, it shifts to low wavenumbers for the pm-Si:H case during light-soaking. It can therefore be stated that the cells are under high compressive stress, as is usually the case for plasma deposited silicon thin films [29]. In other words, the peak position evolution from higher k to lower k (towards 480 cm^{-1}) suggests a relaxation of compressive stress during light-soaking for the pm-Si:H films.

3.3 In-situ AFM studies

Figures 4a−4d show examples of AFM images taken in both the as-deposited and light-soaked states for both a-Si:H and pm-Si:H PIN stacks on glass. For both a-Si:H and pm-Si:H PIN layer stacks we observe a grain structure related to the surface roughness of the samples. Indeed, based on the Raman spectra shown in Figure 2, the grains are amorphous. The Watershed image processing segmentation algorithm was used to extract information about the surface grain area [30, 31] by splitting the images into grains, based on the topology of the image. Figure 4e summarizes the surface grain size change during

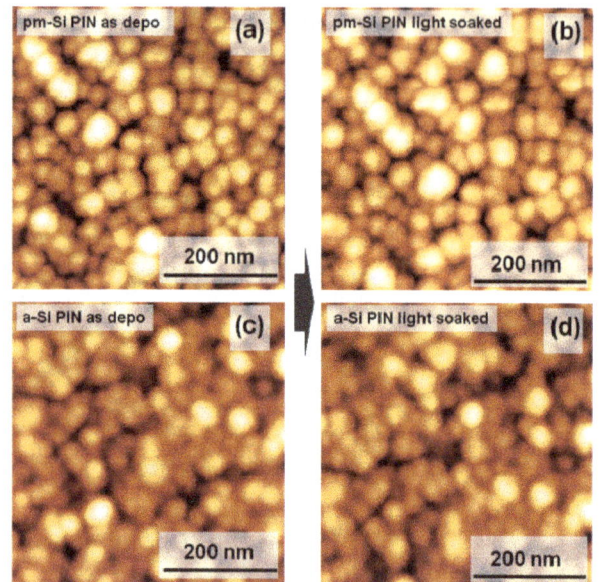

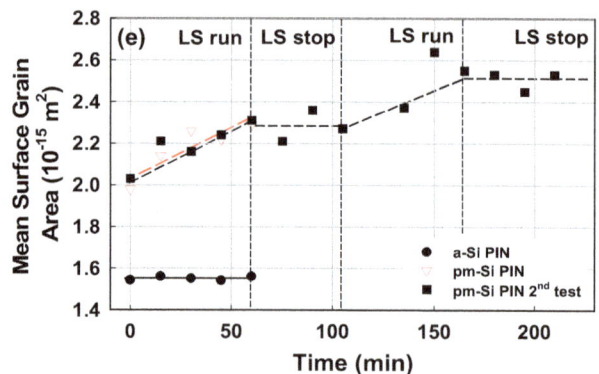

Fig. 4. AFM images of pm-Si:H and a-Si:H solar cells on Corning Eagle glass before (a, c) and after (b, d) light-soaking, and (e) mean surface grain area evolution during light-soaking, as extracted from AFM images. Note that no evolution is observed for the a-Si:H film, and that the evolution for the pm-Si:H cell stops when the light source is off.

light-soaking analyzed from surface grain extraction. One can see that the surface grain area of pm-Si:H PIN increases as a function of light-soaking time, and this expansion stops if light-soaking is stopped, which is to say that the surface expands while light-soaking occurs, and it remains expanded once light-soaking stops. For the a-Si:H PIN, no significant change was observed. As in the case of the Raman measurements, these results were reproduced in multiple trials.

To eliminate the effect of image drift on the AFM scan during light-soaking, individual grain sampling was performed on a selected set of grains. As shown in Figure 5a, a region was first chosen from the AFM images that remained within the field of measurement during the light-soaking, and eight surface-grains from within that region were analyzed. The surface of each grain was extracted by the Watershed algorithm. Figure 5a shows an example

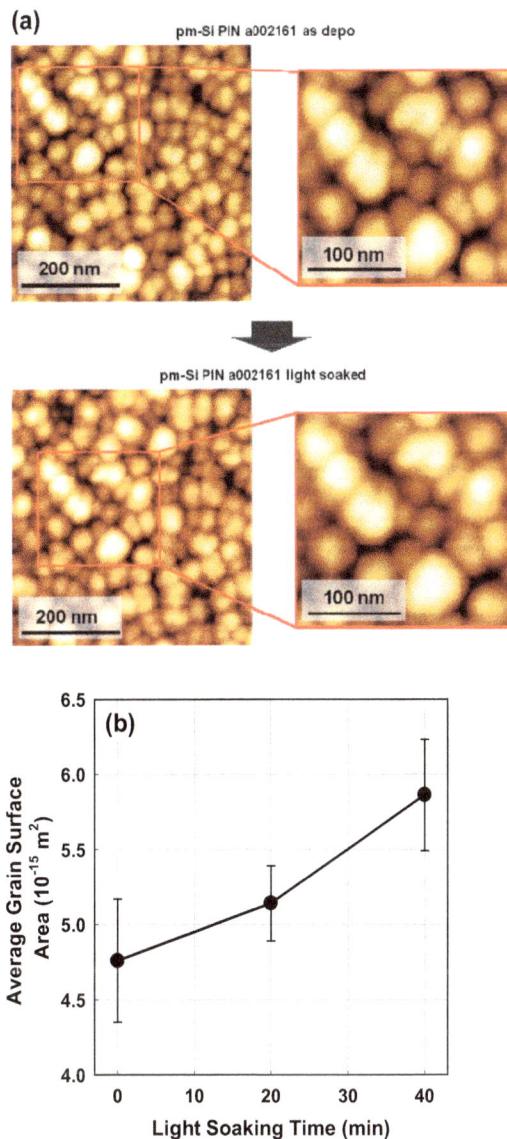

(a) pm-Si PIN a002161 as depo

pm-Si PIN a002161 light soaked

(b)

Fig. 5. (a) Demonstration of zone selection for AFM images to follow changes in area of individual grains, and (b) average surface area evolution of selected grains during light-soaking of pm-Si:H film.

from cropping a sample region, and Figure 5b shows the average area of the eight selected grains as a function of light-soaking time. In spite of the small sample size, the manual grain extraction shows the same trend as the statistical group behavior of surface grains in Figure 4e. Therefore, one can conclude that the "swelling" behavior of the surface grains is not due to an image drift effect, but due to light-soaking induced changes on the thin film topology.

3.4 Infrared absorption studies

Figure 6 shows the absorbance spectra in the stretching region of a-Si:H and pm-Si:H PIN stacks on FZ c-Si substrates extracted from infrared transmission. The stretching region of infrared absorption consists of two peaks centered at 2000 cm^{-1} and 2090 cm^{-1} [32]. The 2000 cm^{-1} mode is commonly associated with isolated Si-H groups in the bulk, related to the saturation of the dangling bonds during growth [33], while the 2090 cm^{-1} mode is attributed to clustered monohydrides and/or dihydrides. Furthermore, there is a third peak, which is detected in pm-Si:H PIN samples, centered at around 2030 cm^{-1}. The stretching mode at 2030 cm^{-1} has been interpreted as clustered Si-H groups in the form of platelet-like configurations [34]. Therefore, this component could come from Si-H bonds at the amorphous-crystalline interface at nanocrystals surface in pm-Si:H.

Infrared absorption behavior during light-soaking is quite distinct for those two materials. a-Si:H PIN shows slow but steady reduction of its stretching modes absorbance during light-soaking. It is believed that Si-H bonds are broken under illumination, creating dangling bonds. However, in pm-Si:H PIN case, there is an increase in its stretching modes absorption at the early stages (up to five hours) of light-soaking. The absorption starts to decrease after light-soaking for 20 h, but the absorption is still higher than as-deposited state. The result shows a good agreement with literature [35]. Few more tests were done for the reproducibility, and the initial absorption increase was found to be reproducible. Multiple infrared transmission spectra were taken to check the error range, which was found to be in ±0.1% of its absolute transmission. Thus the error range is much lower than the change in transmission caused by light-soaking, which is more than ±0.5%. Therefore, one can conclude that there is very small change in total amount of Si-H bonds, both in a-Si:H and pm-Si:H PIN layer stacks, but there is reproducible consistency in their behavior. That is, initial increase of infrared absorption band of pm-Si:H PIN layer stack contrary to a-Si:H, which shows monotonous decrease.

3.5 Hydrogen exodiffusion studies

Figure 7 shows the H_2 partial pressure detected during exodiffusion, normalized to the volume of the sample, as a function of temperature of pm-Si:H PIN layer stack on glass. The exodiffusion spectrum of the as-deposited pm-Si PIN shows two distinct peaks at around 350 °C and 500 °C along with hydrogen spikes at low temperature around 250 °C. The peak at 350 °C is associated with the presence of weakly bonded hydrogen, which characterizes pm-Si:H [34, 36], or it is also considered to be molecular hydrogen in the material [37, 38]. At 500 °C, hydrogen from isolated Si-H bonds effuses, leaving behind dangling bonds [39]. In addition, it should be noted that the low temperature peak at 350 °C can be also come from the fact that we are analyzing a PIN stack instead of just an intrinsic layer. As a matter of fact, it has been shown that there is a quite different effusion behavior between p/i and n/i layer stacks. The p/i structure shows the low temperature effusion peak at 300 °C, while n/i

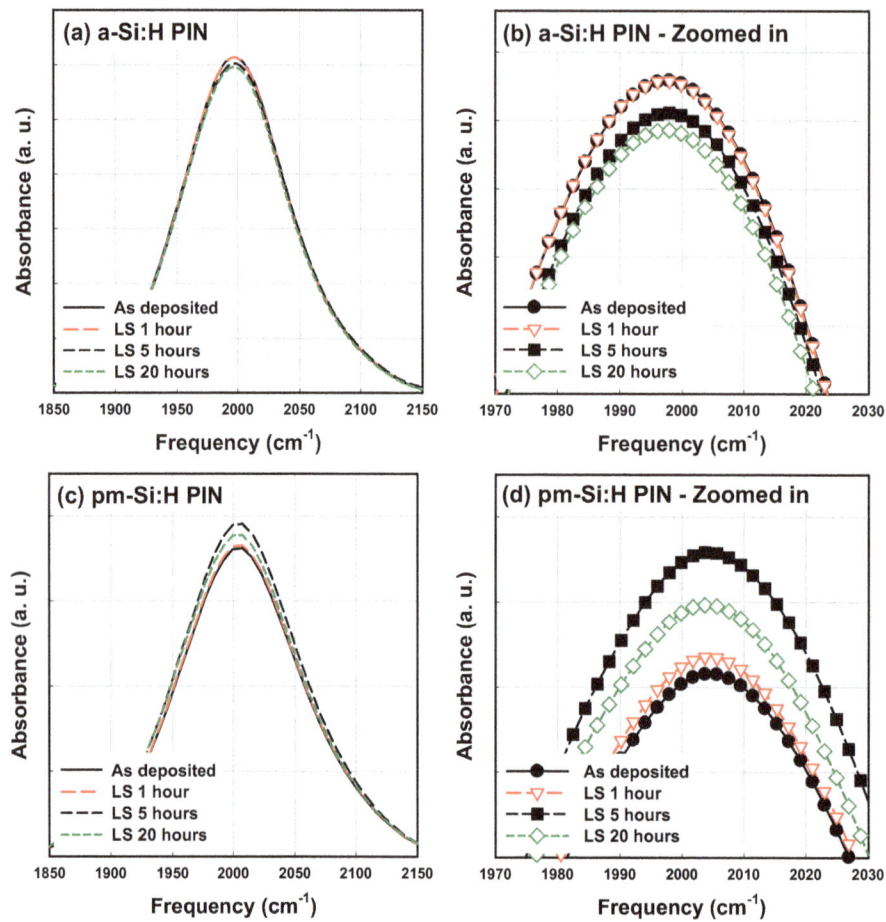

Fig. 6. Absorbance spectra of stretching modes of (a) a-Si:H and (c) pm-Si:H PIN layer stacks on intrinsic FZ wafer and their zoom-in to the peak (b, d). The spectra are extracted from infrared transmission at four light-soaking (LS) states : as-deposited (black), after one hour of LS (red), after 5 h (blue) and after 20 h (green).

structure shows shift of the effusion peak to higher temperature of 450 °C [40]. There are few possible reasons to have a low temperature peak in the structure with *p*-layer at bottom. Boron doped a-SiC:H layers are porous, and have a high diffusion coefficient of hydrogen [41], so atomic hydrogen could incorporate if another layer is deposited on top of the boron doped a-SiC:H layer. Therefore, molecular hydrogen could already exist in *p*-layer or *p*-layer/substrate interface. More detailed comprehensive view with the other results will be dealt in the discussion section.

The most striking result in Figure 7 is that the peak at 350 °C disappeared after light-soaking for one hour. This change is reproducible and is also observed on the sample light-soaked for 20 h. It is in fact surprising because not only the change is very abrupt, but also such change is much stronger than the change in infrared absorption study in previous section. Reminding that only bonded hydrogen is detected by infrared spectroscopy, the origin of the peak at 350 °C should be molecular hydrogen in microvoids or at *p*-layer/substrate interface. Furthermore, the signal at 400 °C increased after light-soaking. It is considered to be related to the initial increase of stretching modes absorbance in previous section (Fig. 6). It implies

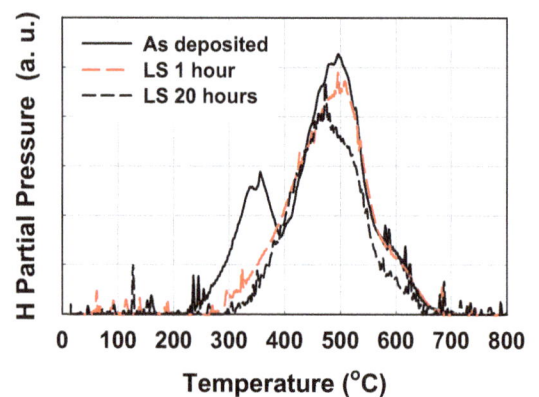

Fig. 7. Hydrogen exodiffusion results of co-deposited pm-Si:H PIN solar cells for three light-soaking (LS) states : as-deposited (black), after one hour of LS (red), and after 20 h (blue). Note the disappearance of the peaks around 350 °C even after 1 h of LS.

that a portion of molecular hydrogen in 350 °C peak may have converted into bonded hydrogen. After light-soaking for 20 h, the signal at 400 °C is still higher than its initial value, even if the high temperature signal (~500 °C) shows

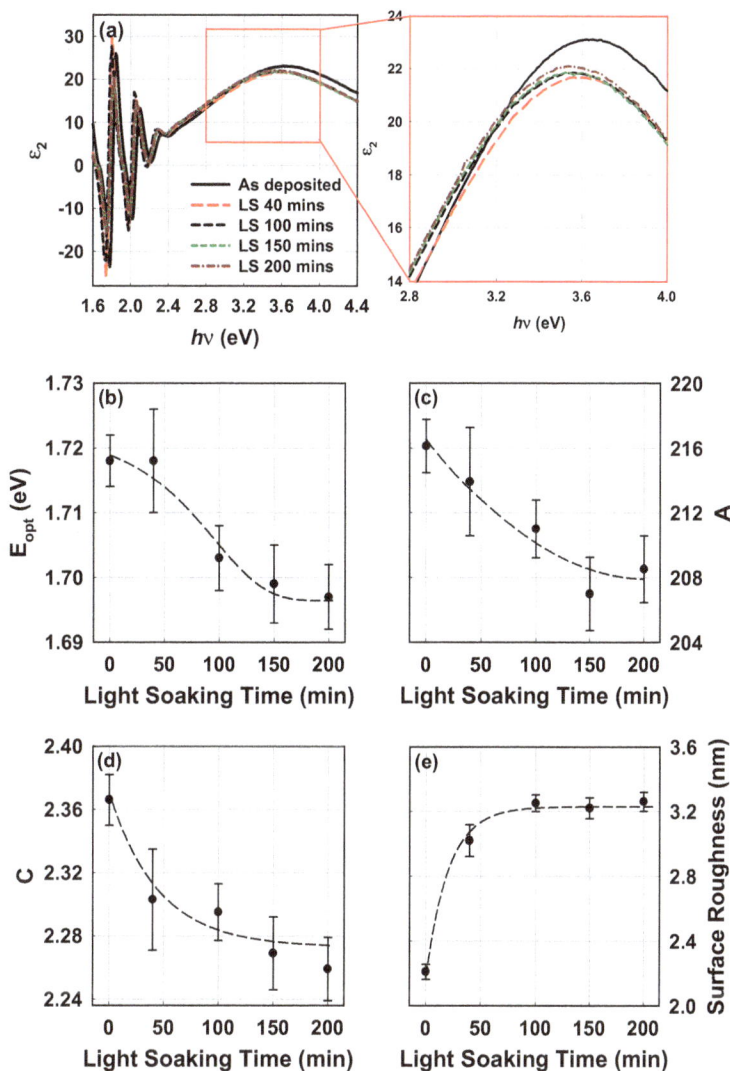

Fig. 8. Evolution of (a) pseudo-dielectric function of a pm-Si:H PIN layer stack on Corning Eagle glass at various stages of light-soaking, as measured by spectroscopic ellipsometry, and of material parameters as extracted from the measured spectra: (b) optical bandgap E_{opt}, (c) density parameter A, (d) disorder parameter C, and (e) surface roughness.

obvious reduction. In order to have more detailed information, the exodiffusion results were deconvoluted into five Gaussian peaks, and the results of such analysis are summarized in Table 1. The exodiffusion result of the as-deposited sample is characterized by five peaks at 320 °C, 353 °C, 469 °C, 516 °C and 591 °C. Note that the spikes were not taken into account in the deconvolution. Table 1 shows that the area of the peak at 460 °C increased after light-soaking for one hour, while the two low temperature peaks (320, 353 °C) disappeared. This gives support to the fact there is not only diffusion of molecular hydrogen, but also reconstruction of Si-H bonding during light-soaking [36], and molecular hydrogen should play an important role in this process [42].

Last but not least, unusual sharp spikes are observed at low temperature (as low as 50 °C) for the light-soaked samples. The detection of these low temperature spikes in the light-soaked samples suggests that a portion of molec-ular hydrogen (350 °C peak) is also transformed to be mobile during the light-soaking and accumulates at internal cavities or at the film/substrate interface. This hydrogen can easily escape during macroscopic cracking events and introduce sharp and rapid increases in hydrogen partial pressure at even lower temperature than 350 °C in the exodiffusion experiments [37].

3.6 Spectroscopic ellipsometry studies

Figure 8 shows the imaginary part of the pseudo-dielectric function of the film measured by spectroscopic ellipsometry, as well as the material parameters deduced from modeling the pm-Si:H PIN stack on glass. Let us recall that E_{opt} is the optical bandgap, whereas A and C are parameters related to the density and the disorder, respectively [7, 43]. It is interesting to note that most of

Table 1. Peak positions and integrated areas extracted from exodiffusion results of Figure 6. Data are presented for co-deposited pm-Si:H cell layer stacks after three light-soaking conditions (as-deposited, 1 h, and 20 h).

As-deposited		Light-soaked 1 h		Light-soaked 20 h	
Peak position (°C)	Area (a.u.)	Peak position (°C)	Area (a.u.)	Peak position (°C)	Area (a.u.)
320	2.43				
353	2.32				
469	9.06	464	12.1	457	9.44
516	2.99	507	2.59	509	1.98
590	2.23	596	1.18	584	1.35
Total	19.03		15.87		12.77

the changes occur during the initial stages of light-soaking, especially during first 100 min. The zoom of the high energy part of the ellipsometry spectra (Fig. 8a) shows that during the first 40 min of light-soaking there is a strong decrease in the amplitude of ε_2, which can be accounted by an increase of the surface roughness of the films (see Fig. 8e), consistent with published results [44] and AFM images (Figs. 4 and 5). At longer times, light-soaking mainly results in a shift of the absorption edge (low energy part in the zoom of ε_2) to lower energy. This shift is associated with the decrease of the E_{opt} (Fig. 8b). There are two possible reasons for the decrease in E_{opt}. It could refer that Si-H is breaking as a consequence of light-soaking, even if our infrared spectroscopy shows initial increase of Si-H bonding absorption, yet E_{opt} showed a bit delayed decrease comparing to the other material parameters. The other reason can be the inter-atomic distance. In-situ Raman and AFM revealed that light-soaking introduces stress-relaxation, which also implies that inter-atomic distance gets larger after light-soaking. Less splitting of orbital energy (band gap) is then expected from larger inter-nuclear distance of atoms. Decrease of the density parameter A suggests that the sample is getting more porous. Such porosity could connect the stress relaxation detected by Raman and AFM measurement to the hydrogen cavity formation in exodiffusion study (sharp low temperature peaks). Furthermore, increase in surface roughness could be also connected to the macroscopic change observed by AFM.

3.7 Long term changes in surface morphology for stacks on transparent conductive oxides

In the previous sections, changes in film morphology during light-soaking were observed for the films deposited on glass or crystalline silicon wafer. We turn now to light-soaking effects on the same PIN layer stacks deposited on transparent conductive oxide substrates. Figure 9 shows SEM and AFM images of the surface of pm-Si:H PINs on flat ZnO:Al both before and after light-soaking for 16 h. Images throughout the entire 1×1 square inch substrate were acquired on the as-deposited state and after light-soaking. While the as-deposited sample shows no clear feature except surface roughness (Fig. 9a), the light-soaked sample shows that the surface morphology is strongly modified (Fig. 9b). A large number of both dark

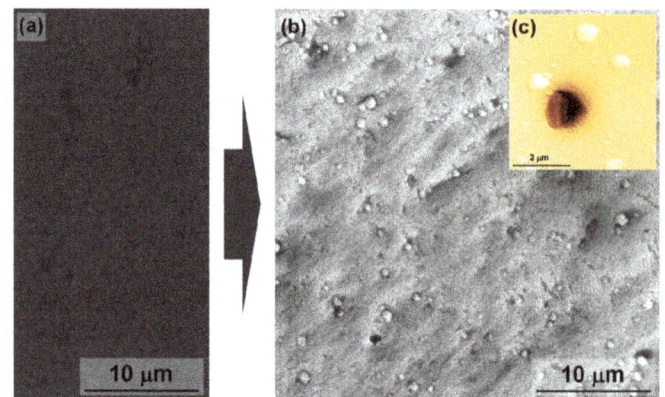

Fig. 9. SEM images of pm-Si:H PIN solar cell surface when deposited on flat ZnO:Al : (a) as-deposited, and (b) light-soaked (LS) for 16 h. (inset, c) AFM image of surface shown in (b).

objects and mound-like objects is observed and the formation of holes takes place all over the surface. The depth of the holes and height of the mounds are estimated from the cross section of the AFM images (Fig. 9c). The depth of the holes varies from 30 to 400 nm, which means that some of the holes correspond to a complete peeling off of the pm-Si:H PIN layer. The height of the mound-like objects is of the order of about 100 nm. For the mound like objects, it is clear that they are not dust or particles because of their surface topology is similar to that of the rest of the surface, as well as the fact that they are not seen in the as-deposited state. In addition, not only circular shaped holes, but also many irregular shaped holes are found in SEM images. It implies that the macroscopic changes are similar to the buckling observed when films are annealed at high temperature [34, 40, 45, 46].

Because the above results could be due to the use of flat ZnO:Al, we performed another light-soaking test on a pm-Si:H PIN layer stack deposited on a textured Asahi (SnO$_2$:F) substrate, which is usually used for PIN solar cells. The PIN layer stack was light-soaked for 100 hours, and again many irregularly shaped large holes were found. Figure 10 shows an AFM image of one of these holes. One can see that the size of the hole is larger ($\sim$20 μm) than the ones in Figure 9 ($\sim$few μm), and its depth corresponds to the film thickness (about 400 nm), suggesting that the film has completely peeled off. Furthermore, many small sized holes ($\sim$few μm) are also seen in Figure 10, similar to those observed on flat ZnO:Al (Fig. 9b). As mentioned

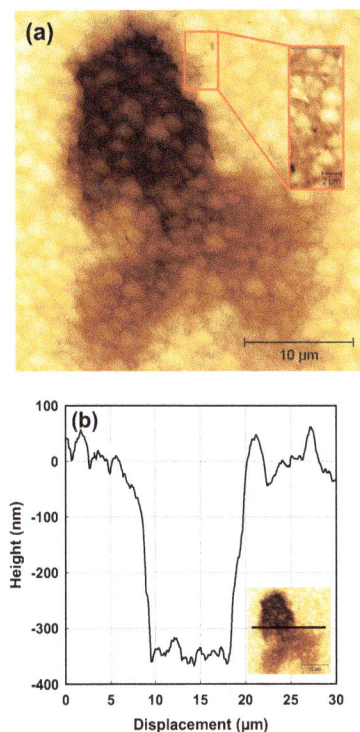

Fig. 10. (a) AFM image of a pm-Si:H PIN on textured Asahi substrate after 100 h of light-soaking and (b) cross-sectional profile measurement through center of hole showing a depth of almost 400 nm.

in the paragraph above, it is believed that the formation of those irregular shaped holes and mounds is linked to the relaxation of the high compressive stress in the as-deposited films.

4 Discussion

With the experimental results presented above, we now aim to understand the particular evolution of the pm-Si:H solar cell parameters (V_{oc}, J_{sc}) shown in Figure 1. In-situ Raman and AFM measurements reveal that light-soaking plastically relaxes mechanical stress. Light-soaking also introduces large structural changes resulting in the diffusion of molecular hydrogen, as deduced from infrared absorption, exodiffusion and spectroscopic ellipsometry studies. For a long exposure time, light-soaking leads to the formation of holes – where the film completely peels off – and mounds, which form with a height up to half of the film thickness.

One could argue that the presented results can be led by illumination induced local heating. However, three observations suggest this is not the case. First, heating of the PIN solar cells by illumination was suppressed by fan cooling. As mentioned above, the temperature stayed under 50 °C even at 2 suns of illumination (well below the deposition temperature of 175 °C). Second, we have observed increasing V_{oc} and decreasing J_{sc}. The opposite

would be seen if pm-Si:H cells were heated. At last, Raman measurements show that the TO peak remained in same position even after cooling down for 20 min. Therefore, we conclude that the macroscopic changes presented here do not rely on heating, but on illumination.

Based on our experimental study, the collected results can be summarized as follows:

1. Light-soaking introduces a fast decrease in J_{sc} and an increase of V_{oc} on pm-Si:H PIN solar cells (Fig. 1).
2. In-situ Raman measurements (Figs. 2 and 3) reveal that light-soaking relaxes the compressive stress in the as-deposited pm-Si:H PIN, and such relaxation is connected to surface expansion as detected by in-situ AFM measurements (Figs. 4 and 5).
3. Infrared absorption and exodiffusion studies (Figs. 6 and 7) demonstrate that light-soaking induces strong changes on the hydrogen distribution in the material. In particular, the disappearance of the exodiffusion peak at 350 °C, the increase in infrared stretching modes absorbance, as well as the increase in the exodiffusion signal at 400 °C and the arise of low temperature sharp spikes after light-soaking. These results suggest that molecular hydrogen in the pm-Si:H PIN stack (350 °C peak in exodiffusion) effuses out during light-soaking, and a portion of the effused hydrogen is converted into bonded hydrogen and into hydrogen filled cavities, in particular at the substrate/p-layer interface.
4. Spectroscopic ellipsometry studies (Fig. 8) also provide evidence of macroscopic changes of pm-Si:H PIN solar cell through the evolution of the density parameter A and the increase of surface roughness.
5. The SEM and AFM results (Figs. 9 and 10) reveal visible macroscopic changes due to light-soaking, which manifest through the formation of macroscopic holes and mounds.

These results lead us to postulate that light-soaking leads molecular hydrogen to effuse from pm-Si:H PIN solar cell, and a fraction of molecular hydrogen (made mobile during light-soaking) forms cavities, probably around the silicon nanocrystals of the pm-Si:H material and at the substrate/p-layer interface. This causes the mounding and peeling behavior observed in the microscope images. Therefore, the question is whether there is a connection between the macroscopic changes and the behavior of the solar cell parameters.

Indeed, there is a general consensus on light-soaking modifying hydrogen incorporations in a-Si:H. Light-induced, long-range H motion has been postulated as the key step in the Branz model of SWE defect formation [47, 48]. According to the Branz model, only a small subset of the Si-H bonds involved in the SWE contributes to metastable dangling bond formation, while most of the broken Si-H bonds are recycled by re-trapping mobile hydrogen. This process could be a reason for the molecular hydrogen diffusion during light-soaking [47–49]. We therefore postulate that the large structural changes reported above are macroscopic manifestations of molecular hydrogen accumulation at the interface between the substrate and the PIN stack.

However, there is still a question remaining about the fast diffusion of hydrogen. As a matter of fact, it is reported that hydrogen diffusion in a-Si:H is more moderate than the result presented here [37,48,49]. Nevertheless, one can still find the difference in the hydrogen diffusivity between pm-Si:H and a-Si:H, recalling the abrupt disappearance of exodiffusion peak at 350 °C (Fig. 7) while infrared absorption measurements show only small changes in Si-H bond concentration (Fig. 6). Therefore, such fast hydrogen diffusion cannot originate from Si-H bond breaking, but from molecular hydrogen in the material. As mentioned above, molecular hydrogen could exist either in microvoids or at the p-layer/substrate interface, and it could have effused out during light-soaking by stress relaxation observed in Raman and AFM studies (Figs. 2 and 3). It has been reported that molecular hydrogen can diffuse faster in a highly inhomogeneous material such as pm-Si:H [38]. In addition, such diffusion would be even accelerated if the material experiences a volume expansion, providing larger space to hydrogen. As a matter of fact, there has been a great deal of research on light-induced volume changes [50]. Various beam bending experiments have been performed using samples consisting of long and narrow pieces of thin glass or quartz substrates coated with a-Si:H [51–57]. These results reveal that light-induced volume expansion follows a stretched exponential behavior, usually showing saturation at $dV/V \sim 10^{-3}$ [29]. This phenomenon has also been connected with hydrogen motion, as light-induced volume changes have been shown to depend on the hydrogen content of the film [29]. "On-the-edge" or mixed phase materials grown under high hydrogen dilution and having improved order result in solar cells with significantly reduced light-induced degradation. Interestingly enough, these materials show the fastest and largest photo-expansion amongst many different a-Si:H materials [29,58]. Their large photo-induced volume change prompts comparison to the macroscopic evolution of our pm-Si:H solar cells. In other words, the volume expansion of the film is a product of the stress being relieved, and is particularly important at the p-layer/substrate interface.

Turning now to the different behavior of the a-Si:H and pm-Si:H solar cells (Fig. 1), one should also consider the different processing conditions of a-Si:H and pm-Si:H materials (notably vastly different hydrogen dilution and ion-bombardment conditions) which could induce different modification of the p-type a-SiC:H layer of the solar cells. Indeed, the deposition conditions of the pm-Si:H layer could strongly modify the hydrogen content and bonding configurations in the p-layer, even if the p-type a-SiC:H layer was deposited under the same conditions for both the a-Si:H and pm-Si:H PIN solar cells. As atomic hydrogen has a high diffusivity in boron doped a-SiC:H [41], it is even more likely that deposition conditions involving high hydrogen dilution will modify the p-type a-SiC:H layer and weaken the interface between the substrate and the p doped layer. The fact that we study PIN stacks and not

just intrinsic layers may also be the reason for the exodiffusion peak at 350 °C (Fig. 7) it has been reported elsewhere [40]. Therefore, we suggest that pm-Si:H deposition conditions may lead to a fragile p-layer/substrate interface, which can be further damaged during light-soaking of the solar cells. This is supported by the low temperature sharp spikes in the exodiffusion result (Fig. 7). Indeed, those sharp spikes evidence molecular hydrogen escaping from cavities at relatively low temperature. The hydrogen evolving from these cavities must have been accumulated during light-soaking. Even if AFM and SEM results in Figure 9 allow us to detect significant macroscopic changes only after quite long light-soaking times (16 hrs), small macroscopic changes could already have taken place at the early stages of light-soaking, as supported by the in-situ Raman and AFM studies shown in Figures 3 and 4.

Another interesting aspect of the macroscopic defect creation can be seen in Figure 10, wherein groups of many small holes are found in the AFM image along with larger ones. The size distribution of the holes shows a high density of small ones ($\sim$few μm), and few larger ones (as large as 20 μm). Considering the fact that small holes are locally grouped, the origin of the large hole might be related to the grouping of small ones. Assuming that the holes are a result of local stress relief, a region where small holes are grouped is most likely less rigid than nearby areas, and when those areas crack and peel-off during light-soaking, compressive stress is relaxed. Large cracked holes could be created through such stress release-crack cycle while more areas are cracked and peeled-off.

The initial infrared absorption increase of pm-Si:H PIN during light-soaking can be explained in the framework of Branz model, where mobile hydrogen is emitted by photocarrier recombination, creating dangling bond. Created mobile hydrogen travels around in the material, and it is captured by another dangling bond and recycled into another Si-H bond [59]. In this cycle, re-trapped hydrogen is assumed to originate from Si-H bond breaking. However, if a large amount of mobile hydrogen exists in the material (for instance molecular hydrogen), and then the number of hydrogen atoms captured by dangling bonds can be larger than the number of hydrogen atoms created by Si-H bond-breaking. The presence of molecular hydrogen in pm-Si:H PIN solar cells, suggested by the exodiffusion spectra (Fig. 7) may cause the fast hydrogen diffusion and initial increase in Si-H bonding (Figs. 6, 7) during light-soaking. Indeed, our result on hydrogen diffusion after light-soaking of pm-Si:H PIN solar cells is found to be much more active than that of a-Si:H in literature [60]. Such an active hydrogen evolution could be also explained by the existence of molecular hydrogen. Furthermore, a multi-body motion model has been proposed in which light-induced molecular hydrogen formation mediated by (another) interstitial molecular hydrogen [42]. In the model, atomic hydrogen comes to passivate a newly created dangling bond left by another atomic hydrogen parted from bond centered position, and two hydrogen atoms form a molecular hydrogen or interstitial hydrogen. This mechanism of bond centered hydrogen creation and

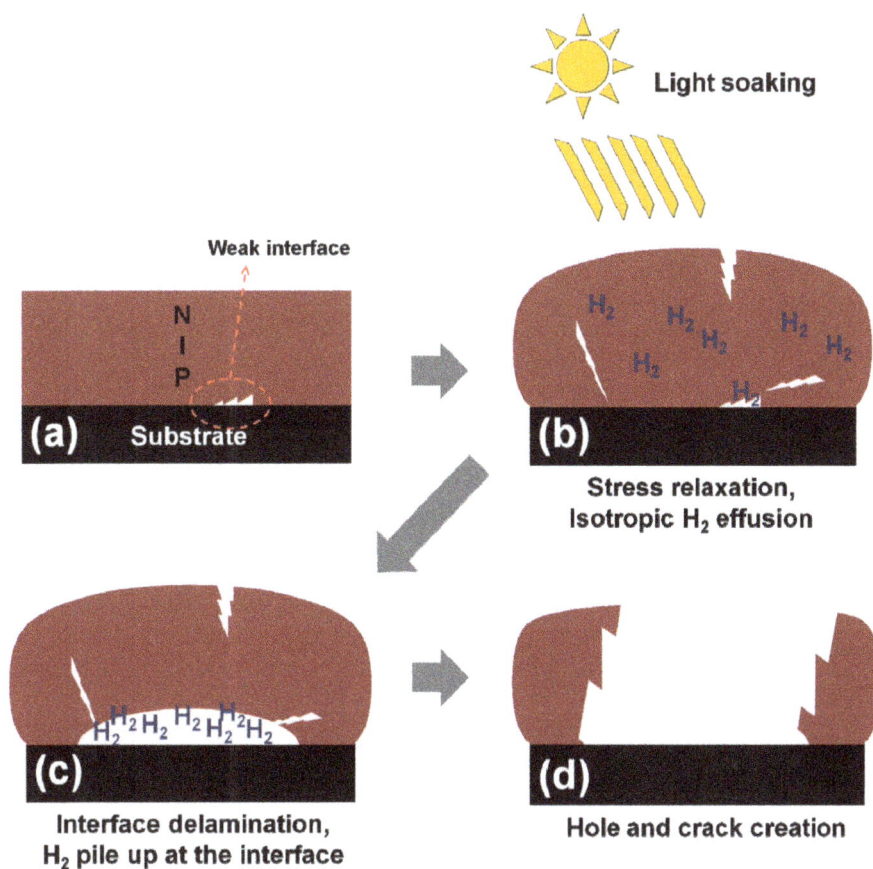

Fig. 11. Diagrammatical representation of proposed mechanism for structural defect formation through localized delamination. See text for description.

annihilation at the very beginning of the light-soaking process indicates that bond centered hydrogen could play the role as an intermediate precursor [35]. It may seem difficult for molecular hydrogen to spontaneously dissociate into atomic hydrogen. Indeed, molecular hydrogen in free space is more stable than Si-H bond. However, the energy level of molecular hydrogen in silicon is much less stable than in free space [61,62]. Moreover, the existence of not only molecular hydrogen, but also diatomic spices in silicon has been reported [63] and recent works have reported on strain-induced dissociation of molecular hydrogen, in particular at vicinity of strained Si-Si bonds [64,65]. As a matter of fact, we have shown that light-soaking on pm-Si:H PIN solar cells is accompanied by a large amount of stress relaxation, as seen in Figures 2–5.

The structural studies can be gathered to propose the macroscopic defect creation scenario shown in Figure 11. In the as-deposited state (Fig. 11a), the pm-Si:H solar cell has a weak film/substrate interface due to the process conditions of the pm-Si:H intrinsic layer. Then, light-soaking (Fig. 11b) introduces volume changes and stress relaxation, molecular hydrogen in the material becomes mobile and isotropically diffuses to the film free surface as well as to the interface with the substrate where it can accumulate and form hydrogen cavities (Fig. 11c). This delamination continues to release built-in "compressive" stress of the

film, and the stress relaxation will eventually result in the formation of mechanical defects such as cracks, mounds and holes (Fig. 11d), as well as the accumulation of hydrogen at the interface. Based on the above scenario we propose that the unusual evolution of pm-Si:H solar cell parameters, particularly rapid J_{sc} decrease (Fig. 1), is related to the diffusion of molecular hydrogen, resulting in a reduction of active solar cell area due to delamination. However such delamination can hardly account for the observed increase in V_{oc}, which in the literature has been related to dissociation of B-H complexes [18], hydrogen evolution from nanocrystals [21], and light-induced defect states re-distribution at p/i interface [22].

At last, it should be mentioned that a-Si:H also shows large macroscopic structural changes when annealed at 350 °C [45,46] and irreversible solar cell degradation when light-soaked under 50 suns at 130 °C [66,67]. These extreme conditions lead to enhanced hydrogen motion at the origin of the reported macroscopic structural changes in both films and a-Si:H solar cells. Interestingly, our results on pm-Si:H solar cells show that such irreversible changes can take in this material under standard light-soaking conditions. We attribute this to the peculiar nanostructure of pm-Si:H which provides good environment for hydrogen diffusion [36], and to the presence of weakly bonded hydrogen in the material (see Fig. 7). Therefore, one can conclude that the reported degradation kinetics of pm-Si:H PIN solar cells is an extreme case of a-Si:H solar cells, which also show hydrogen induced structural changes under extreme conditions.

5 Summary and conclusions

We have performed a detailed study on light-induced changes in a-Si:H and pm-Si:H PIN solar cells deposited on various substrates and characterized these changes by a wide range of techniques. We have observed that light-soaking induces an increase of V_{oc} and a fast drop of J_{sc} in pm-Si:H PIN solar cells. The changes in solar cell parameters are correlated to changes in hydrogen incorporation and the structural properties of the material, such as stress relaxation. The experimental results support the hypothesis that light-soaking results in the formation of cavities at the interface between the substrate and the p-layer which are progressively filled by molecular hydrogen during light-soaking. This process weakens this interface and causes mechanical defects such as partial delamination, cracks and holes. This is an often forgotten aspect of the so-called SWE. Indeed our results show that besides the creation of electronic defects, macroscopic defects related to the delamination of the SnO_2/p-layer interface and H_2 diffusion should be considered. This is particularly true in the case of pm-Si:H solar cells, for which a fast initial drop in J_{sc} has been explained by the delamination of the interface at the SnO_2/p-layer interface.

This work was performed in the TOTAL-LPICM Joint PV Research Team.

References

1. D.L. Staebler, C.R. Wronski, Appl. Phys. Lett. **31**, 292 (1977)
2. M. Vanecek, O. Babchenko, A. Purkrt, J. Holovsky, N. Neykova, A. Poruba, Z. Remes, J. Meier, U. Kroll, Appl. Phys. Lett. **98**, 163503 (2011)
3. J.Y. Ahn, K.H. Jun, K.S. Lim, M. Konagai, Appl. Phys. Lett. **82**, 1718 (2003)
4. P. Roca i Cabarrocas, P. St'ahel, S. Hamma, Y. Poissant, *Proc. 2nd World Conference on Photovoltaic Solar Energy Conversion*, p. 355 (Vienna, Autralia, 1998)
5. P. Roca i Cabarrocas, A. Fontcuberta i Morral, Y. Poissant, Thin Solid Films **403-404**, 39 (2002)
6. P. Roca i Cabarrocas, J. Non-Cryst. Solids **266–269**, 31 (2000)
7. A. Fontcuberta i Morral, P. Roca i Cabarrocas, C. Clerc, Phys. Rev. B **69**, 125307 (2004)
8. Y.M. Soro, A. Abramov, M.E. Gueunier-Farret, E.V. Johnson, C. Longeaud, P. Roca i Cabarrocas, J.P. Kleider, J. Non-Cryst. Solids **354**, 2092 (2008)
9. M. Meaudre, R. Meaudre, R. Butte, S. Vignoli, C. Longeaud, J.P. Kleider, P. Roca i Cabarrocas, J. Appl. Phys. **86**, 946 (1999)
10. Y. Poissant, P. Chatterjee, P. Roca i Cabarrocas, J. Non-Cryst. Solids **299–302**, 1173 (2002)
11. M. Brinza, G.J. Adriaenssens, A. Abramov, P. Roca i Cabarrocas, Thin Solid Films **515**, 7504 (2007)
12. P. Roca i Cabarrocas, J.B. Chevrier, J. Huc, A. Lloret, J.Y. Parey, J.P.M. Schmitt, J. Vac. Sci. Technol. A **9**, 2331 (1991)
13. J. Jang, S.Y. Lee, Appl. Phys. Lett. **52**, 1401 (1988)
14. L. Yang, L. Chen, Mater. Res. Soc. Symp. Proc. **336**, 669 (1994)
15. P. Siamchai, M. Konagai, Appl. Phys. Lett. **67**, 3468 (1995)
16. M. Isomura, H. Yamamoto, M. Kondo, A. Matsuda, *Proceedings of the Second World Conference and Exhibition on Photovoltaic Solar Energy Conversion, 6–10 July 1998* (European Commission, Vienna, 1998), p. 925
17. C. Longeaud, J.P. Kleider, M. Gauthier, R. Brüggemann, Y. Poisant, P. Roca i Cabarrocas, Mater. Res. Soc. Symp. Proc. **557**, 501 (1999)
18. P. St'ahel, P. Roca i Cabarrocas, P. Sladek, M.L. Theye, MRS Symp. Proc. Ser. **507**, 649 (1998)
19. M. Isomura, M. Kondo, A. Matsuda, Jpn J. Appl. Phys. **39**, 3339 (2000)
20. K. Lord, B. Yan, J. Yang, S. Guha, Appl. Phys. Lett. **79**, 3800 (2001)
21. J. Yang, K. Lord, B. Yan, A. Banerjee, S. Guha, Mater. Res. Soc. Symp. Proc. **715**, A26.1.1 (2002)
22. E.V. Johnson, F. Dadouche, M.E. Gueunier-Farret, J.P. Kleider, P. Roca i Cabarrocas, Phys. Stat. Sol. A **207**, 691 (2010)
23. V. Nadazdy, M. Zeman, Phys. Rev. B **69**, 165213 (2004)
24. Y. Wang, O. Matsuda, T. Serikawa, K. Murase, J. Phys. IV **10**, Pr7-259 (2000)
25. J.M. Owens, D. Han, B. Yan, J. Yang, K. Lord, S. Guha, Mat. Res. Soc. Symp. Proc. **762**, A4.5.1 (2003)
26. D.M. Bhusari, A.S. Kumbhar, S.T. Kshirsagar, Phys. Rev. B **47**, 6460 (1993)
27. D. Beeman, R. Tsu, M.F. Thorpe, Phys. Rev. B **32**, 874 (1985)
28. P. Roura, J. Farjas, P. Roca i Cabarrocas, J. Appl. Phys. **104**, 073521 (2008)
29. E. Stratakis, E. Spanakis, P. Tzanetakis, H. Fritzsche, S. Guha, J. Yang, Appl. Phys. Lett. **80**, 1734 (2002)
30. L. Vincent, P. Soillev, IEEE Trans. Pattern Anal. Mach. Intell. **13**, 583 (1991)
31. S. Beucher, F. Meyer, *The watershed transformation in Mathematical Morphology in Image Processing*, edited by E.R. Dougherty (Marcel Dekker, New York, 1992), pp. 433–481
32. M.H. Brodsky, M. Cardona, J.J. Cuomo, Phys. Rev. B **16**, 3556 (1977)
33. W.B. Jackson, C.C. Tsai, Phys. Rev. B **45**, 6564 (1992)
34. S. Lebib, P. Roca i Cabarrocas, Eur. Phys. J. Appl. Phys. **26**, 17 (2004)
35. R. Darwich, P. Roca i Cabarrocas, S. Vallon, R. Ossikovski, P. Morin, K. Zellamam, Philos. Mag. B **72**, 363 (1995)
36. F. Kail, S. Fellah, A. Abramov, A. Hadjadj, P. Roca i Cabarrocas, J. Non-Cryst. Solids **352**, 1083 (2006)
37. A.H. Mahan, W. Beyer, B.L. Williamson, J. Yang, S. Guha, Philos. Mag. Lett. **80**, 647 (2000)

38. F. Kail, J. Farjas, P. Roura, P. Roca i Cabarrocas, Phys. Rev. B **80**, 073202 (2009)

39. J. Zhou, M. Kumeda, T. Shimizu, J. Non-Cryst. Solids **195**, 76 (1996)

40. N. Pham, Y. Djeridane, A. Abramov, A. Hadjadj, P. Roca i Cabarrocas, Mat. Sci. Eng. B **159-160**, 27 (2009)

41. F. Kail, A. Hadjadj, P. Roca i Cabarrocas, Thin Solid Films **487**, 126 (2005)

42. C. Longeaud, D. Roy, O. Saadane, Phys. Rev. B **65**, 085206 (2002)

43. G.E. Jellison, F.A. Modine Jr., Appl. Phys. Lett. **69**, 371 (1996)

44. P. Agarwal, A. Srivastava, D. Deva, J. Appl. Phys. **101**, 083504 (2007)

45. H.R. Shanks, C.J. Fang, L. Ley, M. Cardona, F.J. Demond, S. Kalbitzer, Phys. Stat. Sol. B **100**, 43 (1980)

46. H.R. Shanks, L. Ley, J. Appl. Phys. **52**, 811 (1981)

47. H.M. Branz, Phys. Rev. B **59**, 5498 (1999)

48. H.M. Cheong, S.H. Lee, B. Nelson, A. Mascharenas, S.K. Deb, Appl. Phys. Lett. **77**, 2686 (2000)

49. H.M. Branz, S. Asher, H. Gleskova, S. Wagner, Phys. Rev. B **59**, 5513 (1999)

50. P. Tzanetakis, Sol. Eng. Mater. Sol. Cells **78**, 369 (2003)

51. T. Hatano, Y. Nakae, H. Mori, K. Ohkado, N. Yoshida, S. Nonomura, M. Itoh, A. Masuda, H. Matsumura, Thin Solid Films **395**, 84 (2001)

52. S. Nonomura, N. Yoshida, T. Gotoh, T. Sakamoto, M. Kondo, A. Matsuda, S. Nitta. J. Non-Cryst. Solids **266-269**, 474 (2000)

53. T. Sakamoto, N. Yoshida, H. Harada, T. Kishida, S. Nonomura, T. Gotoh, M. Kondo, A. Matsuda, T. Itoh, S. Nitta, J. Non-Cryst. Solids **266-269**, 481 (2000)

54. K. Shimizu, T. Tabuchi, K. Hattori, H. Kida, H. Okamoto, Mater. Res. Soc. Symp. Proc. **507**, 735 (1998)

55. T. Gotoh, S. Nonomura, M. Nishio, N. Masui, S. Nitta, M. Kondo, A. Matsuda. J. Non-Cryst. Solids **227-230**, 263 (1998)

56. T. Gotoh, S. Nonomura, M. Nishio, S. Nitta, M. Kondo, A. Matsuda, Appl. Phys. Lett. **72**, 2978 (1998)

57. N. Yoshida, Y. Shobajima, H. Kamiguchi, T. Ida, T. Hatano, H. Mori, Y. Nakae, M. Itoh, A. Masuda, H. Matsumura, S. Nonomura, J. Non-Cryst. Solids **299-302**, 516 (2002)

58. H. Fritzsche, Annu. Rev. Mater. Res. **31**, 47 (2001)

59. H.M. Branz, Sol. Eng. Mater. Sol. Cells **78**, 425 (2003)

60. P.V. Santos, N.M. Johnson, R.A. Street, Phys. Rev. Lett. **67**, 2686 (1991)

61. W.B. Jackson, Curr. Opin. Solid State Mater. Sci. **1**, 562 (1996)

62. C.G. Van de Walle, Phys. Rev. B **49**, 4579 (1994)

63. K.J. Chang, D.J. Chadi, Phys. Rev. Lett. **62**, 937 (1989)

64. S.K. Estreicher, J.L. Hastings, P.A. Fedders, Phys. Rev. B **57**, R12663 (1988)

65. S.K. Estreicher, J.L. Hastings, P.A. Fedders, Appl. Phys. Lett. **70**, 432 (1997)

66. D.E. Carlson, K. Rajan, Appl. Phys. Lett. **68**, 28 (1996)

67. D.E. Carlson, K. Rajan, Appl. Phys. Lett. **69**, 1447 (1996)

14

Quo Vadis photovoltaics 2011

A. Jäger-Waldau[a]

European Commission, Joint Research Centre; Renewable Energy Unit, via E. Fermi 2749, TP 450, 21027 Ispra (VA), Italy

Abstract Since more than 10 years photovoltaics is one of the most dynamic industries with growth rates well beyond 40% per annum. This growth is driven not only by the progress in materials knowledge and processing technology, but also by market introduction programmes in many countries around the world. Despite the negative impacts on the economy by the financial crisis since 2009, photovoltaics is still growing at an extraordinary pace and had in 2010 an extraordinary success, as both production and markets doubled. The open question is what will happen in 2011 and the years after as the situation is dominated by huge manufacturing overcapacities and an increasing unpredictability of policy support. How can the PV industry continue their cost reduction to ensure another 10 to 20 years of sustained and strong growth necessary to make PV to one of the main pillars of a sustainable energy supply in 2030. Despite the fact, that globally the share of electricity from photovoltaic systems is still small, at local level it can be already now above 30% of the demand at certain times of the year. Future research in PV has to provide intelligent solutions not only on the solar cell alone, but also on the module and the system integration level in order to permit a 5 to 10% share of electricity in 2020.

1 Introduction

Production data for the global cell production[1] in 2010 vary between 18 GW and 27 GW. The significant uncertainty in the data for 2010 is due to the very competitive market environment, as well as the fact that some companies report shipment figures, others report sales and again other report production figures. In addition, the difficult economic conditions and increased competition led to a decreased willingness to report confidential company data. The previous tight silicon supply situation reversed due to massive production expansions as well as the economic situation. This led to a price decrease from the 2008 peak of around 500 \$/kg to about 50−55 \$/kg at the end of 2009 with a slight upwards tendency throughout 2010 and early 2011.

The presented data, collected from various companies and colleagues were compared to various data sources and thus led to an estimate of 23.5 GW (Fig. 1), representing a doubling of production compared to 2009.

Since 2000, total PV production increased almost by two orders of magnitude, with annual growth rates be-

tween 40% and 80%. The most rapid growth in annual production over the last five years could be observed in Asia, where China and Taiwan together now account for more than 50% of world-wide production.

The change of the market from a supply restricted – to a demand-driven market and the resulting overcapacity for solar modules has resulted in a dramatic price reduction of more than 50% over the last three years. Especially companies in their start-up and expansion phase with limited financial resources and restricted access to capital are struggling in the current market environment. This situation is believed to continue for at least the next few years and put further pressure on the reduction of the average selling prices (ASP). The recent financial crisis added pressure as it resulted in higher government bond yields and ASPs have to decline even faster than previously expected to allow for higher project internal rate of returns (IRRs). On the other hand, the rapid declining module and system prices open new markets, which offer the perspectives for further growth of the industry – at least for those companies with the capability to expand and reduce their costs at the same pace.

For the third year in a row solar power attracted behind wind the second largest amount of new investments into renewable energies [1]. Business analysts are confident that the industry fundamentals as a whole remain strong and that the overall photovoltaics sector will continue to experience a significant long-term growth. Following the stock market decline, as a result of the financial

---

[a] e-mail: `arnulf.jaeger-waldau@ec.europa.eu`

[1] Solar cell production capacities mean: in the case of wafer silicon based solar cells only the cells; in the case of thin-films, the complete integrated module; only those companies which actually produce the active circuit (solar cell) are counted; companies which purchase these circuits and make cells are not counted.

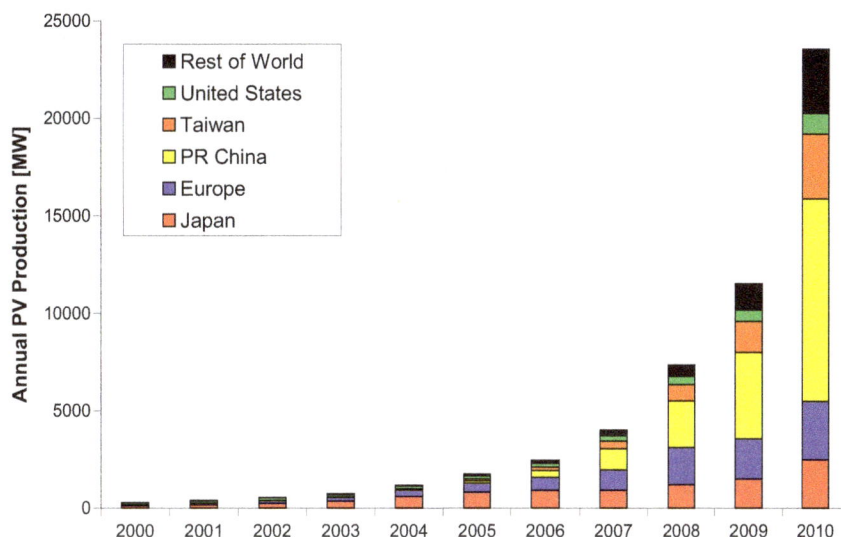

Fig. 1. World PV Cell/Module Production from 2000 to 2010 (data source: Navigant [2], Photon Magazine [3], PV News [4] and own analysis).

turmoil, the PPVX[2] (Photon Pholtovoltaic Stock Index) declined from its high at over 6500 points at the beginning of 2008 to 2095 points at the end of 2008. At the beginning of April 2011 the index stood at 2571 points and the market capitalisation of the 30-PPVX companies[3] was €43.5 billion.

Market predictions for the 2011 PV market vary between 17.3 GW by the Navigant Consulting conservative scenario [5] 19.6 GW by Macquarie [6] and 24.9 GW by iSuppli [7] with a consensus value in the 18 to 19 GW range. Massive capacity increases are underway or announced and if all of them are realised, the world-wide production capacity for solar cells would exceed 50 GW at the end of 2011. This indicates that even with the optimistic market growth expectations, the planned capacity increases are way above the market growth. The consequence would be the continuation of the low utilisation rates and therefore a continued price pressure in an oversupplied market. Such a development will accelerate the consolidation of the photovoltaics industry and spur more mergers and acquisitions.

The current solar cell technologies are well established and provide a reliable product, with sufficient efficiency and guaranteed energy output for at least 25 years of guaranteed power output. This reliability, the increasing potential of electricity interruption from grid overloads, as well as the rise of electricity prices from conventional energy sources, add to the attractiveness of Photovoltaic systems.

About 80% of the current production uses wafer-based crystalline silicon technology. A major advantage of this technology is that complete production lines can be bought, installed and be up and producing within a relatively short time-frame. This predictable production start-up scenario constitutes a low-risk placement with calculable return on investments. However, the temporary shortage in silicon feedstock and the market entry of companies offering turn-key production lines for thin-film solar cells led to a massive expansion of investments into thin-film capacities between 2005 and 2010. More than 200 companies are involved in the thin-film solar cell production process ranging from R&D activities to major manufacturing plants.

Projected silicon production capacities available for solar in 2012 vary between 140 000 metric tons from established polysilicon producers, up to 185 000 metric tons, and including the new producers up to 250 000 metric tons [8,9]. The possible solar cell production will in addition depend on the material use per Wp. Material consumption could decrease from the current 8 g/Wp to 7 g/Wp or even 6 g/Wp, but this might not be achieved by all manufacturers.

Similar to other technology areas, new products will enter the market, enabling further cost reduction. Concentrating Photovoltaics (CPV) is an emerging market. There are two main tracks – either high concentration >300 suns (HCPV) or low to medium concentration with a concentration factor of 2 to approx. 300. In order to maximise the benefits of CPV, the technology requires high Direct Normal Irradiation (DNI) and these areas have a limited geographical range – the "Sun Belt" of the Earth. The market share of CPV is still small, but an increasing number of companies are focusing on CPV. In 2008 about 10 MW of CPV were produced, market estimates for 2010 are in the 10 to 20 MW range and for 2011 about 50 to 100 MW are expected. In addition, dye-cells are getting ready to enter the market as well. The growth of these

---

[2] The PPVX is a non commercial financial index published by the solar magazine "Photon" and "Öko-Invest". The index started on 1 August 2001 with 1000 points and 11 companies and is calculated weekly using the Euro as reference currency. Only companies which made more than 50% of their sales in the previous year with PV products or services are included [Pho 2001].

[3] Please note that the composition of the index changes as new companies are added and others have to leave the index.

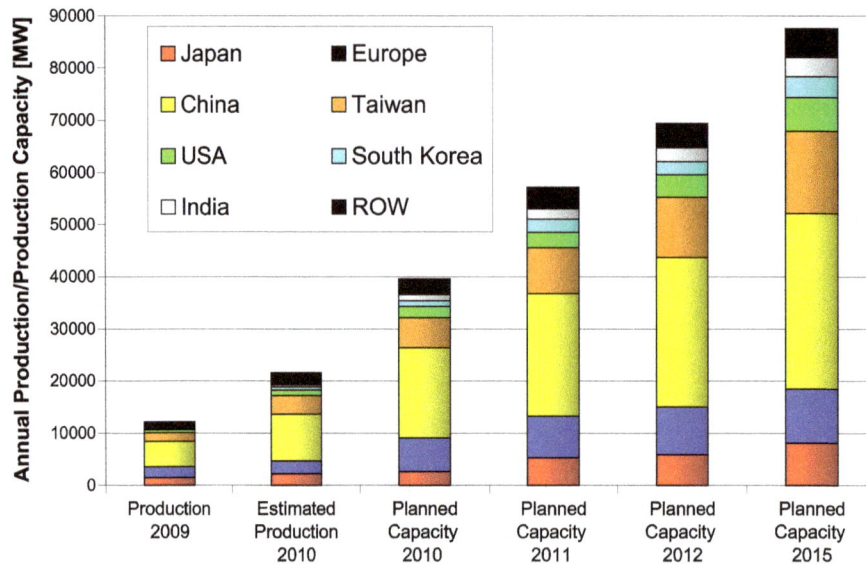

Fig. 2. World-wide PV Production and planned production capacity increases.

technologies is accelerated by the positive development of the PV market as a whole.

It can be concluded that in order to maintain the extremely high growth rate of the Photovoltaic industry, different pathways have to be pursued at the same time:

- Continuation to expand solar grade silicon production capacities in line with solar cell manufacturing capacities;
- Accelerated reduction of material consumption per silicon solar cell and Wp, e.g. higher efficiencies, thinner wafers, less wafering losses, etc.;
- Accelerated ramp-up of thin-film solar cell manufacturing;
- Accelerated CPV introduction into the market, as well as capacity growth rates above the normal trend.

Further photovoltaic system cost reductions will depend not only on the technology improvements and scale-up benefits in solar cell and module production, but also on the ability to decrease the system component costs, as well as the whole installation, projecting, operation, permitting and financing costs.

2 The photovoltaic industry

In 2010 the photovoltaic world market doubled in terms of production to about 23.5 GW. The market for installed systems doubled again and the current estimates are between 16 and 18 GW, as reported by various consultancies. One could guess that this represents mostly the grid-connected photovoltaic market. To what extent the off-grid and consumer-product markets are included is unclear. The difference of roughly 3 to 6 GW has therefore to be explained as a combination of unaccounted off-grid installations (approx. 1–200 MW off-grid rural, approx. 1–200 MW communication/signals, approx. 100 MW off-grid commercial), consumer products (ca. 1–200 MW) and cells/modules in stock.

In addition, the fact that some companies report shipment figures, whereas others report production figures add to the uncertainty. The difficult economic conditions added to the decreased willingness to report confidential company data. Nevertheless, the figures show a significant growth of the production as well as an increasing silicon supply situation.

The announced production capacities – based on a survey of more than 300 companies worldwide – increased despite difficult economic conditions. Despite the fact that a significant number of players announced a scale back or cancellation of their expansion plans for the time being, the number of new entrants into the field, notably large semiconductor or energy related companies overcompensated this. At least on paper the expected production capacities are increasing. Only published announcements of the respective companies and no third source info were used. The cut-off date of the used info was April 2011.

It is important to note, that production capacities are often announced, taking into account different operation models such as number of shifts, operating hours per year, etc. In addition the announcements of the increase in production capacity do not always specify when the capacity will be fully ramped up and operational. This method has of course the setback that (a) not all companies announce their capacity increases in advance and (b) that in times of financial tightening, the announcements of the scale back of expansion plans are often delayed in order not to upset financial markets. Therefore, the capacity figures just give a trend, but do not represent final numbers.

If all these ambitious plans can be realised by 2015, China will have about 38.4% of the world-wide production capacity of 88 GW, followed by Taiwan (18.0%), Europe (11.4%) and Japan (9.3%) (Fig. 2).

All these ambitious plans to increase production capacities at such a rapid pace depend on the expectations that markets will grow accordingly. This, however, is the biggest uncertainty as the market estimates for 2011 vary

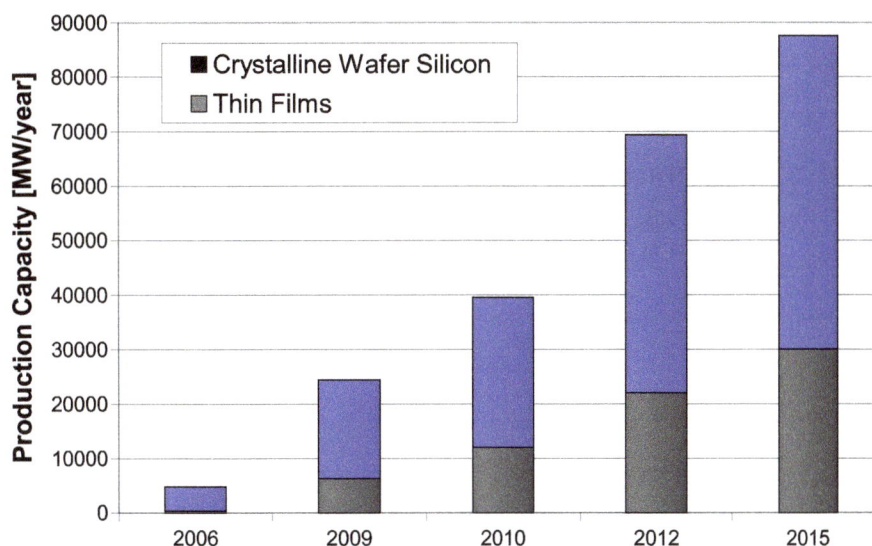

Fig. 3. 2006 and planned PV Production capacities of Thin-Film and Crystalline Silicon based solar modules.

between 17 GW and 24 GW with a consensus value in the 19 GW range. In addition, most markets are still dependent on public support in the form of feed-in tariffs, investment subsidies or tax-breaks.

Already now, electricity production from photovoltaic solar systems has shown that it can be cheaper than peak prices in the electricity exchange. In the first quarter 2011, the German average price index for rooftop systems up to 100 kWp was given with €2546 per kWp without tax [10]. With such investment costs, the electricity generation costs are already at the level of residential electricity prices in some countries, depending on the actual electricity price and the local solar radiation level. But only if markets and competition will continue to grow, prices of the photovoltaic systems will continue to decrease and make electricity from PV systems for consumers even cheaper than from conventional sources. In order to achieve the price reductions and reach grid-parity for electricity generated from photovoltaic systems, public support, especially on regulatory measures, will be necessary for the next decade.

2.1 Technology Mix

Wafer-based silicon solar cells is still the main technology and had around 80% market shares in 2010. Polycrystalline solar cells still dominate the market (45 to 50%), even if the market share has slightly decreased since the beginning of the decade. Commercial module efficiencies are within a wide range between 12 and 20%, with monocrystalline modules between 14%−20%, and polycrystalline modules between 12%−17%. The massive manufacturing capacity increases for both technologies are followed by the necessary capacity expansions for polysilicon raw material.

In 2005, production of thin-film solar modules reached for the first time more than 100 MW per annum. Since then, the Compound Annual Growth Rate (CAGR) of thin-film solar module production was even beyond that of the overall industry, increasing the market share of thin-film products from 6% in 2005 to 10% in 2007 and 16−20 % in 2010.

More than 200 companies are involved in thin-film solar cell activities, ranging from basic R&D activities to major manufacturing activities and over 150 of them have announced the start or increase of production. The first 100 MW thin-film factories became operational in 2007. If all expansion plans are realised in time, thin-film production capacity could be around 22 GW, or 32% of the total 69.4 GW, in 2012 and about 30 GW, or 34%, in 2015 of a total of 87.6 GW (Fig. 3). The first thin-film factories with GW production capacity are already under construction for various thin-film technologies.

One should bear in mind that only one fourth of the over 200 companies with announced production plans have already produced thin-film modules on a commercial scale in 2010.

The majority of the companies are silicon based and use either amorphous silicon or an amorphous/microcrystalline silicon structure. 30 companies announced using $Cu(In,Ga)(Se,S)_2$ as absorber material for their thin-film solar modules, whereas 9 companies use CdTe and 8 companies go for dye and other materials.

Concentrating Photovoltaics (CPV) is an emerging technology which is growing at a very high pace, although from a low starting point. More than 50 companies are active in the field of CPV development and almost 60% of them were founded in the last five years. Over half of the companies are located either in the United States of America (primarily in California) and Europe (primarily in Spain).

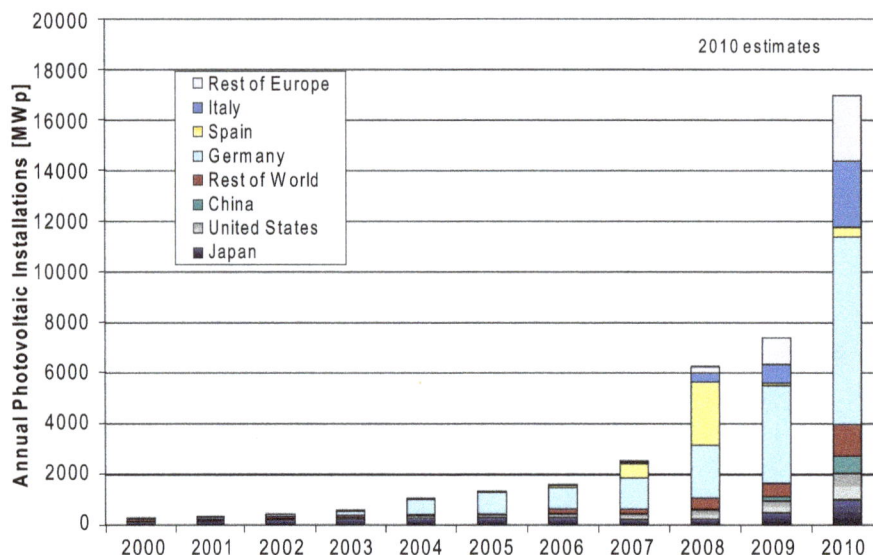

Fig. 4. Annual photovoltaic installations from 2000 to 2010 (data source: EPIA [9], Eurobserver [11] and own analysis).

Within CPV there is a differentiation according the concentration factors[4] and whether the system uses a dish (Dish CPV) or lenses (lens CPV). The main parts of a CPV system are the cells, the optical elements and the tracking devices. The recent growth in CPV is based on significant improvements in all of these areas, as well as the system integration. However, it should be pointed out that CPV is just at the beginning of an industry learning curve with a considerable potential for technical and cost improvements. The most challenging task is to become cost competitive with other PV technologies quickly enough in order to use the window of opportunities for growth.

With market estimates for 2010 in the 10 to 20 MW range, the market share of CPV is still small, but already for 2011 about 50 to 100 MW are expected and there is a wide consensus amongst consultancies and market analysts that CPV will reach a GW market size by 2015.

The existing photovoltaic technology mix is a solid foundation for future growth of the sector as a whole. No single technology can satisfy all the different consumer needs, ranging from mobile and consumer applications with the need for a few watts to multi MW utility-scale power plants. The variety of technologies is an insurance against a roadblock for the implementation of solar photovoltaic electricity if material limitations or technical obstacles restrict the further growth or development of a single technology pathway.

3 The photovoltaic market

In 2010 the world-wide photovoltaic market more than doubled, driven by major increased in Europe. The current estimates are between 17 and 19 GW, as reported by various consultancies (Fig. 4). This represents mostly the grid connected photovoltaic market. To what extent the off-grid and consumer product markets are included is not clear, but it is believed that a substantial part of these markets are not accounted for as it is very difficult to track them. A conservative estimate is that they account for approx. 400 to 800 MW (approx. 1−200 MW off-grid rural, approx. 1−200 MW communication/signals, approx. 100 MW off-grid commercial and approx. 1−200 MW consumer products).

With a cumulative installed capacity of over 29 GW, the European Union is leading in PV installations with a little more than 70 % of the total world wide 39 GW of solar photovoltaic electricity generation capacity at the end of 2010.

3.1 European Union

Market conditions for photovoltaics differ substantially from country to country. This is due to different energy policies and public support programmes for renewable energies and especially photovoltaics, as well as the varying grades of liberalisation of domestic electricity markets. After a tenfold increase of solar photovoltaic electricity generation capacity between 2001 and 2008, the newly installed capacity almost tripled in the last two years and reached 28.6 GW cumulative installed capacity at the end of 2010 [12–14].

Germany had the biggest market with 7.4 GW [13]. The German market growth is directly correlated to the introduction of the Renewable Energy Sources Act or "*Erneuerbare Energien Gesetz*" (EEG) in 2000 [15]. This law introduced a guaranteed feed-in tariff for electricity generated from solar photovoltaic systems for 20 years and already had a fixed build in annual decrease, which was adjusted over time to reflect the rapid growth of the market and the corresponding price reductions. Due to the

---

[4] High concentration >300 suns (HCPV); medium concentration $5 < x < 300$ suns (MCPV); low concentration <5 suns (LCPV).

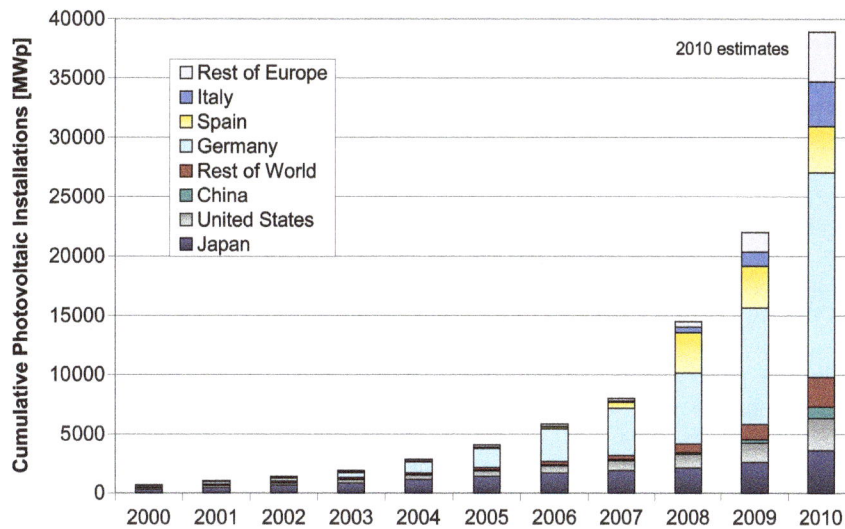

Fig. 5. Cumulative Photovoltaic Installations from 2000 to 2010 (data source: EPIA [9], Euroserver [11] and own analysis).

fact that until 2008 only estimates of the installed capacity existed, a plant registrar was introduced from 1 January 2009 on.

The German market showed to installation peaks during 2010. The first one was in June, when more than 2.1 GW were connected to the grid prior to the 13% feed-in cut which took effect on 1 July 2010. The second peak was in December with almost 1.2 GW just before the scheduled tariff reduction of another 13% on 1 January 2011. Compared to 2009, the feed-in tariff has been reduced by 33 to 36% depending on the system size and classification.

Italy again took the second place with respect to new installations and added a capacity of about 2.5 GW bringing cumulative installed capacity to 3.7 GW at the end of 2010 [14]. The Conto Energia 2011−2013 foresaw a 20% cut in the tariff in 2011 compared to 2010, but the tremendous market growth led to a revision of the support scheme, which outcome is still pending at the time of writing this paper.

The Czech Republic was the third largest market in Europe photovoltaic systems with more than 1.3 GW of new installations in 2010 reaching a cumulative nominal capacity of 1.8 GW and exceeding their own target of 1.65 GW set in the National Renewable Action Plan for 2020. The law on the Promotion of Production of Electricity from Renewable Energy Sources went into effect on 1 August 2005 and guarantees a feed-in tariff for 20 years. The annual prices are set by the Energy Regulator. The Producers of electricity can choose from two support schemes, either fixed feed-in tariffs or market price + Green Bonus. The 2011 feed-in rate in the Czech Republic was cut by almost 50% to CZK 5.5 per kilowatt hour (0.23 €/kWh). In addition, from 1 March 2011, the tariff only applies to roof-top and facade-integrated PV installations with a capacity of up to 30 kW.

Spain is second regarding the total cumulative installed capacity with 3.9 GW. Most of this capacity was installed in 2008 when the country was the biggest market

with close to 2.7 GW in 2008 [9]. This was more than twice the expected capacity and was due to an exceptional race to install systems before the Spanish Government introduced a cap of on the yearly installations in the autumn of 2008. A revised decree (Royal Decree 1758/2008) set considerably lower feed-in tariffs for new systems and limited the annual market to 500 MW with the provision that two thirds are rooftop mounted and no longer free field systems. These changes resulted in a new installed capacity of about 100 MW and about 380 MW in 2010.

In 2010 the Spanish Government passed the Royal Decrees 1565/10 [16] and RD-L 14/10 [17]. The first one limits the validity of the feed-in tariffs to 28 years while the later reduces the tariffs by 10% and 30% for existing projects until 2014. Both bills are "retroactive" and the Spanish Solar Industry Association (ASIF) [18] already announced to take legal actions against them.

France saw another year of massive growth in new photovoltaic system installations with 720 MW of new installations breaking the 1 GW threshold for cumulative installed capacity in 2010. This rapid growth led to a revision of the feed-in scheme in February 2011 setting a cap of 500 MW for 2011 and 800 MW for 2012 [19].

The new tariff levels only apply to rooftop systems up to 100 kW in size. In addition, those installations are divided into three different categories: residential; education or health; and other buildings with different feed-in tariffs depending on the size and type of installation. The tariffs for these installations range between 0.2883 €/kWh and 0.46 €/kWh. All other installations up to 12 MW are just eligible for a tariff of 0.12 €/kWh.

Belgium had another strong market performance in 2010 with new photovoltaic system installations of about 355 MW bringing the cumulative installed capacity to 740 MW according to the Flamish Regulator VREG (Vlaamse regulator van de elektriciteits- en gasmarkt). Most of the installations were done in Flanders, where since 1 January 2006 Green certificates exist which had a value of 0.35 €/kWh for 20 years in 2010. In 2011 Flanders is confronted with a legislation proposal, that will further bring down the value of green certificates for solar PV from the €0.33/kWh, which were set in January 2011 in three-month steps each of € 0.02–0.04/kWh.

Greece introduced a new feed-in-tariff scheme on 15 January 2009. The tariffsremeined unchanged until August 2010 and are guaranteed for 20 years. However, if a grid connection agreement was signed before that date, the unchanged FIT was applied if the system is finalised within the next 18 months. For small rooftop PV systems an additional programme was introduced in Greece on 4 June 2009. This programme covers rooftop PV systems up to 10 kWp (both for residential users and small companies). In 2011 the tariffs decreased by 6.8% to 8.5% depending on the size and location of the installation. In 2010 about 150 MW of new installations were carried out bringing the total capacity to about 180 MW.

The United Kingdom introduced of a new feed-in tariff scheme in 2010, which led to the installation of approximately 55 MW bringing the cumulative installed capacity to about 85 MW. However, in March 2011, the UK government has proposed significant reductions of the tariffs, especially for systems larger 50 kW.

The markets in the remaining European countries are still small and according to the National Renewable Energy Action Plans submitted to the European Commission in 2010 there are now plans to significant growth.

3.2 Asia and Pacific Region

In 2010 the Japanese market experienced a high growth, doubling its volume to 960 MW bringing the cumulative installed PV capacity to 3.6 GW. In 2009 a new investment incentive of ¥70 000 per kW for systems smaller than 10 kW and a new surplus power purchase scheme with a purchase price of ¥48 per kWh from systems smaller than 10 kW was introduced and the discussion about a wider feed-in tariff were started.

In April 2011, METI (Ministry for Economy, Trade and Industry) announced a change in the feed-in tariffs and increased the tariff for commercial installations from ¥20 to 40 per kWh and decreased the tariff for residential installations to ¥42 per kWh.

The Chinese PV market more than tripled again in 2010 with market estimates in the range of 530 to 690 MW bringing the cumulative installed capacity to about 1 GW. Despite this fact, the home market is still less then 10% of total Photovoltaic production. This situation might change because of the revision of the PV targets for 2015 and 2020. According to press reports, the National Energy Administration is seriously considering to rise the 5 year target for PV to 10 GW in 2015 and further up to 50 GW in 2020.

In January 2009, the Korean Government had announced the third National Renewable Energy Plan, under which renewable energy sources will steadily increase their share of the energy mix between now and 2030. The plan

covers such areas as investment, infrastructure, technology development and programmes to promote renewable energy. The new plan calls for a Renewable Energies share of 4.3% in 2015, 6.1% in 2020 and 11% in 2030. In 2010 about 180 MW of new PV installations brought the cumulative capacity to 0.7 GW.

For India 2010 market estimates for solar PV systems vary between 50 to 100 MW. The launching of the Indian National Solar Mission in January 2010 gave impetus to the markets, but the majority of the announced projects will come online from 2011 onwards. The National Solar Mission aims to make India a global leader in solar energy and envisages an installed solar generation capacity of 20 GW by 2020, 100 GW by 2030 and 200 GW by 2050. The short term outlook until 2013 was improved as well when the original 50 MW grid connected PV system target in 2012 was changed to 1000 MW in 2013.

It is worthwhile to mention that the Asian Development Bank (ADB) launched an Asian solar energy initiative in 2010, which should lead to the installation of 3 GW of solar power by 2012. ADB will provide US$ 2.25 billion (€1.73 billion) to finance the initiative, which is expected to leverage an additional US$ 6.75 billion (€5.19 billion) in solar power investments over the period.

3.3 North America

In the USA new grid connected PV installations of 878 MW increasing the cumulative capacity to 2.1 GW were reported by the Solar Energy Industries Association (SEIA) for 2010 [10]. In addition there are another 4 to 500 MW off-grid installations. Over the last 5 years, the market share of grid connected residential PV systems was pretty constant in the 30% range, whereas the utility market showed the largest increase from 1% back in 2005 to 28% in 2010.

Many State and Federal policies and programmes have been adopted to encourage the development of markets for PV and other renewable technologies. These consist of direct legislative mandates (such as renewable content requirements) and financial incentives (such as tax credits). One of the most comprehensive databases about the different support schemes in the US is maintained by the Solar Centre of the State University of North Carolina. The Database of State Incentives for Renewable Energy (DSIRE) is a comprehensive source of information on State, local, utility, and selected federal incentives that promote renewable energy. All different support schemes are described there and it is highly recommended to visit the DSIRE web-site http://www.dsireusa.org/ and the corresponding interactive tables and maps for details.

In 2010 Canada more than tripled its cumulative installed PV capacity to about 420 MW with 300 MW new installed systems. This development was driven by the introduction of a feed-in tariff in the province of Ontario enabled by the "*Bill 150, Green Energy and Green Economy Act, 2009*". On the federal level only a accelerated capital cost allowance exists under the income tax regulations. On a province level, nine Canadian Provinces have

Net Metering Rules with solar photovoltaic electricity as one of the eligible technologies, *Sales Tax Exemptions* and *Renewable Energy Funds* exist in two Provinces and *Micro Grid Regulations* and *Minimum Purchase Prices* each exist in one Province.

4 Conclusions and outlook

New investment in clean energy technologies, companies, and projects increased in 2010 by 30% compared to 2009 and reached $243 billion (€187 billion) [1]. China held largest share of investments with 22.4% followed by Germany with 17% and the USA with 14% [20]. Total 2010 worldwide investment in solar energy reached $79 billion (€61 billion) and was second only to wind with $95 billion (€73 billion).

The Photovoltaic Industry has changed dramatically over the last few years. China has become the major manufacturing place followed by Taiwan, Germany and Japan. Amongst the 15 biggest photovoltaic manufacturers in 2010, only three had production facilities in Europe, namely First Solar (USA, Germany and Malaysia), Q-Cells (Germany and Malaysia) and Solarworld (Germany and USA).

The implementation of the 100 000 roofs programme in Germany in 1990 and the Japanese long-term strategy set in 1994, with a 2010 horizon, were the start of an extraordinary PV market growth. Before the start of the Japanese market implementation programme in 1997, annual growth rates of the PV markets were in the range of 10%, mainly driven by communication, industrial and stand-alone systems. Since 1990 PV, production has increased almost 500-fold from 46 MW to about 23.5 GW in 2010. This corresponds to a CAGR of 36% over the last twenty years. Statistically documented cumulative installations world-wide accounted for over 39 GW in 2010. The interesting fact is, however, that cumulative production amounts to 53 GW over the same time period. Even if we do not account for the roughly 5 GW difference between the reported production and installations in 2010, there is a considerable 9 GW capacity of solar modules which are statistically not accounted for. Parts of it might be in consumer applications, which do not contribute significantly to power generation, but the overwhelming part is probably used in stand-alone applications for communication purposes, cathodic protection, water pumping, street, traffic and garden lights, etc.

The temporary shortage in silicon feedstock, triggered by the high growth-rates of the photovoltaics industry over the last years, resulted in the market entrance of new companies and technologies. New production plants for polysilicon, advanced silicon wafer production technologies, thin-film solar modules and technologies, like concentrator concepts, were introduced into the market much faster than expected a few years ago.

Even with the current economic difficulties, the increasing number of market implementation programmes world-wide, as well as the overall rising energy prices, the need to re-evaluate the validity of a nuclear option after the tragic events in Fukujima, Japan, in March 2011 and the pressure to stabilise the climate, will continue to keep the demand for solar systems high. In the long-term, growth rates for photovoltaics will continue to be high, even if the economic frame conditions vary and can lead to a short-term slow-down. This view is shared by an increasing number of financial institutions, which are turning towards renewables as a sustainable and secure long-term investment. Increasing demand for energy is pushing the prices for fossil energy resources higher and higher. Already in 2007, a number of analysts predicted that oil prices could well hit 100 $/bbl by the end of 2007 or early 2008 [21]. After the spike of oil prices in July 2008, with close to 150$/bbl, prices have decreased due to the worldwide financial crisis and hit a low around 37 $/bbl in December 2008. However, the oil price has rebounced and fluctuates in the 70 to 90 $/bbl range since August 2009. It is obvious that the fundamental trend of increasing demand for oil will drive the oil price higher again. Already in March 2009, the IEA Executive Director, Nobuo Tanaka, warned in an interview that the next oil crisis with oil prices at around 200 $/bbl due to a supply crunch, could be as close as 2013 because of lack of investments in new oil production.

Over the last 20 years, numerous studies about the potential growth of the photovoltaic industry and the implementation of photovoltaic electricity generation systems were produced. In 1996 the Directorate General for Energy of the European Commission published a study "Photovoltaics in 2010" [22]. The medium scenario of this study was used to formulate the White Paper target of 1997 to have a cumulative installed capacity of 3 GW in the European Union by 2010 [23]. The most aggressive scenario in this report predicted a cumulative installed PV capacity of 27.3 GW world-wide and 8.7 GW in the European Union for 2010. This scenario was called "*Extreme scenario*" and it was assumed that in order to realise it a number of breakthroughs in technology and costs as well as continuous market stimulation and elimination of market barriers would be required to achieve it. The reality check reveals that even the most aggressive scenario is lower than what we expect from the current developments. A cumulative installed capacity of about 39 GW world-wide and 29 GW in Europe was estimated as cumulative installations of PV systems at the end of 2010.

According to investment analysts and industry prognoses, solar energy will continue to grow at high rates in the coming years. The different Photovoltaic Industry Associations, as well as Greenpeace, the European Renewable Energy Council (EREC) and the International Energy Agency, have developed new scenarios for the future growth of PV. Table 1 shows the different scenarios of the Greenpeace/EREC study, as well as the different 2008 IEA *Energy Technology Perspectives* scenarios.

These projections show that there are huge opportunities for the photovoltaics industry in the future if the right policy measures are taken, but we have to bear in mind that such a development will not happen by itself. It will

Table 1. Evolution of the cumulative solar electrical capacity scenarios until 2050 [24–26].

Year	2010 [GW]	2020 [GW]	2030 [GW]	2050 [GW]
Actual Installations	**39**			
Greenpeace* (reference scenario)	14	80	184	420
Greenpeace* ([r]evolution scenario)	18	335	1036	2968
Greenpeace* (advanced scenario)	21	439	1330	4318
IEA Reference Scenario	10	30	<60	non competitive
IEA ACT Map	22	80	130	600
IEA Blue Map	27	130	230	1150
IEA PV Technology Roadmap	27	210	870	3155

* 2010 values are extrapolated as only 2007 and 2015 values are given.

require the constant effort and support of all stakeholders to implement the envisaged change to a sustainable energy supply with photovoltaics delivering a major part. The main barriers to such developments are perception, regulatory frameworks and the limitations of the existing electricity transmission and distribution structures.

The above-mentioned scenarios will only be possible if new solar cell and module design concepts can be realised, as with current technology the demand for materials like silver would exceed the available resources within the next 30 years. Research to avoid such kind of problems is underway and it can be expected that such bottle-necks will be avoided.

The photovoltaic industry is developing into a fully-fledged mass-producing industry. This development is connected to an increasing industry consolidation, which presents a risk and an opportunity at the same time. If the new large solar cell companies use their cost advantages to offer lower-priced products, customers will buy more solar systems and it is expected that the PV market will show an accelerated growth rate. However, this development will influence the competitiveness of small and medium companies as well. To survive the price pressure of the very competitive market situation, and to compensate the advantage of the big companies made possible by economies of scale that come with large production volumes, they have to specialise in niche markets with high value added in their products. The other possibility is to offer technologically more advanced and cheaper solar cell concepts.

Despite the fact that Europe – especially Germany – is still the biggest world market, the European manufacturers are losing market shares in production. This is mainly due to the rapidly growing PV manufacturers from China and Taiwan and the new market entrants from companies located in India, Malaysia, Philippines, Singapore, South Korea, UAE, etc. Should the current trend in the field of world-wide production capacity increase continue, the European share will further decrease, even with a continuation of the growth rates of the last years. At the moment, it is hard to predict how the market entrance of the new

players all over the world will influence future developments of the markets.

A lot of the future market developments, as well as production increases, will depend on the realisation of the currently announced world-wide PV programmes and production capacity increases. During 2009 and 2010, the announcements from new companies which want to start a PV production, as well as established companies to increase their production capacities, continued to increase the expected overall production capacity. If all these plans are realised, thin-film production companies will increase their total production capacities even faster than the silicon wafer-based companies and increase their market share from the 2007 market share of 10% to about 30% in 2015. However, the number of thin-film expansion projects which are caught between the fact that margins are falling, due to decreasing module prices and the need to raise additional capital to expand production in order to lower costs, is increasing.

Already for a few years, we have now observed a continuous rise of oil and energy prices, which highlights the vulnerability of our current dependence on fossil energy sources, and increases the burden developing countries are facing in their struggle for future development. On the other hand, we see a continuous decrease in production costs for renewable energy technologies as a result of steep learning curves. Due to the fact that external energy costs, subsidies in conventional energies and price volatility risks are generally not taken into consideration, renewable energies and photovoltaics are still perceived as being more expensive in the market than conventional energy sources. Nevertheless, electricity production from photovoltaic solar systems have already proved now that it can be cheaper than peak prices in the electricity exchange in a wide range of countries and if the new EPIA and SEIA visions can be realised, electricity generation cost with photovoltaic systems will have reached grid parity in most of Europe and the USA by 2020. In addition, renewable energies are, contrary to conventional energy sources, the only ones to offer a reduction of prices rather than an increase in the future.

References

1. World Economic Forum, April 2011, Green Investing 2011 – Reducing the Cost of Financing, Ref. 200311
2. P. Mints, *Manufacturer Shipments, Capacity and Competitive Analysis 2009/2010*. Navigant Consulting Photovoltaic Service Program, Palo Alto, CA.
3. Photon Interanational, March 2011
4. PV News, published by The Prometheus Institute, ISSN 0739-4829
5. P. Mints, *Global PV Demand 2011 and Beyond*, Webinar: Vote Solar, January 12, 2011
6. Mercom, Market Intelligence Report, 4 April 2011
7. iSupply, PV Perspectives, February 2011
8. Bundesverband Solarwirtschaft, Preisindex Photovoltaik http://www.solarwirtschaft.de/preisindex
9. European Photovoltaic Industry Association, Global Market Outlook for Photovoltaics until 2015, 2011
10. Solar Energy Industry Association, *March 2011, US Solar Market Insight^{TM} , 2010 Yearin review*, Executive Summary
11. Photovoltaic Energy Barometer, Systèmes Solaires, le journal du photovoltaique n° 3 – 2011, April 2011, ISSN 0295-5873
12. Photovoltaic Energy Barometer, Systèmes Solaires, le journal du photovoltaique n° 5 – 2011, April 2011, ISSN 0295-5873
13. German Federal Network Agency (Bundesnetzagentur), Press Release 21 March 2011
14. Gestore Servici Energetici, Press Release, 15 February 2011
15. Gesetz für den Vorrang erneuerbarer Energien vom 29. März 2000; Bundesgesetzblatt Jahrgang 2000 Teil I Nr.13 S. 305 ausgegeben am 30. März 2000
16. Royal Decree 1565/10, published on 23 November 2010 http://www.boe.es/boe/dias/2010/11/23/pdfs/BOE-A-2010-17976.pdf
17. Royal Decree RD-L 14/10, published on 24 December 2010 http://www.boe.es/boe/dias/2010/12/24/pdfs/BOE-A-2010-19757.pdf
18. Asociación de la Industria Fotovoltaica (ASIF), http://www.asif.org/principal.php?idseccion=565
19. Ministère de l'Économie, de l'industrie et de l'emploi, Press Release, 24 February 2011
20. The PEW Charitable Trust, 2011, *Who's Winning the Clean Energy Race? G-20 Investment Powering Forward*
21. International Herald Tribune, 24 July 2007 http://www.iht.com/articles/2007/07/24/bloomberg/bxoil.php
22. European Commission, Directorate-General for Energy, Photovoltaics in 2010, Office for Official Publications of the European Communities, 1996, ISBN 92-827-5347-6
23. Energy for the Future: Renewable sources of energy; White Paper for a Community Strategy and Action Plan, COM(1997)599 final (26/11/97)
24. Greenpeace International, European Renewable Energy Council (EREC), 2010, energy [r]evolution, June 2010
25. International Energy Agency, 2008, Energy Technology Perspectives – Scenarios & Strategies to 2050, ISBN 9789264041424
26. International Energy Agency, 2010, PV Technology Roadmap

15

Performance potential of low-defect density silicon thin-film solar cells obtained by electron beam evaporation and laser crystallisation

J. Dore[1,2], S. Varlamov[2,a], R. Evans[1], B. Eggleston[1,2], D. Ong[1], O. Kunz[1], J. Huang[2], U. Schubert[1], K.H. Kim[1,2], R. Egan[1], and M. Green[2]

[1] Suntech R&D Australia, Pty., Ltd. 82-86 Bay St., Botany, NSW 2019, Australia
[2] University of NSW Sydney, NSW 2052, Australia

Abstract A few microns thick silicon films on glass coated with a dielectric intermediate layer can be crystallised by a single pass of a line-focused diode laser beam. Under favorable process conditions relatively large linear grains with low defect density are formed. Most grain boundaries are defect-free low-energy twin-boundaries. Boron-doped laser crystallised films are processed into solar cells by diffusing an emitter from a phosphorous spin-on-dopant source, measuring up to 539 mV open-circuit voltage prior to metallisation. After applying a point-contact metallisation the best cell achieves 7.8% energy conversion efficiency, open-circuit voltage of 526 mV and short-circuit current of 26 mA/cm^2. The efficiency is significantly limited by a low fill-factor of 56% due to the simplified metallisation approach. The internal quantum efficiency of laser crystallised cells is consistent with low front surface recombination. By improving cell metallisation and enhancing light-trapping the efficiencies of above 13% can be achieved.

1 Introduction

Crystalline silicon (c-Si) wafer-based solar cells dominate the photovoltaic (PV) market due to the mature and constantly improving technology and a decreasing manufacturing cost. One of the major contributors into cost reduction is the use of thinner cells to lower the consumption of the material. A number of approaches exist to produce thin c-Si wafers and layers [1]. c-Si films on supporting substrates such as glass is one of such approaches and a few technologies have been developed to produce solar cells from such films [2–6]. The highest module efficiency of c-Si thin-film solar cells on glass is 10.5% [4]. Further improvement in the cell performance is limited by the high density of intragrain defects and related poor electronic quality of the material [7, 8] typically obtained by solid-phase crystallisation (SPC) or epitaxy (SPE) at relatively low temperatures compatible with the glass substrate. A new approach has recently emerged that exploits zone-melt-like liquid-phase crystallisation (LPC) of precursor silicon films by an electron beam to produce large-grained and low-defect density c-Si films on glass with the superior electronic quality compared to SPC silicon [9, 10]. The open circuit voltage (V_{OC}) of 545 mV and the efficiency of 4.7% were demonstrated for a cell with a LPC absorber and an a-Si:H heterojunction emitter. This paper presents initial results for a similar approach but where a diode laser is used for the LPC to obtain high crystallographic and electronic quality Si films. These films are processed into 7.8% efficient cells with a diffused homojunction emitter.

2 Experiment

Planar Schott Borofloat33 glass was used as a substrate for film deposition. It was coated with an intermediate dielectric layer, such as SiC_x, SiO_x, SiN_x or their combination. These layers, intrinsic or boron-doped, were prepared by RF magnetron sputtering or co-sputtering using dielectric and pure boron targets. Then, 10 μm thick Si films, undoped or boron doped at 1E16 cm^{-3}, were deposited by electron-beam evaporation at 650 °C on boron-doped or undoped intermediate layer respectively. The Si films were crystallised by a single pass of a line-focus beam from an 808 nm CW diode laser, LIMO450-L12 $\times$ 0.3-DL808-EX937. The laser beam has a Gaussian profile in

[a] e-mail: s.varlamov@unsw.edu.au

the short (scan) direction (FWHM 0.170 mm) and a top-hat profile in the long direction (FWHM 12 mm). The laser conditions were 15–25 kW/cm^2 and 10–15 ms exposure time at sample stage pre-heat temperature in the range of 550–700 °C. Solar cells of 1.7×0.6 cm^2 size were fabricated in the middle of the crystallised strips. The cell emitter was formed by phosphorous diffusion at about 900 °C from a spin-on source. The hydrogen plasma passivation and metallisation processes applied to the cells are similar to those used for SPC Si thin-film solar cells [4–6]. Metallisation relies on forming point contacts on the rear side of the device by inkjet printing holes in a resin layer and then etching down to the emitter and absorber layers in sequence. Sputtered Al is then used to form contacts to the n- and p-type openings. This metallisation is very similar to the one described in [11] but simplified by reducing the contact density and no heavy Si doping under the p-type contacts to the absorber to accommodate and facilitate quicker cell development. Current research-level planar cells rely on a pigmented rear diffuse reflector for modest light-trapping.

The Si film crystal quality was characterised by SEM and TEM imaging. The dopant concentration in the films was estimated from SIMS measurements. The electronic film quality and the cell performance was characterised by Hall measurements, Suns-V_{oc}, Light I-V and spectral response techniques.

3 Results and discussion

A few micron thick e-beam evaporated Si films can be crystallised in a quick, zone-melt like process by scanning with a high power line-focus diode laser beam. Continuous lateral crystal growth can be achieved whereby the growth front is seeded by the preceding crystallised region, forming long parallel grains in the direction of scanning (Fig. 1). Factors having significant effects on the crystallisation process include an intermediate layer between glass and Si, laser beam parameters and the substrate temperature during crystallisation.

3.1 Intermediate layer

The intermediate layer has to be stable at the Si melting point of 1414 °C and it serves a few important functions: enables molten Si to wet substrate sufficiently to avoid balling up; blocks impurity diffusion from glass; acts as a dopant source; provides an antireflection (AR) effect; and passivates the front device surface (the Si-glass interface). Transparent dielectric films, SiO$_x$, SiC$_x$, and SiN$_x$, and their different combinations were tested as the intermediate layers, along with their interactions with laser beam parameters. Properties of different intermediate layers and their effects on cell performance are described in detail elsewhere [11,12]. No single layer alone can perform all required functions. SiC$_x$ (refractive index $(n) = 2.9$) is the best wetting layer providing the widest laser process window but it allows impurity diffusion from glass, absorbs too much of the short wavelength light <500 nm, and

it does not passivate the Si-glass interface well (Sect. 3.4). SiO$_x$ is the best impurity diffusion barrier and the surface passivating layer resulting in the highest V_{OC} but because its refractive index is very similar to that of the superstrate glass ($n = 1.51$) does not offer any additional AR effect to that already provided by the glass alone and it has the smallest laser process window. SiN$_x$ ($n = 2.11$) offers the best AR effect and a reasonably wide laser parameter widow but it leads to pinholing and the poorest cell performance. Only a combination of different dielectrics is proved to deliver a satisfactory process and cell performance. Results presented in this report are obtained using a three layer stack of SiO$_x$/SiC$_x$/SiO$_x$ (O/C/O, 80 nm, 14 nm, 15 nm respectively). The oxide layers in the stack are boron-doped at a level of $4 \sim 5E19$ cm^{-3} (estimated from SIMS measurements) such that after boron diffusion that occurs during laser melting the sheet resistance of crystallised Si films is about 2 000 Ω/sq and the boron concentration of about 1E16 cm^{-3}.

3.2 Laser process

The range of laser parameters that allows a successful crystallization process is limited by the resulting Si crystal structure at the lower energy-dose end, which has to avoid amorphous and/or microcrystalline material, and by dewetting of a Si film from the substrate at the high dose end. Without any intermediate wetting layer, and without substrate heating dewetting and/or delamination occurs at energy doses comparable with the crystallisation limit and no process is feasible.

With the intermediate layer, a large-grained crystalline silicon material without dewetting or delamination is obtained at substrate temperatures exceeding the glass softening point of about 550 °C and the minimum laser energy dose of 220 J/cm^2. The experimentally determined process window for the dose is about 250, 220 and 190 J/cm^2 wide for the SiC$_x$, SiN$_x$ and SiO$_x$, intermediate layer respectively. The cells on the O/C/O intermediate layer coated glass presented in this report typically require the energy dose of 240–260 J/cm^2.

3.3 Structural quality

The optimised laser crystallisation process results in lateral growth of linear silicon grains which extend through the whole film thickness of 10 μm and they are a few millimetres long and 50–500 μm wide (Fig. 1). According to electron backscatter diffraction (EBSD) most of the linear grains have [101] orientation in direction normal to the film plane [13]. TEM cross-sectional images and their corresponding diffraction patterns (Figs. 2, 3) confirm complete crystallisation with [110] grain orientation in direction parallel to the film. It is consistent with the EBSD images from [13] as EBSD and TEM are measured at approximately 90° relative to each other. The silicon grains contain regions with the varying defect density which was evaluated by counting dislocation lines

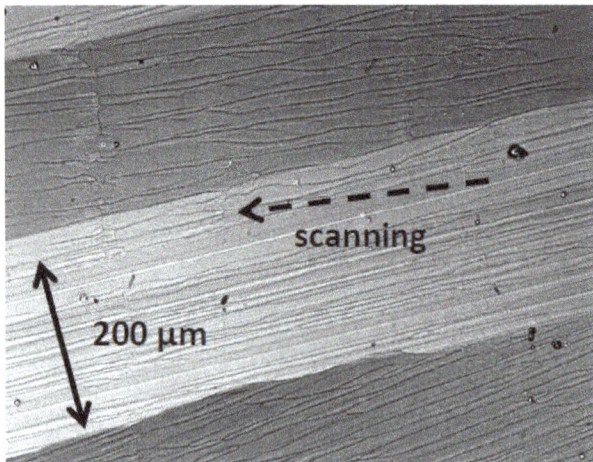

Fig. 1. Backscatter SEM image of lateral grains in diode laser crystallised Si film. ~200 μm-wide linear grains are shown in the same shade of grey; wavy lines within each grain are surface texture developed during crystallisation while straight parallel lines are twin boundaries.

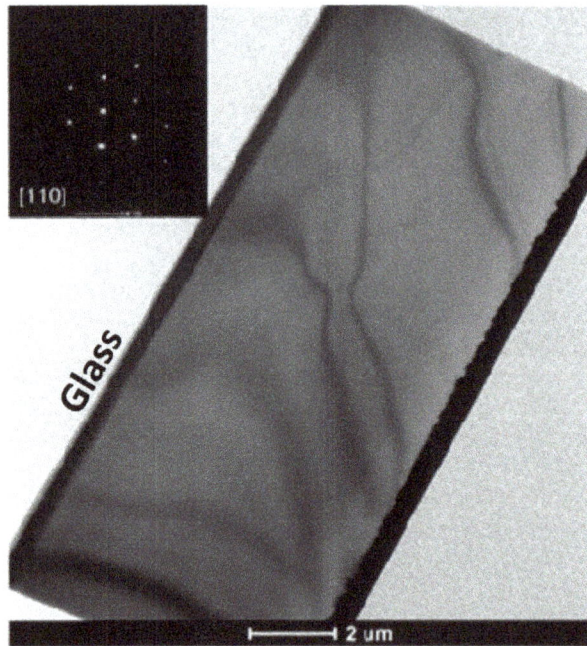

Fig. 2. TEM image and diffraction pattern of defect-free grain with [110] orientation in diode laser crystallised Si film.

in Weak-Beam-Dark-Field (WBDF) images. Most laterally grown grains are nearly defect-free on the TEM scale which translates into the defect density of 1E5 cm^{-2} or lower (Fig. 2) while the most defect-rich grains can contain dislocations with the density up to 1E9 cm^{-2} (Fig. 3). The most commonly observed boundaries between linear grains are defect-free low-energy Σ3 twin-boundaries which are electrically inactive, i.e. they do not significantly contribute into carrier recombination [10, 14].

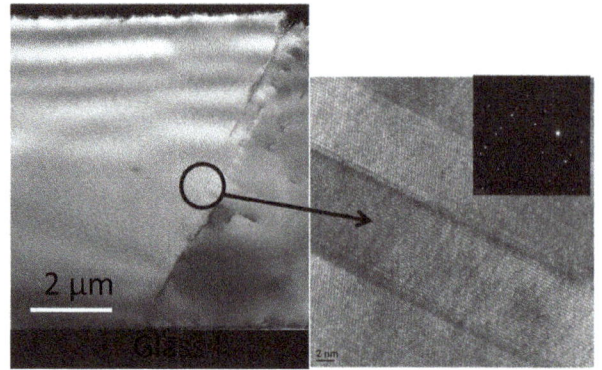

Fig. 3. (Left) TEM image of diode laser crystallised Si grains with twin boundary (circle) and dislocations; (right) high resolution TEM image of defect-free twin boundary.

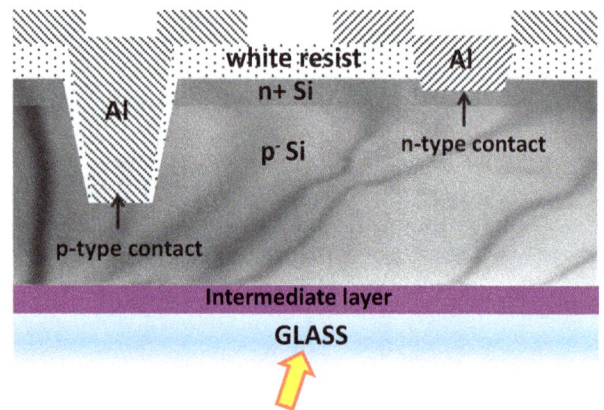

Fig. 4. The structure of a metallised solar cell. The light enters the cell through the front, glass side.

3.4 Electronic properties and cell performance

The carrier mobility, concentration and bulk resistivity of laser crystallised boron-doped Si films used for cell fabrication were calculated from Hall effect measurements using the Van der Pauw cross pattern. Depending on a particular location the mobility typically varies between 300 and 470 cm^2/V s, the carrier concentration ~10E16 cm^{-3}, and resistivity 1–3 Ω cm. These values are similar to those measured for a reference c-Si wafer: 414 cm^2/V s, 1.6E16 cm^{-3}, 0.94 Ω cm respectively, indicating a high electronic quality of the laser crystallised silicon film, which is a lot better than the quality of a reference SPC Si film with a similar boron concentration and a mobility of 50–120 cm^2/V s, which is in agreement with the mobilities of 40–90 cm^2/V s reported elsewhere [15].

The structure of a metallised solar cell is shown in Figure 4.

The cell voltages depend on the intermediate dielectric layers, which is discussed elsewhere [11, 12]. The cells on SiO$_x$ achieve V_{OC} in a range of 520–540 mV, while the cells on SiN$_x$ have the poorest V_{OC}, below 500 mV. These effects are not well understood but it can be speculated that the oxide provides the best silicon-glass interface passivation which is consistent with high blue response for cells on SiO$_x$ (Fig. 5, right). For cells on SiN$_x$ SIMS data indicate

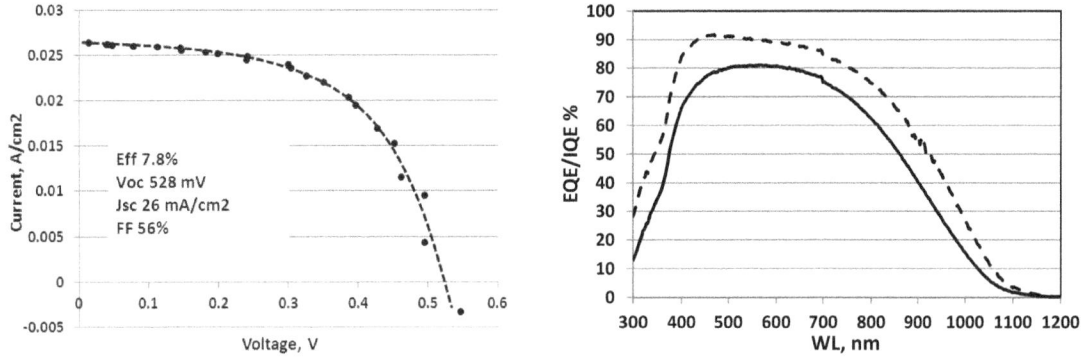

Fig. 5. (Left) Light I-V curve and (right) EQE (solid) and IQE (broken) of 7.8% efficient laser crystallised Si thin-film cell.

presence of nitrogen contamination which can be negatively affecting the voltages. The best performing metallised 7.8%-efficient cell on the O/C/O triple layer stack has V_{OC} of 526 mV, similar to V_{OC} of the cells on the oxide only but an improvement in cell antireflection properties due to presence of a very thin SiC_x layer on the front leads to a significant current gain, 26 mA/cm^2 versus 24 mA/cm^2 for the cell on SiO_x [11].

The light I-V curve of the best, 7.8% efficient cell is shown in Figure 5 (left).

The series and shunt resistances (R_s and R_{sh}) of 4.5 and 317 Ω cm^2 respectively, were estimated by measuring the inverse slope of the I-V curve at the voltage and current axes. High R_s is due to poor metal contacts that are made to lightly-doped ($\sim$1E16 cm^{-3}) silicon. It is also possible that a Schottky diode is formed at such contacts, which has a negative contribution to the cell voltage [16]. A source of relatively low R_{sh} is not yet identified. Both high R_s and low R_{sh} result in the low FF of 56%. It can be shown [17] that R_s of 4.5 Ω cm^2 results in the FF reduction of about 13%, from nominal 70%, typical for poly-Si thin-film cells with a similar metallisation scheme and R_s of 1.3 Ω cm^2 [4,6], to 57%, in a good agreement with the measured FF value. It is also estimated that R_{sh} of 317 Ω cm^2 can be responsible for only about 3% reduction in the FF. Both these issues can be addressed by improving cell metallisation, for example, by selectively doping the contacts to the absorber. With the FF of at least 70%, which is typical for optimised contact schemes of this type [18], the cell efficiency would approach 10%.

The EQE, and IQE curves of the best cell are shown in Figure 5 (right). The IQE is adjusted to take into account parasitic absorption in the SiC_x layer, which was measured prior to silicon deposition, as presented in [11]. The IQE peaks at around 90% and it stays around this value for most of the visible wavelength range. For a rear-junction cell it suggests the low front surface recombination velocity and the minority carrier diffusion length greater than the cell thickness of 10 μm.

3.5 Performance potential

In order to estimate the performance potential of the laser crystallised silicon material the cell IQE was fitted

by using PC1D modelling. This allows indicative values of bulk lifetime (τ) and front surface recombination velocity (S_F) to be extracted. The rear-surface recombination velocity was found to have no impact on the IQE in this model, which is likely due to the low minority carrier concentration at the heavily doped n+ rear surface.

The best 7.8% efficient cell is difficult to model because of complexity of its optical properties in the short-wavelength region due to presence of the triple O/C/O intermediate layer stack on the front. Instead, a similar cell but on the single SiO_x layer was used. It is justified because both cells have similar V_{OC} and similarly high IQE in the visible region (Figs. 5 and 6). The oxide film and the glass were treated as one 3.3 mm thick layer with the refractive index of 1.5.

Possible ranges for τ and S_F can be estimated by looking at two extreme cases. The case when recombination is dominated by the front surface can be modelled by setting τ very high (e.g. 100 μs). In this case, a good IQE fit with maximum of 93% is achieved when $S_F = 1900$ cm/s or less (blue dotted line in Fig. 6). Similarly, the case dominated by bulk recombination is modelled by setting S_F very low (e.g. to 100 cm/s). In this case, a good IQE fit is obtained when $\tau = 260$ ns or more (purple dotted line in Fig. 6). Thus, we can conclude from using this model that τ is at least 260 ns, and that S_F is at most 1900 cm/s. At present, there is no indication as to which case, bulk or surface recombination, limits the cell performance. The parameters used to obtain the best fits are listed in Table 1.

Even at the minimum value of 260 ns, the lifetime would be sufficient for significantly higher efficiencies, should all other device properties be optimised. This is instructive because it shows that the achieved cell performance is not limited by the laser crystallised material quality and thus identifies device optimisation as the immediate development priority leading to better cell performance, rather than the silicon material itself. For example, an ideal cell with R_s of 1–1.5 Ω cm^2 (e.g. with selectively doped absorber contacts), no shunts, optimum antireflection coating and internal reflectance of over 90% (e.g. via light-trapping by surface texture as described in [19]) would be potentially capable of an efficiency of over 13%.

Table 1. PC1D model parameters.

Thickness [μm]	10
p-type background doping [cm^{-3}]	7E15
Peak n-type emitter doping [cm^{-3}]	1.5E19
Bulk lifetime [μs] (bulk-limited; $S_F = 100$ cm/s)	0.26
Rear-surface recombination velocity [cm/s]	1E4
Front-surface recombination velocity [cm/s] (surface-limited, $\tau = 100$ μs)	1900
Front internal reflectance [%]	51
Rear internal reflectance [%]	58

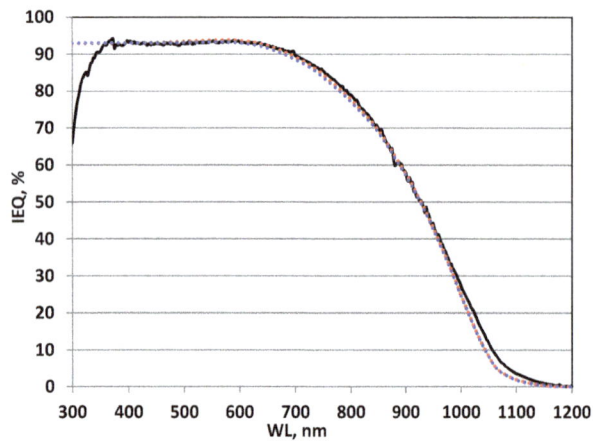

Fig. 6. Experimental (black-solid line) IQE curve and front-surface (blue-dotted line) and bulk recombination (red-dotted line) limited model fits of the cell on the oxide layer.

4 Conclusions

It is demonstrated than line-focused diode laser crystallisation of a few micron thick silicon films on glass can produce a high crystal and electronic quality material. Such a material consists of a few millimetre long and 50–500 micron wide grains with low defect density. The majority of grains boundaries are defect free low-energy $\Sigma 3$ twin boundaries. Laser process parameters, crystal and electronic film and cell properties are influenced by an intermediate dielectric layer between glass and silicon. Solar cells with a diffused emitter that are fabricated on the SiO$_x$ layer achieve the highest V_{OC} up to 539 mV. The cells on the SiO$_x$/SiC$_x$/SiO$_x$ triple stack layer achieve the best efficiency up to 7.8%. The performance is not limited by the quality of the laser crystallised Si material but by simplified device design and poor light-trapping. PC1D modelling suggests a bulk lifetime of at least 260 ns, which in an otherwise optimised device, would be compatible with efficiencies

of over 13%. Future work will focus on device optimisation to determine how much of this improvement is achievable.

References

1. F.J. Henley, in *Proceedings of the IEEE Photovoltaic Specialist Conference, Honolulu, USA, 2010*, p. 1184
2. H.M. Branz, C.W. Teplin, M.J. Romero, I.T. Martin, Q. Wang, K. Alberti, D.L. Young, P. Stradin, Thin Solid Films **519**, 4545 (2011)
3. I. Gordon, L. Carnel, D. Van Gestel, G. Beaucarne, J. Poortmans, Prog. Photovolt.: Res. Appl. **15**, 574 (2007)
4. M. Keevers, T. Young, U. Schubert, M. Green, in *Proceedings of the 22 European Photovoltaic Solar Energy Conference, Milan, Italy, 2007*
5. O. Kunz, Z. Ouyang, S. Varlamov, A. Aberle, Prog. Photovolt.: Res. Appl. **17**, 567 (2009)
6. R. Egan et al., in *Proceedings of the 24 European Photovoltaic Solar Energy Conference, Hamburg, Germany, 2009*
7. J. Wong, J. Huang, B. Eggleston, M.A. Green, O. Kunz, R. Evans, M. Keevers, R.J. Egan, J. Appl. Phys. **107**, 123705 (2010)
8. J. Wong, J. Huang, S. Varlamov, M. Green, M. Keevers, Prog. Photovolt.: Res. Appl. (2011), DOI: 10.1002/pip.1154
9. D. Amkreutz, J. Muller, M. Schmidt, T. Hanel, T.F. Schulze, Prog. Photovolt.: Res. Appl. **19**, 937 (2011)
10. W. Seifert, D. Amkreutz, T. Arguirov, M. Krause, M. Schmidt, Solid State Phenomena **178-179**, 116 (2011)
11. J. Dore, R. Evans, U. Schubert, B. Eggleston, D. Ong, K. Kim, J. Huang, O. Kunz, M. Keevers, R. Egan, S. Varlamov, M. Green, Prog. Photovolt.: Res. Appl., DOI: 10.1002/pip.2282
12. J. Dore, R. Evans, B. Eggleston, S. Varlamov, M. Green, MRS Online Proceedings Library **1426** (2012), Doi:10.1557/opl.2012.866
13. B. Eggleston, S. Varlamov, J. Huang, R. Evans, J. Dore, M. Green, MRS Online Proceedings Library **1426** (2012), Doi:10.1557/opl.2012.1260
14. J. Chen, T. Sekiguchi, D. Yang, F. Yin, K. Kido, S. Tsurekawa, J. Appl. Phys. **96**, 5490 (2004)
15. T. Noguchi, Jpn J. Appl. Phys. **32**, L1584 (1993)
16. S.W. Glunz, J. Nekarda, H. Mäckel, A. Cuevas, in *Proceedings of the 22 European Photovoltaic Solar Energy Conference, Milan, Italy, 2007*, p. 849
17. M.A. Green, *Solar Cells* (University of NSW, 1992), p. 97
18. S. Partlin, N. Chang, R. Egan, T. Young, D. Kong, R. Evans, D.A. Clugston, P. Lasswell, A. Turner, J. Dore, T. Florian, in *Proceedings of the 25 European Photovoltaic Solar Energy Conference, Valencia, Spain, 2010*, p. 3568
19. Q. Wang, T. Soderstrom, K. Omaki, A. Lennon, S. Varlamov, Energy Procedia **15**, 220 (2012)

16

Optical evaluation of doping concentration in SiO$_2$ doping source layer for silicon quantum dot materials

T. Zhang[1,a], I. Perez-Wurfl[1], B. Berghoff[2], S. Suckow[2], and G. Conibeer[1]

[1] The University of New South Wales, UNSW, Sydney, NSW 2052, Australia
[2] Institute of Semiconductor Electronic, RWTH Aachen University, Aachen, Germany

Abstract We have investigated and proposed a simple method to correlate optical absorption with high B doping concentrations in thin SiO$_2$ films that offer a potential doping source for Si quantum dots. SiO$_2$ films with boron and phosphorus were deposited using a computer controlled co-sputtering system. By assessing the absorption coefficients, it was observed that the doping can dramatically increase the absorption of the transparent SiO$_2$. Additionally, the highly doped SiO$_2$ films have a very broad Urbach like absorption tail and the absorption corresponds well with the doping level.

1 Introduction

The use of silicon quantum dots (QD) in photovoltaic has been proposed as a means to develop an all-silicon tandem solar cell. This third generation solar cell concept, involving depositing superlattice materials with Si quantum dots embedded in dielectric materials (SiO$_2$, Si$_3$N$_4$ or SiC), has been previously investigated [1]. A p-i-n structure has already been fabricated demonstrating a device with rectifying characteristics [2]. Doping is one critical requirement for device fabrication. Using dopant rich SiO$_2$ layers as the doping source for diffusing boron or phosphorous into the superlattice is found to change the conductivity of layers containing QDs [3]. This method has an advantage over the typical in-situ way of doping during film deposition because in-situ doped B or P in SRO layer influence the quantum dot formation and QD size control becomes difficult [4,5]. Therefore research interest is drawn to these highly doped SiO$_2$ film focusing on controlling the concentration and the diffusion as well as its optical properties.

Due to the insulating property of thin SiO$_2$ films, the doping level of this material cannot be evaluated simply through electrical measurement. Meanwhile, although there are many advanced material analysis approaches for thin film layers, such as SIMS and RBS, they are not suitable for quick examination of the doping concentration. In this work we attempt to correlate the optical properties and doping concentration to find a possible method to analyze doped SiO$_2$ films quickly.

To achieve this goal, we deposited SiO$_2$ films on quartz substrates with by co-sputering boron and phosphorus with various concentrations. This work could be helpful for researchers working on quantum dot materials and those who may want to utilize doped insulator layers as doping sources in any other material system.

2 Experimental

In this work, the SiO$_2$ doping source layers with various boron and phosphorus concentrations were deposited using a computer-controlled AJA ATC-2200 co-sputtering system. The films were sputtered from a 99.999% silica 4 inch target at a nominal rate of 1.5 nm/min using a 135 W of a 13.56 MHz AC in an Ar atmosphere at 1.5 mTorr. Doping was achieved by varying the RF power applied to 2 inch dopant targets. We investigated 48 W, 98 W and 196 W on a Boron target (Metal Boron) and 18 W, 36 W and 72 W on a phosphorus target (P$_2$O$_5$). The concentration of dopants is proportional to the RF power used based on the deposition rate of B and P$_2$O$_5$. In order to keep the constant SiO$_2$ matrix, the power on SiO$_2$ target was fixed. The SiO$_2$ stoichiometry change will be very small so this effect is only minor effect on absorption comparing to highly doping. X-ray Reflectivity (XRR) was used to accurately measure the film thickness [6]. The equipment used for XRR is a PANanalytical X'Pert MRD, which has a primary X-ray source of CuKα (λ = 0.154 nm), which is collimated with a divergence slit of $1/4°$ (divergence of 0.27°). Incident and reflected beams are respectively collimated by two Soller slits with 2.3° divergence. A $1/32°$ anti-scatter slit was used to improve the resolution of the detection by PIXcel3D detector configured in "receiving slit" mode.

The optical characterization was carried out measuring the reflectance and transmittance of the films

[a] e-mail: `tianz@student.unsw.edu.au`

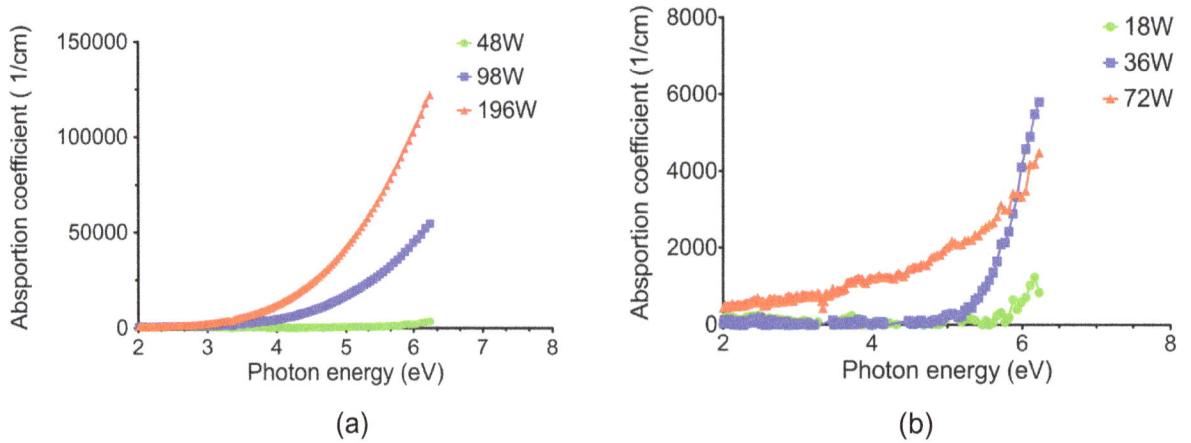

(a)   (b)

Fig. 1. Absorption results of Boron (a) and Phosphorus (b) doped samples are different. Boron has clear absorption relationship with doping level while Phosphorus loses the trend in higher power doping.

Table 1. Thickness measured by XRR and refractive index determined by optical simulation.

Sample	Thickness (nm)	Refractive Index (n)
Boron 48 W	37.6	1.442
Boron 98 W	40.7	1.468
Boron 196 W	48.3	1.617
Phosphorus 18 W	40.0	1.444
Phosphorus 36 W	45.8	1.451
Phosphorus 72 W	60.8	1.474

on quartz substrate using a Lambda1050 PerkinElmer UV/VIS Spectroscopy. A photomultiplier Tube (PMT) detector was used for the 200 nm to 860 nm range and an InGaAs photodiode detector was used for the 800 nm to 2000 nm range. To determine the absorbance of the films, R and T were measured and the absorption coefficient was found as suggested by Maley [7], so that the absorption coefficient can be evaluated simply without iterative procedures. The Cauchy model, usually used for transparent semiconductor, was applied to evaluate the refractive index of the film by modeling T only.

Finally, two samples were examined by XPS. One sample is doped with Boron using 98 W on the Boron target and the other is phosphorus doped using 36 W on the P_2O_5 target. The measurement was carried out using an ESCALAB220i-XL system with an Al Kα X-ray source.

3 Results and discussion

The films thickness was designed to be around 40 nm. However due to the high boron and phosphorus sputtering rates the thickness increases proportionally with the power used on the dopant targets. In Table 1, the refractive index at 600 nm from simulation with Cauchy model shows that the Boron doped samples have a stronger change than phosphorus doped samples. The Cauchy model was applied for simulating T for the non-absorption region of the spectrum. The accuracy of the simulation was improved by using the XRR measured thickness, avoiding the need

to also fit this parameter. It manifests that refractive index increases with doping level.

The absorption coefficient results were evaluated according to N. Maley's approach [7]. The optical properties of the substrate (quartz) are known therefore the R and T on the surface can be easily calculated by using the known index of refraction "n". The equations [7] used for calculating the absorption coefficient are,

$$\begin{cases} T = \frac{T_{sa}T_f}{(1-R_{sa}R_f)} \\ R = R_{sa} + \frac{T_{sa}^2 R_f}{(1-R_{sa}R_f)} \\ \frac{T_f}{1-R_f} = \frac{(1-R_{fa})e^{-ad}}{(1-R_{fa}^2 e^{-2ad})}. \end{cases} \quad (1)$$

Where T and R are the transmittance from the film side and reflectance from the substrate side respectively. T_{sa} and R_{sa} represent the transmittance and reflectance from the substrate/air interface, which can be calculated from the known "n" of quartz. R_{fa} represents the film/air reflectance. But R_{fa} has a minor effect on the absorption coefficient result in practice according to Maley here we used the modeling results at 600 nm for calculation.

According to the absorption results in Figure 1, both boron and phosphorus doped samples exhibit higher absorption starting from the photon energy over 4 eV compared to that of bare SiO_2 (9 eV). The absorption increases proportionally with sputtering power, which is proportional to the doping level. By comparing two doping types, the boron doped samples have much higher absorption than the phosphorus doped ones. The XPS results showed that boron sample (98 W) has 10 at% concentration is much higher than phosphorus (36 W) with 2 at.%. So boron samples must have relatively higher concentration, which could explain the higher absorption since doping concentration is basically proportional to the sputtering power. What should be noticed in the Boron doped sample's absorption curves is that they have a very regular increment with increasing sputtering power. On the other hand, among the phosphorus doped samples the trend deviates when the power reaches 72 W. But it is still clear that the absorption increases significantly over the whole

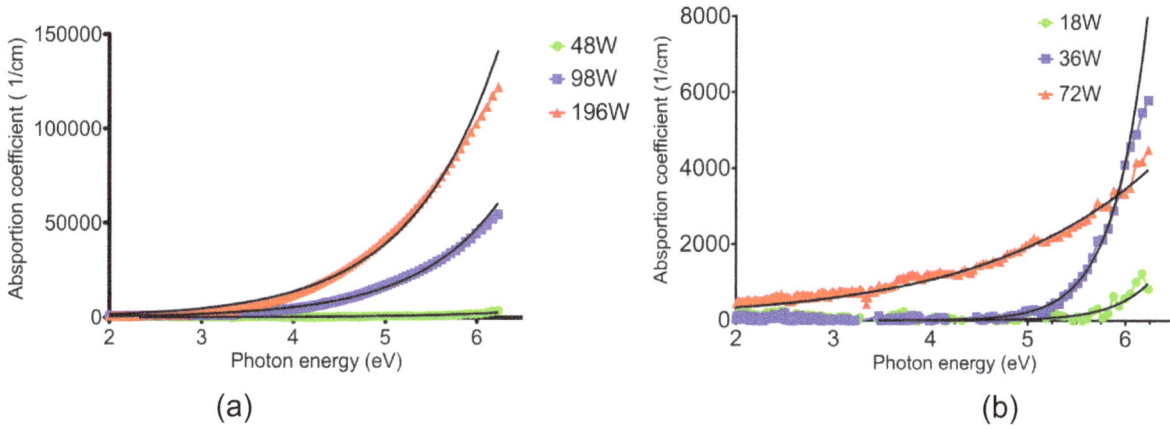

Fig. 2. Graph (a) is B doped samples' absorption with good fitting in exponential curve. Graph (b) is P doped samples only for 18 W and 36 W. 72 W P doped sample has very different absorption curve. It is hard to get good exponential fitting for Phosphorus doped samples because the absorption is quite weak.

Table 2. Doping concentration predicted according to the relation.

Sample	α	k	Tauc edge
Boron 48 W	5.2	1.00	4.7 eV
Boron 98 W	67.8	1.09	5.1 eV
Boron 196 W	203.4	1.05	4.8 eV
Phosphorus 18 W	8.28e-5	2.6	5.5 eV
Phosphorus 36 W	9.44e-5	2.9	5.4 eV
Phosphorus 72 W	101.7	0.59	4.6 eV

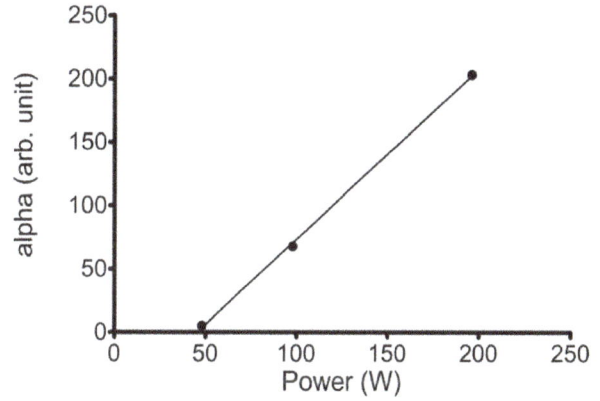

Fig. 3. Boron doped samples exhibit linear relationship between sputtering power and α.

spectrum. It should be mentioned that the PerkinElmer UV/VIS Spectroscopy gave negative absorption values therefore the absorption coefficient is upward shifted by 1000 cm^{-1} offset. The negative value is due to a systematic error in R/T measurement from the PerkinElmer UV/VIS Spectroscopy. Additionally, the absorption in Phosphorus doped SiO$_2$ samples is much smaller than T and R since the film is only 40$-$60 nm. However, the trend is still obvious that absorption is proportional to doping level.

The spectrum range from 4 eV to 6 eV for boron and 5.5 eV to 6 eV for phosphorus were selected for curve fitting in order to overcome lower absorption region that gives more error points making it hard to fit curves. In the case of boron, we can fit curves well by using an exponential function $Y = \alpha e^{kx}$ (Fig. 2(a) and Table 2), however fitting accuracy decreases if we include higher photon energy spectrum. Phosphorus doped samples (Fig. 2(b)) did not show very good fittings (lower R-square) due to the weak absorption which has larger noise. It can be seen that for all the spectrums $E > 6$ eV, the fitting curve deviate from data points. This indicates that Tauc region should be over 6 eV. The P 72 W doped sample has a very broad absorption region compared to the samples doped using lower sputtering power. The reason for this effect is not clear yet This change for high P concentration might be caused by that more defects generated and different kinds of defects might exist.

Comparing α values from fitting, which represents the intensity, it is not possible to correlate concentration with absorption in P doped samples because they have very different k values. On the contrary, Boron doped samples exhibits similar k values thus there is obvious trend in α values (Table 2). We can find a linear relationship between sputtering power and α (Fig. 3). Therefore it might be possible to simply correlate α with doping concentration that is proportional to the sputtering power for B doped SiO$_2$ But it is not effective for P doped samples especially for higher doping level. In order to quantify the doping centration, RBS or SIMS will be necessary.

It is obvious that higher doping B and P into SiO$_2$ dramatically enhance the absorption. Generally it is believed that the dopants will form distinct energy levels as Giusy et al. observed in PL/PLE measurement on phosphorus doped silica [8] and Dong et al. also tried to simulate energy levels for different III/V dopants by using first principle calculation on substitutional dopants in SiO$_2$ materials [9]. However in this experiment it is not likely to be the case that the dopants will form constant defect energy levels within the bandgap. Because if we try

(a)   (b)

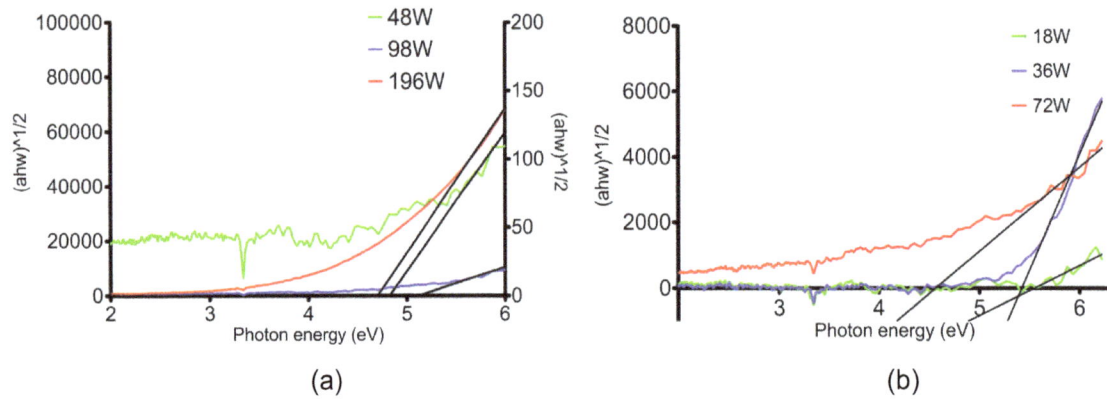

Fig. 4. Tauc plot for Boron (a) and Phosphorus (b) doped samples. 5−6 eV was used for linear fitting. It is clear that X intercepts are very different. Boron 48 W has very low absorption therefore use secondary Y-axis.

to consider that the high energy region has a Tauc plot shape $(ah\omega)^n$, the doped samples should show similar absorption edges independent of doping level. But according to our measurements this is not the case. Here we use $n = 2$ as generally used value, the Tauc absorption edge varies from 5.1 eV to 4.7 eV for boron-doped samples and from 5.5 eV to 4.6 eV for the phosphorous-doped samples (Figs. 4(a) and 4(b)). But it might be necessary for us to extend the optical absorption spectrum to higher energy region to show more accurate Tauc plot and also apply PL measurement to see whether the defect levels really exist.

Urbach tail region following Tauc region is usually manifest in amorphous materials and it is usually related to disorder, impurity and defects. The absorption curves in our case have very Urbach-like shape so it is helpful to interpret the overall trend in the absorption by applying Urbach theory, such as doping induced defects [10]. Additionally, from the fact that P doped samples have very different k values rather than similar k values as B doped samples, which usually represents Urbach energy, it might be the clue that P forms more different and complex bonds network or defects than B. In order to investigate more details on this disordered material from doping, it requires more accurate characterization methods such as by FTIR, PDS or even EPR [11]. Such characterization techniques are beyond what this paper is aiming to present. Here a very simple model could be used to explain the trends we observed.

But it might be easier to apply a simple model to explain the higher absorption with increasing doping level. If we consider the amorphous material mixed with dopants as "condensed solution" we can then apply Beer's law ($\alpha = \varepsilon C d$) that can concisely represent the relationship between the concentration and absorption. C represents defects concentration, d represents film thickness and ε represents absorption cross-section. The absorption is proportional to the defects which means more absorption sites and it is believed to be proportional to doping level.

4   Conclusion

In this paper, we proposed and examine one rapid way to examine the high doping level in SiO_2 thin films by simply evaluating optical absorption. It was found that boron doped SiO_2 exhibits a well proportional relationship between doping level and absorption. However phosphorus doped samples did not show this relationship. More accurate measurements on absorption and doping concentration will be further needed to verify the results in this paper.

The authors acknowledge the support from The School of Photovoltaic and Renewable Energy Technology, The University of New South Wales and The Institute of Semiconductor Electronic, RWTH Aachen University, Aachen, Germany.

References

1. G. Conibeer, M. Green, et al., Thin Solid Films **516**, 6748 (2008)
2. I. Perez-Wurfl, X. Hao, et al., Appl. Phys. Lett. **95**, 153506 (2009)
3. L. Ma, D. Lin, et al., Physica Status Solidi (c) **8**, 205 (2011)
4. X.J. Hao, E.C. Cho, et al., Sol. Energy Mat. Sol. Cells **93**, 273 (2009)
5. X.J. Hao, E.C. Cho, et al., Sol. Energy Mat. Sol. Cells **93**, 1524 (2009)
6. A. Bender, Th. Gerber, et al., Thin Solid Films **229**, 29 (1993)
7. N. Maley, Jpn. J. Appl. Phys. **31**, 768 (1992)
8. Giusy, Fabrisio, et al., Phys. Rev. B 80, 205208 (2009)
9. Dong Han, D. West, et al., Phys. Rev. B **82**, 155132 (2010)
10. M. Offengerg, R. Meyer, et al., J. Vac. Sci. Technol. A **4**, 1009 (1986)
11. M. Stutzmann, D.K. Biegelsen, et al., Phys. Rev. B **35**, 5666 (1987)

17

Thin film pc-Si by aluminium induced crystallization on metallic substrate

F. Delachat[1,a], F. Antoni[1], P. Prathap[1], A. Slaoui[1], C. Cayron[2], and C. Ducros[2]

[1] CNRS-UdS InESS, Strasbourg, France
[2] CEA-Liten, Grenoble, France

Abstract Thin film polycrystalline silicon (pc-Si) on flexible metallic substrates is promising for low cost production of photovoltaic solar cells. One of the attractive methods to produce pc-Si solar cells consists in thickening a large-grained seed layer by epitaxy. In this work, the deposited seed layer is made by aluminium induced crystallization (AIC) of an amorphous silicon (a-Si) thin film on metallic substrates (Ni/Fe alloy) initially coated with a tantalum nitride (TaN) conductive diffusion barrier layer. Effect of the thermal budget on the AIC grown pc-Si seed layer was investigated in order to optimize the process (i.e. the quality of the pc-Si thin film). Structural and optical characterizations were carried out using optical microscopy, μ-Raman and Electron Backscatter Diffraction (EBSD). At optimal thermal annealing conditions, the continuous AIC grown pc-Si thin film showed an average grain size around 15 μm. The grains were preferably (001) oriented which is favorable for its epitaxial thickening. This work proves the feasibility of the AIC method to grow large grains pc-Si seed layer on TaN coated metal substrates. These results are, in terms of grains size, the finest obtained by AIC on metallic substrates.

1 Introduction

Compared to classical photovoltaic technologies based on bulk crystalline silicon, thin film approach enables to reduce the material consumption drastically. Furthermore, it allows large area deposition on low-budget foreign substrates. Thus, thin film polycrystalline silicon (pc-Si) solar cells on non-silicon substrates are interesting to reduce the cost of photovoltaic electricity provided high quality silicon is produced and efficient optical confinement is applied.

As substrate's candidates, ceramic or glass ceramic insulating materials have been previously suggested [1–3] Nevertheless, metallic foils can offer similar and additional advantages. They are cheap, thermally stable and flexible thus they can be rolled. Several material technologies have been proposed to obtain pc-Si thin films on metallic substrate. Some technologies consist in using buffer and template layers on an metallic substrate for the hetero-epitaxial growth of large-grain (20-50 μm) silicon thin films (2-10 μm) [4,5]. However, these approaches do not take profit of the metal substrate conductivity. Hence, the theoretical efficiency of such cells will be strongly limited by an important shadowing effect due to the inter-digitated contact configuration. Another technology consists in the direct deposition of crystalline silicon films

produced by standard plasma processes at low temperature. The metal substrate directly plays the role of the back-contact in a double-side contact configuration cell. Efficiency up to 5.8% has been reported [6]. However, the size of the Si crystallites are relatively small (<1 μm) which limited the improvement of the electronic quality of the Si thin films thus obtained. One attractive method to produce pc-Si solar cells consists in thickening by epitaxy a large-grained seed layer. Large grains p-type pc-Si can be obtained by aluminum induced crystallization (AIC) [7]. Thin-film polycrystalline-silicon solar cells based on AIC and thermal CVD with 8% efficiency has already been reported on alumina substrate [8].

In the present work, aiming to exploit simultaneously the advantages of the AIC method and metallic substrates, the feasibility of AIC process on a metallic substrate is investigated. The AIC method is a relatively simple process, which enables a low thermal budget and a much shorter crystallization time compared to solid phase crystallization activated by classical annealing. To avoid any diffusion of metallic impurities from the substrate during the process, a tantalum nitride (TaN) conductive diffusion barrier layer has been used. The annealing temperature and time are the key parameters of the Al/Si exchange process. Thus, the thermal budget influence on the pc-Si film quality was studied for optimizing the AIC process on metallic substrate coated by TaN.

[a] e-mail: `florian.delachat@icube.unistra.fr`

Table 1. AIC thermal budget parameters.

Sample id.	Temperature (°C)	Time (h)
S.1	450	16
S.2	475	16
S.3	500	8
S.4	525	4
S.5	550	2

2 Experiment details

Metallic foils of 5×5 cm^2 made of ferritic steel (from APERAM Inc.) were used as substrates. Because the substrate's roughness can modify the nucleation rate [5, 6], it has a major influence on the crystallographic quality of the AIC grown pc-Si layers. Therefore, the substrates with the minimal roughness achievable were used. Surface maps have been realized by WYKO NT9100 surface profiler on the substrates prior to the deposition of TaN layers. An average roughness of 255 nm has been measured on a representative sample. This controlled surface roughness is favorable to the AIC process [9]. The metallic substrates were coated with a 1 μm thick cubic face-centered TaN layer deposited by sputtering. Afterwards, as a precursor in the AIC process, aluminium layer (200 nm thick) was deposited by e-beam evaporation (EBE). The samples were then exposed to air for 1 week prior to amorphous silicon (a-Si) deposition in order to achieve an AlO$_x$ permeable membrane. This membrane is essential for a successful layer exchange [10]. Afterward, the AlO$_x$/Al layers were coated with radiofrequency magnetron sputtered a-Si (400 nm).

Then, the samples were annealed in a quartz tube furnace under nitrogen gas flow. The investigated temperature range was chosen above the threshold crystallization temperature of a-Si in contact with aluminium and below the eutectic temperature (Al/Si; $T_{eu} = 577$ °C). By considering that for lower temperature annealing the incubation time (characteristic time needed to form the first nuclei) is longer than for higher temperature, the thermal budget was compensated by extending the annealing time. However, a maximum time of 16 h was fixed in a partial way in order to limit the duration of the whole process. The parameters used to optimize the AIC annealing are reported Table 1.

After annealing, a residual layer composed of aluminium and silicon islands on the top of the pc-Si surface is formed. This layer was removed by an appropriate chemical etching or mechanical polishing. A schematic illustration of the AIC process on these samples is reported in Figure 1.

The resulting p-type (Al) pc-Si thin films obtained were analyzed with a Renishaw RAMASCOPE 2000 μ-Raman spectrometer using the 633 nm excitation wavelength of a HeNe laser. The crystalline fractions were evaluated from the optical microscope observations. The crystal orientation and grain size analysis were carried out by SEM LEO 1530 using the electron backscatter diffraction (EBSD) configuration.

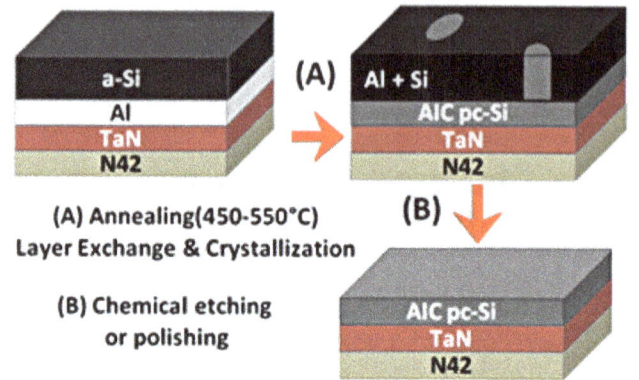

Fig. 1. Schematic illustration of the AIC process on metallic substrate.

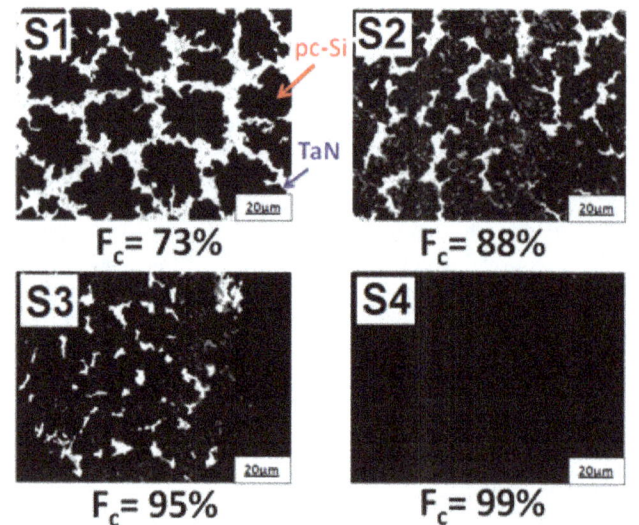

Fig. 2. Optical microscope images of AIC grown pc-Si layer after annealing and chemical etching of the residual Al/Si layers. The images were processed by a grey level converter in order to estimate the crystallized fraction (F_c).

3 Results and discussions

3.1 Grains surface coverage

After annealing, the residual Al/Si top layer was removed by chemical etching. Hence, the samples were analyzed by optical microscope. The surface images of the AIC-grown pc-Si are reported in Figure 2. The images were consecutively converted in high contrasted gray level and filtered in order to estimate the crystallized fraction (F_c).

On these images, the surface coverage of the silicon crystallites on the top of the TaN layer is clearly visible. The crystallized fraction can be described by the analysis of the grains surface coverage of the optical microscope images. The resulting values are reported under each image on Figure 2. The crystallized fractions increase from S1 to S5. It reaches 73% for S1 while it is closed to 100% for S4 and S5 (S5 is not shown here as it is similar to S4). The crystallization is almost completed

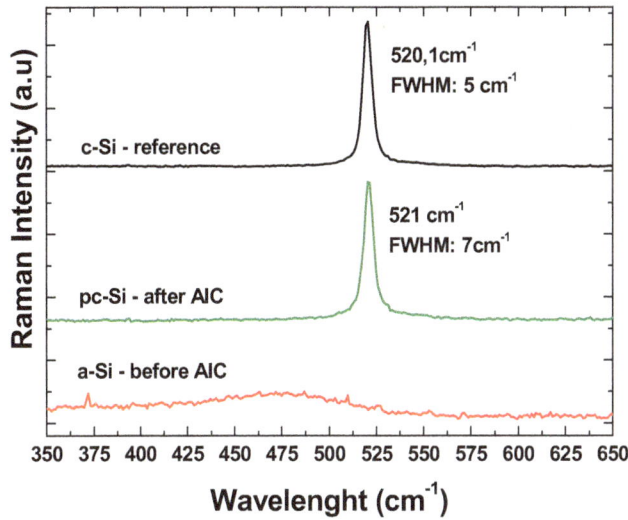

Fig. 3. Raman spectra of a representative sample (S3) before and after AIC annealing. A c-Si μ-Raman spectrum is also plot for reference purpose.

for sample S3 where F_c reaches 95%. A continuous pc-Si layer is formed for the following annealing conditions (525 °C – 4 h and 550 °C – 2 h).

3.2 Optical and structural analyses of the pc-Si layers

All the samples were analyzed by μ-Raman spectroscopy, before and after annealing. The system has been calibrated first with the transversal optic (TO) phonon band of a mono-crystalline silicon (c-Si) wafer. Representative spectra of the measurement is shown in Figure 3. The reference spectrum of the c-Si is also reported in the same graph for comparison.

Before AIC annealing, the amorphous phase of the as-deposited silicon is witnessed by a large band at around 480 cm^{-1}, attributed to the transverse optical (TO) phonons related band of a-Si [11]. After AIC annealing, a typical Lorentzian-like band centered at 521 cm^{-1} is attributed to the transverse optical (TO) band of crystalline silicon. On this spectrum, the band attributed to the amorphous silicon has totally disappeared which indicates the full crystallization of the pc-Si layer. The full half width maximum (FHWM) of the Raman peak is equal to 7 cm^{-1}, which is slightly higher than the 5 cm^{-1} of the c-Si reference. This value is similar for all samples, which means that the degree of crystallinity is relatively acceptable for all samples.

On the other hand, if we consider the Raman peak position versus the thermal budget as reported in Figure 4, we observe that the thermal budget has a considerable influence on the stress that affects the pc-Si layer.

Indeed, compressive stress induces a shift of the Raman peak to higher wavelengths. The correlation between the Raman shift and the stress can be roughly determined from the following formula [12]:

$$\sigma(\text{MPa}) = -250\,(\omega_{\text{s}} - \omega_0)$$

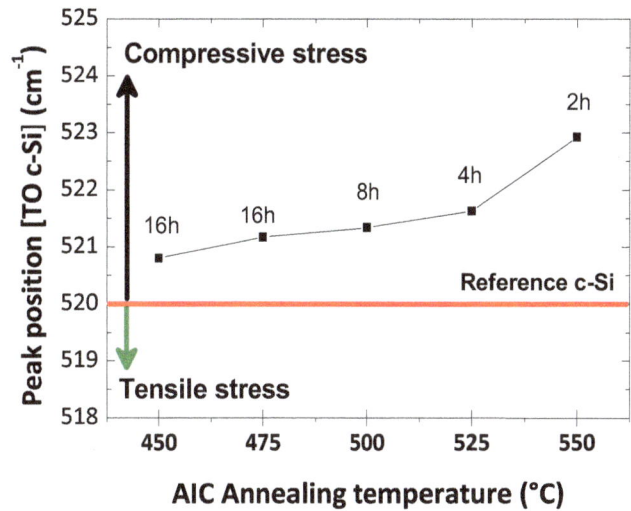

Fig. 4. Evolution of the Raman peak position as a function of the thermal treatment applied.

Table 2. Evaluation of the stress in the pc-Si layer according to the thermal treatment received.

Sample id.	Raman shift (cm^{-1})	σ (MPa)
S1	0.7	−175
S2	1.1	−275
S3	1.2	−300
S4	1.5	−375
S5	2.8	−700

where ω_0 is the wavenumber of the stressed free c-Si and ω_s is the wavenumber of the stressed pc-Si layer. The stress deduced value for each sample is reported in Table 2.

As the thermal treatment increases, the compressive stress observed in the layer is increasing too. A relative constant stress growth is observed from S1 to S4, from 175 up to 375 MPa respectively. For the sample S5, the stress is then raising strongly to reach 700 MPa. The stress in the layer can be induced by the grain boundaries which are known to compressively stress the pc-Si films [13]. The increase of the compressive stress is then correlated to the decrease of the grains size.

In order to investigate the crystallographic quality of the pc-Si layer more deeply, cross-sectional EBSD analyses were carried out. Prior to EBSD, a cross-sectional scanning electron microscopy observation of the sample was realized. This observation enables notably to check the surface quality of the polished surface. A picture of a representative sample is reported in Figure 5.

On this sample, the residual layer composed of aluminium and silicon islands on the top of the pc-Si surface has not yet been removed. The porous nature of this residual layer is clearly visible on this instructive picture. Thank to this cross sectional observation, each component of the multilayer can be clearly identified (metal/barrier layer/pc-Si/residual layer). Thus, the good quality of the layer exchange is clearly witnessed. The thickness of the pc-Si layer is about 200 nm, in accordance with the precursor aluminium layer's thickness.

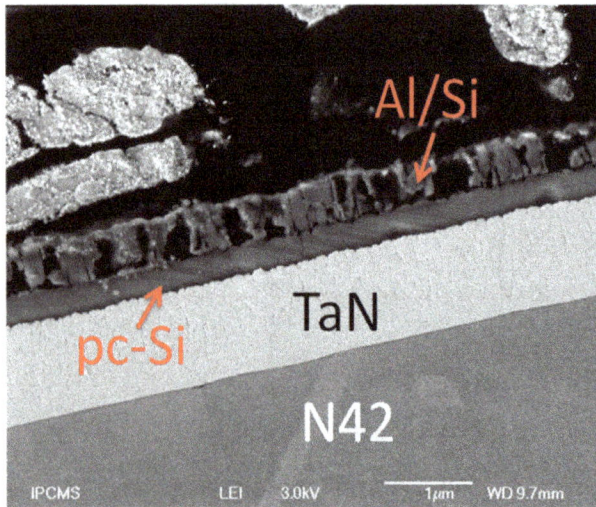

Fig. 5. Cross sectional scanning electron microscopy of a representative sample after AIC processing.

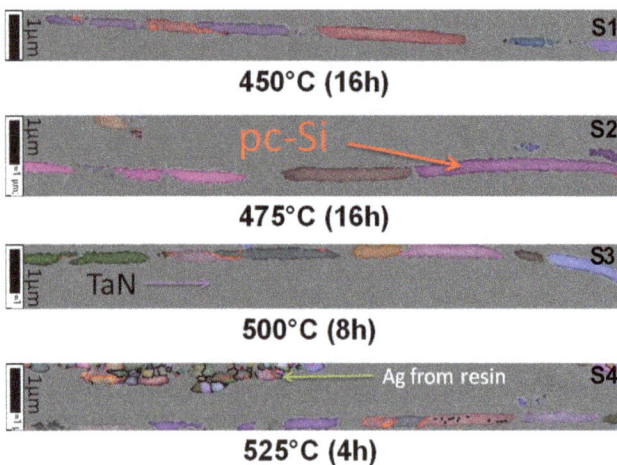

Fig. 6. Cross sectional orientation map deduced from EBSD observation (from top to bottom S1, S2, S3, S4 respectively). Each color in these images indicates a specific crystallographic direction.

The EBSD images of sample 1 to sample 4 are reported in Figure 6. Each color in these images indicates a specific crystallographic direction. The influence of the annealing conditions on the grain size can be therefore investigated. The portion of layers analyzed is too small to give reliable statistical values on the grain size; but it gives a good idea to define the optimization direction. Indeed, we can distinguish that the low annealing temperature results in much larger grain size (up to 5 μm for S1 and S2). This is clearly due to a lower nucleation rate at low temperature [14]. However, even if the grain size of the crystallites is higher, the pc-Si is not continuous for both of these samples, even though the longest annealing time of 16 h has been used. For S3, the pc-Si layer is not completely continuous ($F_c = 95\%$), thus a longer time of annealing will certainly lead to a complete exchange of the layers. For S4, the apparent grain size is much smaller than for S3 but the layer is continuous ($F_c = 99\%$).

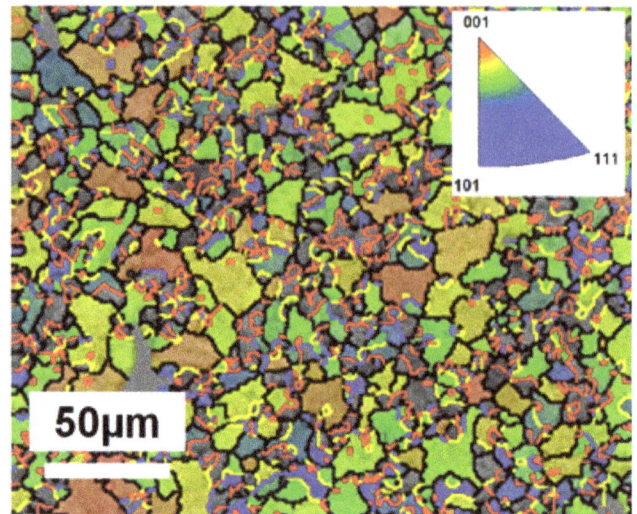

Fig. 7. Orientation map deduced from EBSD analysis. The lines shows the grain boundaries (twins: red = $\Sigma3$, yellow = $\Sigma9$, blue = $\Sigma29$, black = randomly orientated). The inverse pole figure in the z-axis in reported on the inset (red color = high density, blue = low density).

As a summary, from the grain size estimation versus the exchange annealing conditions we can observe that the lowest the temperature conditions (S1 and S2) result in the biggest grains. However, the process is not completed after 16 h, which lead to non-continuous pc-Si layers. For the highest temperature conditions (S4 and S5), the pc-Si is continuous, but the grain size is limited to a few micrometers. In addition, the induced stress in the pc-Si is increasing with the temperature augmentation. Hence, the best compromise we choose to fit our requirements is to process the layer at 500 °C. At this temperature a consequent grain size is achievable. In order to complete the AIC process at this temperature the annealing time was increased up to 16 h.

4 Optimization results

Consequently to the previous experiments, a new sample was annealed at 500 °C during 16 h. After this process, the residual Al/Si layer was removed by chemical etching in order to achieve an EBSD plan view analysis. The plan view EBSD analysis is reported in Figure 7.

This observation is in good agreement with the expected optimization's results. Indeed, the EBSD orientation map shows a continuous pc-Si layer formed by AIC at 500 °C. The inverse figure pole reported in the inset Figure 6 indicates that there is preferential orientation along the $\langle 100 \rangle$ direction as usually reported for AIC layers [15]. From the orientation map, the grain size can be evaluated. For that purpose, a representative area of 1.44 mm² was analyzed. The extracted grain size distribution is reported in Figure 8.

The diameter of each grain is deduced by assuming disk geometry of a similar area. The grain size is comprised

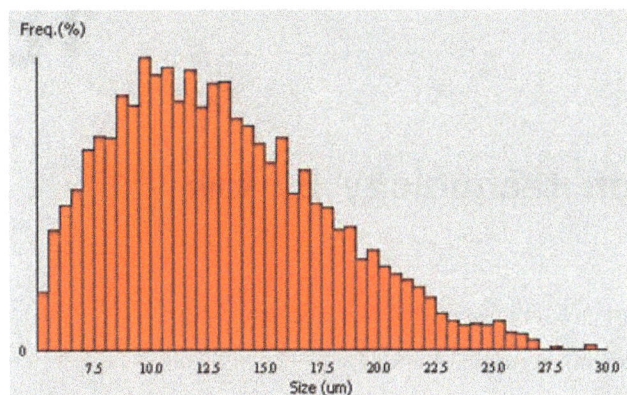

Fig. 8. Grain size distribution of the pc-Si thin films on metal substrate obtained by AIC after optimization.

between 3 to 30 μm with an average size around 15 μm. This result is, from our knowledge the best result obtained by AIC on metallic substrate up to now.

5 Conclusion

In this work, a-Si thin films were crystallized by AIC on a metallic substrate (Ni/Fe alloy) coated with a conductive diffusion barrier layer (TaN). The effects of the thermal budget on the pc-Si obtained were investigated in order to optimize the AIC process on such TaN coated metal substrates. The consecutive optical and structural analyses enabled us to achieve a continuous p-type pc-Si layer composed of the large grains ($\approx$15 μm on average) with limited residual stress. This work proves that a good crystallization can be obtained by AIC on metallic substrates (N42) coated with a TaN conductive barrier layer. The AIC method applied to metal substrate enables to produce pc-Si formed by larger grains than those achievable by a thermal annealing of a-Si or by CVD direct deposition of pc-Si. This result is promising is for the fabrication of thin film solar cell on a metallic substrate by the epitaxial thickening of the seed layer thus obtained. The electronic quality of the AIC seed layer is under evaluation by employing it into a solar cell.

The authors gratefully acknowledge G. Ferblantier for the technical assistance with the magnetron sputtering system, J. Faerber for the SEM observations. The authors would like also to thank S. Roques, S. Schmitt and J. Bartringer for their valuable contributions. This work was partially funded by the National Research Agency (ANR-Habisol) in the project SILASOL.

References

1. G. Beaucarne, S. Bourdais, A. Slaoui, J. Poortmans, Thin Solid Films **403**, 229 (2002)
2. Ö. Tüzün, Y. Qiu, A. Slaoui, I. Gordon, C. Maurice, S. Venkatachalam, S. Chatterjee, G. Beaucarne, J. Poortmans, Sol. Energy Mater. Sol. Cells **94**, 1869 (2010)
3. A. Slaoui, E. Pihan, A. Focsa, Sol. Energy Mater. Sol. Cells **90**, 1542 (2006)
4. A.T. Findikoglu, W. Choi, V. Matias, T.G. Holesinger, Q.X. Jia, D.E. Peterson, Adv. Mater. **17**, 1527 (2005)
5. C.W. Teplin, M.P. Paranthaman, T.R. Fanning, K. Alberi, L. Heatherly, S.-H. Wee, K. Kim, F.A. List, J. Pineau, J. Bornstein, Energ. Environ. Sci. **4**, 3346 (2011)
6. A. Torres Rios, Epitaxial Growth of Crystalline Silicon on N42 Alloys by PECVD at 175 °C for Low Cost and High Efficiency Solar Cells, in *Proceeding 27th E.U. PVSEC*, Vol. Thin Film Crystalline Silicon Solar Cells, No. 3DO.7.1 (2012)
7. P. Prathap, O. Tuzun, D. Madi, A. Slaoui, Sol. Energy Mater. Sol. Cells **95**, S44 (2011)
8. I. Gordon, L. Carnel, D. Van Gestel, G. Beaucarne, J. Poortmans, Prog. Photovolt.: Res. Appl. **15**, 575 (2007)
9. E. Pihan, *Élaboration et caractérisations de silicium poly-cristallin par cristallisation induite par aluminium de silicium amorphe : Application au photovoltaïque*, thesis, Université de Strasbourg, 2005
10. S. Gall, Polycrystalline Silicon Thin-Films Formed by the Aluminum-Induced Layer Exchange (ALILE) Process, in *Crystal Growth of Silicon for Solar Cells*, edited by K. Nakajima, N. Usami (Springer Berlin Heidelberg, 2009), Vol. 14, pp. 193–218
11. M.H. Brodsky, M. Cardona, J.J. Cuomo, Phys. Rev. B **16**, 3556 (1977)
12. V. Paillard, P. Puech, M.A. Laguna, P. Temple-Boyer, B. Caussat, J.P. Couderc, B. de Mauduit, Appl. Phys. Lett. **73**, 1718 (1998)
13. Ö. Tüzün, *Polycrystalline Silicon Films by Aluminium Induced Crystallization and Epitaxy: Synthesis, Characterizations and Solar Cells* (UdS-CNRS, 2009)
14. Ö. Tüzün, A. Slaoui, C. Maurice, S. Vallon, Appl. Phys. A **99**, 5361 (2009)
15. D. Van Gestel, I. Gordon, L. Carnel, K. Van Nieuwenhuysen, J. D'Haen, J. Irigoyen, G. Beaucarne, J. Poortmans, Thin Solid Films **511-512**, 3540 (2006)
16. I. Gordon, D. Van Gestel, K. Van Nieuwenhuysen, L. Carnel, G. Beaucarne, J. Poortmans, Thin Solid Films **487**, 113 (2005)
17. E. Pihan, A. Slaoui, A. Focsa, P.R.I. Cabarrocas, Polycrystalline silicon films on ceramic substrates by aluminium-induced crystallisation process, in *Proceedings of 3rd World Conference on Photovoltaic Energy Conversion, Osaka, Japan* (2003), Vol. 2, pp. 1182–1185

18

Power change in amorphous silicon technology by low temperature annealing

Ankit Mittal[1,2,a], Marcus Rennhofer[1], Angelika Dangel[1], Bogdan Duman[1], and Victor Schlosser[2]

[1] Photovoltaics System, Austrian Institute of Technology, 1220 Vienna, Austria
[2] Department of Physics, University of Vienna, 1010 Vienna, Austria

Abstract Amorphous silicon (a-Si) is one of the best established thin-film solar-cell technologies. Despite its long history of research, it still has many critical issues because of its defect rich material and its susceptibility to degrade under light also called as Staebler-Wronski effect (SWE). This leads to an increase in the defect density of a-Si, but as a metastable effect it can be completely healed at temperatures above 170 °C. Our study is focused on investigating the behavior of annealing of different a-Si modules under low temperature conditions below 80 °C indicated by successive change of module power. These conditions reflect the environmental temperature impact of the modules in the field, or integrated in buildings as well. The power changes were followed by STC power rating and investigation of module-power evolution under low irradiance conditions at 50 W/m^2. Our samples were recovered close to their initial state of power, reaching as high as 99% from its degraded value. This shows the influence of low temperature annealing and light on metastable module behavior in a-Si thin-film modules.

1 Introduction

The demand of amorphous silicon in the photovoltaic industry was growing quite rapidly for several years, which gave this technology a substantial share in the market [1]. Marginal cost of amorphous silicon (a-Si) is playing its decent role in the green energy market, since in many parts of the world has the short fall in energy supply [2]. Further research to make the a-Si technology cheaper with higher efficiency is therefore needed. Under diffuse light conditions, this technology is found to be good and can yield better performance than crystalline silicon [3, 4]. Therefore, it is an optimal technology for applications in the tropical climates, low AM, high turbidity and tends to give better performance behavior in blue light than in the red spectrum of the sun [4–9].

It is also a suitable technology for installing the modules in facades as it has low temperature coefficients for power and power gain due to thermal annealing effect during its operation outdoors [3].

The SWE has been known for many decades. It leads to a reduction in photoconductivity and dark conductivity of the a-Si solar cells [10]. Although the effect has been

studied for many decades a commonly agreed reason for the cause of this effect is still unknown. Therefore, degradation effects are central focus for understanding the fully behavior and formation of this PV material. A-Si technology has a unique property which leads in reversibility of the degraded photoconductivity, which is unique under all PV technologies. It can be attained by heating the cell at 170 °C [11] which leads to annihilation of all the defects.

Many defect models have been proposed to explain this phenomenon. The Bond breaking model was the first widely accepted model to explain this phenomenon in terms of electronic properties [12]. Thereafter, the Hydrogen collision model came [13] and then cluster phase model [14].

In this paper, all the experiments were carried out at the commercial module level in order to analyze the impact of low temperature annealing and low light irradiance on the modules maximum power. This was chosen in order to reflect out-door conditions. The experimental section will explain about the procedures followed during the low temperature annealing of the modules. The results and discussion section will show the recovery of the a-Si modules initial power at low temperature annealing and further discuss the influence of metastabilities due to pretreatment effect under light and in dark.

[a] e-mail: `ankit.mittal.fl@ait.ac.at`

Table 1. Modules used during the experiment.

Modules	Annealing temp. [°C]	Initial power [Wp]	Stabilization
sample 1	60–70	116	outdoor
sample 2	65–70	118	indoor
sample 3	60–70	2	indoor

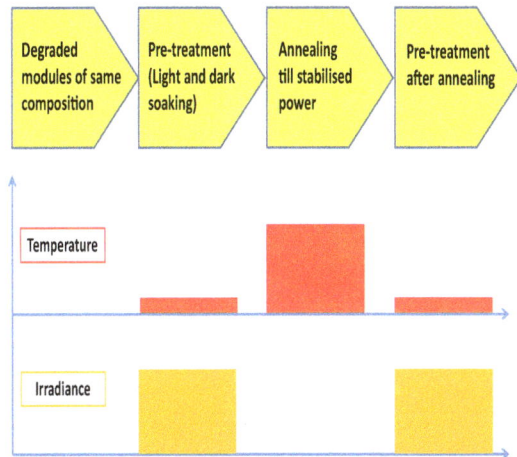

Fig. 1. Experimental scheme to analyze the effect of temperature and irradiance on a-Si modules.

2 Experimental details

Various commercial modules of double junction a-Si, stabilized indoors and outdoors were used for the experiments. The chosen a- Si modules are tabulated in Table 1. The samples 1 and 2 were the commercial modules (1.1 m × 1.3 m) of double junction a-Si on a glass substrate. The sample 3 was a-Si double junction on a polymer substrate type with a size of 15 cm × 45 cm.

Sample 1 was held outdoors operating at close to short circuit current (exactly at $V = 1/3 \times V_{OC}$, STC) for around 3 years. Samples 2 and 3 were stabilized indoors using a BAA-class static sun simulator.

2.1 Measurement protocol

The basic measurement protocol to see the effect of annealing and low irradiance is shown in Figure 1. Following scheme was:

a. degradation of the modules;
b. pre-treatment cycle, as described in Section 2.4;
c. annealing at low temperature, as described in Section 2.2;
d. pre-treatment.

After step (a), steps (b)–(d) were repeated until the module power was stabilized.

2.2 Low temperature annealing

All the chosen modules were put in a climatic chamber at a particular temperature for a certain time until the power got stabilized at that temperature i.e. when the power from the two two successive annealing steps were almost equivalent to each other. Low temperature annealing of the modules were initiated at 60 °C (except for sample 2), 65 °C and 70 °C. All modules were annealed completely in dark. Sample 2 was annealed in light and dark both. This was done by annealing it at a temperature T_i under illumination until stabilization, which means less than ±1% change in power from the last measurement point, and then the measurement protocol was followed (see Sect. 2.1, (b)–(d) again at T_i).

2.3 Power rating

During the annealing period a successive IV characterization was done after 30 h to 65 h of annealing in a pulsed solar simulator for each module, respectively. All measurements were done according to the IEC 60904 standard (procedure for IV characteristics measurement) [15] and corrected according to the IEC 60891 for the temperature and irradiance correction [16] at STC (Standard test conditions i.e. 1000 W/m^2, 25 °C, Air mass 1.5) [17].

2.4 Illumination under low-light conditions

To measure the effect of the annealing on the power rating in more detail prior to the STC power rating as described in Section 2.3, a pre-treatment was carried out. Before the start of pre-treatment, the modules were stored in dark overnight to eliminate any effects from light. In pre-treatment, the modules were illuminated ("light soaking" – LS) for 2 h and then immediately stored in dark ("dark soaking" – DS) for 2 h. The modules were exposed to an irradiance of ∼50 W/m^2 during LS, and for successive time intervals an IV-curve was measured in a pulsed solar simulator at STC conditions for both LS and DS cycle. The detailed procedure for the thin film pre-treatment as done here is described elsewhere [18–20]. A pre-treatment cycle was performed to show that annealed modules have a much stable behavior regarding power changes from low light illumination as they appear at lab manipulation and handling (approx. 50 W/m^2).

2.5 Arrhenius' law

The Arrhenius equation is a simple and accurate formula for the temperature dependence of the reaction rate constant. It is valid, if the basic activated mechanism is following the thermal Boltzmann statistics of energy distribution. Here it is used as an indicator for thermal activation of the changes in the modules resulting from the annealing procedure.

An approach to the evaluation here was a standard Arrhenius behavior in which the relationship between defects and other intrinsic parameters and the power change with the temperature. Here the temperature T dependent parameter tested was the modules power P-(T), while the

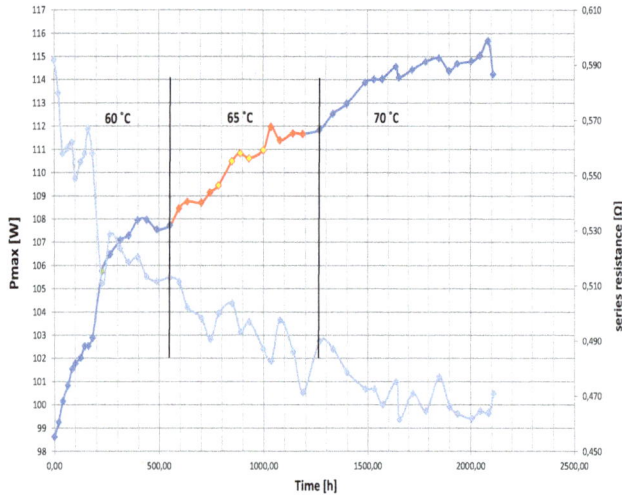

Fig. 2. Power evolution vs. series resistance of sample 1, annealed in dark. The lines are for guidance to the eyes.

intrinsic constants c_0 and c_1 linking the materials properties changing in the modules cells and the power are still unknown and will be measure of experiments to follow.

$$c_0 \Delta P(T) = c_1 \Delta P_0 e^{\frac{-E_a}{kT}}, \qquad (1)$$

where E_a is the activation energy of the driving process in [J], k is the Boltzmann constant and T is the temperature in [K].

3 Results and discussion

3.1 Experimental results

3.1.1 Impact on module power

All modules annealed showed an increase in their STC power. The increase depends on the history of the module and the annealing temperature. The power of sample 1 (cf. Tab. 1), which was prior degraded outdoor, increased 16.5% from its degraded state as shown in Figure 2 along with its series resistance. In Table 2, the relative and absolute increase in power from annealing is also shown.

Series resistance is generally caused by the intrinsic resistance of the a-Si cell and all successive transition resistances (e.g. between metal contact and silicon). Its increase causes loss in performance of the module. In sample 1 a decrease of the series resistance from 0.51 Ω to 0.46 Ω was registered, as shown in Figure 2. Also, the shunt resistance is caused by the manufacturing defects. Low shunt resistance in a solar cell, cause loss in power by providing an alternate current path. Here, the shunt resistance was increased from 42.8 Ω to 64.3 Ω along with power, as shown in Figure 3.

The changes in series and shunt resistances are small, but they are shown in graphs correlating to the power change with each annealing cycle.

In Figure 2, the first part for about 550 h represent the 60 °C annealing. The module has stabilized after

Table 2. Power evolution for sample 1.

Temp. [°C]	Power [W]	Power [%]	116 Wp = 100%	
25	98.6	100	85	
60	107.7	109.2	92.9	+7.9
65	111.7	113.2	96.3	+3.4
70	114.9	116.5	99.0	+2.7

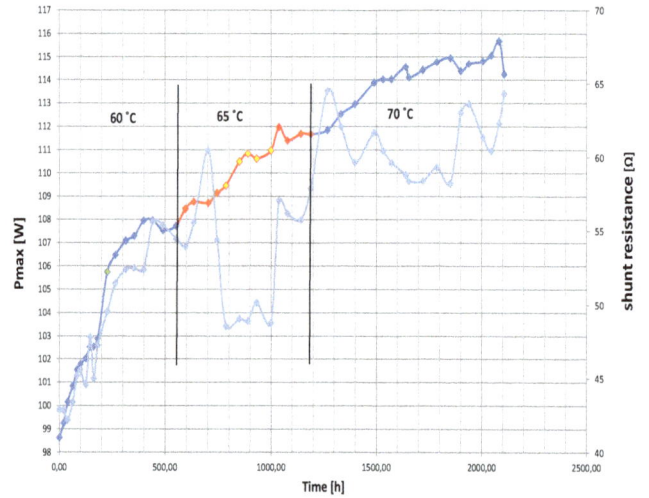

Fig. 3. Power evolution vs. shunt resistance of sample 1, annealed in dark. The lines are for guidance to the eyes.

about 553 h starting at 98.6 ($\pm$0.2%) Watt peak (Wp) and a final stabilized value of 107.7 Wp ($\pm$0.2%) of power. This results in a performance improvement of 9.1 W (9.2%). At 65 °C cycle, the stabilized final value was reached after 635 h with 111.7 Wp ($\pm$0.2%), resulting in a further increase of 4 W (4%). At 70 °C the module finally reached 114.9 Wp ($\pm$0.2%) after 920 h. The power increase from 65 °C to 70 °C cycle was 3.2 W (3.3%). Thus, the overall increase for three runs is 16.3 Wp, i.e. 16.5% from the degraded value.

In Figure 4, the relative increase in power of the module for each temperature step is shown. All temperature steps lead to an increase of power. The initial power value after manufacturing (data sheet value of 116 Wp) was not reached.

A different behavior was found for the flexible thin film module, sample 3 as given in Figure 5. The performance at the first measuring point was 1.82 Wp (stabilized value). After about 553 h at 60 °C the power increased to 1.93 Wp ($\pm$0.8%). This results in an increase of 0.11 Wp (6.2%). During the 65 °C and 70 °C run, no further change in power was registered.

In Table 3 the maximum power change due to annealing is summarized.

3.1.2 Impact on low light behavior

Before starting with annealing, a pre-treatment cycle was performed, as explained in Section 2.4. After

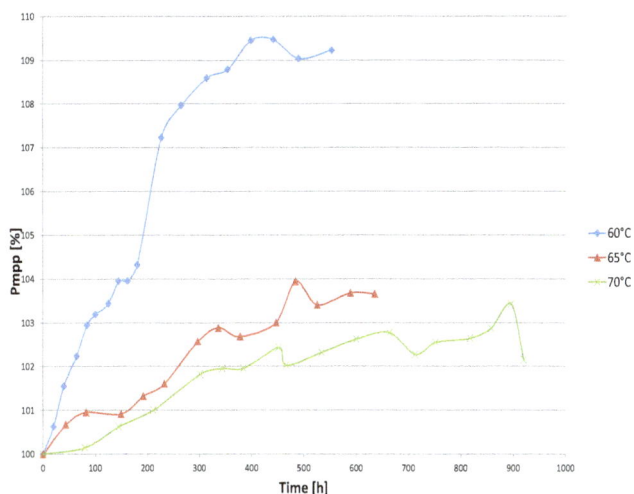

Fig. 4. Power curve normalized to its initial state at each annealing temperature of sample 1.

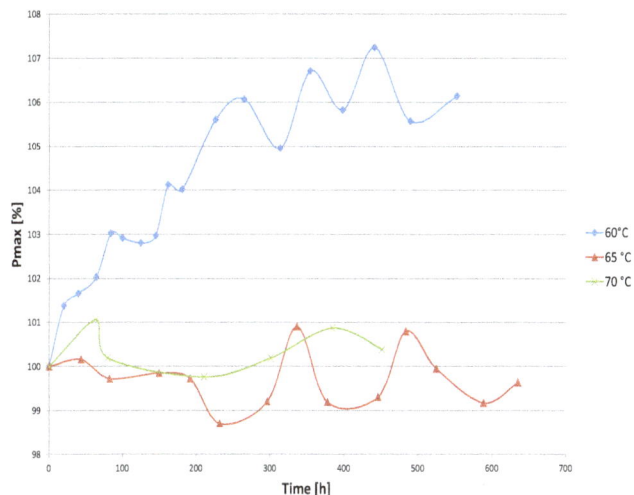

Fig. 5. Power curve normalized to its initial state at each annealing temperature of sample 3.

Table 3. Power evolution for sample 3.

Temp. [°C]	Power [W]	Power [%]	2 Wp = 100%	
25	1.82	100	91	
60	1.93	106.3	96.5	+6.3
65	1.92	105.9	96.0	−0.4
70	1.93	106.2	96.5	+0.3

the annealing procedure was finished, a subsequent pre-treatment cycle was again carried out. All IV characterizations were measured at STC (standard test conditions). The obtained results were compared with the evolution characteristics found for the degraded (i.e. initial) state. Figure 6 shows the evolution of module power at STC during the pre-treatment cycle for the degraded state (Fig. 6a) and the annealed state (Fig. 6b). In Figures 6a and 6b the power was normalized to the initial

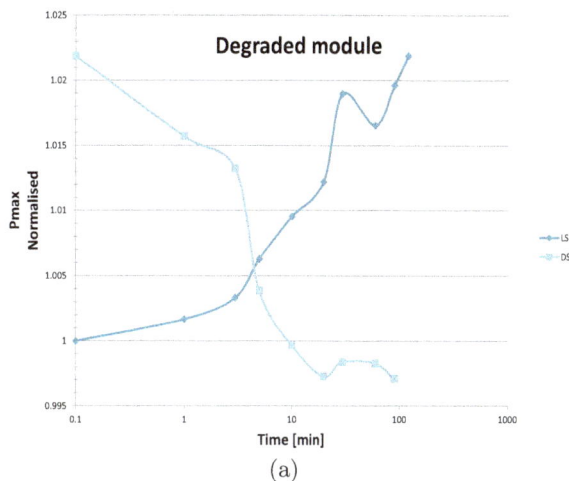

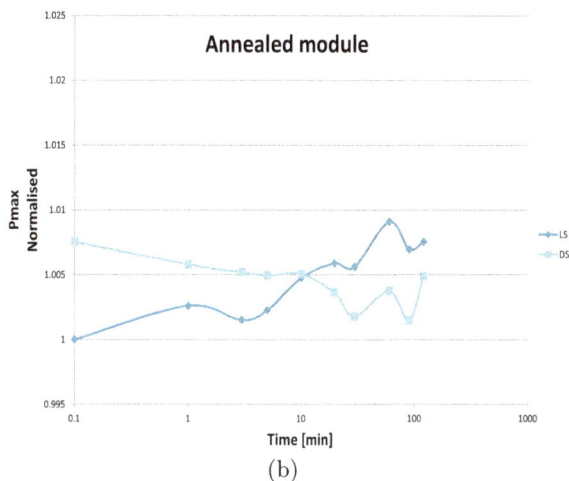

Fig. 6. Evolution of the normalized power of a double junction a-Si module (sample 2) before (a) and after (b) annealing.

measurement point of LS i.e. the IV curve of the module stored in dark overnight.

The last point ($t = 120$) of LS curve is the starting point ($t = 0$) of DS cycle and the end point of the DS cycle shows the completion of pre-treatment. The power change from the initial point of LS was compared with the final point of DS.

During the pre-treatment of the degraded module the output power, as shown in Figure 6a, gradually increases by +2.2% under illumination (LS) followed by a relaxation with reversed dynamics while holding in dark (DS). The initial power of the degraded module was recovered after 10 min in the dark while the module power saturation (−2.5%) was reached after 20 min at a value being slightly below the initial value before start of the LS cycle. Application of the pre-treatment procedure on annealed modules yields a remarkable change of the amplitude and dynamics of the metastable behavior. There, the magnitude of the power change due to illumination was much smaller during the period of the LS (>0.8%) and DS (<0.3%) treatments, respectively (see Fig. 6b).

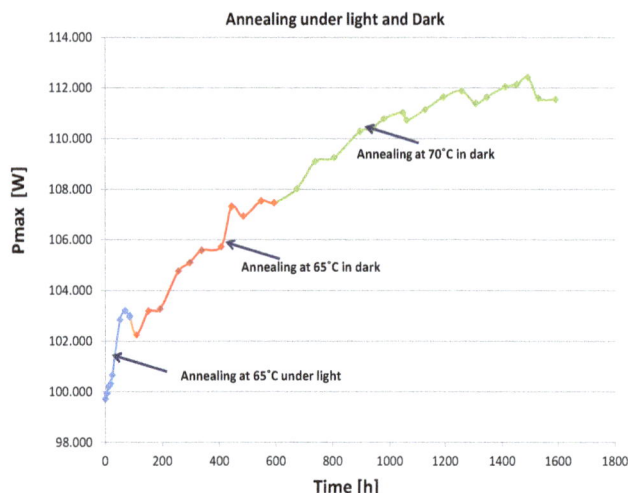

Fig. 7. Annealing curve of sample 2, low temperature annealing with light and without light.

Table 4. Power evolution for sample 2.

Temp. [°C]	Power [W]	Power [%]	118 Wp = 100%	
25	99.726	100	84.5	
65	102.97	103.3	87.3	+2.8
65	107.491	107.8	91.1	+3.8
70	111.562	111.9	94.8	+3.7

3.1.3 Annealing under illumination

A single module was used to follow a different annealing/illumination path. Sample 2 was used in order to analyze the correlation between illumination and annealing in combination, compared to the dark annealed modules (see Fig. 7).

During annealing in light at 65 °C and an irradiance of approx. 830 W/m^2 a stabilised power of 102.97 Wp with an absolute increase of 3.2 W after 85 h were obtained. According to the graph, the portion of annealed power with light was considered stable from the final three consecutive measurement points, in degree of their deviation in power of ±0.1 W. Following this, annealing in dark for around 509 h at 65 °C, lead to a further increase to 107.4 Wp with a ΔP of 5.2 W. The power increase from 65 °C to 70 °C was 112 Wp with an increase in 4.6 W and overall increase in power after the final annealing step was 112 Wp (12%) from its degraded state of 99.7 Wp.

All results for the measured power values stayed stable at room temperature i.e. the module state did not change even after storing the modules in dark for some hours and the IV characterization gave reproducible results.

3.2 Discussion

3.2.1 Annealing

The results of the annealing cycles show that a huge part of the initial (manufactured) power can be recovered

by annealing in the dark even at low temperatures. This is interesting for two reasons. First it is definitely below the temperature damaging the module structure. Second it is at temperatures measured as operation temperatures in the field. As it has been observed as a seasonal effect seen in building integrated a-Si technologies, where the power of the PV modules improves during summer and decreases during winter season [21]. Therefore, it was interesting to use outdoor operating temperatures of PV modules in dark to observe the effect of low temperature on the recovery of power. The annealing was done for a-Si double junction technologies, glass-encapsulated and fully plastic encapsulated modules.

Interestingly, the series resistance was reduced and the shunt resistance was raised, after each annealing cycle. The change in shunt resistance represents the recombination losses in the bulk. This causes reduction in the fill factor. Through annealing a change in shunt resistance also improved the fill factor, for example in sample 1, the fill factor was increased from 63.4% to 68.9%. This shows a reduction in the recombination losses through low temperature annealing. Also, with each pre-treatment the modules showed a decrease in metastabilities. This also suggests that there is some kind of stabilization introduced in the structure of the a-Si technology.

Finally, presented measurement also shows how to differentiate between the spectral and thermal annealing effects as a-Si technologies as shown in Figure 7, where the difference between annealing in light and annealing in dark can be seen, where it exhibit a power recovery dependence on the incident spectrum [22].

Retain in power also suggests that some change in metastabilities is occurring inside the a-Si:H technology. The overall change was almost the full module degraded power from the manufactured state.

3.2.2 Arrhenius behavior

The annealing procedure was done for a long duration of time – around 2000 h, but our purpose was to investigate the main evolution of power correlated to defect annihilation. The energy barriers of the metastable state from the ground state as found in literature have a variation from 0.9 eV to 1.3 eV [12,23].

Since defect annealing is a thermal activated process, it must show an Arrhenius behavior. The stabilized power at each temperature was plotted vs. (1000/temperature) for sample 1. It follows a perfect Arrhenius curve, as shown in Figure 8.

Annealing might have changed some important parameter in the a-Si:H structure, this could be the Si-H$_2$ which is according to the experimental research suggesting its increase from 50 °C [24]. The di-hydride bonds require much less activation energy than the single Si-H bonds. This is also proven by these researchers from [24]. This might be an essential factor in increasing the efficiency of the module. Also, as the Si-H$_2$ lies generally in the shallow traps, this suggests that it requires low activation to transport in the material.

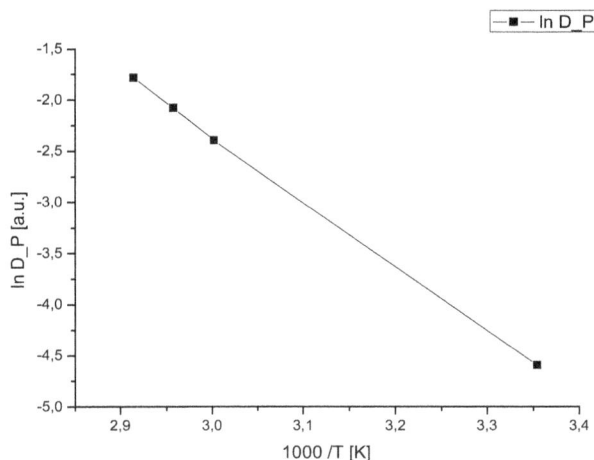

Fig. 8. Shows the Arrhenius curve for sample 1, the annealed power showing the correlation to the defect densities.

Nevertheless, from the graph in Figure 8, no activation energies could be extracted as described in Section 2.2. Although it is just an assumption we think that the power changes which were measured in the presented experiments may origin from defect states with similar activation energies. Some of the modules were able to reach their initial state (quoted by the manufacturer) of power even after getting degraded below their nominal power. Recover of power to its initial state refers to change in the metastable behavior.

Therefore, there might be majority of metastable defects in a-Si:H caused by low energy defects rather than higher activation energy. A detail study at the cell level must be carried out to allow deciding these questions, which cannot be solved on the module level. This will lead to naming the potential candidates for the changes in photoconductivity, as discussed here.

Finally it should be mentioned that, in field degradation is happening at the high irradiance and at around temperature of around 60 to 70 °C. Therefore, in outdoor conditions always the case of annealing like shown in Figure 7 for sample 2 (for the part- annealing at 65 °C in light) will be valid.

There are still many open questions from the low temperature annealing since no measurements could be performed at the material level. Optical characterization technique, electroluminescence imaging, did not show any changes in the treated modules, when compared with its initial degraded state to the fully annealed state. This says something about no damages and changes being done externally but in the internal structure of the module.

4 Conclusion

Low temperature annealing caused several changes in the modules behavior: (a) an increase of power was recorded for all annealed modules power, (b) the short term metastable behavior was decreased while the overall power rose and (c) the difference between annealing under

illumination and in the dark was shown. For the underlying driving process the thermal nature could be shown as the results follow Arrhenius behavior. This observation suggests, that the presence and the dynamics of metastabilities in double-juncion a-Si:H modules are related to optically/thermally induced changes of the intrinsic defect density.

The almost full power recovery and activation energy suggests that the defects which cause degradation might be just of low activation energy. This findings and conclusions were all done on module level, which lowers their value for reason of model character. Nevertheless, we hope, that these kind of experiments can also help to understand full scale mass marked modules in their behavior much better.

References

1. Q.Y. Research, Market Research Report on Global and China Amorphous Silicon Thin Film Solar Cell Industry, June, 2014
2. European Commission, Communication from the commission to the European parliament and the council Renewable Energy: Progressing towards the 2020 target, 2011
3. D. Craciun et al., in *Proceedings of 27th PVSEC, Frankfurt, 24–28 September, 2012*, pp. 3074–3081
4. J. Wagner et al., in *Proceedings of the 26th European PVSEC and Exhibition, Poster, Hamburg, 5–9. September, 2011*
5. S. Krauter, A. Priess, in *Proceedings of 25th PVSEC, Valencia, 6–10 September, 2011*, p. 3141
6. V. Helmbrecht et al., in *Photovoltaics International 15th Edition, April 2012*, pp. 194–201
7. J.A. del Cueto, B. von Roedern, Prog. Photovolt.: Res. Appl. **7**, 101 (1999)
8. R. Ruther et al., in *33rd IEEE Photovoltaic Specialists Conference, 2008, PVSEC 08* (2008), pp. 1–5
9. M. Rennhofer et al., Correlation of weather and spectral changes for evaluation of thin-film PV module performance, in *Proceedings of the "Thin Film Conference", München, 2011*, pp. 56–62
10. L. Staebler, C.R. Wronski, Appl. Phys. Lett. **31**, 292 (1977)
11. M. Stutzmann, in *Amorphous and Microcrystalline Semiconductor Devices*, edited by J. Kanicki (Artech House, Boston, 1992), Vol. II, p. 129
12. M. Stutzmann, W.B. Jackson, C.C. Tsai, Phys. Rev. **32**, 23 (1985)
13. H. Branz, Solid State Commun. **105**, 387 (1998)
14. S. Zafar, E.A. Schiff, Phys. Rev. B **40**, 5235 (1989)
15. IEC 60904-1 Ed.2.b, photovoltaic devices – part 1, measurement of photovoltaic current-voltage characteristics, 2010
16. IEC 60891 Ed. 2, Procedures for temperature and irradiance corrections to measured I-V characteristics of photovoltaic devices, 2008
17. IEC 61646 Ed. 2.0, Thin-film terrestrial photovoltaic (PV) modules – Design qualification and type approval, 2008

18. A. Dangel, Staebler Wronski effect in Photovoltaic modules, Master thesis, FH Technikum, Wien, 2014

19. S. Novalin, M. Rennhofer, J. Summhammer, Thin Solid Films **535**, 261 (2013)

20. A. Mittal, Short-term metastable effect in a-Si technology, Master thesis, Austrian Institute of Technology, 2012

21. A. Virtuani, L. Fanni, Prog. Photovolt.: Res. Appl. **22**, 208 (2012)

22. Y. Hirata, T. Tani, Solar Energy **55**, 463 (1995)

23. A. Kolodzeij, Opto-electronics Review **12**, 21 (2004)

24. P.K. Lim et al., J. Phys.: Conf. Ser. **61**, 708 (2007)

19

Diode laser crystallization processes of Si thin-film solar cells on glass

Jae Sung Yun[1,a], Cha Ho Ahn[1], Miga Jung[1], Jialiang Huang[1], Kyung Hun Kim[1,2], Sergey Varlamov[1], and Martin A. Green[1]

[1] University of New South Wales, NSW, 2033, Kensington, Australia
[2] Suntech R&D Australia, Pty., Ltd. 82-86 Bay St., BSW 209, Botany, Australia

Abstract The crystallization of Si thin-film on glass using continuous-wave diode laser is performed. The effect of various processing parameters including laser power density and scanning speed is investigated in respect to microstructure and crystallographic orientation. Optimal laser power as per scanning speed is required in order to completely melt the entire Si film. When scan speed of 15–100 cm/min is used, large linear grains are formed along the laser scan direction. Laser scan speed over 100 cm/min forms relatively smaller grains that are titled away from the scan direction. Two diode model fitting of $Suns$-V_{oc} results have shown that solar cells crystallized with scan speed over 100 cm/min are limited by grain boundary recombination ($n = 2$). EBSD micrograph shows that the most dominant misorientation angle is $60°$. Also, there were regions containing high density of twin boundaries up to $\sim 1.2 \times 10^{-8}/\mathrm{cm}^2$. SiO_x capping layer is found to be effective for reducing the required laser power density, as well as changing preferred orientation of the film from $\langle 110 \rangle$ to $\langle 100 \rangle$ in surface normal direction. Cracks are always formed during the crystallization process and found to be reducing solar cell performance significantly.

1 Introduction

Polycrystalline silicon thin-film solar cells on glass are strong candidate for next generation photovoltaic technology as it combines advantages of both wafer Si solar cells technology and thin-film solar cells technology. Si material has advantages of nontoxicity and low-cost. Commercial polycrystalline silicon thin-film solar cells have been fabricated by CSG Solar which achieved photovoltaic conversion efficiency of 10.4% in 2007 [1]. In this solar cell, 2 μm thick a-Si thin-films on borosilicate glass were crystallized using solid phase crystallization (SPC) process which produced grain sizes in the range of 1–2 μm. However, high density of intragrain defects are generated during the SPC process and it was found to be lifetime limiting recombination pathway which greatly limit the open-circuit voltage [2]. Alternate to SPC, it was reported that continuous-wave (CW) diode laser crystallization of Si thin-film on glass can form defect-free grains with very large grains size up to few tenths of millimeters in length [3]. Performance potential of this solar cells has shown that photovoltaic conversion efficiency above 13% can be achieved with a diffused homo-junction emitter [4].

In this work, the diode laser crystallization processes of Si thin-film on glass are reported in detail. Grain microstructure and crystallographic orientation were investigated in terms of the laser parameters. Then, effect of SiO_x capping layer is investigated in terms of the crystallization parameters and crystal orientation. Finally, cracks in the film are discussed and their influence on solar cell performance is evaluated.

2 Experiment

On a 3.3 mm thick planar borosilicate glass (Schott Borofloat33) with an area of 50×50 mm, SiO_x 100 nm thick barrier layer was deposited by plasma-enhanced chemical vapor deposition (PECVD) at 200–300 °C, below 1 mT at deposition rates of 5–30 nm/min. Then, the SiO_x layer was subjected to a dehydrogenation annealing step at 500 °C for 2 h under N_2 flow. 10 μm thick Si films were deposited by e-beam evaporation at 350–650 °C. During the deposition, in-situ boron doping was performed to realize a boron concentration of $\sim 5 \times 10^{15}$ to $\sim 2 \times 10^{16}$ cm^3. The Si films were placed on pre-heated stage at 650 °C for 3 min. Subsequently, CW diode laser (FWHM, 0.17 cm wide and 12 cm long) was scanned perpendicular to the 5 cm long axis for crystallization. After

---
[a] e-mail: j.yun@student.unsw.edu.au

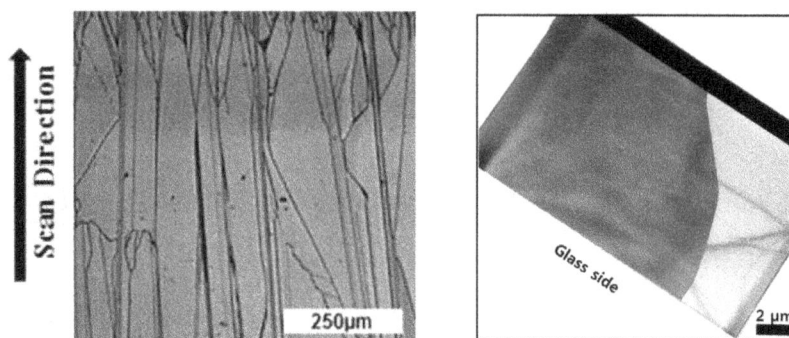

Fig. 1. (Left) Optical microscope image and (right) TEM image of diode laser crystallized Si film (40 cm/min with 15 kW/cm^2) with complete melting of entire film.

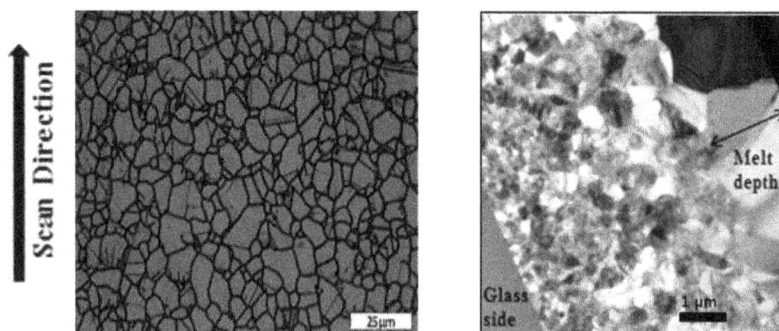

Fig. 2. (Left) Optical microscope image and (right) TEM image of diode laser crystallized Si film (40 cm/min with 9.7 kW/cm^2) with partial melting of film.

removal of native oxide formed during the crystallization by HF dip, phosphorous dopant source (P508, Filmtronics, Inc.) was spin-coated. Subsequently, rapid thermal process (RTP) was conducted at 870 °C for 2 min. Junction depth around 500 nm was achieved and emitter sheet resistance of 300–500 ohms/□ was obtained. The hydrogen passivation was applied through remote plasma passivation at ~650 °C for 20 min. Point contact metallization method is applied as described in reference [5].

The crystallographic orientation was analyzed by X-ray diffraction (XRD). Philips X'Pert MRD 4-circle diffractometer using Cu-K$_\alpha$ radiation. Receiving slit of 7 mm^2 was used and (100), (110), (111), and (422) pole figures were measured with background subtraction. Samples were placed in the center of an open three axes goniometer in Bragg orientation up to 75° tilting angle. Measurement time per step was between 5 and 10 s, step size for both rotation and tilting was 5°. The data was corrected and analyzed using X'pert Texture (Version 1.1a) software which can generate pole figures and inverse pole figure maps. For high resolution electron beam scattering diffractometer (EBSD) measurement, a NordlysF detector EBSD analytical system interfaced to a Carl Zeiss AURIGA® CrossBeam® workstation was used.

3 Results and discussion

3.1 Liquid phase crystallization of Si film

Figures 1 and 2 are showing two distinctly different microstructures after Secco etching and the corresponding

TEM images of the laser crystallized Si thin-films. In our diode laser system, laser scan speed (cm/min) and power density (W/cm^2) are two major parameters that control the crystallization process. Appropriate laser power density is required to completely melt the entire film [3]. Figure 1 represents the desired microstructure with large linear grains grown parallel to the laser scan direction when scan speed of 40 cm/min with laser power density of 15 kW/cm^2 is used. Corresponding TEM image shows defect-free grains that were formed by complete melting of the film. Grain size can be up to few tenths of millimeters in length and several nanometers to several hundred microns in width. On the other hand, if the power density is insufficient to melt the entire film, only surface melting occurs and generates columnar grains several microns in size as shown in Figure 2. In this case, scan speed of 40 cm/min with laser power density of 9.7 kW/cm^2 is used. Corresponding TEM image of partially melted Si film shows the formation of columnar grains at the surface which extends few microns into the film. If the power density is higher than the required power to achieve full-melting, film delamination takes place.

Figure 3 is EBSD micrographs and corresponding misorientation charts of the completely melted films that were scanned at 40 cm/min at 15 kW/cm^2. Due to the top-hat profile of the laser beam, intensity at the edges of the laser traces is not sufficient to achieve full-melting and form small columnar grains as shown in Figure 3a. The corresponding misorientation chart is showing that there are variety misorientation angles with relatively higher number of 60° misorientation. As the intensity of the beam

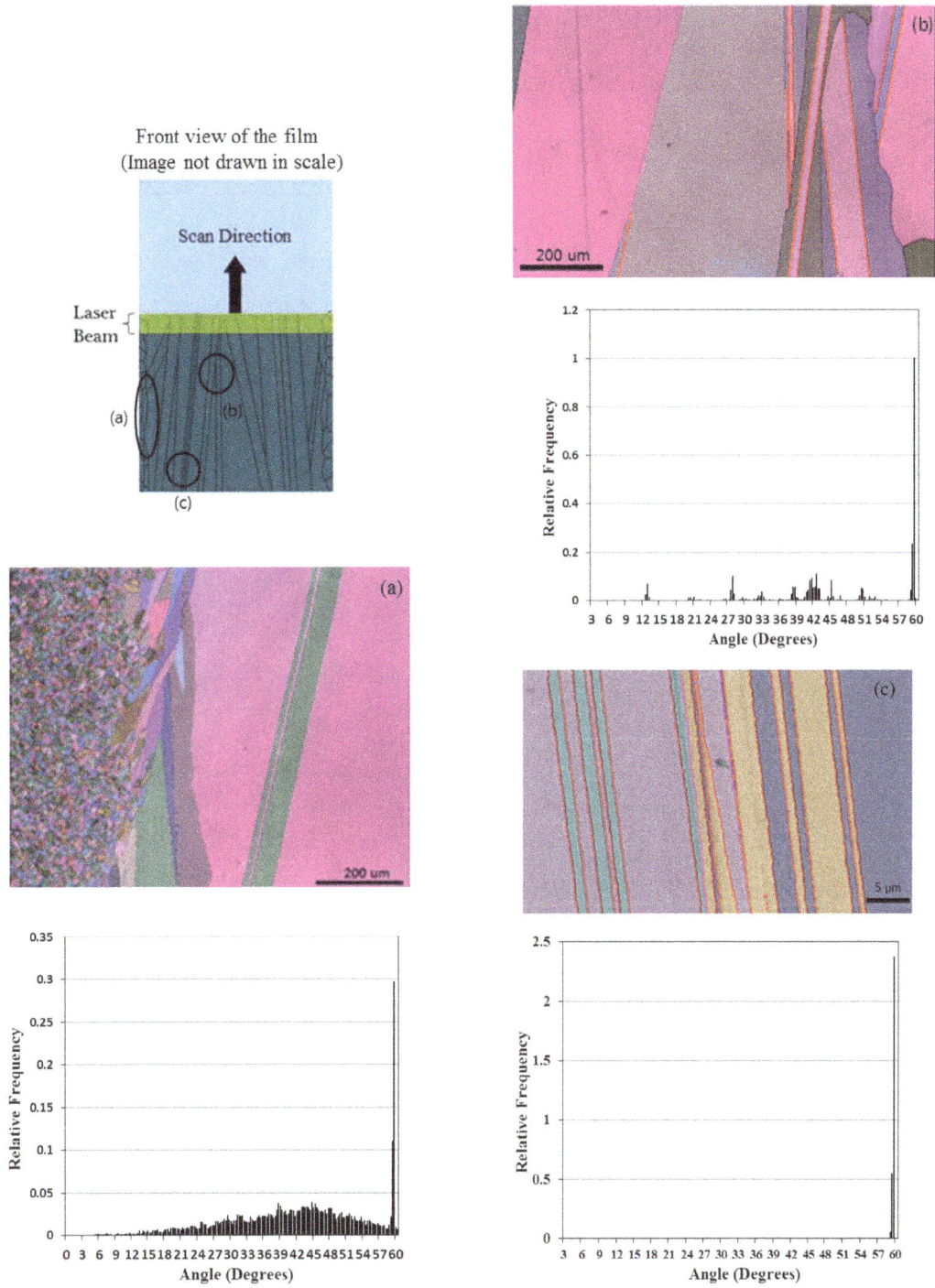

Fig. 3. EBSD micrographs and corresponding charts showing the relative frequency corresponding to each misorientation value of the laser crystallized films (40 cm/min with 15 kW/cm^2) showing (a) small columnar grains at left edge; (b) parallel grains meet each other; and (c) twin boundaries. Note that red lines represent twin boundaries.

gets stronger towards the middle region, the melt depth increases and the film becomes liquid phase. As the laser scans, this liquid Si solidifies along with, it forming large linear grains along the scan direction. Figure 3b is showing termination of growth of such linear grains when adjacent grains meet each other. If the grains do not converge, they can grow continuously until the scan stops. Such parallel grains have 60° misorientation angles which are marked in red lines. Frequently, there are regions where high density of parallel grain boundaries, up to $\sim 1.2 \times 10^{-8}$/cm^2, are formed, as shown in Figure 3c. All of these boundaries are found to be 60° misorientation as shown in corresponding relative misorientation chart and large number of these boundaries are found to be first order $\Sigma 3$ twin boundaries.

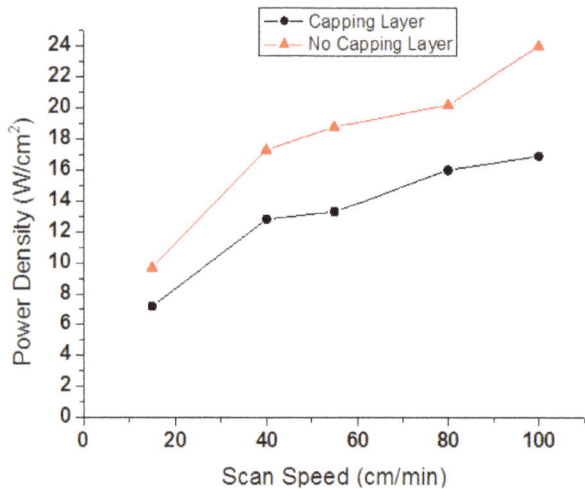

Fig. 4. Required laser power density to achieve full-melting of Si film vs. laser scan speed with and without capping layer.

3.2 Effect of SiO$_x$ capping layer

In laser processing of Si film, few hundred nanometers thick SiO$_x$ capping layer is often deposited on top of the Si film in order to enlarge grain size of the film [6]. When 100 nm thick SiO$_x$ capping layer is deposited by PECVD, the required power density for full-melting of the film is reduced due to an antireflection effect. Figure 4 is an empirically obtained graph showing the required laser power density to achieve full-melting as a function of scan speed with and without capping. As can be seen from this graph, the required power density is lower when the capping layer is used. This can be explained by an antireflection characteristic of the SiO$_2$ layer which enhances the absorption of the laser power. Our measurement shows that absorption at 808 nm is increased from 56% to 65%. In addition to the enhanced absorption, the capping layer stores some of generated heat during the laser irradiation and the heat flows back into the Si film which elevates temperature of the film [7].

Pole figure measurements are performed using XRD and Figure 5 is showing x (surface normal), y (scan direction), and z (plane normal) inverse pole figure maps of completely melt Si film with and without the capping layer. As can be seen, when there is no capping layer, (110) preferential orientation in z direction is present. Also, there is a maximum intensity present at (112) orientation for both x and y directions. Although this type of orientations is not uniformly distributed all over the crystallized area, it is a very common crystallographic orientation distribution in our film. When the capping layer is applied, (100) preferential orientation is found in z direction. Highest intensity at (110) orientation is formed along the x direction while highest intensity is found at (111) orientation in y direction. According to Atwater et al. [8], the interfacial energy at the interface between SiO$_2$ intermediate layer and Si film determines the (100), (110), and (111) preferential orientation in plane normal of the Si film. The interfacial energy depends on heat flow

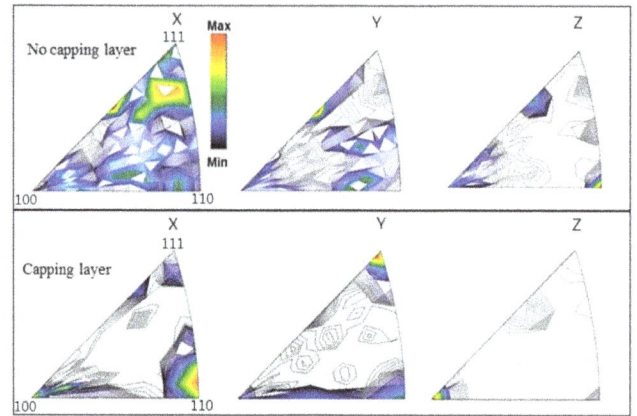

Fig. 5. Inverse pole figure maps of completely melt Si film at scan speed of 80 cm/min with and without the capping layer.

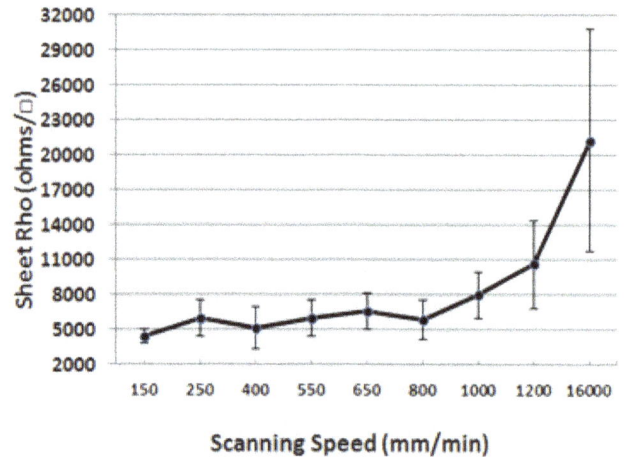

Fig. 6. Sheet resistance as function of scan speed.

during crystallization process and (100) orientation requires longest melt duration [9]. As mentioned earlier, the capping layer elevates temperature of the film and thereby it has effect of extending duration of liquid state. As a result, the (100) preferred orientation formation could have been allowed by extended melt duration in our case.

3.3 Effect of laser scan speed

Figure 6 is a graph of sheet resistance as function laser scan speed. The sheet resistance is almost comparable for the scan speed between 15–80 cm/min. However, the sheet resistance starts to increase from 80 cm/min and further more rapidly from 100 cm/min. Microstructure of the laser crystallized films scanned at 40 cm/min and 160 cm/min is shown in Figure 7. As can be seen, grains are severely tilted away from the scan direction when 160 cm/min is used. Figure 7 is a schematic illustration of these two grain growth behaviors. Zone 1 is liquid phase and Zone 2 is a phase transition region where liquid Si is in cooling stage. Zone 3 refers to a solid phase region with very high surface temperature. As can be seen, area of the Zone 2 is much larger in Type I compared to Type II. Due to excessively

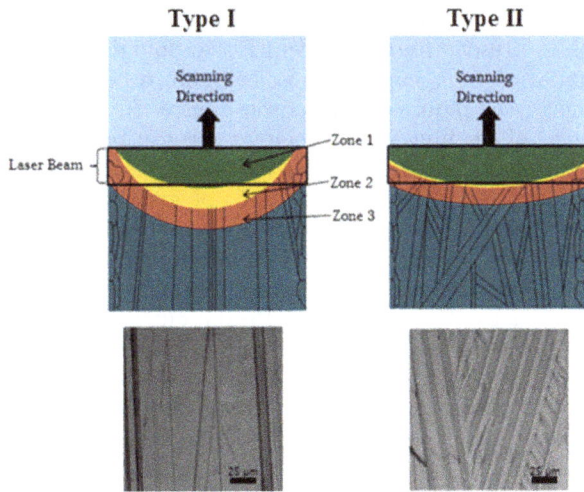

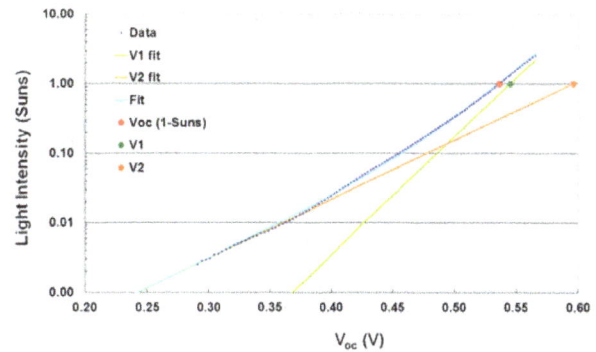

(a)

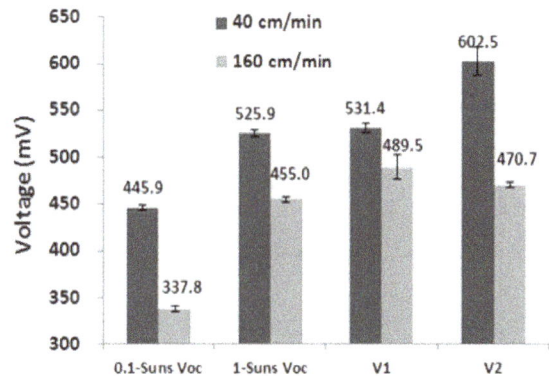

(b)

Fig. 7. Schematic illustration of grain growth behavior on two distinct scan speed regimes. Type I refers to scan speed of 15–100 cm/min and Type II refers to scan speed over 100 cm/min.

Fig. 8. (a) $Suns$-V_{oc} measurement of the laser crystallized solar cells (40 cm/min with 15476 W/cm^2) with fitted curve. (b) $Suns$-V_{oc} results of scan speed of 40 and 160 cm/min.

fast scan speed in Type II, temperature in this region decreases rapidly. Therefore, molten Si quickly transforms into solid phase. As a result, grain growth cannot follow the scan direction and thus forms in random directions. Whereas, Type I provides long enough time for grain formation thereby grain growth can proceed along the scan direction.

Emitter was formed on these two types of microstrucures and $Suns$-V_{oc} was measured after removing some portion of emitter to make contact with the absorber. Figure 8a is showing the obtained data and fitted curve of solar cells that were crystallized at scan speed of 40 cm/min with 15 kW/cm^2. $Suns$-V_{oc} curves are useful to identify diode properties such as open-circuit voltage, pseudo fill factor, diode ideality factors, etc., and to determine the effects of bulk and depletion region recombination [10]. Obtained data is fitted to two-diode model ($n = 1$ and $n = 2$) in order to obtain parameters V_{oc}, V_1, V_2 and R_{sh}. Detailed fitting analysis is described in elsewhere [11]. Two charactersitic voltages V_1 and V_2 are obtained at the intersection of the two fitted curves for the $n = 1$ and $n = 2$ diodes correspondingly with the horizontal line at 1-Sun light intensity. Generally, the $n = 2$ diode ($V_1 > V_2$) accounts for SRH (Shockley-Read-Hall) recombination in the junction space charge region and at grain boundaries, whereas the $n = 1$ diode ($V_1 < V_2$) accounts for bulk and surface recombination.

As depicted in Figure 8b, 1 and 0.1 $Suns$-V_{oc} were 525.9 mV and 445.9 mV for the laser crystallized solar cells at 40 cm/min and 160 cm/min, respectively. It is also shown that $n = 1$ recombination ($V_1 < V_2$) is dominant for the 40 cm/min sample, while $n = 2$ recombination ($V_1 > V_2$) is dominant for the 160 cm/min sample. Presumably, the grain boundaries tilted away from scan direction formed in 160 cm/min act as a carrier recombination center and limits the solar cell performance.

3.4 Solar cell performance and effects of film cracks

It is clear that large cracks, width up to several hundred microns and length up to several millimeters are generated along the scan direction as shown in Figure 9. The position of the crack formation is random. In some case, few centimeters long cracks could be generated in right middle or near to the edges of the crystallized area. The cracks are always generated during the laser crystallization process and likely responsible for shunting problem when p-n junction is formed and metallized. It is shown that grain boundaries continue even after the crack which implies that the crack occurred after the film is solidified. Also, structural defects are often observed near the crack which is not surprising since the crack is likely to be formed by excessive tensile stress in the film. The stress could be responsible from high undercooling rate as well as the spatial non-uniformity of the undercooling rate. Solar cell device is fabricated on absorber layer crystallized at scan speed of 80 cm/min by RTA emitter diffusion. After hydrogen passivation, the material is metallized using the point contacts. Figure 10 compares two types of cells, one with few severe cracks running through the middle of entire active area of the cells and one that the active area is made in the region of no crack. As can be seen, shunt resistance of the cell without cracks is around two times higher for the cell with cracks. Also, V_{oc} and J_{sc}, and FF

adjacent grains meet. High density of parallel grain boundaries is present frequently which was found to be twin boundaries. SiO_x capping layer is found to be effective in reducing the required laser power density to achieve full-melting of the film. Also, preferred orientation is changed to (100) from (110) in surface normal direction when the capping layer is used. Two types of growth behaviors, Type I and Type II, are identified in respect to laser scan speed. In Type II (over 100 cm/min) growth behavior the solid and liquid phase region was shortened compared to that of Type I (15–100 cm/min). As a result, grains growth could not follow the scan direction and tilted away from the scan direction. $Suns$-V_{oc} measurement showed that band to band recombination was dominant for Type I while grain boundary recombination was dominant for Type II. Cracks are generated during the laser crystallization and it was found to reduce solar cell efficiency from 7.1% to 5.9%.

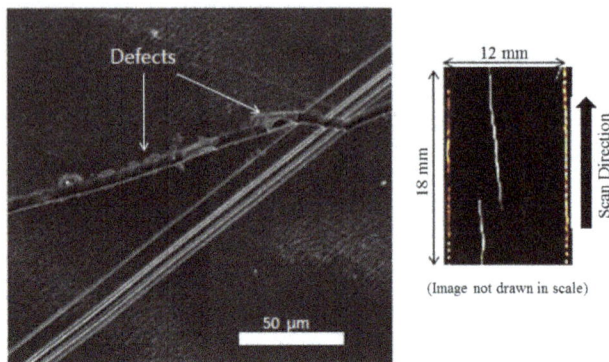

Fig. 9. (Left) SEM image of cracked region and (right) real image of cracks.

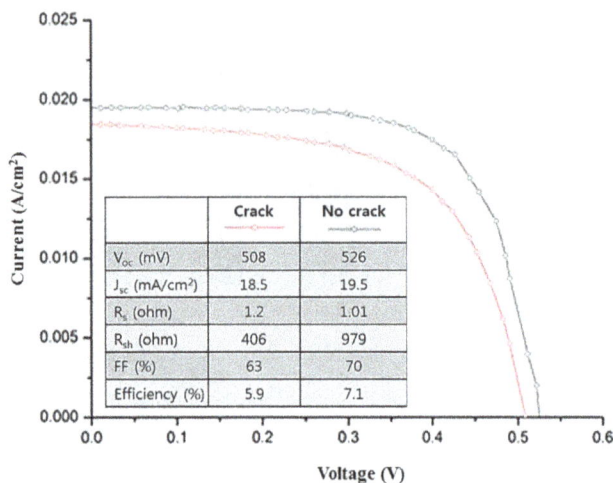

Fig. 10. Light I-V curve comparison of solar cells with cracks and without cracks.

are lower for the cell with cracks. Overall efficiency is also reduced from 7.1% to 5.9% due to the cracks. Film cracking presents a major challenge for the laser crystallization of thin-film on glass.

4 Conclusions

High quality polycrystalline Si thin-film on glass was fabricated using CW diode laser crystallization. It was necessary to melt the entire silicon thin-film to obtain desired microstructure. When scanning speed was between 150 to 100 cm/min, large linear grains were formed, up to few tenths of millimeters in length and up to several hundred microns in width. Grain growth ceased when

This program has been supported by the Australian Government through the Australian Renewable Energy Agency (ARENA). The Australian Government, through ARENA, is supporting Australian research and development in solar photovoltaic and solar thermal technologies to help solar power become cost competitive with other energy sources. The views expressed herein are not necessarily the views of the Australian Government, and the Australian Government does not accept responsibility for any information or advice contained herein. Jae Sung Yun acknowledges support from the Australian Government through the Australian Solar Institute ARENA scholarship.

References

1. M.J. Keevers et al., in *22nd European Photovoltaic Solar Energy Conference, Milan, 2007*
2. J. Wong et al., J. Appl. Phys. **107**, 123705 (2010)
3. B. Eggleston et al., in *MRS Online Proc. Library* (Cambridge University Press, 2012), Vol. 1426
4. J. Dore et al., EPJ Photovoltaics **4**, 40301 (2013)
5. J. Dore et al., Progr. Photovolt.: Res. Appl. **21**, 1377 (2013)
6. W. Yeh, M. Matsumura, Jpn J. Appl. Phys. **41**, 1909 (2002)
7. R. Vikas et al., Jpn J. Appl. Phys. **45**, 4340 (2006)
8. H. Atwater, C.V. Thompson, H.I. Smith, J. Mater. Res. **3**, 1232 (1988)
9. D. Witte et al., J. Vacuum Sci. Technol. B **26**, 2455 (2008)
10. S.I. Sulaiman, in *Proceedings of the IEEE International Conference on Semiconductor Electronics, 2004*
11. O. Kunz, Ph.D. thesis, University of New South Wales, 2009

20

Thin metal layer as transparent electrode in n-i-p amorphous silicon solar cells

Martin Theuring[a], Stefan Geissendörfer, Martin Vehse, Karsten von Maydell, and Carsten Agert

NEXT ENERGY – EWE Research Centre for Energy Technology at Carl von Ossietzky University, Carl-von-Ossietzky-Straße 15, 26129 Oldenburg, Germany

Abstract In this paper, transparent electrodes, based on a thin silver film and a capping layer, are investigated. Low deposition temperature, flexibility and low material costs are the advantages of this type of electrode. Their applicability in structured n-i-p amorphous silicon solar cells is demonstrated in simulation and experiment. The influence of the individual layer thicknesses on the solar cell performance is discussed and approaches for further improvements are given. For the silver film/capping layer electrode, a higher solar cell efficiency could be achieved compared to a reference ZnO:Al front contact.

1 Introduction

Transparent electrodes are used in many electro-optical devices such as displays and solar cells. The main requirements of these electrodes are a low electrical resistance and a high optical transmittance. However, both properties are interdependent, e.g. by the amount of free charge carriers, and improving one usually comes at the expense of the other. For an optimum trade-off between both properties, various different technologies can be applied. The most common approach is the use of degenerate metal oxide semiconductors (TCO: transparent conductive oxide) such as indium tin oxide (ITO). Furthermore, graphene, carbon nanotubes and various types of metal nanostructures are currently under investigation. However, in terms of fabrication those are more complex technologies (review papers on transparent electrodes can be found e.g. from Granqvist [1] and Ellmer [2]).

Solar cells are large area mass products. To be price competitive to other energy technologies, high priced materials and costly fabrication steps have to be avoided. Hence, ITO and sophisticated nanostructures are not suitable for the transparent electrode in these devices. Further cost reductions and new applications can be created with roll-to-roll processing and flexible devices. Yet, this leads to additional requirements for the individual layers in the cell, such as bendability. Temperature-sensitive substrates or absorber layers also limit the possible processing techniques.

However, a transparent electrode is required in thin film solar cells for charge carrier collection. In silicon thin film solar cells highly doped zinc oxides (e.g. AZO: aluminum doped zinc oxide) or tin oxides are often applied for this purpose [3]. To achieve a sufficient conductivity for solar cell applications, thick layers (up to several μm) are needed, which are less flexible. Furthermore, for sputtered AZO an optimized performance will be obtained if the material is deposited on a heated substrate [4]. But a high temperature process can degenerate the underlying layers by decomposition or cracking due to differences in their temperature coefficients.

Amongst others (see [5]), Sahu et al. [6] have shown that the incorporation of a few nanometers of silver in thin TCO layers strongly reduces the sheet resistance, while still providing high optical transmittance in the visible part of the spectrum. Higher mechanical stability and a simple low temperature fabrication process are the advantages of this design [7].

The objective of this paper is to demonstrate the integration of thin metal layers as a transparent electrode in n-i-p amorphous silicon (a-Si) solar cells. In this work, we focus on thin metal layers adjacent to the p-layer of the cell. As an anti-reflection (AR) and protection layer, sputtered AZO is compared to an AZO/SiO_x multilayer. We analyze the influence of the individual layers' thicknesses and deduce design rules for an optimized performance.

2 Experimental details

As substrates for the solar cells, we use glass slides with a layer of chemically wet etched, structured AZO. The latter was coated with a 30 nm layer of aluminum followed

---

[a] e-mail: `martin.theuring@next-energy.de`

by 50 nm of silver for opaqueness. Both metal layers were deposited by electron beam evaporation. Subsequently, a 70 nm layer of AZO was sputtered on the back contact. Hydrogen-passivated amorphous silicon layers were fabricated in the n-i-p configuration (deposition order: n-doped/intrinsic/p-doped) by plasma-enhanced chemical vapor deposition (PECVD). The cell structure is depicted on the right sides of Figures 1, 3 and 5. SiH_4, H_2 and PH_3 were used as process gases for the n-doped a-Si layer, SiH_4 and H_2 for the intrinsic a-Si layer (i-layer) and SiH_4, CH_4, H_2 and B_2H_6 for the p-doped a-SiC layer, respectively. SiO_x AR-layers were fabricated in a PECVD process at 180 °C substrate temperature. SiH_4, H_2 and CO_2 were used with a dilution rate of CO_2 to SiH_4 of approximately 18.3 to 1, which led to highly transparent, SiO_2-like layers (refractive index n of SiO_x in the range of 1.5–1.6).

The silver layer in the front contact and all AZO layers were deposited by DC magnetron sputtering. We estimate that the deposition rate for the silver layer is approximately 1 nm/s. This value was obtained by measuring a silver layer on a flat substrate by spectroscopic ellipsometry and subsequent modeling. However, the actual layer thickness might be different on structured substrates, such as the solar cells in our experiments. In the following, we therefore only give the deposition times for the silver layers as a rough indication for the layer thicknesses. For the AZO layers, we used a ZnO sputter target with a 2wt% concentration of Al_2O_3. Before the deposition of AZO and silver no substrate heating was used, which is crucial for the thin silver layers. Island formation was observed with heating temperatures exceeding 80 °C, depending on the layer thickness. However, the 700 nm AZO reference layer was deposited on a substrate at 300 °C. All solar cells were annealed at 160 °C for 30 min after fabrication.

We defined the active cell area with the size of the front contact. A marker pen was used to draw square openings with a size of ≈0.65 cm^2 before front contact deposition. The liftoff process was carried out by sonication in an acetone bath. Short circuit current densities (J_{SC}) were obtained from external quantum efficiency (EQE) measurements by convolution with the AM1.5G spectrum. Current density-voltage (JV) curves were measured with a solar simulator under standard test conditions and scaled to the J_{SC} values calculated from the EQE data. Sheet resistances were obtained by four point measurement method.

The simulation results were calculated with the software Scout/Code (by W. Theiss Hard- and Software), which solves the electromagnetic wave equation for the one dimensional case. The layer stack of the solar cell used in the simulation was matched to the experimental data. Simulated EQEs were obtained from the computed absorption in the intrinsic layer. Optical material parameters were partly (Ag, SiO_2) taken from reference [8], partly obtained by spectroscopic ellipsometry and subsequent modeling (all other materials).

3 Results and discussion

3.1 Optimization of the silver layer thickness

As already outlined at the beginning of this paper, the incorporation of a thin metal layer in TCO strongly influences the sheet resistance of the structure and of course, the thicker metal layers lead to lower resistances. However, an increase of the metal layer thickness decreases the transmittance of the device due to higher reflectivity and absorption from the metal [6]. On the other hand, for thinner metal layers the sheet resistance increases nonlinearly with the thickness: scattering at grain boundaries and surface boundaries play a more dominant role [9]. Depending on what kind of metal is used, its thickness and the substrate type, island formation of the metal is observed, which leads to interrupted current paths (see inset of the graph in Fig. 1) [1]. Furthermore, discontinuities in the metal layer allow the excitation of localized surface plasmon polaritons [10]. This reduces the optical transmittance due to an increased absorption in the metal and its proximities.

On the left side of Figure 1, the JV-characteristics of four n-i-p a-Si solar cells are shown (cell composition sketched on the right side of the same figure). In the Figure legend, the corresponding sheet resistances of the front contacts are specified. The cells differ in the deposition time of the silver layer in the transparent front electrode. For 4 s of deposition (red curve) a sheet resistance of $R_{Sq} \approx 1800$ Ω was measured, which is in the range of the sheet resistance of the AZO capping layer. This can be explained with the discontinuity of the silver caused by island formation (SEM image of silver islands is shown in the inset of the graph). This behavior is also confirmed by the cells' JV-characteristics: the red JV-curve is dominated by the high ohmic resistance of the front contact. By increasing the deposition time, the islands connect, resulting in a much lower sheet resistance (e.g. orange curve). When the silver layer thickness is increased beyond an optimal value (optimized sample roughly represented by the green curve), more and more light is reflected by the cell and the maximum value of extractable charge carriers decreases (blue curve). Further research should be directed towards applicable adhesion layers for the metal, which could prevent island formation for thinner silver films [11].

3.2 Influence of anti-reflection capping layer

The application of an AR layer on a thin metal film is essential for a high transmittance of the transparent electrode [7]. The dotted blue line in Figure 2 represents the simulated EQEs of a solar cell stack with only a thin Ag layer as transparent front contact. In comparison the solid green line shows the data from the same device but with an additional AZO layer of optimized thickness for the antireflective coating. Both, the EQE curves as well as the calculated J_{SC} values given in the table inset demonstrate the necessity of the capping layer. The data from

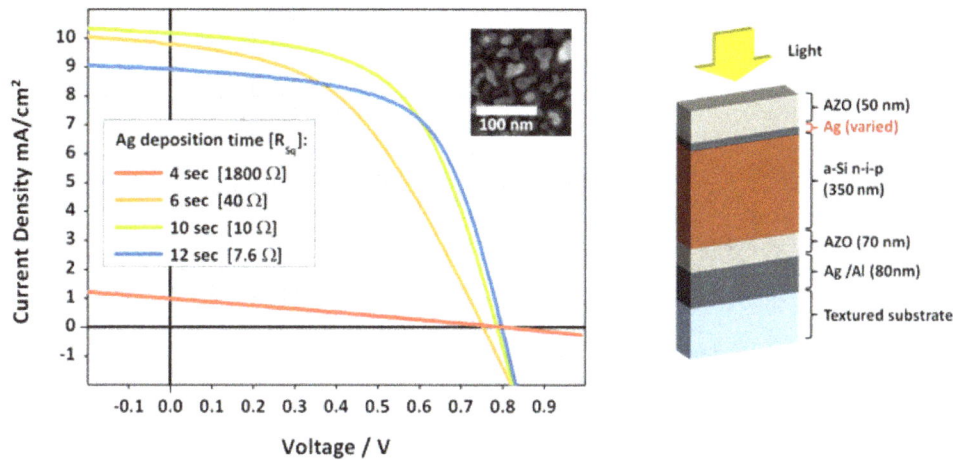

Fig. 1. Variation of the silver layer thickness of the front contact. Left: JV-characteristics of the solar cells (approximate front contact sheet resistance in parentheses; inset: example for island formation in thin Ag layers [scanning electron microscope image]). Right: Design of the solar cell stack (individual layer thicknesses in parentheses).

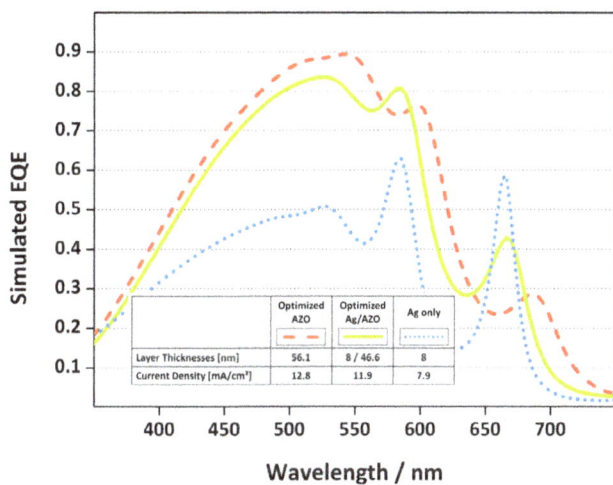

Fig. 2. Simulated external quantum efficiencies for solar cells with different front contacts. Table inset contains layer thicknesses of the different front contacts and the calculated current density of the devices.

a solar cell stack with only a thin AZO layer optimized for antireflective quality is also shown in the same graph. Without the silver layer, the obtained J_{SC} is almost identical to the Ag/AZO sample. For this cell, the high sheet resistance of the front electrode would not allow a good current extraction.

The thickness of the AR-coating material and its optical density play a key role for the solar cell performance. The computed results clearly show the influence of the front contact on the optical properties of the solar cell. However, the results obtained from the 1D-simulation cannot directly be adapted to the experiment. Here, light trapping mechanisms are employed, which augment the spectral response of the solar cell for longer wavelengths. Also the optical cavity of the thin film stack, which causes

the optical interference peaks visible in the simulation results, is weakened due to the light scattering at the textured back contact. The curves in Figure 3 show the EQE and 1-total cell reflectivity (1-R) values for a variation of the AZO capping layer thickness in 10 nm steps. Towards the red spectral band the absorption coefficient of a-Si decreases. For wavelengths above 750 nm, almost no charge carriers are generated in the absorber layer (i-layer). The light is transmitted to the back electrode, where it is mainly absorbed at the structured silver back contact [12].

With increasing AZO layer thickness, the reflectivity peak around 350 nm grows larger. This behavior is directly translated to the EQE curves of the cells: the EQE for shorter wavelengths is reduced. But even for good AR-properties (40 nm AZO sample), a comparably low efficiency is found. In this spectral range, the light is partly absorbed, before reaching the absorber layer. In Figure 4, the simulated absorption distribution in a solar cell is depicted for the 1D case. A large part of the parasitic absorption occurs in the doped layers, where most excited electrons recombine due to high defect densities. Further optimization of these layers helps to overcome such problems, which however, is not the objective of this study.

A full understanding of parasitic absorption in the investigated solar cells is still not given. The simulation results suggest that considerable parasitic absorption can also be attributed to the thin metal film. For thin film optics, the mode distribution in the individual layers plays a key role. Additionally, the influence of the surface roughness on the thin metal film and the light interaction with non-flat metal layers require further investigations. But also the AZO layer is responsible for parasitic absorption. Especially for lower deposition temperatures, which are required for this application, the AZO's transmittance decreases for shorter wavelengths [4].

Therefore, we applied an advanced capping layer, consisting of an AZO/SiO$_x$ double structure. The AZO is required to prevent island formation of the silver layer, as

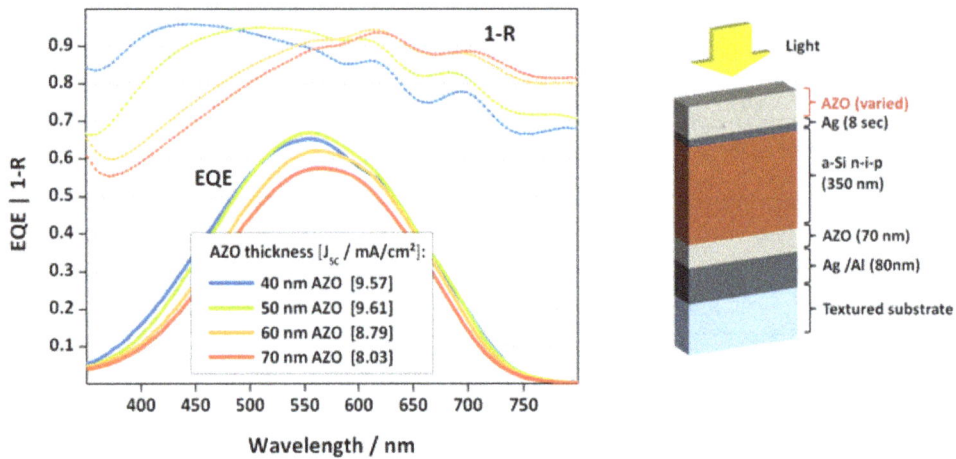

Fig. 3. Thickness variation of the aluminum zinc oxide (AZO) layer in the front contact. Left: External quantum efficiency and 1-total cell reflectivity (short circuit current densities of the corresponding cells in parentheses). Right: Design of the solar cell stack (individual layer thicknesses in parentheses).

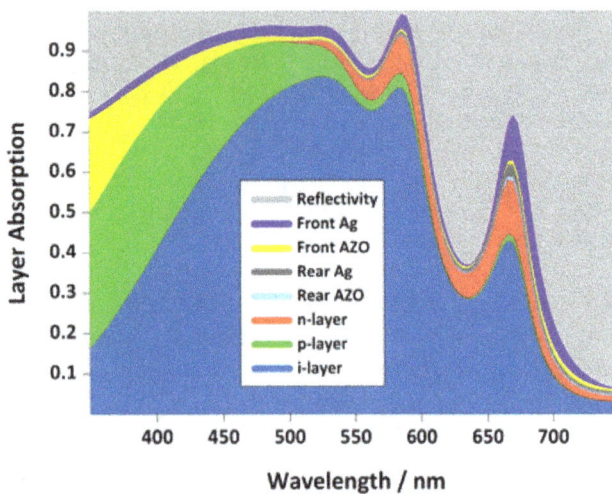

Fig. 4. Simulated absorption distribution in the individual layers of a n-i-p solar cell with Ag/AZO front contact.

the SiO_x was deposited at elevated temperatures. In order to investigate the influence of the capping layer, we performed a thickness variation of the SiO_x layer. The results of this experiment are depicted on the left side of Figure 5. A similar behavior compared to the AZO variation is found. An increase of the SiO_x layer thickness results in a red shift of the AR properties. The overall reflectivity of the cell is found to be higher with the SiO_x capping layer compared to the Ag/AZO contact (the best performing samples with Ag/AZO and Ag/AZO/SiO_x front contact, respectively, are shown in Fig. 6). However, a gain in efficiency for wavelengths up to approximately 475 nm can be found in direct comparison. For further optimizing the performance, a low-absorptive dielectric with higher n (such as Al_2O_3 or TiO_2) should be considered as capping layer material. Also the use of a different deposition method with lower temperatures could be beneficial. This

would allow omitting the remaining 15 nm AZO cover layer and further reduce the parasitic absorption.

Figure 6 also displays the measurement results from a cell with a 700 nm AZO front contact. From Table 1 one can see, that for our experiment, both, the Ag/AZO and the Ag/AZO/SiO_x front contact show a higher efficiency compared to the 700 nm AZO contact. The explanation for this behavior can be found in both, the electrical and optical performance of the front electrode. The sheet resistance of the AZO layer is twice as for the silver-based electrodes. Accordingly, JV-curve and thus the fill factor are influenced negatively. For a lower sheet resistance, a thicker AZO is required which on the other hand would further increase the parasitic absorption in this layer. However, the parasitic loss is already higher for the AZO-only sample. Especially for shorter wavelengths, the EQE is lower while the anti-reflective properties of the contacts are in a comparable range. We, therefore, achieve a higher efficiency with the silver film based transparent electrodes.

4 Conclusion

In this paper it was demonstrated that transparent electrodes based on thin silver films are applicable in structured a-Si solar cells in n-i-p configuration. We narrowed down the possible thicknesses of the silver layer for good optical and electrical performance. The capping layer is crucial for the optical properties of the transparent electrode. Its antireflective properties can be tuned with a thickness or material variation. In this way a higher efficiency can be obtained than with a solar cell with a 700 nm AZO front contact.

The authors would like to thank Volker Steenhoff and Samuel Gage for discussion and Regina Nowak and Tim Möller for discussion and ellipsometry measurements. We would like to

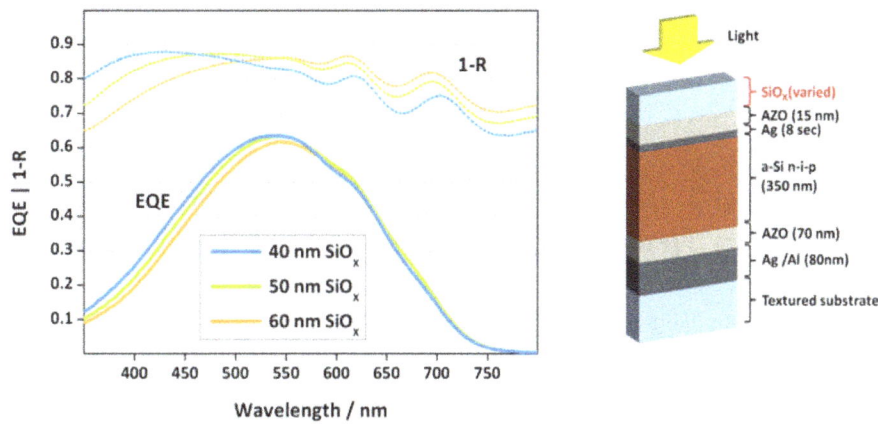

Fig. 5. Variation of the SiO_x layer thickness. Left: External quantum efficiency and 1-total cell reflectivity. Right: Design of the solar cell stack (individual layer thicknesses in parentheses).

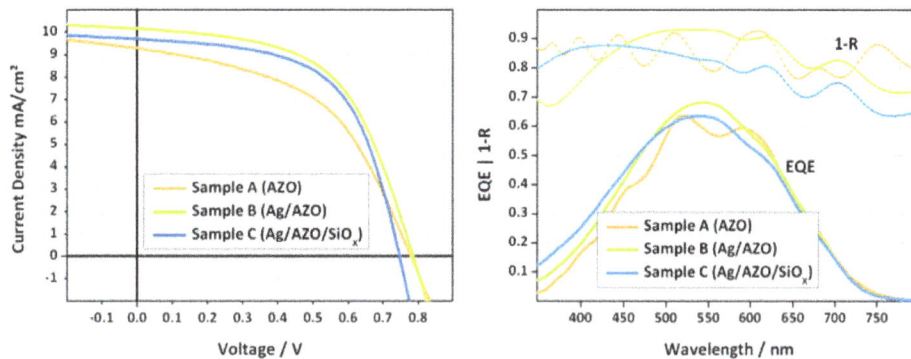

Fig. 6. Comparison of different front contacts (figures of merit see table below). Left: JV-characteristics of the solar cells. Right: External quantum efficiency and 1-total cell reflectivity.

Table 1. Measured cell performance of a-Si solar cells with different front contact. Open circuit voltage (V_{OC}), short circuit current density (J_{SC}), fill factor (FF), sheet resistance of the front contact (R_{Sq}) and cell efficiency (η).

Front Contact (approximate layer thicknesses)	V_{OC} [mV]	J_{SC} [mA/cm^2]	FF [%]	R_{Sq} [Ω]	η [%]
Sample A (700 nm AZO)	779	9.3	49.5	19	3.6
Sample B (8 s Ag/50 nm AZO)	785	10.2	55.7	10	4.5
Sample C (8 s Ag/15 nm AZO/15 nm SiO$_x$)	747	9.7	58.9	8.3	4.3

thank the BMBF for supporting this work in the framework of the project SiSoFlex (FKZ 03SF0418B).

References

1. C.G. Granqvist, Sol. Energy Mater. Sol. C **91**, 1529 (2007)
2. K. Ellmer, Nat. Photon. **6**, 809 (2012)
3. J. Müller, B. Rech, J. Springer, M. Vanecek, Sol. Energy **77**, 917 (2004)
4. O. Kluth, G. Scho, B. Rech, R. Menner, M. Oertel, K. Orgassa, H. Werner, Thin Solid Films **502**, 311 (2006)
5. C. Guillén, J. Herrero, Thin Solid Films **520**, 1 (2011)
6. D. Sahu, S. Lin, J. Huang, Appl. Surf. Sci. **252**, 7509 (2006)
7. D.S. Ghosh, T.L. Chen, N. Formica, J. Hwang, I. Bruder, V. Pruneri, Sol. Energy Mater. Sol. C **107**, 338 (2012)
8. E.D. Palik, *Handbook of Optical Constants of Solids* (Academic Press, Burlington, 1997)
9. J.R. Sambles, Thin Solid Films **106**, 321 (1983)
10. M. Theuring, M. Vehse, I. Noureddine, K. von Maydell, C. Agert, in *MRS Online Proc. Library* (Cambridge University Press, 2012), Vol. 1426
11. V.J. Logeeswaran, N.P. Kobayashi, M.S. Islam, W. Wu, P. Chaturvedi, N.X. Fang, S.Y. Wang, R.S. Williams, Nano Lett. **9**, 178 (2009)
12. F.-J. Haug, T. Söderström, O. Cubero, V. Terrazzoni-Daudrix, C. Ballif, J. Appl. Phys. **104**, 064509 (2008)

21

Impact of deposition parameters on the material quality of SPC poly-Si thin films using high-rate PECVD of a-Si:H

Avishek Kumar[1,2,3,a], Per Ingemar Widenborg[1], Goutam Kumar Dalapati[3], Gomathy Sandhya Subramanian[3], and Armin Gerhard Aberle[1,2]

[1] Solar Energy Research Institute of Singapore, National University of Singapore, 7 Engineering Drive 1, 117574 Singapore, Singapore
[2] Department of Electrical and Computer Engineering, National University of Singapore, 117583 Singapore, Singapore
[3] Institute of Materials Research and Engineering, A*STAR (Agency for Science, Technology and Research), 3 Research Link, 117602 Singapore, Singapore

Abstract The impact of the deposition parameters such as gas flow (sccm) and RF plasma power density (W/cm^2) on the deposition rate of a-Si:H films is systematically investigated. A high deposition rate of up to 146 nm/min at 13.56 MHz is achieved for the a-Si:H films deposited with high lateral uniformity on 30×40 cm^2 large-area glass substrates. A relationship between the SiH$_4$ gas flow and the RF power density is established. The SiH$_4$ gas flow to RF power density ratio of about 2.4 sccm/mW cm^{-2} is found to give a linear increase in the deposition rate. The influence of the deposition rate on the material quality is studied using UV-VIS-NIR spectrophotometer and Raman characterisation techniques. Poly-Si thin film with crystal quality as high as 90% of single-crystalline Si wafer is obtained from the SPC of high rate deposited a-Si:H films.

1 Introduction

SPC poly-Si thin-film material is a promising semiconductor for the PV industry, combining the robustness of c-Si with the advantages of the thin-film approach [1]. SPC poly-Si thin films are obtained from the annealing of a-Si:H films in a N$_2$ atmosphere for 12 h at 610 °C [2]. Plasma-enhanced chemical vapour deposition (PECVD) is widely used in industry for large-area deposition of a-Si:H films. However, the low deposition rate of a-Si:H films (25–35 nm/min) in traditional PECVD processes significantly adds to the cost of poly-Si thin-film solar cells [3]. There are other deposition techniques such as e-beam evaporation and hot wire chemical vapour deposition (HWCVD) which are capable of a-Si:H film depositions at a very high rate ($\gg$100 nm/min). However, the electronic properties of SPC poly-Si films obtained by these deposition techniques are inferior to those obtained from PECVD, which makes them difficult to commercialise. The a-Si films deposited by the e-beam evaporation technique have a high degree of thickness non-uniformity over large areas, which leads to problems at the device level (such as laterally non-uniform short-circuit current densities of solar cells, or pinholes and cracks that are detrimental to the solar cell performance [4]). On the other hand, a-Si:H films deposited by HWCVD have a very high nucleation rate, leading to SPC poly-Si films with much smaller grains compared with films deposited by PECVD [5].

Some results on a-Si:H films deposited with the high-rate PECVD technique can be found in the literature [6–11]. However, most of these deposited films suffered from internal stress, large microvoids and high dangling bond densities. The possible reason for this could be the fact that the substrate temperature (T_s) used for these a-Si:H depositions was restricted to below 250 °C. Recently, Jin [12] demonstrated in his doctoral thesis that an increase in the substrate temperature to 400 °C yields a high-quality poly-Si film prepared from the SPC of an a-Si:H film deposited at significantly higher rate using PECVD. The performance of the poly-Si thin-film solar cell obtained from the SPC of a-Si:H films deposited at high rate was also found to be comparable to that of the standard low-rate deposited cells [12]. In addition, it was shown that there was a decrease in the contamination level in the absorber layer when the a-Si:H film deposition rate was increased [12]. In a traditional PECVD,

---
[a] e-mail: `avishek.kumar@nus.edu.sg`

contaminants such as carbon (C), nitrogen (N) and oxygen (O) are controlled in the a-Si:H films by keeping the base deposition pressure extremely low in the range of 10^{-8} Torr. This is generally achieved by the use of expensive turbo pumps. Thus, by controlling the contaminants, high-rate PECVD provides the flexibility to deposit at higher base pressure. This makes high-rate PECVD very interesting for large-scale industrial applications where the use of expensive turbo pumps is often not economical. In addition, a much thicker absorber layer is possible through the use of high-rate deposition, which otherwise is virtually impossible with a low deposition rate (<35 nm/min). A thicker absorber layer is required to enhance the current of poly-Si thin-film solar cells, and hence their PV efficiency. Furthermore, the electronic properties of a SPC poly-Si thin film largely depend on the post-deposition treatments, which gives us the possibility to extensively explore the process parameter space of a-Si:H deposition for SPC poly-Si thin-film solar cell applications far beyond of what has been investigated for the a-Si:H solar cell field.

In this paper, we investigate the effects of the main deposition parameters (RF power density (mW/cm^2), SiH$_4$ gas flow (sccm), process pressure, etc.) on the deposition rate and thickness uniformity of a-Si:H films deposited on 30×40 cm^2 large-area glass substrates. Furthermore, the paper presents a detailed study about the impact of the deposition parameters of the a-Si:H films on the structural quality of the resulting SPC poly-Si films. A high deposition rate of up to 146 nm/min at 13.56 MHz is achieved for the a-Si:H films deposited with high lateral uniformity on 30×40 cm^2 large-area glass substrates.

2 Experimental details

An $\sim$70 nm thick SiN$_x$ barrier coating was deposited onto $300 \times 400 \times 3.3$ mm^3 planar glass sheets (Schott, Borofloat) in a PECVD chamber (MVSystems, USA) at a temperature $\sim$460 °C. The SiN$_x$ film acts as an antireflection coating as well as a diffusion barrier for impurities in the glass substrate. The SiN$_x$ coated glass was then loaded into a special PECVD chamber with an electrode spacing of 23 mm. The a-Si:H films were then deposited using different combinations of SiH$_4$ gas flow rate and RF power density, at the conventional plasma generation frequency of 13.56 MHz. The complete recipe used to deposit a-Si:H films is summarised in Table 1. The deposited a-Si:H films were then annealed (Nabertherm, N 120/65HAC furnace, Germany) at 610 °C in a N$_2$ atmosphere for a duration of 12 h to achieve solid phase crystallisation of the film. A rapid thermal anneal (RTA, CVD Equipment, USA) for 1 minute at a peak temperature of 1000 °C in N$_2$ atmosphere was then used to remove crystallographic defects from the SPC poly-Si thin-films and to activate the dopants. Each 30×40 cm^2 poly-Si coated glass sheet was then cut into 12 equal small pieces of size 10×10 cm^2. Optical reflectance measurements in the 250–1500 nm wavelength range were performed on each of the 12 pieces of the poly-Si film using an UV-VIS-NIR spectrophotometer (PerkinElmer, Lambda 950, UV/VIS Spectrometer). The

Table 1. Experimental details used for the PECVD of the p^- a-Si:H films.

Process condition	p^- a-Si layer
SiH$_4$ (sccm)	60–400
100 ppm B$_2$H$_6$:H$_2$ (sccm)	0.2
Substrate temperature (°C)	380
Pressure (Pa)	107
Time (min)	5–20

thickness of the poly-Si thin film was then calculated by curve fitting of the reflectance data with an optical simulation programme (WVASE). The thickness uniformity of the poly-Si thin film over the 30×40 cm^2 glass sheet was determined by calculating the thickness of each of the 12 pieces of the poly-Si film obtained from the glass sheet. The thicknesses of selected a-Si:H samples were also determined by ellipsometry and were found to be in good agreement with the thickness data of the poly-Si films obtained from WVASE. The deposition rate was then obtained by dividing the average thickness of the poly-Si film by the deposition time. Finally, the material quality of the poly-Si thin film was determined using UV reflectance measurements [13–15] and Raman spectroscopy [16, 17] measurements (Witec Alpha 300R confocal Raman microscope equipped with a 532 nm Nd:YAG lase), whereby the samples were always measured from the air side.

3 Results and discussion

3.1 Effect of SiH$_4$ gas flow rate and RF power density on the deposition rate of a-Si:H films

Figure 1 shows the deposition rate of a-Si:H films as a function of the SiH$_4$ gas flow rate. The plasma power density, process pressure and the substrate temperature were kept constant at 67 mW/cm^2, 107 Pa and $\sim$380 °C, respectively, while the SiH$_4$ gas flow rate was systematically increased. It can be clearly seen that the deposition rate of the a-Si:H films increases from 17 to 75 nm/min when the SiH$_4$ gas flow rate is increased from 60 to 225 sccm. Detailed observation of figure 1 reveals that the deposition rate approaches to a saturation value when the SiH$_4$ gas flow rate is increased beyond about 200 sccm. The saturation in deposition rate at high SiH$_4$ gas flows could be from the limitation of plasma power density to ionize the gas present in the chamber. Thus, further experiments were carried out to understand the effect of the plasma power density on the deposition rate of a-Si:H films at high SiH$_4$ gas flow rate.

Figure 2 shows the deposition rate of a-Si:H films as a function of the RF plasma power density. The SiH$_4$ gas flow rate, process pressure and the substrate temperature were kept constant at 200 sccm, 107 Pa and $\sim$380 °C, respectively, while the plasma power density was systematically increased. At 200 sccm of SiH$_4$ gas flow, there is a significant increase in the a-Si:H deposition rate from 71 to 93 nm/min when the plasma power density is increased from 67 to 100 mW/cm^2. The a-Si:H deposition

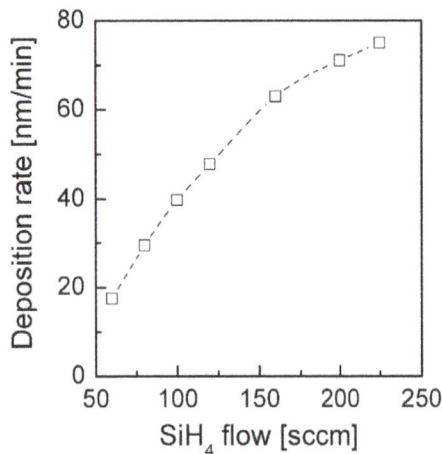

Fig. 1. Deposition rate of a-Si:H films as a function of SiH_4 flow.

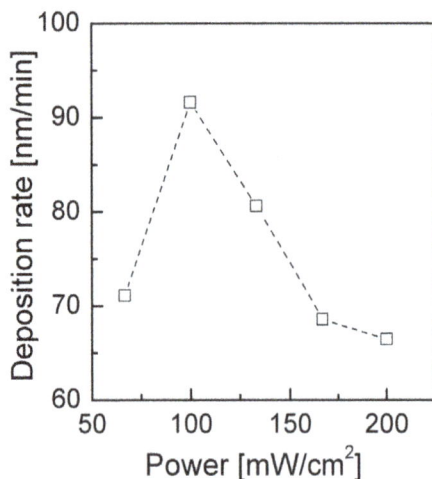

Fig. 2. Deposition rate of a-Si:H films as a function of the RF power density. The dotted lines are guides to the eye.

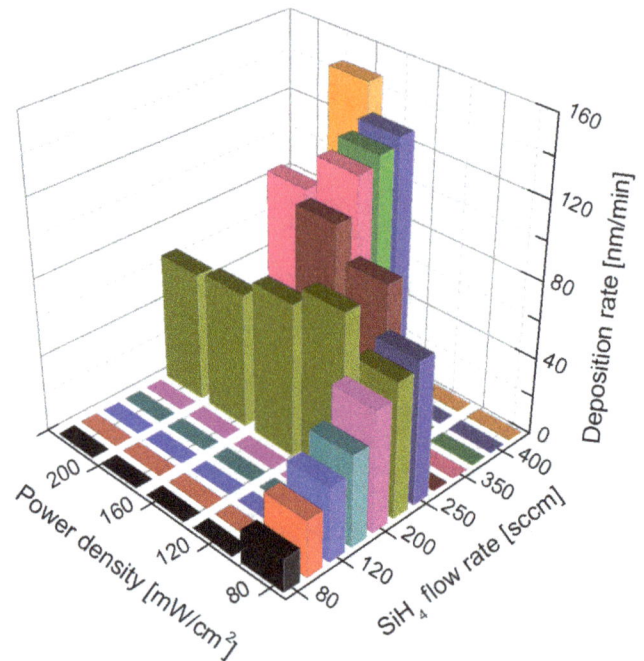

Fig. 3. Deposition rate of the a-Si:H films as a combined function of the plasma power and the SiH_4 flow rate.

films. Further experiments were carried out to study the combined effect of the plasma power density and the SiH_4 gas flow rate on the deposition rate of the a-Si:H films. Figure 3 shows the a-Si:H deposition rate as a combined function of the SiH_4 gas flow rate and the RF plasma power density. From Figure 3 it can be clearly seen that, at a RF power density of 67 mW/cm^2, the a-Si:H deposition rate starts to saturate when the SiH_4 gas flow rate increases beyond 200 sccm. A very different trend in the deposition rate is observed when the RF power density is varied. There is an increase in the a-Si:H deposition rate from 71 to 91 nm/min when the power density is increased from 67 to 100 mW/cm^2. The deposition rate then decreased with the further increase in the RF power density until it saturated at a significantly lower value of 67 nm/min. Further analysis of Figure 3 reveals that a sustainable increase in the a-Si:H deposition rate can be achieved through the simultaneous control of the SiH_4 gas flow and the power density. In this work, we found that a SiH_4 gas flow to RF power density ratio of around 2.4 sccm/mW cm^{-2} gives a sustainable increase in the a-Si:H deposition rate. A high deposition rate of 146 nm/min was obtained through the control of the SiH_4 gas flow to RF power density ratio.

3.2 Impact of deposition rate on thickness uniformity of the a-Si:H films over the 30×40 cm^2 glass

Thickness uniformity over large areas is one of the key issues for any film deposition technique and is very critical for many semiconductor applications. The PECVD technique is renowned for its conformal deposition over large areas at low deposition rates. However, very few results are

rate then decreases with a further increase of the plasma power density. It seems that the further increase of the plasma power density led to the saturation of the active species (responsible for the deposition rate) present in the plasma region at 200 sccm of SiH_4 gas flow rate [18]. In addition, the increase in the plasma power density also results in the formation of dust or small particles [18]. The combined effects of dust generation and saturation of the active species (positive ions) might be responsible for the decrease in the a-Si:H deposition rate. Furthermore, the dust particles generated at high power density plasma get accumulated and form a sheath near the throttle valve connected to the pump. This accumulated layer of dust particles impacts the flow of gas in the chamber, which also causes a laterally more non-uniform deposition (to be discussed in detail in Sect. 3.2). From the above discussion, it was observed that the increase in the deposition rate cannot be sustained by adjusting the power and gas flow rate individually. It is thus, desired to establish a relationship between the RF power density and the SiH_4 gas flow rate to have a sustained growth rate of the a-Si:H

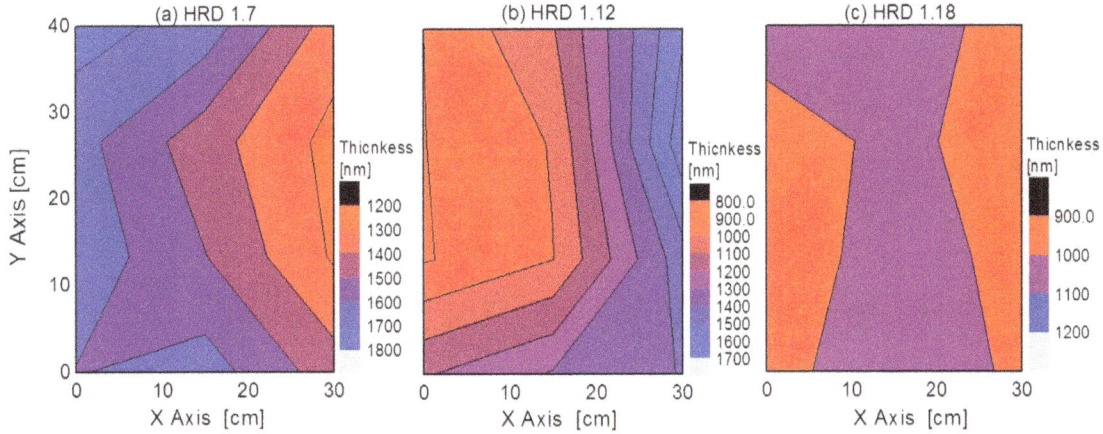

Fig. 4. Contour maps for a-Si:H thickness non-uniformity over the $30 \times 40 \text{ cm}^2$ glass sheet at a deposition rate of (a) 75 nm/min, (b) 67 nm/min and (c) 146 nm/min.

available in the literature for high-rate conformal PECVD deposition over large area. Thus, we further investigate the thickness uniformity of high-rate deposited a-Si:H films over the large-area ($30 \times 40 \text{ cm}^2$) glass sheet. The thickness non-uniformity is quantified using the following relation:

$$\text{non-uniformity} \% = \left[\frac{\text{max}_{\text{thickness}} - \text{min}_{\text{thickness}}}{\text{max}_{\text{thickness}} + \text{min}_{\text{thickness}}} \right] \times 100. \tag{1}$$

Figure 4 shows the contour maps for the a-Si:H thickness non-uniformity over the $30 \times 40 \text{ cm}^2$ glass sheet at an a-Si:H deposition rate of (a) 75 nm/min, (b) 67 nm/min and (c) 146 nm/min. It is found that the a-Si:H films deposited at 75 nm/min have a thickness non-uniformity of $\pm 16\%$ over the $30 \times 40 \text{ cm}^2$ glass sheet, whereas the a-Si:H films deposited at a slightly lower deposition rate of 67 nm/min are found to be highly non-uniform (thickness non-uniformity of $\pm 30\%$). However, the a-Si:H films deposited at a very high deposition rate of 146 nm/min are found to be highly uniform over the $30 \times 40 \text{ cm}^2$ glass sheet, with a thickness non-uniformity value of less than $\pm 6\%$ (see Fig. 4c). Further analysis of the thickness non-uniformity contour maps and the deposition parameters reveals that a highly uniform a-Si:H deposition is possible at high deposition rate, provided that the ratio of SiH_4 gas flow rate to RF power density is kept at around $2.4 \text{ sccm/mW cm}^{-2}$. We also observed that SiH_4 gas flow to RF power density ratios of below $1.6 \text{ sccm/mW cm}^{-2}$ or above $3 \text{ sccm/mW cm}^{-2}$ lead to a very high lateral thickness variation of the a-Si:H film on the $30 \times 40 \text{ cm}^2$ glass sheet. The samples HRD 1.7 and HRD 1.12 (see Figs. 4a and 4b) were obtained with a SiH_4 gas flow to RF power density ratio of $3.3 \text{ sccm/mW cm}^{-2}$ and $1.0 \text{ sccm/mW cm}^{-2}$, respectively. A large amount thickness non-uniformity in the a-Si:H film over the large-area ($30 \times 40 \text{ cm}^2$) glass sheet leads to the formation of cracks during the SPC process. The cracks are mainly formed by the stress produced in the film during the heat treatment process. The large thickness variation in the a-Si:H film on the $30 \times 40 \text{ cm}^2$ glass sheet area is one of the main causes for the stress generation. The cracks formed during the

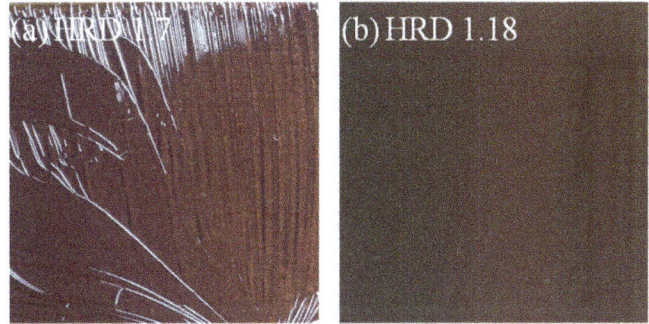

Fig. 5. Photograph of a poly-Si film obtained from SPC of a-Si:H films deposited with a SiH_4 gas flow to RF power density ratio of (a) $3.3 \text{ sccm/mW cm}^{-2}$ and (b) $2.4 \text{ sccm/mW cm}^{-2}$.

SPC process open up during the RTA process and cover the entire film, as can be seen in Figure 5.

Figure 5 shows photographs of selected poly-Si thin-film samples obtained from SPC of a-Si:H films deposited using a SiH_4 gas flow to RF power density ratio of (a) $3.3 \text{ sccm/mW cm}^{-2}$ and (b) $2.4 \text{ sccm/mW cm}^{-2}$. The photos shown here were taken after the RTA process and depict a small sample area of $10 \times 10 \text{ cm}^2$ from the centre of $30 \times 40 \text{ cm}^2$ glass sheet. It can be clearly seen that the poly-Si film prepared from SPC of laterally highly non-uniform a-Si:H films (HRD 1.7) has cracks all over the surface (see Fig. 5a). In contrast, no cracks are observed after the RTA process for the poly-Si film obtained from SPC of a highly uniform a-Si:H film (see Fig. 5b). The high density of cracks makes the poly-Si film unsuitable for solar cell applications and thus needs to be avoided.

3.3 Effect of deposition rate on the crystal quality of the poly-Si thin film

The material quality of the poly-Si thin film is one of the key parameters that define its electrical properties. Thus, it is desirable to obtain a high crystal quality of the poly-Si film while trying to achieve a high-rate

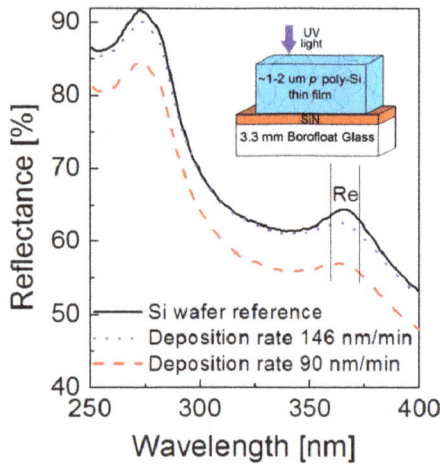

Fig. 6. Hemispherical UV reflectance measured on two poly-Si films obtained by SPC of a-Si:H films deposited at 90 and 146 nm/min, respectively. Also shown (solid line) is the UV reflectance measured on a polished singlecrystalline Si wafer.

deposition. Thus further investigation was undertaken to study the effect of a high deposition rate on the crystal quality, using the UV reflectance and Raman characterization techniques. Figure 6 shows the total hemispherical reflectance at short wavelengths (UV region, 250–400 nm) measured on two different poly-Si thin films prepared by SPC of high-rate deposited a-Si:H films, as well as on a polished FZ c-Si (100) wafer. Two characteristic peaks resulting from direct optical transitions in c-Si [19] were observed in the UV reflectance spectrum at ~360 nm and ~275 nm ($Re1$ and $Re2$), respectively. Structural disorder and defects in the surface region of the probed poly-Si film lead to a decrease and broadening of the peaks [20]. It can be clearly seen that the poly-Si thin film fabricated by SPC of an a-Si:H film deposited at 146 nm/min has less structural defects than the poly-Si film fabricated by SPC of an a-Si:H film deposited at 90 nm/min. In addition, the reflectance spectrum obtained for the poly-Si film is quite close to that of a single-crystal Si wafer. Furthermore, the UV reflectance spectra obtained for the c-Si wafer and the poly-Si thin films were quantified in terms of the crystal quality (Q_r) using the following relationship given in references [14, 15]:

$$Q_r = \frac{1}{2}\left[\frac{Re1}{Re1.\text{c-Si}} + \frac{Re2}{Re2.\text{c-Si}}\right]. \qquad (2)$$

The UV-Vis-NIR measurement at wavelengths below 300 nm has a small signal-to-noise ratio. Therefore, we exclude the $Re2$ to get the following equation:

$$Q_r = \frac{Re1}{Re1.\text{c-Si}}. \qquad (3)$$

The crystal quality factor Q_r quantifies how closely the UV reflectance of a single-crystalline Si wafer is mimicked by a particular poly-Si thin-film sample and thus provides a qualitative measure of its area-averaged crystal quality [14].

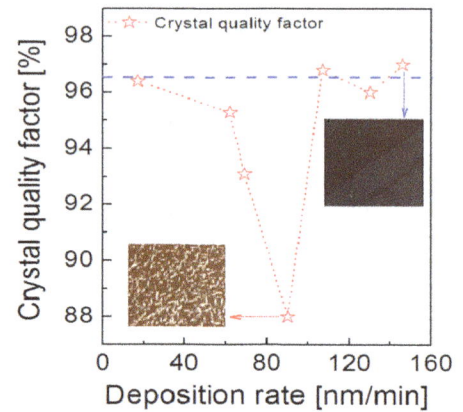

Fig. 7. Crystal quality of the SPC poly-Si thin films calculated from UV reflectance as a function of the a-Si:H deposition rate. The dotted lines are guides to the eye.

Figure 7 shows the calculated crystal quality factor for selected poly-Si thin films obtained from a-Si:H films deposited at various deposition rates. It can be clearly seen that few poly-Si thin-film samples fabricated by SPC of a-Si:H films deposited at high rate have a better UV crystal quality than those obtained from a-Si:H deposited at lower rate. Detailed analysis of the graph (see Fig. 7) and the experimental data revealed that the SPC poly-Si films with the best crystal quality were fabricated from a-Si:H films deposited using a SiH$_4$ flow to RF power density ratio of about 2.4 sccm/mW cm^{-2}. In contrast, the SPC poly-Si films with poor crystal quality were found to be fabricated from a-Si:H films using a SiH$_4$ gas flow to RF power density ratio of less than 1.6 sccm/mW cm^{-2} or greater than 3 sccm/mW cm^{-2}. It was also observed that few of the poly-Si thin films with low UV crystal quality had an extremely high number of cracks (see Fig. 7). These cracks might have been generated during the SPC and RTA processes due to the large a-Si:H thickness variation across the large area of 30×40 cm^2, as discussed earlier (see Sect. 3.2). The UV reflectance merely probes the surface region (~5 nm thickness) of the poly-Si film and thus the obtained UV crystal quality could just be a representation of the surface quality of the poly-Si film. The poor UV crystal quality observed in some of the poly-Si films with an extreme number of cracks could well be due to the effects of the cracks and thus does not represent the true effect of the a-Si:H deposition rate on the crystal quality of the SPC poly-Si film. Hence, selected poly-Si thin film samples were further characterised from the air side by Raman spectroscopy [16] (depth resolution ~400 nm), to better understand the effect of the a-Si:H deposition rate on the structural quality of the SPC poly-Si film.

Figure 8 shows Raman spectra acquired from the visible (532 nm) laser line for selected poly-Si thin films fabricated by SPC of a-Si:H films deposited at 17 and 146 nm/min. As a reference, the Raman spectrum was also obtained for a single-crystal high-resistivity FZ Si (100) wafer (solid line). A strong peak at a frequency ω_0 of about 522 cm^{-1} is observed for the c-Si wafer. This peak position value of c-Si may slightly vary from experiment to

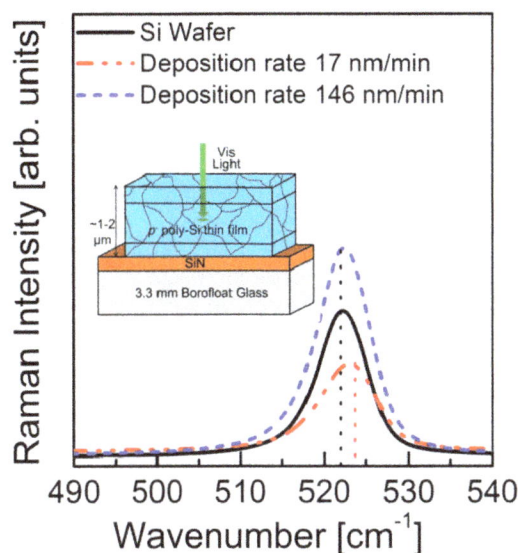

Fig. 8. Raman spectra of poly-Si films deposited at two different deposition rates of 17 and 146 nm/min, respectively. Also shown (solid line) for comparison is the Raman spectrum measured on a polished single-crystal Si wafer.

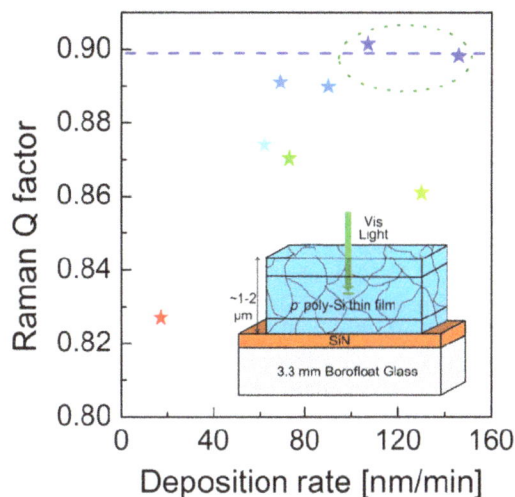

Fig. 9. Raman quality factor (R_Q) of SPC poly-Si thin films as a function of the a-Si:H deposition rate.

experiment, depending on the calibration of the spectrometer and monochromator. Furthermore, the Raman spectra for poly-Si film obtained from SPC of a-Si:H films deposited at 17 nm/min reveal that there is a shift in the peak position towards higher wave numbers with respect to c-Si, indicating the presence of compressive stress in the poly-Si film [21]. However, it is worth to note that the poly-Si film fabricated by SPC of a-Si:H deposited at a high deposition rate of 146 nm/min is found to have no stress in the film.

From the analysis of the Raman spectra and the process parameters revealed that stress in the SPC poly-Si film can be reduced by the control of the SiH$_4$ gas flow to RF power density ratio. Furthermore, the Raman spectra reveal that the full width at half maximum (FWHM) of the poly-Si thin films varies with the deposition rate of the a-Si:H film. The FWHM is an excellent indicator of the crystal quality of the poly-Si film. An increase in the defect density and disorder in Si thin films leads to a broadening of the peak (FWHM) [16,22,23]. A Raman quality factor (R_Q) is defined here as the ratio between the FWHM of single-crystal Si to that of the poly-Si film ($R_Q = \frac{FWHM_{\text{c-Si}}}{FWHM_{\text{poly-Si}}}$) to quantify the defects in the poly-Si film relative to a (stress-free) single-crystal Si wafer.

Figure 9 shows the calculated Raman quality factor of the poly-Si thin film obtained by SPC of a-Si:H films as a function of the deposition rate. It can be clearly seen that few poly-Si films obtained by SPC of a-Si:H films deposited at higher rate had better crystal quality than those obtained from a-Si:H deposited at low deposition rate. The interpretation of poly-Si thin film crystal quality obtained from Raman measurements is in good agreement with the crystal quality factor obtained from UV reflectance measurements for most of the poly-Si thin film reported here.

4 Conclusion

In conclusion, a high deposition rate of 146 nm/min was achieved through the control of the SiH$_4$ gas flow and the RF power density. A laterally highly uniform deposition of a-Si:H over the large glass sheet area of 1200 cm^2 using high-rate PECVD was achieved. A relationship between the SiH$_4$ gas flow and the RF power density was established. A linear increase in the deposition rate up to 146 nm/min was achieved by keeping the SiH$_4$ gas flow to RF power density ratio constant at about 2.4 sccm/mW cm^{-2}. The SiH$_4$ gas flow to RF power density ratio was also found to affect the thickness uniformity of a-Si:H films and the material quality of the SPC poly-Si films obtained from the a-Si:H films. A very high SPC poly-Si crystal quality with a thickness non-uniformity of less than ±6% over 30 × 40 cm^2 was obtained. A further increase in the deposition rate to about 250 nm/min seems possible through the control of the SiH$_4$ gas flow and the RF power density, while maintaining good thickness uniformity and a high crystal quality.

The Solar Energy Research Institute of Singapore (SERIS) is sponsored by the National University of Singapore (NUS) and the National Research Foundation (NRF) of Singapore through the Singapore Economic Development Board (EDB). This research was supported by the NRF, Prime Minister's Office, Singapore under its Clean Energy Research Programme (Award No. NRF2009EWT-CERP001-046). A.K. acknowledges Ph.D. scholarships from Singapore's Energy Innovation Programme Office.

References

1. A.G. Aberle, Thin Solid Films, **517**, 4706 (2009)
2. A. Kumar, P.I. Widenborg, H. Hidayat, Q. Zixuan, A.G. Aberle, in *MRS Proceedings, 2011*, Vol. 1321

3. P.A. Basore, in *Proceedings of the 4th World Conference on Photovoltaic Energy Conversion, Hawaii, 2006*, pp. 874–876

4. Z. Ouyang, O. Kunz, M. Wolf, P. Widenborg, G. Jin, S. Varlamov, in *18th International Photovoltaic Science and Engineering Conference, Kolkata, 2009*, p. 023

5. D.L. Young, P. Stradins, Y. Xu, L. Gedvilas, B. Reedy, A.H. Mahan, H.M. Branz, Q. Wang, D.L. Williamson, Appl. Phys. Lett. **89**, 122103 (2006)

6. T. Takagi, K. Takechi, Y. Nakagawa, Y. Watabe, S. Nishida, Vacuum **51**, 751 (1998)

7. T. Takagi, R. Hayashi, G. Ganguly, M. Kondo, A. Matsuda, Thin Solid Films **345**, 75 (1999)

8. V. Kirbs, T. Drusedau, H. Fiedler, J. Phys.: Condens. Matter **4**, 10433 (1992)

9. M. Estrada, A. Cerdeira, I. Pereyra, S. Soto, Thin Solid Films **373**, 176 (2000)

10. B.A. Korevaar, G.J. Adriaenssens, A.H.M. Smets, W.M.M. Kessels, H.Z. Song, M.C.M. van de Sanden, D.C. Schram, J. Non-Cryst. Solids **266-269**, 380 (2000)

11. J. Hautala, Z. Saleh, J.F.M. Westendorp, H. Meiling, S. Sherman, S. Wagner, in *MRS Proceedings, 1996*, Vol. 420

12. G. Jin, Ph.D. Thesis, The University of New South Wales, 2010

13. G. Harbeke, L. Jastrzebski, J. Electrochem. Soc. **137**, 696 (1990)

14. A. Straub, P.I. Widenborg, A. Sproul, Y. Huang, N.P. Harder, A.G. Aberle, J. Cryst. Growth **265**, 168 (2004)

15. P.I. Widenborg, A.G. Aberle, J. Cryst. Growth **306**, 177 (2007)

16. S. Nakashima, M. Hangyo, IEEE J. Quantum Electron. **25**, 965 (1989)

17. M. Holtz, W.M. Duncan, S. Zollner, R. Liu, J. Appl. Phys. **88**, 2523 (2000)

18. G. Bruno, P. Capezzuto, G. Cicala, Chemistry of Amorphous Silicon Deposition Processes: Fundamentals and Controversial Aspects, in: *Plasma Deposition of Amorphous Silicon-Based Materials*, edited by G. Bruno, P. Capezzuto, A. Madan (Academic Press, San Diego, 1995), pp. 1–62.

19. P.Y. Yu, M. Cardona, *Fundamentals of semiconductors: physics and materials properties* (Springer, Berlin, 1999)

20. T. Kamins, *Polycrystalline silicon for integrated circuit applications* (Kluwer Academic Publishers, 1988)

21. K. Kitahara, A. Hara, in *Crystallization – Science and Technology*, edited by M.R.B. Andreka (Intech, Rijeka, 2012)

22. A. Kumar, H. Hidayat, C. Ke, S. Chakraborty, G.K. Dalapati, P.I. Widenborg, C.C. Tan, S. Dolmanan, A.G. Aberle, J. Appl. Phys. **114**, 134505 (2013)

23. R.C. Teixeira, I. Doi, M.B.P. Zakia, J.A. Diniz, J.W. Swart, Mater. Sci. Eng. B **112**, 160 (2004)

22

UV and IR laser induced ablation of Al$_2$O$_3$/SiN:H and a-Si:H/SiN:H

T. Schutz-Kuchly[1,a], A. Slaoui[1], J. Zelgowski[2], A. Bahouka[2], M. Pawlik[3], J.-P. Vilcot[3], E. Delbos[4], M. Bouttemy[5], and R. Cabal[6]

[1] Laboratoire des sciences de l'Ingénieur, de l'Informatique et de l'Imagerie (ICUBE) UMR 7357, UdS/CNRS, 23 rue du Loess, BP 20 CR, 67037 Strasbourg Cedex 2, France
[2] IREPA-Laser Pole API – Parc d'innovation, 67400 Strasbourg, France
[3] Institut d'Électronique, de Microélectronique et de Nanotechnologie (IEMN) UMR 8520, Université Lille 1 Sciences et Technologies, CS 60069, 59652 Villeneuve d'Ascq, France
[4] KMG Group, 45 avenue des États-Unis, 78035 Versailles, France
[5] Institut Lavoisier de Versailles UMR 8180, Université de Versailles-St-Quentin en Yvelines, 45 avenue des États-Unis, 78000 Versailles, France
[6] CEA-INES, 50 avenue du Lac Léman, 73375 Le Bourget du Lac, France

Abstract Experimental work on laser induced ablation of thin Al$_2$O$_3$(20 nm)/SiN:H (70 nm) and a-Si:H (20 nm)/SiN:H (70 nm) stacks acting, respectively, as p-type and n-type silicon surface passivation layers is reported. Results obtained using two different laser sources are compared. The stacks are efficiently removed using a femtosecond infra-red laser (1030 nm wavelength, 300 fs pulse duration) but the underlying silicon surface is highly damaged in a ripple-like pattern. This collateral effect is almost completely avoided using a nanosecond ultra-violet laser (248 nm wavelength, 50 ns pulse duration), however a-Si:H flakes and Al$_2$O$_3$ lace remain after ablation process.

1 Introduction

In today's photovoltaic (PV) industry, laser processing is mostly limited to edge isolation [1] for producing standard silicon solar cells. Owing to its versatility, cost-effectiveness and high throughput, laser processing is well adapted for the PV industry need. For instance, laser doping has entered into production since 2009 for the fabrication of selective emitter solar cells [2]. Next step would be the use of laser induced ablation of dielectrics towards the fabrication of selective emitter silicon solar cells by structuring narrow lines into the antireflective coating. This will lead to increase the short circuit current by reducing electrode shadowing loss.

Selective laser ablation is particularly interesting when it is combined to self aligned metallization technologies such as nickel (Ni) electro-less deposition [3] and copper (Cu) plating. This combination can be seen as an attractive alternative to the standard screen-printed metallization which becomes a bottleneck in the PV industry because its use of expensive silver. Moreover, since the Ni-Cu metallization implies lower thermal budget than conventional screen printing route, surface passivation using, for example, amorphous silicon will be feasible.

N-type bifacial solar cells [4] are passivated with an aluminum oxide/hydrogenated silicon nitride (Al$_2$O$_3$/SiN:H) stack on their front side (P-type emitter) and an SiN:H on their rear side (N-type). Laser ablation of one single SiN:H layer has been studied by several research groups and is now well understood [5–8]. The ablation of the Al$_2$O$_3$/SiN:H stack has already been studied by Jaffrennou et al. [9] and Jin et al. [10] using UV picosecond lasers. However, as Al$_2$O$_3$ shows no significant optical absorption coefficient for this wavelength range, its correct ablation seems to be problematic, whatever it is stacked or not with another material film.

So, we present hereby experimental results concerning laser induced ablation of two dielectric stacks that are used for the passivation of n-type bifacial solar cells; an Al$_2$O$_3$/SiN:H stack for the passivation of the boron front emitter and an a-Si:H/SiN:H stack for the passivation of the phosphorus rear back surface field (BSF). The results obtained using two different laser sources, in terms of laser

a e-mail: `tschutzk@gmail.com`

Table 1. UV and IR laser characteristics.

Laser	Wavelength (nm)	Pulse duration	Profile	Frequency
UV	248	50 ns	Flat-hat	1−10 Hz
IR	1030	300 fs	Gaussian	0−2 MHz

wavelength and pulse duration, are compared. The first laser source is in the infra-red (IR) domain with a pulse duration of 300 fs and the second laser source is in the ultra-violet (UV) domain with a pulse duration of 50 ns.

The impact of the laser ablations on the silicon wafer is characterized in terms of carrier lifetime mappings. Scanning electron microscope (SEM) observations are used to visualize the ablation quality and the surface damages.

2 Experimental details

2.1 Laser characteristics

The characteristics of the two laser sources that were used are given in Table 1.

The first one is a femtosecond IR laser (1030 nm) with pulse duration of 300 fs from AMPLITUDE Systems, the model is Tangerine®. Minimum achievable spot size is close to 40 μm and the repetition frequency can reach up to 2 MHz which is an interesting feature for industrial purposes.

The second laser source is a Lambda Physics COMPex 201 UV Krypton Fluorine (KrF) excimer laser. The pulse duration is 50 ns and the repetition frequency is low, about 10 H. The typical spot size is around 3×3 mm^2 and physical masks will be used in order to pattern the substrates.

2.2 Wafer processing and characterization tools

The samples used in this work are 5×5 cm^2 squares issued from 125 PSQ Czochralski (Cz) N-type silicon wafers. The base resistivity is ~ 3 Ω cm and the thickness is ~ 200 μm. The samples were either KOH/IPA texturized or KOH polished.

The a-Si:H and the SiN:H layers were deposited in an electron-cyclotron-resonance plasma enhanced chemical vapor deposition (ECR-PECVD) reactor from Roth and Rau company. The a-Si:H and SiN:H thickness is 20 nm and 70 nm, respectively.

The Al$_2$O$_3$ layer was deposited in a plasma assisted atomic layer deposition (PA-ALD) reactor (BeneqTFS 200); its thickness is 20 nm. A SiN:H layer was also deposited on top.

The SEM analyses were performed using a SEM-FEG (Scanning Electron Microscope enhanced by a Field Emission Gun) equipment (JSM 7001-F, JEOL). Lifetime mappings were performed with a Semilab WT-2000 in Micro-Wave-Photoconductance-Decay

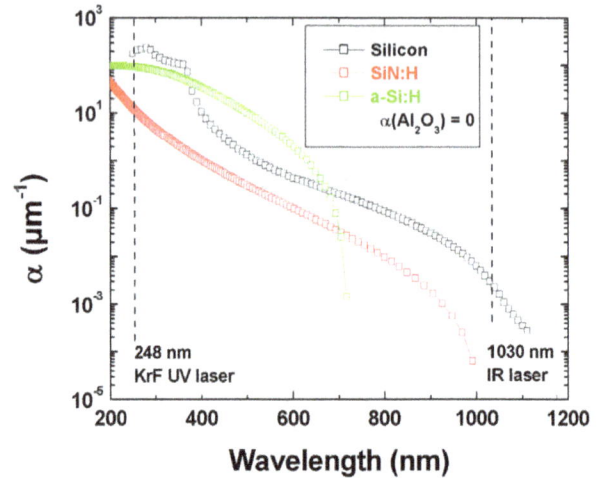

Fig. 1. Absorption coefficient (μm^{-1}) as a function of wavelength for silicon, SiN:H, a-Si:H and Al$_2$O$_3$.

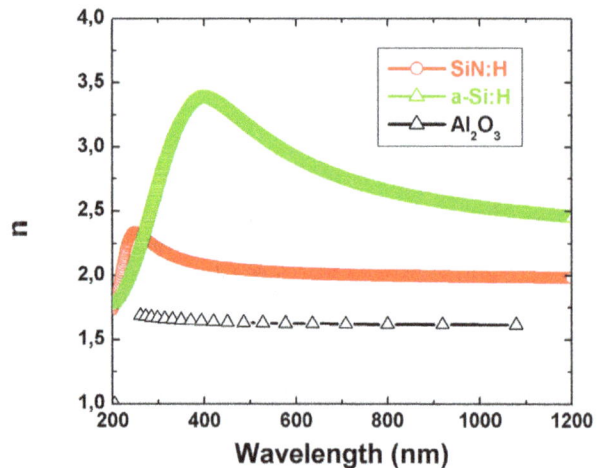

Fig. 2. Refractive index as a function of wavelength for SiN:H, a-Si:H and Al$_2$O$_3$.

(μW-PCD) mode. XPS (X-ray Photoelectron Spectroscopy) (Thermo Electron K-Alpha spectrometer) analyses and local nano-Auger measurements (Jamp 9500F, JEOL) were carried out over the ablated areas of the Al$_2$O$_3$/SiN:H coated wafers. Optical characterizations were performed with a Horiba Jobin-Yvon ellipsometer.

3 Results and discussions

The optical properties of the different materials were firstly measured versus the wavelength. Absorption coefficient and refractive index are shown in Figures 1 and 2, respectively.

In the IR wavelength range of interest, the absorption coefficient of a-Si:H and SiN:H is negligible compared to silicon. In the UV wavelength range, the absorption coefficient of a-Si:H is similar to the one of silicon. Concerning Al$_2$O$_3$, the absorption coefficient remains almost 0, or in any case immeasurable, whatever the wavelength range is.

Table 2. Thermodynamic properties [11].

Name	c-Si	a-Si	Si_3N_4	Al_2O_3
T_m (°C)	1414	~1200	1877	2072
T_b (°C)	3217	3217	N.A.	2977
k_{th} (W/(m °C))	150	<5	25	18

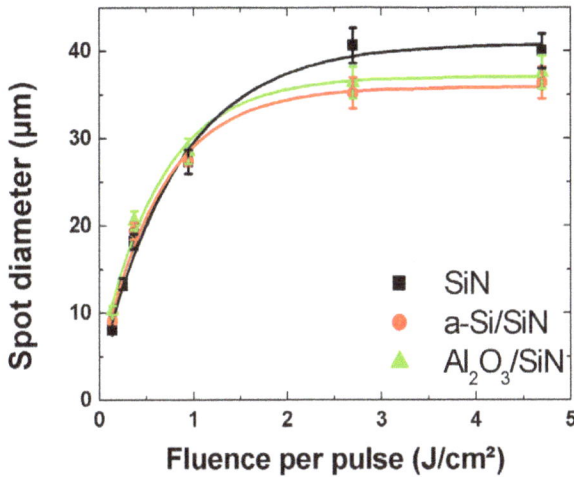

Fig. 3. Diameter of ablated spot versus fluence for SiN, a-Si:H/SiN:H, Al_2O_3/SiN:H materials.

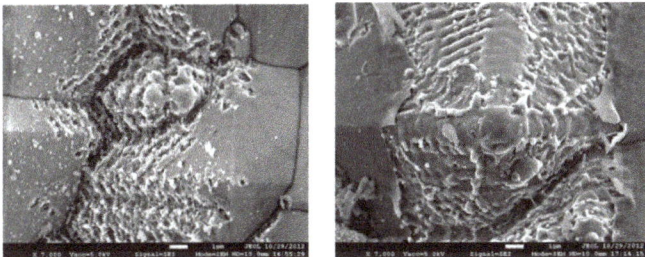

Fig. 4. SEM micrographs of ablated areas using the IR laser for a randomly texturized wafer covered with Al_2O_3/SiN:H. Left: Fluence close to ablation threshold. Right: High fluence value.

When performing ablation experiments, the thermodynamic properties of each material have to be kept in mind. Table 2 shows the melting temperature (T_m), the boiling temperature (T_b) and the thermal expansion coefficient (k_{th}) for silicon, Si_3N_4, a-Si, and Al_2O_3.

3.1 Ablation results using femtosecond IR laser

Based on optical microscope observations, Figure 3 shows the size of the ablated area as a function of the fluence value for three different layers: SiN:H, a-Si:H/SiN:H stack and Al_2O_3/SiN:H stack. The ablation threshold occurs at 0.14 J/cm² in any case. The ablated spot diameter reaches a plateau of ~40 μm, for pulse energy densities greater than 2 J/cm².

Whatever it concerns an Al_2O_3/SiN:H (Fig. 4) or an a-Si:H/SiN:H (Fig. 5) stack deposited on a KOH texturized silicon surface, a ripple-like pattern can be observed

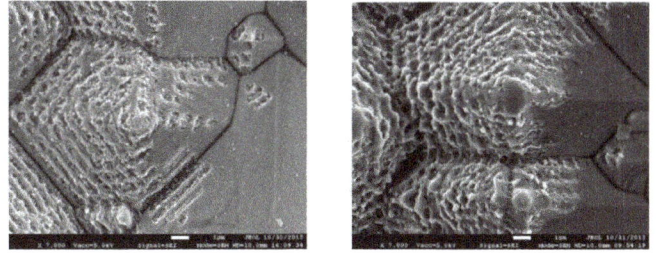

Fig. 5. SEM micrographs of ablated areas using the IR laser for a randomly textured wafer covered with a-Si:H/SiN:H. Left: Fluence close to ablation threshold. Right: High fluence value.

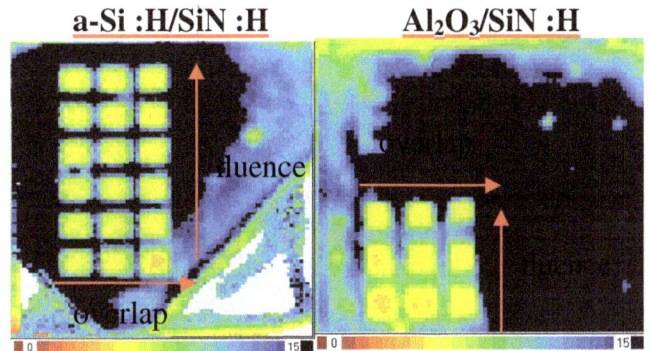

Fig. 6. Lifetime (μ s) mappings of re-passivated wafers. The areas irradiated by IR laser are 5×5 mm². Left: a-Si:H/SiN:H coated wafer. Right: Al_2O_3/SiN:H coated wafer.

over the ablated areas whatever the energy density is. This ripple-like pattern is caused by a modulated intensity pattern on the sidewalls of pyramids that is created by interference patterns during laser irradiation [12].

Based on the respective material absorption coefficients given in Figure 1, it can be seen that IR wavelength is mainly absorbed into the silicon substrate since a-Si:H, SiN:H and Al_2O_3 layers are transparent or quasi-transparent at 1030 nm. From the SEM observations (Figs. 4 and 5), we can speculate that the ripples are the resulting pattern of melted/re-crystallized silicon. As it seems that all the dielectric layers have been efficiently removed, it is reasonable to state that the difference in thermal expansion coefficient of silicon and of dielectric materials (see Tab. 2) was high enough for peeling the dielectric layers off.

In order to evaluate the substrate damages, lifetime mappings using μ W-PCD were performed (Fig. 6). Before performing these mappings, the wafers were HF dipped until the complete removal of Al_2O_3/SiN:H or a-Si:H/SiN:H stacks, and then re-passivated on both sides with a SiN:H film.

Figure 6 clearly shows that lifetime is degraded at the ablated areas. The decrease in lifetime is approximately the same when IR ablations are done on a-Si:H/SiN:H or Al_2O_3/SiN:H stack. Small variation of the lifetime value is observed on the ablated areas when the beam overlapping is increased. Same observation can be made when the fluence value is increased.

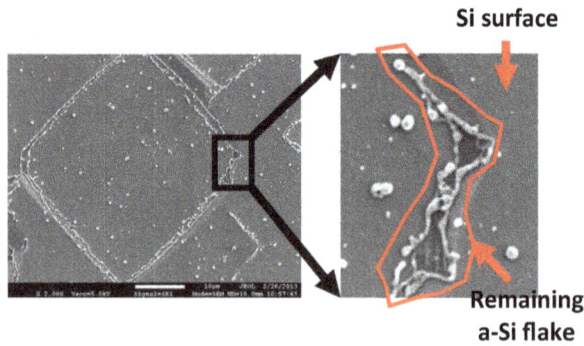

Fig. 7. SEM micrograph of an ablated area using UV laser on KOH polished wafer covered with a-Si:H/SiN:H film. Optimal fluence condition.

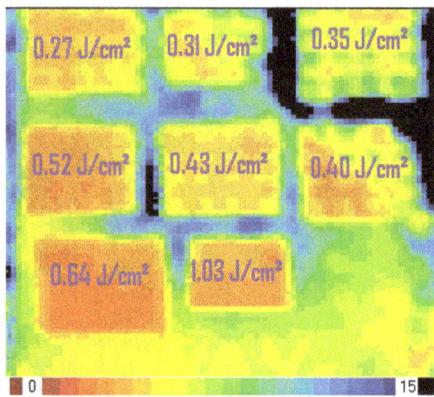

Fig. 8. Lifetime (μs) mapping of re-passivated a-Si:H/SiN:H wafer irradiated by UV laser.

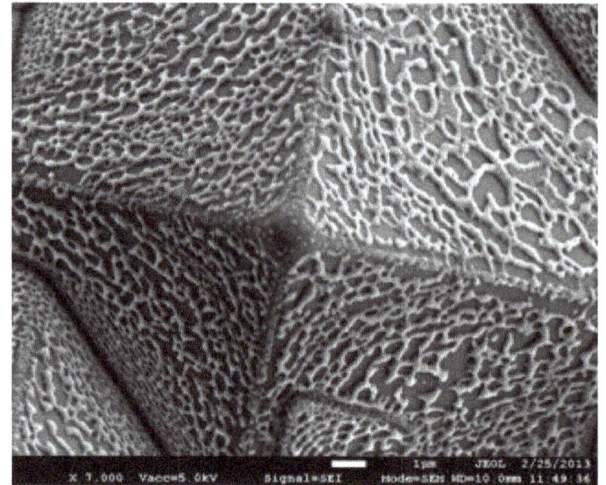

Fig. 9. SEM micrograph of an ablated area using UV laser on KOH textured wafer covered with Al_2O_3/SiN:H. Optimal fluence condition.

Table 3. XPS measurement results of the silicon surface after UV laser ablation of the Al_2O_3/SiN:H stack.

SiN	SiO_x	Metallic Si	Al_2O_3	Metallic C	CO
8.2%	58%	17.2%	2.5%	11.9%	4.2%

Table 4. Nano-Auger results after UV laser ablation of AL_2O_3/SiN:H stack.

C	O	Si	N	Al
11.3%	34.3%	32.8%	17.7%	3.9%

3.2 Ablation results with nanosecond UV laser

3.2.1 a-Si:H/SiN:H stack

Same ablation experiments were done by using a KrF excimer laser. Ablation fluences were varied from ~0.3 J/cm^2 to 1.0 J/cm^2. Figure 7 shows a SEM picture of an ablated a-Si:H/SiN:H stack on a KOH-polished surface using an optimal fluence value. It can be seen that a-Si:H flakes are still present on the border of the truncated pyramids. It seems that a-Si:H material was melted and re-crystallized thus forming randomly distributed balls on the ablated surface.

As observed in the SEM micrograph, as no SiN:H remains and as the silicon wafer surface stays intact, we can reasonably state that the laser beam was absorbed in the a-Si:H/SiN:H stack and silicon wafer extreme surface. The increase in temperature due to the laser beam absorption induced the melting of the a-Si:H layer. The temperature increase was high enough to induce decomposition of the SiN:H layer [7].

Figure 8 shows the lifetime mapping of a re-passivated wafer that was a-Si:H/SiN:H coated. An expected decrease in lifetime is observed when the laser fluence increases. The increase between 0.31 J/cm^2 and 0.43 J/cm^2 (right top corner) is due to a local better wafer lifetime and not to less ablation damages.

3.2.2 Al_2O_3/SiN:H stack

Figure 9 shows a SEM micrograph of the remaining Al_2O_3 material after ablation of an Al_2O_3/SiN:H stack deposited on a KOH textured silicon wafer. Al_2O_3 has been melted/re-crystallized resulting in a "lace" over the silicon surface. As for the a-Si:H/SiN:H case above reported, the heating during laser ablation induces the decomposition of SiN:H. As Al_2O_3 is transparent to the UV wavelength, its melting is induced by the heating of the silicon surface. As it can be seen on the pyramid tip (Fig. 9), the extreme silicon surface was melted as well.

The presence of the silicon surface in between the voids of the Al_2O_3 lace has been confirmed by XPS analysis, this is shown in Table 3. Local nano-Auger measurements show that this silicon surface has been doped with Al atoms from the Al_2O_3 layer (Tab. 4).

Figure 10 shows that lifetime is degraded on the ablated areas when the fluence is increased. It means that the silicon extreme surface is progressively damaged with increasing fluence value.

4 Conclusion

IR femtosecond and UV nanosecond lasers were investigated for the ablation of thin Al_2O_3 (20 nm)/SiN:H

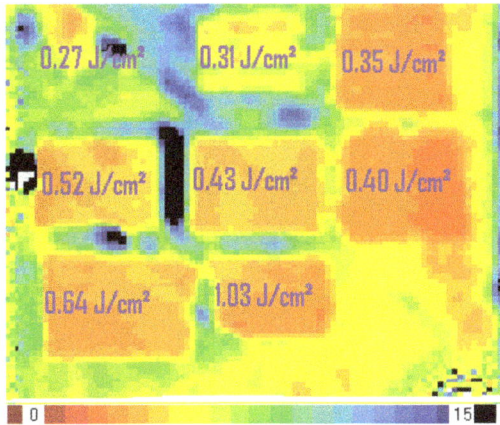

Fig. 10. Lifetime (μs) mapping of re-passivated Al_2O_3/SiN:H wafer irradiated by UV laser.

(70 nm) and a-Si:H (20 nm)/SiN:H (70 nm) stacks deposited on silicon.

The stacks are efficiently removed with the IR laser but the resulting ripple-like pattern resulting from the ablation process indicates that the silicon surface is highly affected and damaged. As a consequence, the emitter doping homogeneity can be affected and the risk of shunting is increased. The IR laser seems not well matched to this application.

We can also evoke the cost and the novelty of this kind of laser system for industry. The process speed is not adapted for long term processes and a big amount of cells to treat.

Concerning the ablation experiments done with the UV laser, the SiN:H is efficiently removed from the stacks. There are however a-Si flakes remaining on the silicon surface after ablation of the a-Si:H/SiN:H stack. On the other hand, Al_2O_3 is difficult to be removed due to its optical transparency to the UV wavelength and a rather high level of waste is present after the ablation process.

As the ablation process will be followed by Ni electroless deposition, a coming paper will show that Ni can be selectively deposited on locally laser ablated areas despite incomplete ablation of the Al_2O_3/SiN:H.

The authors would like to thank the ANR project BIFASOL ANR-11-PRGE-004 for financial support.

References

1. M. Doering, K. Meyer, A. Kaps, H.-J. Krokoszinski, H. Eschrich, in *Proceedings of 25th European Photovoltaic Solar Energy Conference/5th World Conference on Photovoltaic Energy Conversion, Valencia, Spain, 2010* (WIP, München), p. 1778
2. S.K. Chunduri, Photon Int., 158 (2010)
3. L. Tous, D.H. van Dorp, R. Russell, J. Das, M. Aleman, H. Bender, J. Meersschaut, K. Opsomer, J. Poortmans, R. Mertens, Energy Procedia **21**, 39 (2012)
4. A. Carr, *NPV Workshop, Konstanz, Germany, 2011*
5. A. Grohe, C. Harmel, A. Knorz, S.W. Glunz, R. Preu, G.P. Willeke, in *Proccedings IEEE of 4th World Conference on Photovoltaic Energy Conversion, Waikoloa, Hawai, 2006*, p. 1399
6. P. Engelhart, S. Hermann, T. Neubert, R. Grischke, N.P. Harder, R. Brendel, in *Proceedings of the 17th Workshop on Crystalline Silicon Solar Cells and Modules: Materials and Processes, Vail, Colorado, USA, 2007*, edited by B.L. Sopori (NREL, Golden, 2007), p. 57
7. S.A.G.D. Correia, J. Lossen, M. Wald, K. Neckermann, M. Bähr, in *Proceedings of 22nd European Photovoltaic Solar Energy Conference, Milano, Spain, 2007*
8. R. Preu, S.W. Glunz, S. Schafer, R. Ludemann, W. Wettling, W. Pfleging, in *Proceedings of the 16th European Photovoltaic Solar Energy Conference, Glasgow, UK, 2000* (WIP, München), p. 1181
9. P. Jaffrennou, A. Uruena, J. Das, J. Penaud, M. Moors, A. Rothschild, B. Lombardet, J. Szlufcik, in *Proceedings of 26th European Photovoltaic Solar Energy Conference, Hamburg, Germany, 2011* (WIP, München), p. 2180
10. Y. Jin, P.W. Yoon, C. Park, J. Jang, J. Kim, G. Shim, Y. Choe, J.-W. Jeong, in *Proceedings of 26th European Photovoltaic Solar Energy Conference, Hamburg, Germany, 2011* (WIP, München), p. 507
11. E.D. Palik, in *Handbook of Optical Constants of Solids*, edited by E.D. Palik (Elsevier, 1997), Vol. 1
12. A. Knorz, M. Peters, A. Grohe, C. Harmel, R. Preu, Prog. Photovolt.: Res. Appl. **17**, 127 (2009)

Effects of process parameters on μc-Si$_{1-x}$Ge$_x$:H solar cells performance and material properties

Nies Reininghaus[a], Martin Kellermann, Karsten von Maydell, and Carsten Agert

NEXT ENERGY, EWE Research Centre for Energy Technology at the University of Oldenburg, Carl-von-Ossietzky-Str. 15, 26129 Oldenburg, Germany

Abstract On our way to develop a very thin and highly efficient triple-junction thin-film solar cell in a-Si:H/μc-Si:H/μc-SiGe:H configuration and μc-SiGe:H single cell samples were prepared and characterized using an industrial relevant 13.56 MHz 0.5 nm/s process on an industrial like 30×30 cm^2 PECVD tool. To attain a better understanding of the μc-SiGe:H absorber we varied process pressure, germane flow, dilution and silane flow while looking at the electrical and material properties. By realizing a total absorber thickness less than 2 μm for high efficiency cell concepts in triple technology, our intention is to develop an industrial relevant process with attractive fabrication times by benefiting from the enhanced absorption of μc-SiGe:H compared to μc-Si:H.

1 Introduction

Beside low production costs and the usage of abundant raw materials, silicon based thin-film solar cells have the advantage to be built up as multi junction devices like tandem or triple junction solar cells. Comparing silicon solar cells based on wafer technology and thin film silicon solar cells, the advantages of the thin film variant are low energy payback time, better temperature coefficient, large area fabrication and mechanical stability. Therefore developing cheap thin film silicon solar cells with high efficiencies plays a key role.

Typical absorber thicknesses of a-Si:H single junction solar cells are 300–400 nm having very low production costs. However, amorphous silicon degrades under illumination (Steabler-Wronski effect, light-induced degradation (LID)) limiting the efficiency of single junction solar cells. To circumvent this effect, a-Si:H is used in tandem configurations as top cell paired with μc-Si:H as bottom cell i-layer material that does not degrade. Typical absorber thicknesses for tandem cell configurations are 200–250 nm for a-Si:H top cells and 900–1200 nm for μc-Si:H bottom cells.

For a-Si:H/μc-Si:H tandem cell configurations, stabilized efficiencies exceeding 12% were already demonstrated [1]. Recently, Kim et al. published a new efficiency record for silicon based thin-film solar cells which exhibits a stabilized efficiency of 13.4% and is based on the triple junction technology consisting of amorphous and microcrystalline silicon absorber layers [2].

Because of the poor absorption of μc-Si [3], the bottom absorber of triple junction solar cells has to be deposited very thick (>2500 nm) resulting in quite cost intensive industrial application due to corresponding long layer deposition times. There are different approaches to reduce the thickness of the absorber layer. Optimizing light trapping due to scattering structures at the front contact and thereby increasing the path length of light through the device is one of such options. Using novel back reflectors at the back interface is another. A different approach is to substitute conventional doped layers with more transparent oxygenated doped layers where more light reaches the absorber layer and increases the cell current densities enabling a reduced layer thickness. Another approach is to increase the absorption coefficient of the absorber material by alloying silicon with germanium. Matsui et al. introduced the usage of microcrystalline silicon germanium alloys (μc-SiGe:H) with increased absorption coefficient compared to μc-Si:H [4,5].

This contribution addresses the development of μc-SiGe:H absorbers and their application as bottom absorber in triple junction configuration. It is part of a triennial research project based in the Photovoltaic Innovation Alliance established by the BMU (German Federal Ministry for the Environment) and BMBF (German Federal Ministry of Education and Research), that started in July 2011. Partners of Next Energy in that joint research project are the solar module manufacturer

---

[a] e-mail: `nies.reininghaus@next-energy.de`

Inventux Solar GmbH, Hüttinger Elektronik GmbH & Co. KG, a manufacturer of generators for plasma deposition processes, and the Competence Centre Thin-Film and Nanotechnology for Photovoltaics Berlin (PvcomB). In this paper, we will present a comprehensive study of the effects process parameters of the intrinsic layer in a single μc-SiGe:H cell have on the material properties in terms of germanium content and crystallinity while also looking at the device performance. Finally we will compare our triple-junction a-Si:H/μc-Si:H/μc-Si:H solar cells with the triple solar cell type in a-Si:H/μc-Si:H/μc-SiGe:H configuration.

2 Experimental

All μc-Si:H and μc-SiGe:H bottom absorber and μc-SiGe:H single junction solar cells were deposited in pin sequence in a PECVD reactor from Leybold Optics at NEXT ENERGY for 30×30 cm^2 substrate size at substrate temperatures below 250 °C. The solar cells were deposited using a 13.56 MHz plasma excitation frequency at narrow electrode gap under high pressure. The intrinsic μc-Si:H and μc-SiGe:H layers were deposited under high power conditions with deposition rates above 0.5 nm/s.

10×10 cm^2 ZnO:Al coated glass served as substrate for the single junction cells. Ag back contact pads prepared by electron beam evaporation define test cell areas of 1 cm^2. For triple junction solar cells we applied a combination of a 80 nm thick ZnO:Al layer and an Ag layer as back reflector. ZnO:Al layers were prepared by magnetron sputtering.

After deposition all solar cells were annealed for 30 min at 160 °C, followed by IV-measurements. To keep the defined cell area at 1 cm^2 for the triple-junction cells, excess ZnO:Al was removed by etching with HCl.

Raman spectroscopy was performed on single μc-SiGe:H cells. The wavelength of the used laser light was 633 nm. IV-curves of solar cells under illumination (AM 1.5, 100 mW/cm^2, 25 °C) were measured under a continuous light (DC) sun simulator of class A (WACOM). For determining the external quantum efficiency (EQE) of μc-SiGe:H solar cells, monochromatic probe beams using RERA System equipment with wavelength from 300 to 1100 nm were used.

As presented in previous contributions [6,7], we developed a high-quality process for a μc-SiGe:H i-layers on a 15×15 cm^2 PECVD reactor and transferred this process to a small scale industrial PECVD tool which can deposit on 30×30 cm^2 substrates at higher tact times. To benchmark the obtained μc-SiGe:H i-layer quality on the 30×30 cm^2 PECVD tool, μc-SiGe:H single junction cells with an absorber thickness of 1.2 μm were deposited. As substrate rough ZnO substrates from Solayer were used. The p-layer consisted of 50 nm μc-Si:H doped with diborane. The n-layer consisted of 40 nm a-Si:H doped with phosphine. The cell area of 1 cm × 1 cm was determined by evaporating silver pads. The germanium content in μc-SiGe:H layer was determined by measuring the raman crystallinity and using the correlation of the shift of the 520 cm^{-1} LO-TO

phonon resonance to smaller wave numbers to the germanium content in the layer as suggested by Alonso and Winer [8] using equation (1) below:

$$\omega_{\text{Si–Si}} = 520 - 70x.$$

This was found to stay in good accordance to EDX measurements of μc-SiGe:H films [9]. The optimal germanium content in μc-SiGe:H i-layers was found to be 10% because incorporating more germanium leads to the formation of acceptor like states induced by germanium dangling bond defects that decrease the solar cell performance [10].

To gain a better understanding of the relation between the process parameters and the performance of the μc-SiGe:H absorber, different process parameters were varied and the impact on IV cell performance, germanium content and crystallinity was evaluated. The ultimate goal is to reach μc-SiGe:H absorber performances that are comparable to μc-Si:H absorbers in terms of fill factor and V_{oc} but excel in terms of generated current density.

3 Results and discussion

3.1 Germane flow variation

In the first study the germane flow of the μc-SiGe:H absorber was varied from 1.0 sccm to 1.6 sccm in steps of 0.1 sccm while all other parameters were kept constant. As seen in Figure 1, the increased flow of germane leads to an increased germanium content in the absorber by a factor of 6.3%/sccm. This steep increase can be attributed to the high gas utilization because of the lower dissociation energy of germane molecules compared to silane molecules. At the same time the crystallinity decreases from 80% to 71%. When increasing the germane flow the increased germanium contend introduces stress into the lattice and thereby reduces the crystallinity. Looking to the V_{oc} values of the cells in Figure 1, one can see a decrease in V_{oc} for a germanium flow up to 1.2 sccm. The crystallinity stays nearly constant in this range, but the germanium content increases. As alloying of μc-Si:H with germanium narrows the band gap and increases the carrier recombination rate it may be responsible for the change in V_{oc} [4].

The very steep change in crystallinity in the variation between 1.2 < germane flow < 1.4 sccm could account for the increase in V_{oc} values. A shift to a more amorphous phase and thus higher band gap over-compensates the effect of V_{oc} loss due to higher germanium content. At high and low germane flows the reverse happens: Raman-crystallinity stays constant, germanium content increases and the V_{oc} values drop. Since state of the art μc-Si:H single junction cells exhibit V_{oc} values of 500 mV or more, V_{oc} values for the μc-SiGe:H absorbers have to optimized with other process parameter variations in order to build efficient triple junction cells.

3.2 Silane flow variation

Increasing the silane flow from 21.5 sccm to 25.5 sccm (while keeping all other parameters constant) reduces

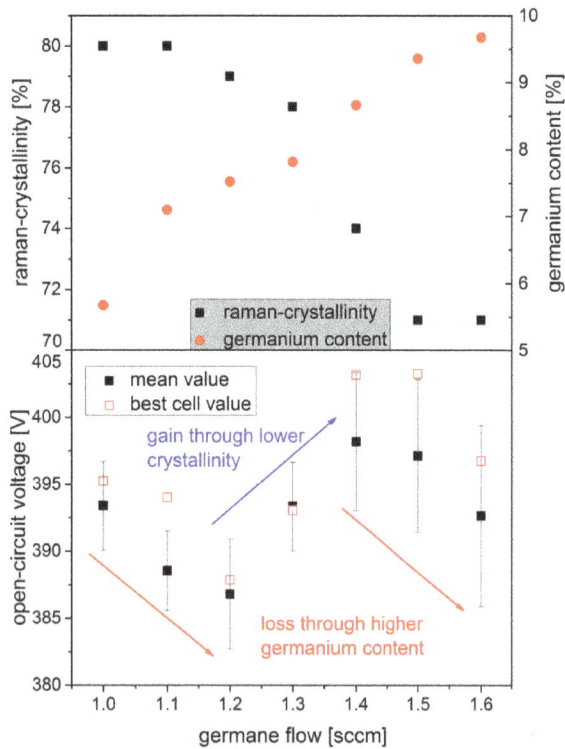

Fig. 1. Raman crystallinity, germanium content and V_{oc} values in relation to the germane flow of single μc-SiGe:H cells. Best cell values were taken from the best cells in terms of efficiency (true for all figures).

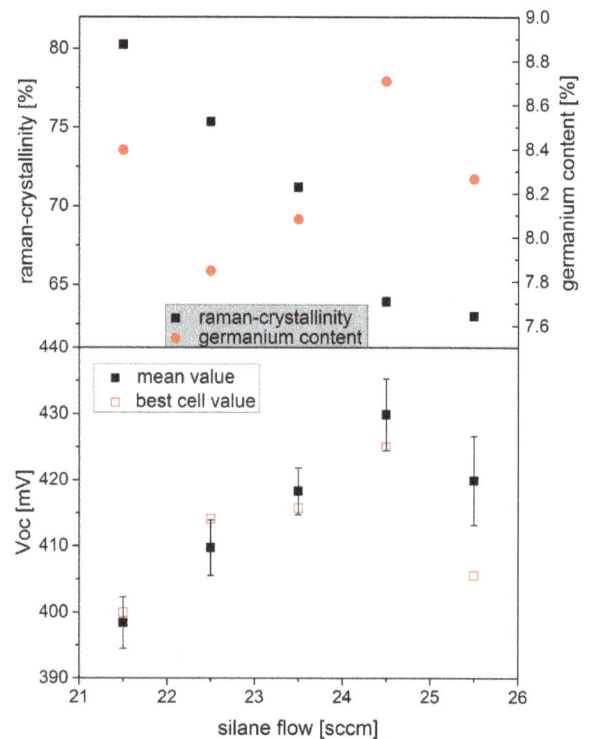

Fig. 2. Raman crystallinity, germanium content and V_{oc} values in relation to the silane flow of single junction μc-SiGe:H cells.

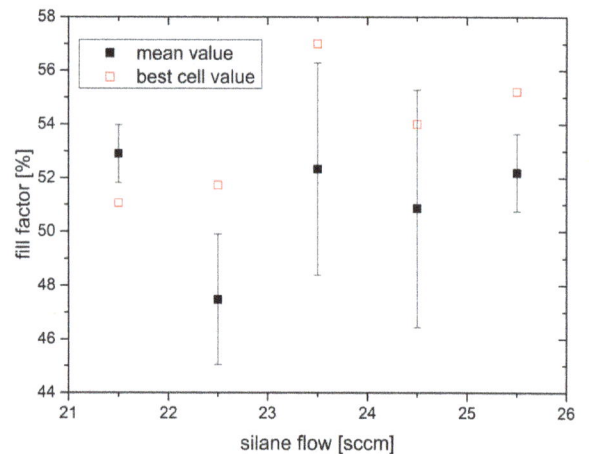

Fig. 3. Fill factor values in relation to the silane flow of single μc-SiGe:H cells.

the raman crystallinity from 80% to 62% due to a larger amorphous volume fraction in the absorber. Comparing Figures 1 and 2 it should be remarked that changing the germane flow had a similar effect on the crystallinity but changing the silane flow did not have the same effect on the germanium content of the layer. The reason for the almost constant germanium content despite changes in crystallinity might be found in the low dissociation energy of germane. Splitting two hydrogen atoms from silane takes twice the energy compared to splitting two hydrogen atoms from germane [11]. Because of the high energy plasma used in microcrystalline depositions all available germane is dissociated and deposited. Changing the relation between hydrogen and silane in the plasma alters the crystallinity while still incorporating all germanium.

The linear decrease in crystallinity depicted in Figure 2 leads to a linear increase in V_{oc} values. The higher amorphous volume fraction shifts the band gap to higher energies without the additional influence of a great variation in germanium content.

Similar to the fill factor values of the germane flow variation (not shown) the fill factor values do not change with increasing silane flow (Fig. 3) and remain around 52%. The low value can be explained by the presence of germanium dangling bonds that act as acceptor like states. Since the presented values were deposited with a not optimized p-layer part of the low values can also be attributed to a reduced interface quality and insufficient internal field.

3.3 Pressure variation

In addition to the germane and silane flow, the process pressure is a key parameter in μc-SiGe:H depositions.

As can be seen in Figure 4, the pressure was varied from 8.5 to 11 mbar. The germanium content is almost constant and varies only slightly between 8.3% and 8.7%. As the pressure increases, the crystallinity declines from 77% to 61%.

Comparing Figures 2 and 4 one can see that increasing the silane flow shows the same effect on crystallinity and germanium content as increasing the pressure: crystallinity decreases but germanium content stays constant.

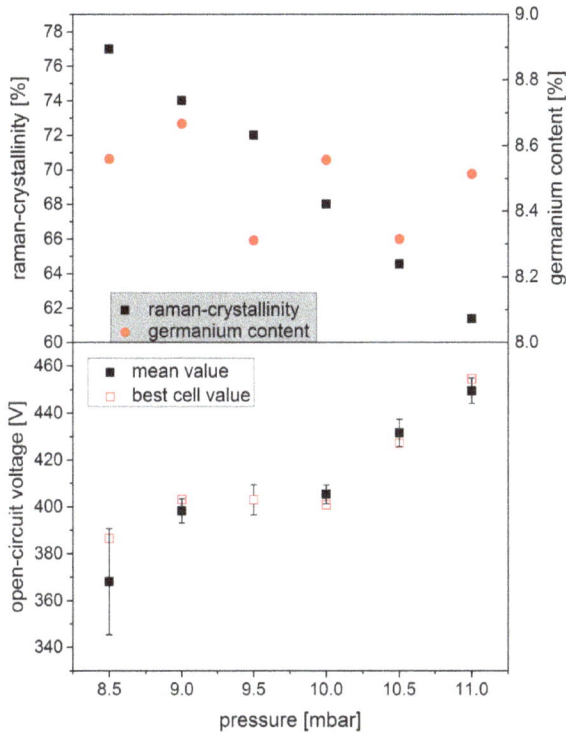

Fig. 4. Raman crystallinity, germanium content and V_{oc} values of single μc-SiGe:H cells in relation to the intrinsic layer process pressure.

The reduction of crystallinity has an enhancing effect on the open circuit voltage.

Because bad material quality will influence the shunt and series resistance, fill factor values are a good gage to estimate the material quality. Figure 5 shows a linear increase of fill factor values with increasing process pressure. Similar to the silane flow variation this parameter does not change the germanium content in the absorber. Comparing Figures 3 and 5, one can see that changes in the crystallinity due to process pressure variations are crucial to good fill factor values while a change in crystallinity through variation in the silane flow has no impact. The increased pressure leads to compact material and thereby reduces defect density of germanium dangling bonds [12].

3.4 Triple junction cells

To show the total cell thickness reduction potential in triple solar cells by the application of μc-SiGe:H bottom cells, we fabricated an a-Si:H/μc-Si:H/μc-Si:H triple-junction cell with a total cell thickness of 4.4 μm. As shown in Figure 6, the V_{OC} performance of the stacked cell is good with a value of 1.8 V resulting in a cell efficiency of 11.0%. Bringing such a device into industrial production is not very feasible, because production costs attributed to 4.4 μm total cell thickness and the corresponding deposition time would be very cost intensive.

Figure 6 also shows the IV-curve under illumination of an a-Si:H/μc-Si:H/μc-SiGe:H triple-junction solar cell

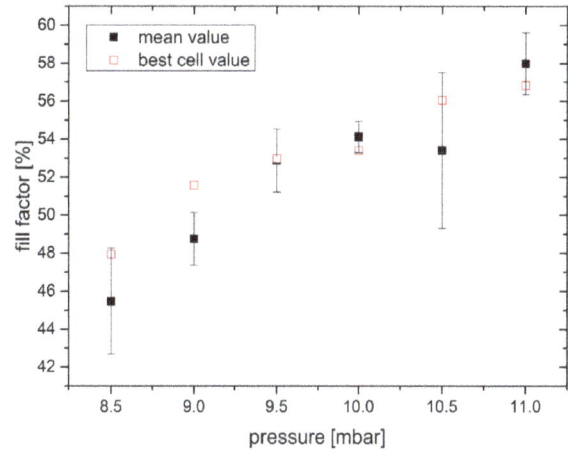

Fig. 5. Dependency of fill factor values on the increase of process pressure.

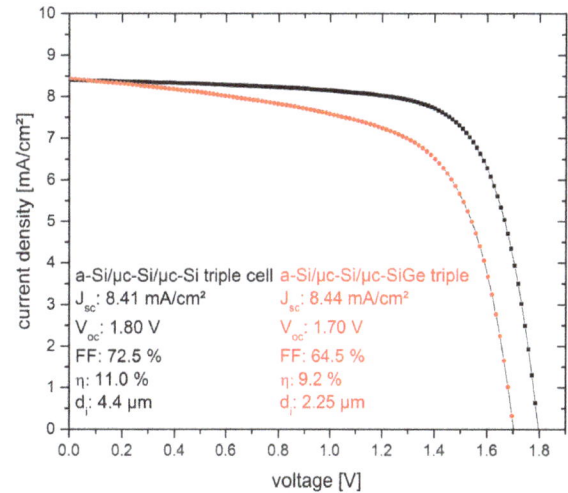

Fig. 6. Comparison of IV-curves of a triple-junction solar cellc in an a-Si:H/μc-Si:H/μc-Si:H and a-Si:H/μc-Si:H/μc-SiGe:H configuration.

with a reduced total cell thickness by a factor of 2 compared to the a-Si:H/μc-Si:H/μc-Si:H reference solar cell. Though the fill factor is only at 64.5%, the remarkable fact is that we managed to produce the same short-circuit current density with 58% of the absorber thickness.

Since the deposition of those triple junction cells were done before we did the extensive material study, we were able to use the knowledge to reduce the difference in V_{oc} values to 60 mV. The next goal will be to achieve fill factor values that are comparable to the μc-Si:H counterpart. By reducing the deposition time of the triple junction technology it should enable the industry to bring triple junction cells with efficiencies greater 14% to the market.

4 Conclusions

The presented work shows the current status of the BMU joined research project SiliziumDS12plus concerning the development of a-Si:H/μc-Si:H/μc-SiGe:H triple-junction solar cells. First we gave a comprehensive overview of the impact of different process parameters on the electrical performance, germanium content and crystallinity of single μc-SiGe:H cells. Since this two phase, two component material is very difficult to fabricate at high qualities, this paper gives the opportunity to gain insight into the parameters needed to reach a good deposition regime. The material quality of the μc-SiGe:H absorber has to be increased further in order to rise the triple cell fill factor above 70%. By using the knowledge gained, we were able to reduce the V_{oc} difference between μc-Si:H and μc-SiGe:H absorbers to 60 mV. A remarkable result is that the same short-circuit current density can be achieved with a 44% thinner triple-junction cell compared to an a-Si/μc-Si/μc-Si reference triple-junction cell by using μc-SiGe:H bottom cells.

The authors would like to thank the BMU for financial support for our work on a-Si:H/μc-Si:H/μc-SiGe:H triple-junction solar cells within the project SiliziumDS-12plus under contract number FKZ0325317. Help from the PV group at NEXT-ENERGY is gratefully acknowledged.

References

1. M.A. Green, K. Emery, Y. Hishikawa, W. Warta, E.D. Dunlop, Progr. Photovolt.: Res. Appl. **21**, 11 (2013)
2. S. Kim, J.W. Chung, H. Lee, J. Park, Y. Heo, H.M. Lee, Sol. Energy Mater. Sol. Cells **119**, 26 (2013)
3. M. Berginski, B. Rech, J. Hüpkes, G. Schöpe, M.N. van den Donker, W. Reetz, T. Kilper, M. Wuttig, ZnO film with tailored material properties for highly efficient thin-film silicon solar modules, in *21st European Photovoltaic Solar Energy Conference, 2006*, p. 1539
4. T. Matsui, H. Jia, M. Kondo, Progr. Photovolt.: Res. Appl. **18**, 48 (2010)
5. T. Matsui, K. Ogata, M. Isomura, M. Kondo, J. Non-Cryst. Solids **352**, 1255 (2006)
6. P. Cuony, D.T.L. Alexander, I. Perez-Wurfl, M. Despeisse, G. Bugnon, M. Boccard, T, Söderström, A. Hessler-Wyser, C. Hébert, C. Ballif, Adv. Mater. **24**, 1182 (2012)
7. T. Kilper, K. Grunewald, M. Kellermann, P. Klement, O. Sergeev, K.V. Maydell, C. Agert, Development of μc-SiGe:H solar cells for the application as bottom cell in triple solar cells, *27st European Photovoltaic Solar Energy Conference, 2012*, p. 2608
8. M.I. Alonso, K. Winer, Phys. Rev. B **39**, 10056 (1989)
9. K.V. Maydell, K. Grunewald, M. Kellermann, O. Sergeev, P. Klement, N. Reininghaus, T. Kilper, Microcrystalline SiGe absorber layers in thin film silicon solar cells, in *Proceedings of the E-MRS spring meeting, 2013*
10. C.W. Chang, T. Matsui, M. Kondo, J. Non-Cryst. Solids **354**, 2365 (2008)
11. C. Boehme, Quantenchemische Berechnungen von Übergangsmetallkomplexen mit Liganden aus den ersten vier Hauptgruppen, Dissertation, 2004
12. Tian-wei Li, Jian-jun Zhang, Yu Cao, Zhen-Hua Huang, Jun Ma, Jian Ni, Ying Zhao, Optoelectron. Lett. **10**, 202 (2014)

24

Photovoltaic yield: correction method for the mismatch between the solar spectrum and the reference ASTMG AM1.5G spectrum

Thomas Mambrini[1,2,3,a], Anne Migan Dubois[2,3,4], Christophe Longeaud[2,3], Jordi Badosa[5], Martial Haeffelin[5], Laurent Prieur[6], and Vincent Radivoniuk[6]

[1] Université Paris-Sud 11, UMR 8507, LGEP, Bâtiment 301, 91405 Orsay Cedex, France
[2] SUPELEC, LGEP, UMR 8507, 3 rue Joliot-Curie, Plateau de Moulon, 91192 Gif-sur-Yvette Cedex, France
[3] CNRS, LGEP, UMR 8507, 11 rue Joliot-Curie, Plateau de Moulon, 91192 Gif-sur-Yvette Cedex, France
[4] Sorbonne Universités, UPMC Univ. Paris 06, UMR 8507, LGEP, 5 Place Jussieu, 75005 Paris Cedex, France
[5] LMD, Institut Pierre-Simon Laplace, CNRS, Ecole Polytechnique, 91128 Palaiseau Cedex, France
[6] SOLEÏS Technologie, 4 allée Jean-Paul Sartre, 77186 Noisiel, France

Abstract We propose a method for a spectral correction of the predicted PV yield and we show the importance of the spectral mismatch on the solar cell. Indeed, currently predicted PV yield are made considering solar irradiation, ambient temperature, incidence angle and partially (or not) the solar spectrum. However, the solar spectrum is not always the same. It varies depending on the site location, atmospheric conditions, time of the day... This may impact the photovoltaic solar cells differently according to their technology (crystalline Silicon, thin film, multi-junctions...) This paper presents a method for calculating the correction of the short-circuit current of a photovoltaic cell due to the mismatch of the solar spectrum with the reference ASTM AM1.5G spectrum, for a specific site, throughout the year, using monthly data of AERONET (AErosol RObotic NETwork established by NASA and CNRS) and the model SMARTS (simple model for atmospheric transmission of sunshine) developed by the NREL. We applied this correction method on the site of Palaiseau (France, 48.7°N, 2.2°E, 156 m), close to our laboratory, just for comparison and the example of Blida (Algeria, 36°N, 2°E, 230 m) is given for one year. This example illustrates the importance of this spectral correction to better estimate the photovoltaic yield. To be more precise, instead of modeling the solar spectral distribution, one can measure it with a spectro-radiometer, and then, derive the spectral mismatch correction. Some of our typical measurements are presented in this paper.

1 Introduction

Nowadays, the rated performance of photovoltaic (PV) modules is evaluated under standard test conditions (STC), that is to say at an irradiance of 1000 W m^{-2}, a PV cell temperature of 25 °C and a spectral energy distribution according to the reference spectrum AM1.5. This reference AM1.5 solar spectral distribution, adopted by the International Electrotechnical Commission (IEC), corresponds to a global irradiation (direct and diffuse) of incident sunlight on a flat surface facing the sun, tilted at 37° to the horizontal, under precise atmospheric conditions [1, 2]. In outdoor conditions, the solar spectral distribution varies continuously mainly due to (i) the path length through the atmosphere, named Air Mass (AM),

(ii) the amounts of atmospheric water vapor (w) and (ii) the aerosol volume (AOD). All those parameters depend on the considered site and the date.

The solar spectrum has a direct impact on the portion of sunlight absorbed by PV cells, more specifically, on the short-circuit current (I_{sc}). The spectral photoresponse of a PV cell depends on the band gap energy, which is technology dependent. Thus, the effects of the solar spectral distribution should be different according to the PV technology considered [3,4]. Then, it may be useful to calculate the correction in I_{sc} due to the spectral mismatch (M) of the real solar spectrum compared to the reference solar spectrum for a given site and the considered PV cells. Currently, predictions of PV yield are made considering solar irradiation, ambient temperature, tilt and orientation of the panels. A few models take into account the effect of solar spectrum (e.g., PVsyst) [5] and when it is the case, not

[a] e-mail: `thomas.mambrini@lgep.supelec.fr`

for all the PV technologies. Furthermore, the only parameter taken into account is AM. But the solar spectrum not only varies with AM and can be very different according to the site (pollution for instance). We propose a method for a spectral correction of the predicted PV yield.

This paper proposes a method for estimating the correction of I_{sc} due to the spectral mismatch, of any PV module knowing the spectral response of its unit cells. We explain how to reconstruct the solar spectrum with precise atmospheric data provided by AERONET (AErosol RObotic NETwork) [6], if available, or directly measure it with a spectro-radiometer.

The first part of this paper presents the tools used to study the solar spectrum and to recover the solar spectrum from atmospheric data. We also simulated the influence of atmospheric conditions on the solar spectrum through key parameters such as AM, w and AOD. Then, we study the spectral mismatch correction for 3 years (2008–2010) on the site of Palaiseau (France, 48.7°N, 2.2°E, 156 m) for 2 PV technologies (crystalline Silicon (c-Si) and thin film Silicon (a-Si:H)). For comparison, the same exercise is made for Blida (Algeria, 36°N, 2°E, 230 m) during one year (2009). Then, we present some solar spectrum measurements and derived mismatch factor M. The last part proposes a method, non PV technology dependant, to derive M from atmospheric data.

2 Tools for studying the solar spectral distribution

In this part, we present first, a mathematical index, the average photon energy (APE), calculated to quantify the amount of energy available in the solar spectrum. Then, to take into account the spectral response of the PV cell, we define the spectral mismatch correction in I_{sc} (M). Next, the modelization of the solar spectrum is described, based on the SMARTS model. We need geographical and atmospheric input parameter for this model. The AERONET atmospheric data are presented in the last paragraph of this part.

2.1 Average photon energy

APE has been first introduced by Betts et al. [7]. This index is defined as the sum of the light energy in the spectrum divided by the total density of the photon flux:

$$APE = \frac{\int_a^b E(\lambda)\, d\lambda}{q \int_a^b \phi(\lambda)\, d\lambda}. \tag{1}$$

q is the electronic charge (C), E is the solar spectral irradiance (W m^{-2} nm^{-1}), and Φ is the spectral photon flux density (photon m^{-2} s^{-1} nm^{-1}). In this study, a and b are equal to 300 nm and 1100 nm, respectively, corresponding to the lower and upper wavelength of the spectral response of most PV cells.

The main advantages of this tool are:

– easy indication of the "color" of the sunlight: $APE >$ 1.895 for blue-rich sky; $APE <$ 1.895 for red-rich sky;
– non device-dependent indicator.

2.2 Spectral mismatch correction

The second tool is the spectral mismatch correction developed in equation (2):

$$M = \frac{\int_a^b E_{ref}(\lambda) S_{ref}(\lambda)\, d\lambda}{\int_a^b E_t(\lambda) S_{ref}(\lambda)\, d\lambda} \frac{\int_c^d E_t(\lambda) S_t(\lambda)\, d\lambda}{\int_c^d E_{ref}(\lambda) S_t(\lambda)\, d\lambda} \tag{2}$$

where E_t and E_{ref} are the test and reference spectra (W m^{-2} nm^{-1}), S_t and S_{ref} are the spectral response (A W^{-1}) of the test and reference cell, a and b, and c and d, are the wavelength limits (nm) of the spectral response of the reference cell and of the test cell, respectively. In this study, an ideal pyranometer is considered as reference, with a flat response over the entire solar spectrum, that is to say that $S_{ref} = 1$.

Moreover, we can derive I_{sc} by convoluting the spectral response of the PV cell with the incident solar spectrum, using the following equation:

$$I_{sc} = A \int_0^{\lambda(E_g)} S(\lambda)\, E(\lambda)\, d\lambda \tag{3}$$

where I_{sc} is the short-circuit current (A), A the cell area (m^2), $S(\lambda)$ the spectral response of the PV cell (A W^{-1}), $E(\lambda)$ the irradiance (W m^{-2} nm^{-1}), λ the wavelength (nm) and E_g the band gap energy (eV) of the PV cell.

In our study, equation (2) becomes:

$$M = \frac{E(AM1.5)}{E(t)} \times \frac{I_{sc}(t)}{I_{sc}(AM1.5)} \tag{4}$$

where $E(AM1.5)$ and $E(t)$ are the irradiation of the reference spectrum and test spectrum, between 300 nm and 1100 nm, respectively. $I_{sc}(AM1.5)$ and $I_{sc}(t)$ are the short-circuit current for the considered PV cell for the AM1.5 spectrum and for the test spectrum, respectively.

The spectral mismatch correction adjusts the I_{sc} of a PV cell, for a given spectrum, from the rated I_{sc} measured under STC.

This tool is device dependent, which means that the spectral response of the PV cell must be known.

2.3 SMARTS solar spectrum model

The FORTRAN code "simple model for atmospheric transmission of sunshine" (SMARTS) [8–12] is the program used to model the solar spectrum under clear sky conditions by injecting atmospheric and geographic input parameter. It covers the whole wavelengths range of the solar spectrum (280 nm to 4000 nm). It is the program used by the American Society for Testing and Materials

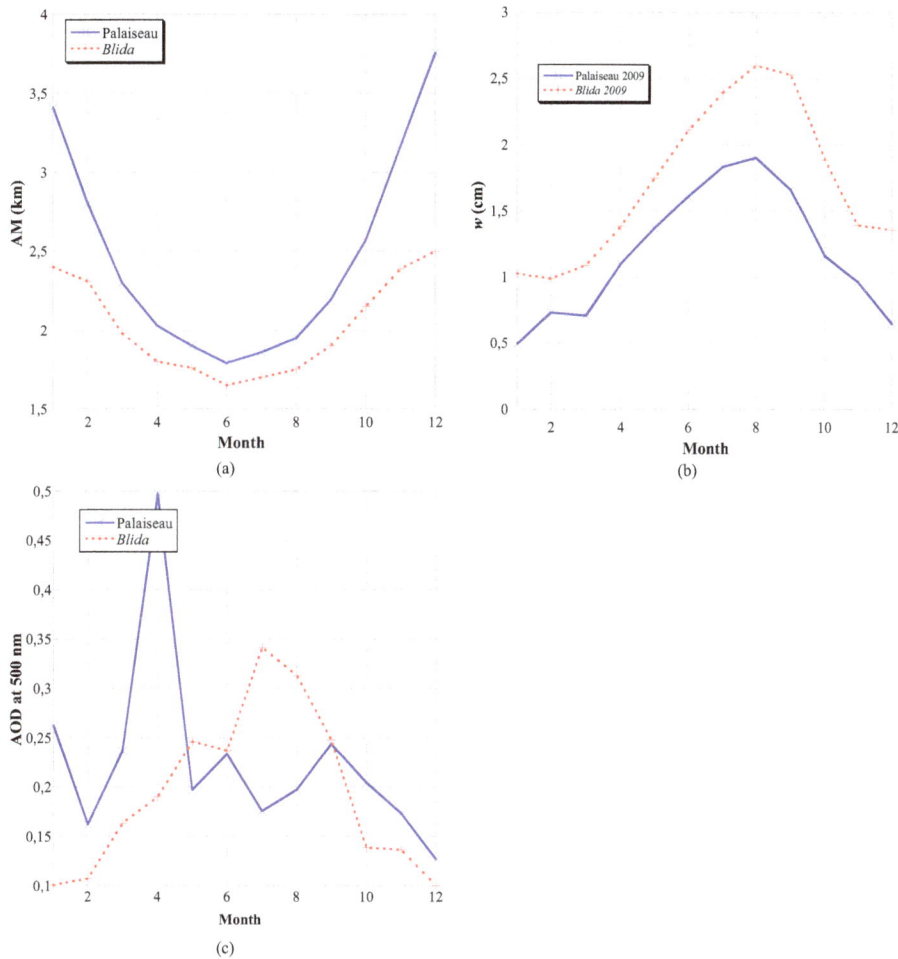

Fig. 1. Monthly average calculated AM (a), w (b) and AOD at 500 nm (c) derived from AERONET data for Palaiseau and Blida during 2009.

(ASTM) to model the reference solar spectrum AM1.5, ASTM G 173-03 [2]. This model is based on parameterizations of transmittance and absorption function for atmospheric constituents include molecular (Rayleigh), ozone, water vapor, mixed gases, trace gases and aerosol transmittances.

2.4 Atmospheric data of AERONET

Sun photometer measurement-derived parameters from the AERONET database are considered in this study as input to the model. In particular we consider:

- aerosol optical depth (AOD) at 500 nm;
- angstrom's alpha parameter for two wavelength intervals (380–500 nm and 500–870 nm);
- integrated water vapor (w).

The single scattering albedo and the asymmetry parameters are set constant to 0.9 and 0.7 in the model, respectively, as typical values for a semi-rural site [13].

These data are entered into the SMARTS model to recover the solar spectrum distribution at a given time and

for a considered site. Moreover, the AM must be calculated from the zenith angle derived from the solar equation.

In this paper, only less than 78.5° zenith angles are considered, leading to AM under 5. In one hand, the equation used to calculate AM [14] is not accurate and, on the other hand, the solar irradiation is low and not relevant for PV applications.

Figure 1 shows monthly averaged AM, AOD at 500 nm and w values (for 2009) for the two considered sites, Palaiseau and Blida.

Those geographic and atmospheric data have a great influence on the solar spectral distribution.

3 Influence of atmospheric conditions on the solar spectrum simulated with SMARTS

The parameters that affect the most the solar spectrum are the path length through the atmosphere (AM), the amount of water vapor (w) and aerosols (AOD) [15]. To evaluate their impacts, we model with the SMARTS

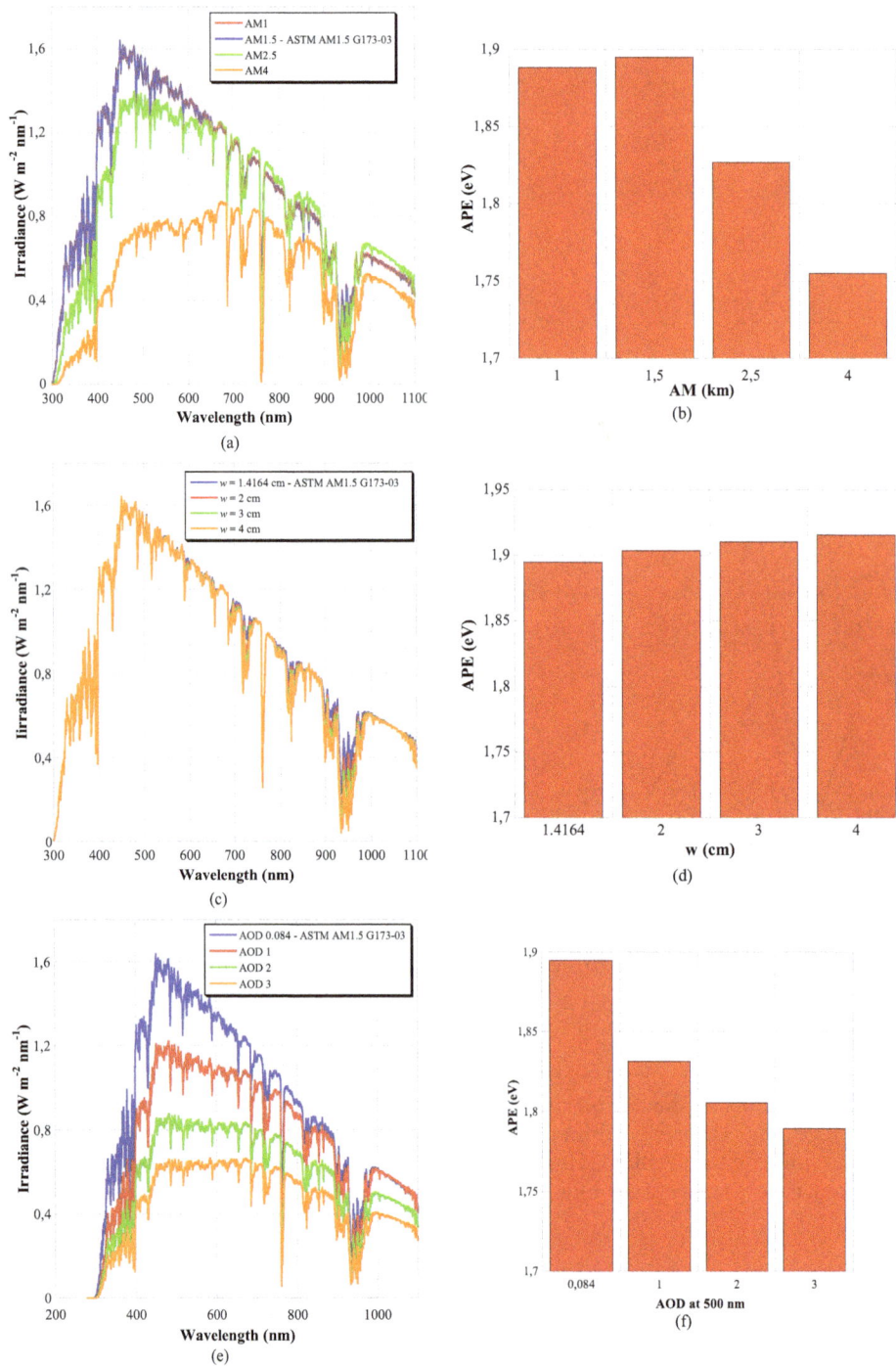

Fig. 2. Simulated spectra for different AM (a), different precipitable water vapor (c) and different AOD at 500 nm (e). Calculated APE associated with the spectra, for different AM (b), different precipitable water vapor (d) and different AOD at 500 nm (f).

program, the solar spectrum with the same inputs as the ASTM AM1.5 G173-03 hemispherical spectrum excepting one parameter, such as AM, water vapor or aerosols. Default values are $AM = 1.5$ km and $w = 1.4164$ cm and AOD $= 0.084$.

The spectral range considered begins at 300 nm and ends at 1100 nm because of the spectral response of all PV cells.

Figures 2a and 2b represent the variations of the simulated solar spectrum and the calculated APE, for AM varying from 1 to 4. Figures 2c and 2d are the same but for w varying from 1 cm to 4 cm. At last, Figures 2e and 2f consider the variation of AOD at 500 nm between 0.084 to 3. The AOD variation is mainly due to the pollution as the pollution peak observed at Palaiseau during March 2014 which leaded to a 0.8 AOD value or during major

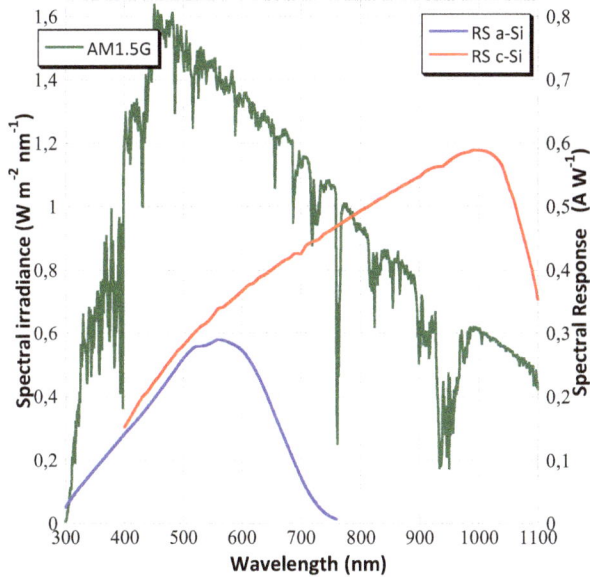

Fig. 3. Reference ASTM AM1.5G spectrum (left axis) and spectral response of a crystalline Silicon PV cell and of a thin film Silicon PV cell (right axis).

short-term pollution events as the wildfire during summer 2010 at Moscow which leaded to a 4.3 AOD value [16].

In Figure 2a, we observe that changing the path length modifies the solar spectrum intensity and distribution and in Figure 2b, we can notice that when AM increases, APE decreases. That means that the more the path length through the atmosphere is long (the sun is low in the sky, for example in early in the morning or late in the evening), the more the sunlight is red-rich. It is understandable because when the path length is thick, it is the short wavelengths that are scattered.

Water vapor absorption bands are centered at 724 nm, 824 nm and 938 nm, in the considered wavelength range as one can see in Figure 2c. Figure 2d shows that when the water vapor increases, the APE increases too.

To take into account the turbidity of the atmosphere, the spectral effects of changing AOD at 500 nm are plotted in Figures 2e and 2f. It can be studied as the main parameter to explain the radiative extinction by aerosols. However, the aerosols spectral effects depend also on the properties of the aerosols, such as the particle size and absorption properties. Increasing AOD have the impact of solar irradiance decreasing with larger impact for shorter wavelengths, so APE decreases. Choosing on other particle size or type also has an effect on the solar spectrum.

Taking into account the variation of the solar spectral distribution is of the most important in prediction of the photovoltaic yield because of the spectral response proper to each technology of PV cell like underscored in Figure 3.

Figure 3 shows that the spectral response for the amorphous cell is more selective than that of the crystalline cell. A change in the solar spectrum acts then differently respect to the cell technology because of its spectral response.

4 Calculating method for the spectral mismatch correction (M) during one year, for a given site

4.1 Modeling the solar spectrum

The SIRTA Atmospheric Research Observatory (www. sirta.fr), located at Palaiseau (France: 48.8°N, 2°E, 156 m) has an AERONET station. The idea of this work is to calculate the spectral mismatch correction during one year for crystalline Silicon PV cells and thin film Silicon PV cells. We take monthly mean values of atmospheric conditions (AOD, w, size of the aerosols below and above 500 nm) and monthly mean values of AM. In other words, we model a typical spectrum for each month of the year, for the site of Polytechnique and we derive M as a function of the spectral response of the considered PV cell.

The results of this study are presented in Figure 4.

From Figure 4a, we can notice that for the thin film Silicon PV cell, there may be a difference of 9.7% between I_{sc} predicted during the winter and during the summer months. I_{sc} is close to its rated value during summer months and lower for winter months. In summer, AM decreases and w increases, as shown in Figure 3 where we can see that w is four times higher in summer than in winter. This leads, in summer, to a more blue-rich solar spectrum ($APE > 1.895$ as shown in Fig. 4c), which is better suitable to the thin film PV cell spectral response (maximum absorbance around 550 nm). Whereas in winter the sun is lower in the sky and AM increases, which leads to a more red-rich sky ($APE < 1.895$ as shown in Fig. 4c).

The opposite effect occurs for crystalline Silicon PV cells that absorb also in the near infrared. Moreover, we can see in Figure 3 that there are two important water bands of absorption close to crystalline Silicon PV cells maximum of absorption (900–1000 nm). This means that increasing w in summer has a strong effect on crystalline Silicon PV cells yield. In Figure 4b, we observe a difference due to the season of 6.6% for crystalline Silicon PV cells. They are less sensitive to the solar spectrum distribution than the thin film PV cells because of their wide spectral response [17].

On the 3 studied years in Palaiseau, the reference solar distribution spectrum overestimated I_{sc} of a thin film Silicon PV cell of about 3.5% and underestimated that of a crystalline Silicon PV cell of about 2.4%.

In Figure 5, the equivalent exercise is presented, for the site of Blida (Algeria: 36°N, 2°E, 230 m) in 2009.

For the Blida site, the average APE is 1.876 eV, the average M for thin film Silicon PV cells is 0.985 and the average M for crystalline Silicon PV cells is 1.011 during 2009. Compared to the first studied site Palaiseau, w is higher and AM is lower as resumed in Table 1. This leads to a more blue-rich sky for Blida than Palaiseau. We can say that, with respect to the spectral response point of view, crystalline Silicon PV cells are more efficient for high latitude site compared to thin film PV cells.

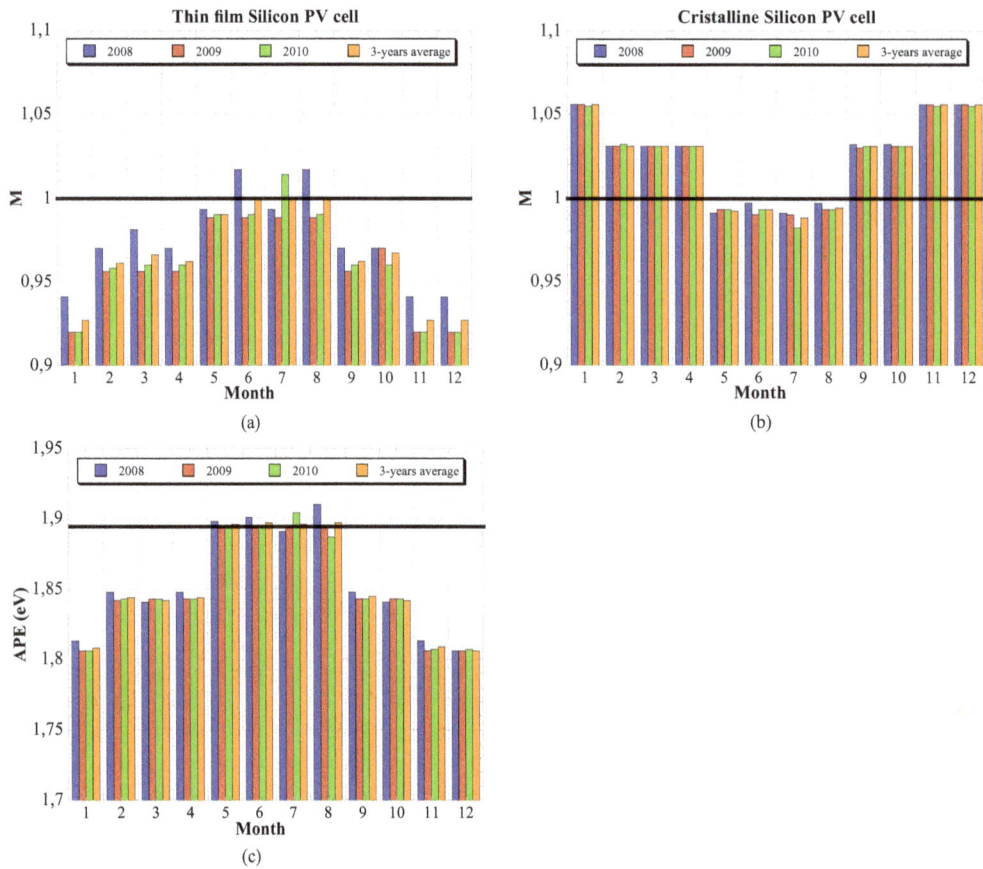

Fig. 4. M calculated for thin film Silicon PV cell (a) and crystalline Silicon PV cell (b) and APE (c) corresponding to the modeled solar spectrum. All the calculated values used monthly mean atmospheric data for years between 2008 and 2010, at Polytechnique (Palaiseau, France: 48.8°N, 2°E, 156 m).

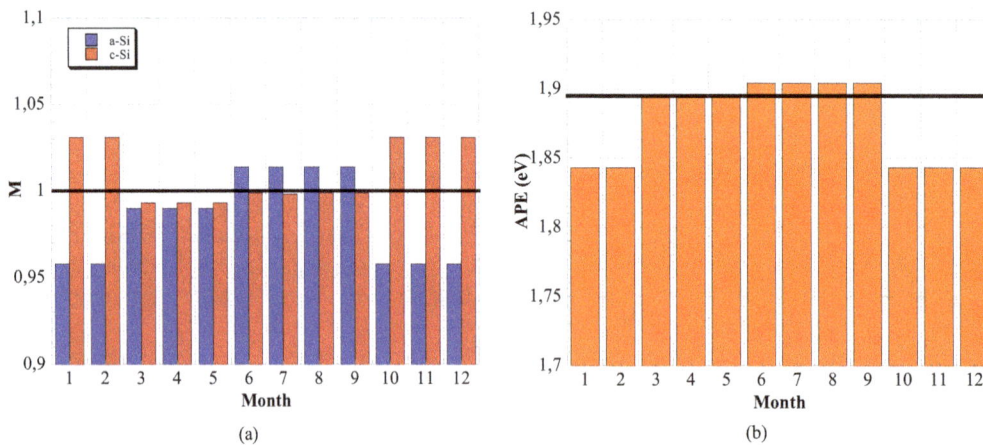

Fig. 5. M calculated for thin film and crystalline Silicon PVs cell (a) and APE (b) corresponding to the modeled solar spectrum. All the calculated values used monthly mean atmospheric data 2009, at Blida (Algeria: 48.8°N, 2°E, 230 m).

Table 1. Average AM, w and APE for Palaiseau, Blida and ASTM AM1.5 G173-03 in 2009.

Site	AM (km) Moy.	w (cm) Moy.	APE (eV) Moy.	M (thin film) Min.	Max.	Moy.	M (crystalline) Min.	Max.	Moy.
Palaiseau	2.478	1.246	1.852	0.920	0.988	0.959	0.990	1.056	1.024
Blida	2.024	1.712	1.876	0.958	1.014	0.985	0.993	1.031	1.011
ASTM AM1.5	1.5	1.416	1.895			1			1

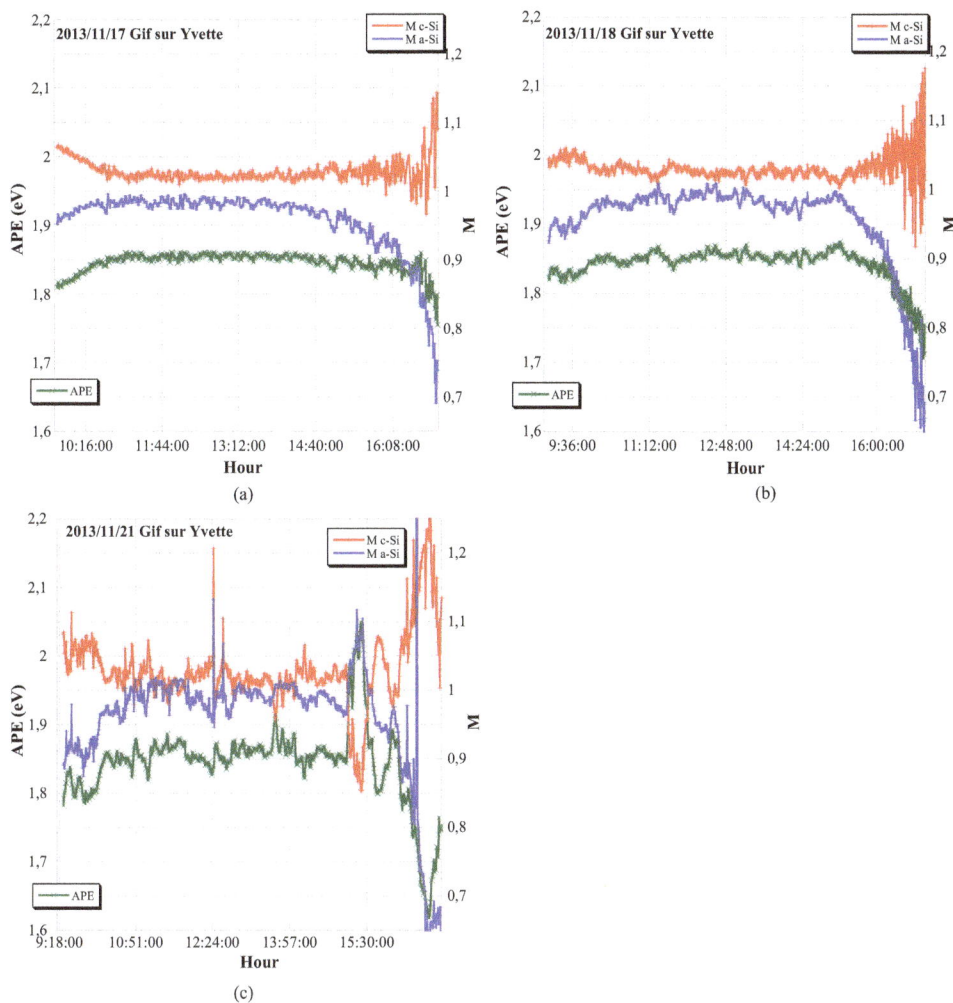

Fig. 6. Calculated APE and M during three days at LGEP, Gif-sur-Yvette (near Palaiseau), 2013/11/17 (a), 2013/11/18 (b) and 2013/11/21 (c).

4.2 Measuring the solar spectrum

Another way, more accurate, to estimate the spectral mismatch correction is to measure the solar spectrum with a USB2000+ Spectrometer with Enhanced Sensitivity (from 210 to 1035 nm) produced by Ocean Optic. Measures of the solar spectrum are taken every 30 s from the sunrise to the sunset and we derived, in Figure 6, the values of APE and M for crystalline Silicon PV cells and thin film Silicon PV cells. We can observe that at the beginning and the end of the day, thin film Silicon PV cell has a lower M contrary to the crystalline Silicon PV cell. This can be mainly explained by the fact that the AM is higher during these periods and then the spectrum is less blue-rich. This is confirmed by looking at the APE which is higher around noon than at the beginning or at the end of the day.

One goal of that kind of study would be to make more accurate models used for PV plant production. For instance, the very used PVsyst software or the Sandia model, utilize an empirical function $f_1 (AM_a)$ to quantify the influence of variation in the solar spectrum, on the

short-circuit current of the PV module [18]:

$$I_{sc} = I_{sc0} \times f_1 (AM_a) \times f_E (\text{AOI}) \times f_T (T) \quad (5)$$

$$f_1 (AM_a) = a_0 + a_1 AM_a + a_2 (AM_a)^2 \\ + a_3 (AM_a)^3 + a_4 (AM_a)^4. \quad (6)$$

f_1 is a 4th order polynomial function of absolute air mass AM_a, and is called the air mass modifier, where a is a vector of coefficients that are determined from module testing.

If we model or measure M during one year in one site, then substitute $f_1 (AM_a)$ by the measured M in these programs should reduce uncertainty respect to the solar spectrum.

5 Conclusion

The solar spectrum variations differently impact the short-circuit current of solar cells, according to their technology. A spectral mismatch correction (M) must be applied to I_{sc} estimated under reference spectral conditions

(ASTM G 173-03) to evaluate precisely the yield of a PV module, for a given site.

This paper gives a method to calculate the spectral mismatch correction according to a PV cell, in a given site. The needed information to take into account is the AERONET monthly data or the solar spectrum measured by a spectrometer. Thus, we can derive the APE and M, and the associated I_{sc}. This can be particularly useful for decrease the uncertainty in PV plant prediction models like PVsyst or the Sandia model.

Application for the site of Polytechnique (Palaiseau) in France, is realized using data issued from the SIRTA during 3 years (2008–2010). We found a difference of 9.7% for I_{sc}, throughout one year for thin film Silicon PV cells and 6.6% for crystalline Silicon PV cells. That highlights the importance to consider the spectral mismatch correction. For comparison, we made the same study in the site of Blida in Algeria.

We also emphasize that the reference AM1.5 solar spectrum is blue-rich respect to the mostly real solar spectrum for high latitudes.

The authors acknowledge the support from ADEME project POLYSIL and from the collaboration with the JEI SOLEÏS Technologie, France.

References

1. IEC 60904-3, *Photovoltaic devices-Part 3: Measurement principals for terrestrial photovoltaic (PV) solar devices with reference spectral irradiance data* (2008)
2. American Society for Testing and Materials, ASTM G 173-03, *Standard Tables for Reference Solar Spectral Irradiances: Direct Normal and Hemispherical for 37° Tilted Surface* (American Society for Testing and Materials, West Conshohocken, 2003)
3. K. Akhmad, A. Kitamura, F. Yamamoto, H. Okamoto, H. Takakura, Y. Hamakawa, Sol. Energy Mater. Sol. Cells **46**, 209 (1997)
4. T. Minemoto, M. Toda, S. Nagae, M. Gotoh, A. Nakajima, K. Yamamoto, H. Takakura, Y. Hamakawa, Sol. Energy Mater. Sol. Cells **91**, 120 (2007)
5. A. Mermoud, T. Lejeune, *Performance assessment of a simulation model for PV modules of any available technology, 25th European Photovoltaic Solar Energy Conference, Valencia, Spain* (2010)
6. B.N. Holben et al., Rem. Sensing Environ. **66.1**, 1 (1998), http://aeronet.gsfc.nasa.gov/new_web/system_descriptions.html.
7. T.R. Betts, C.N. Jardine, R. Gottschalg, D.G. Infield, K. Lane, *3rd World Conference on Photovoltaic energy Conversion, Osaka, 2003*
8. C. Gueymard, Development and performance assessment of a clear sky spectral radiation model, *Solar'93-22nd Annual Conf., American Solar Energy Society, Boulder, Washington, 1993*
9. C. Gueymard, Updated transmittance functions for use in fast spectral direct beam irradiance models, in *Solar '94 Conf., American Solar Energy Society, Boulder, San Jose, 1994*
10. C. Gueymard, SMARTS2, *Simple model of the atmospheric radiative transfer of sunshine: Algorithms and performance assessment*, Rep.FSEC-PF-270-95 (Florida Solar Energy Center, Cocoa, 1995)
11. C. Gueymard, Sol. Energy **71**, 325 (2001)
12. C. Gueymard, D. Myers, K. Emery, Sol. Energy **73**, 443 (2002)
13. M. Iqbal, *An introduction to solar radiation* (Elsevier, 1983)
14. F. Kasten, A.T. Young, Appl. Opt. **28**, 4735 (1989)
15. G. Litjens, *Investigation of Spectral Effects on Photovoltaic Technologies by Modelling the Solar Spectral Distribution* (Universiteit Utrecht, 2013)
16. A. Donkelaar, R.V. Martin, R.C. Levy, A.M. Da Silva, M. Krzyzanowski, N.E. Chubarova, E. Semutnikova, A.J. Cohen, Atmospheric Environment **45**, 6225 (2011)
17. D.L. King, W.E. Boyson, J.A. Kratochvil, *Photovoltaic Array Performance Model* (Sandia National Laboratories, Albuquerque, 2004)
18. Y. Hirata, T. Tani, Sol. Energy **55**, 463 (1995)

25

Use of hexamethyldisiloxane for p-type microcrystalline silicon oxycarbide layers

Prabal Goyal[1,2,3,a], Junegie Hong[1,3], Farah Haddad[2,3], Jean-Luc Maurice[2,3], Pere Roca i Cabarrocas[2,3], and Erik Johnson[2,3]

[1] Air Liquide, Centre de Recherche Paris Saclay, 78354 Jouy-en-Josas, France
[2] Laboratoire de Physique des Interfaces et des Couches Minces, CNRS, Ecole Polytechnique, 91128 Palaiseau, France
[3] LPICM- LPICM, CNRS, Ecole polytechnique, Université Paris-Saclay, 91128 Palaiseau, France

Abstract The use of hexamethyldisiloxane (HMDSO) as an oxygen source for the growth of p-type silicon-based layers deposited by Plasma Enhanced Chemical Vapor Deposition is evaluated. The use of this source led to the incorporation of almost equivalent amounts of oxygen and carbon, resulting in microcrystalline silicon oxycarbide thin films. The layers were examined with characterisation techniques including Spectroscopic Ellipsometry, Dark Conductivity, Fourier Transform Infrared Spectroscopy, Secondary Ion Mass Spectrometry and Transmission Electron Microscopy to check material composition and structure. Materials studies show that the refractive indices of the layers can be tuned over the range from 2.5 to 3.85 (measured at 600 nm) and in-plane dark conductivities over the range from 10^{-8} S/cm to 1 S/cm, suggesting that these doped layers are suitable for solar cell applications. The p-type layers were tested in single junction amorphous silicon p-i-n type solar cells.

1 Introduction

The improvement of thin-film doped layers deposited by plasma enhanced chemical vapor deposition (PECVD) – used in thin film silicon and heterojunction silicon solar cells – provides a great opportunity for increased device performance and cost reduction. These films are necessary to provide an electron-hole pair separation mechanism in the absorber layer, but they are also a source of parasitic absorption [1,2]. However, these layers can provide greater control over optical reflection if properly designed. The most common materials for doped amorphous or microcrystalline layers in thin film silicon solar cells are silicon, silicon-carbon alloys [3] and silicon-oxygen alloys [4]. In particular, doped microcrystalline silicon oxide (μc-SiO$_x$:H) is a material which has gained scientific interest because of it having a reduced optical absorption and low refractive index, while still generating sufficient built-in potential across the absorber layer for electron-hole separation [5–8]. It has been demonstrated that it is possible to control the refractive index (n) of undoped and doped μc-SiO$_x$:H in the range from 1.8–3.6 [9–11] and that these layers can be designed for several functions in thin film silicon solar cells, namely as a doped layer or as an intermediate reflector layer (IRL) [12].

[a] e-mail: prabal.goyal@polytechnique.edu

Table 1. Values of refractive index (n) in literature for different μc-SiO$_x$:H type materials.

Material	Function	n
p-μc-SiO$_x$:H	p-type layer	2.97–3.5 [5,13]
n-μc-SiO$_x$:H	n-type layer + IRL	1.8–2.6 [9,10,14,15]
SiO$_x$	IRL	2.0 [12,16]

Table 1 gives a summary of n values reported in the literature for such films.

The electrical conductivity of doped μc-SiO$_x$:H is attributed to the presence of doped microcrystalline silicon filaments in an amorphous silicon oxide matrix, which results in an out-of-plane conductivity of the layers above 10^{-5} S/cm, while their in-plane conductivities are below 10^{-10} S/cm [9]. The amorphous silicon oxide matrix consists of Si-Si, Si-O-H as well Si-O-Si bonds [17]. Carbon-dioxide (CO$_2$) mixed with silane (SiH$_4$) and hydrogen (H$_2$) (with p-type or n-type doping gas) is the most common chemistry for depositing doped μc-SiO$_x$:H. In this work, we explore the use of hexamethyldisiloxane (C$_6$H$_{18}$OSi$_2$, HMDSO) as an alternative oxygen and carbon source for p-type microcrystalline silicon oxycarbide layers. Figure 1 shows the chemical structure of hexamethyldisiloxane (HMDSO). In such molecules, the desired Si-O-Si bonds already pre-exist, whereas for CO$_2$, dissociation in the plasma produces atomic O [9] which is

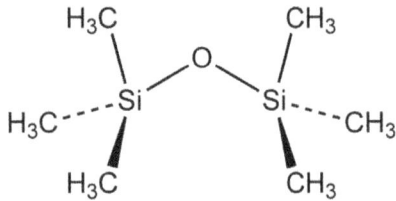

Fig. 1. Structure of hexamethyldisiloxane (HMDSO).

then incorporated into bonds at the film growth surface. This may result in the less desirable Si-O-H configuration, which is a deep electron trap [18, 19]. HMDSO has already been used in the literature to deposit thermally stable SiO_2 layers [20] and dielectric barrier layers [21]. It was suggested that the Si-C bonds dissociate first and that the Si-O-Si bond remains intact during HMDSO decomposition in a plasma [22].

2 Experimental method

HMDSO is a liquid precursor usually stored in a canister. The canister was installed in a commercial liquid-source gas distribution system (Mini CANDI^{TM}) and the liquid was kept heated at 60 °C to increase the vapor pressure. The resulting gaseous HMDSO was injected in a PECVD chamber and was mixed with hydrogen (H_2), silane (SiH_4) and diborane (1% dilution in Ar). The PECVD reactor is a Plassys CVD 300 operated at a fixed RF frequency of 13.56 MHz. The distance between the electrodes is 25 mm. The layers were deposited on Corning Glass substrates and crystalline silicon float zone (FZ) wafers. To study the influence of HMDSO on the film properties, the HMDSO flow rate was varied from 0 to 1.2 sccm while other process parameters were kept constant. The RF power was fixed at 30 W ($\sim$0.17 W/cm^2), pressure at 2.5 Torr, substrate temperature at 150 °C, and gas flows in the ratio $H_2/SiH_4/B_2H_6 = 2000/4/1.6$.

The layers deposited on glass substrates were characterized with an ex-situ spectroscopic ellipsometer (UVISEL from Horiba Jobin Yvon) in the range from 1.5 eV (827 nm) to 4.5 eV (276 nm). The surface profile of the layers was analyzed by a conductive probe atomic force microscope (Resiscope II) which can measure resistances in the range of 10^2 to 10^{12} Ω. The in-plane dark conductivity (σ) of the layers was measured using a co-planar aluminium electrode configuration in a temperature (T) range from 40 °C to 125 °C. From the slope of the $\ln(\sigma)$ plotted versus $1/T$, activation energy of the conductivity of the layers (E_a) of the layers was estimated using $\ln \sigma = -E_a/kT + \ln \sigma_o$ where k is the Boltzmann constant and σ_0 is the conductivity prefactor.

The composition of gas species in the plasma was studied using a Fourier Transform Infrared (FT-IR) spectrometer installed downstream of the dry process pump. The spectral range of this gas-phase FT-IR spectrometer is 1100–4000 cm^{-1}. The molecular bond configuration in the layers was studied with another FT-IR spectrometer (Nicolet 6700 with a Mercury Cadmium Telluride detector cooled by liquid nitrogen). The FT-IR spectra were

acquired on the layers deposited on FZ c-Si substrates over a range from 500–4000 cm^{-1}. The elemental composition of the layers was checked using secondary ion mass spectroscopy (SIMS). The local characterisation of the phase (μc-Si:H and a-Si:H) was performed by transmission electron microscopy (TEM) using selected area electron diffraction (SAED) and high-resolution TEM (HRTEM).

Finally, these layers were used as the p-type doped layer in single junction p-i-n solar cells. ZnO:Al (1 μm) was sputtered on Corning Glass (sheet resistance $\sim$10 Ω/square) as a front contact and was etched in HCl solution (0.5%) for 40 s. More details about the preparation of such substrates can be found in reference [23]. The p-type layers were deposited with an HMDSO flow rate in a range from 0 to 1.2 sccm. The thickness of the i-layer (a-Si:H) is 300 nm and that of the n-layer (n-μc-SiO$_x$:H) was 30 nm. To isolate the influence of the p-layers, all of the i-layers and n-layers were co-deposited, and therefore were the same on all substrates. However, this necessitates a vacuum break at the p/i interface. Furthermore, due to the lack of an n-type dopant source in the HMDSO equipped reactor, a vacuum break was also necessary at the i/n interface as the samples were transferred to another reactor. To check for any disproportionate impact of a vacuum break on HMDSO layers, standard p-a-Si:H and p-μc-SiO$_x$:H (deposited using CO_2) layers were used as references during the same i-layer and n-layer deposition. The back contact (ZnO/Ag) was deposited by sputtering. The solar cells were then annealed at 150 °C for 30 min. For each p-type layer, six solar cells were fabricated and the active area of a solar cell is about 0.125 cm^2. The J-V characteristics were measured under 1 sun at 25 °C, and the external quantum efficiency (EQE) of the p-i-n solar cells was measured under short circuit conditions, and certified using a calibrated diode.

3 Results

3.1 Ellipsometry

Figure 2 shows the imaginary part of the pseudo-dielectric function, $\langle \varepsilon_2 \rangle$ of the layers deposited using different HMDSO flow rates. The thickness of the layers is around 30 nm.

In Figure 2, we see that as HMDSO flow rate increases, the intensity of the $\langle \varepsilon_2 \rangle$ spectra at high energy ($\sim$3.7 eV) decreases. The fitting of the ellipsometry spectra was done with the Delta-Psi2 software using a two layer model composed of a bulk layer and a roughness layer. In this optical modeling, we used an effective medium approximation based on Bruggeman's theory [24] to calculate the optical dielectric functions of each layer. The bulk layer model consists of three phases: amorphous silicon, small grain microcrystalline silicon and SiO_2. The percentage of small grain microcrystalline silicon in the model indicates the crystalline volume fraction (X_c) in the layer, and has been verified to be consistent with Raman scattering (not shown). The roughness layer consists of a 50/50 mix of

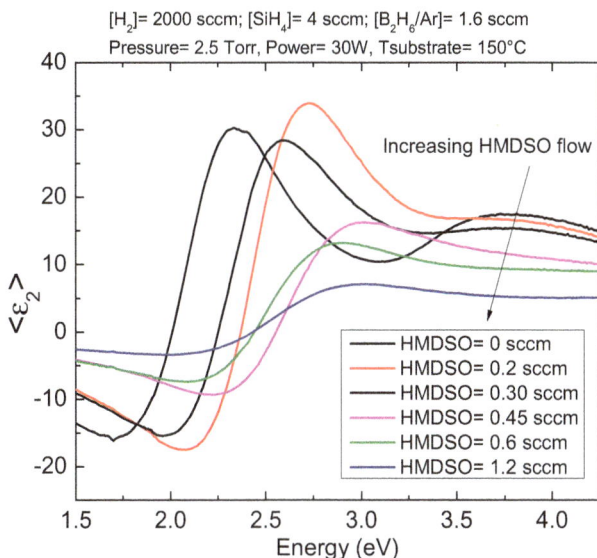

Fig. 2. Imaginary part of pseudo-dielectric function of layers deposited using different HMDSO flow rates, measured by spectroscopic ellipsometry (SE).

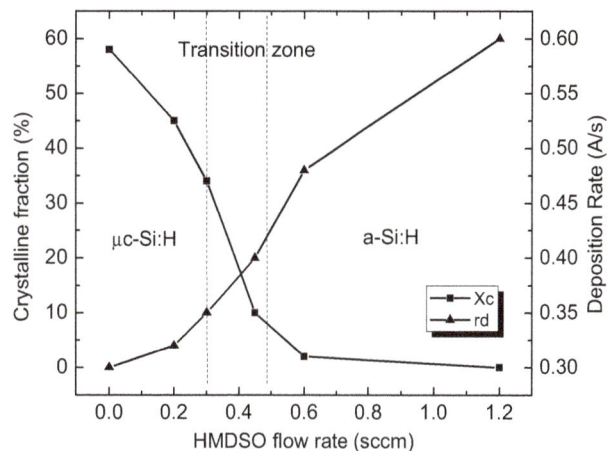

Fig. 3. Crystalline fraction (Xc) and deposition rate (r_d) of HMDSO layers as a function of HMDSO flow rate, as obtained by ellipsometric modeling.

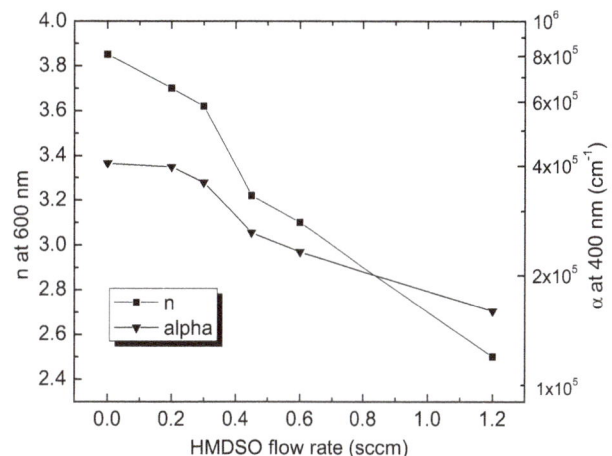

Fig. 4. Optical parameters n (600 nm) and α (400 nm) of the HMDSO layers.

voids and microcrystalline silicon. More details about the optical modeling can be found in reference [8]. The fitting of the bulk layer allows one to estimate the thickness (or deposition rate) and the optical parameters (n, α) of the deposited materials. Figure 3 shows the crystalline fraction (X_c) and deposition rate (r_d). The refractive index (n) at 600 nm and absorption coefficient (α) at 400 nm (extracted from the ellipsometric modelling) as a function of the HMDSO flow rate are shown in Figure 4. Although the model does not explicitly include an a-SiC:H component, it is assumed that the inclusion of a-Si:H in the optical model allows one to account for the incorporation of oxygen and carbon into the amorphous phase of the film. Regardless, the accuracy in the thickness, n, and α would not be affected by the details of the model as long as the fit to the data remains good. However, the extracted value of X_c is more strongly influenced by inaccuracies in the model. Nevertheless, based on comparison with Raman measurements, we can assert that the value of X_c obtained from ellipsometry is accurate to within $\pm 4\%$ absolute for these samples.

Figure 3 shows that as the HMDSO flow rate increases from 0 to 1.2 sccm, the deposition rate (r_d) increases from 0.3 Å/s to 0.6 Å/s, while the crystalline fraction (X_c) decreases. When the HMDSO flow rate is zero, the layer is microcrystalline with $X_c \sim 60\%$. However, as the incorporation of oxygen and carbon in the films inhibit microcrystalline growth, at high flow rate (1.2 sccm) of HMDSO, X_c is almost zero. For p-type nc-SiO$_x$ layers deposited with CO$_2$ as the O source under similar conditions [5], X_c determined by Raman spectroscopy also gave values in the range of 40% to 75%. It has been suggested that for p-μc-SiO$_x$:H, the best performance of solar cells is obtained when using p-type layers deposited at the transition region between μc-Si:H and a-Si:H [13].

Figure 4 shows that as the HMDSO flow rate increases, the refractive index (n at 600 nm) decreases from 3.85 to 2.5. It is possible to obtain values of n down to 1.8 with even higher flows of HMDSO (above 1.2 sccm) but the in-plane conductivity of these layers is less than 10^{-12} S/cm which is out of the range of our in-plane conductivity measurement set up. These films are considered as to be SiO$_x$-rich dielectric films with a high band gap (E_g above 3 eV) and are not within the scope of this paper. Also we note in Figure 4 that α decrease from 4×10^5 cm^{-1} to 2×10^5 cm^{-1} as the HMDSO flow rate increases from 0 to 1.2 sccm. Standard p-μc-SiO$_x$:H layers deposited with CO$_2$ as the O precursor under similar conditions also have $\alpha \sim 3 \times 10^5$ cm^{-1} [5]. The sharp decrease in n and α in the range of 0.3 sccm to 0.45 sccm of HMDSO could also be attributed to the change in X_c from 35% to 10% (Fig. 3). Furthermore, it should be noted that in comparison with p-μc-SiO$_x$:H layers reported in reference [14] with similar crystalline fractions of 35%, our p-μc-SiO$_x$C$_y$:H layers display much higher values of n (3.6 vs. 3.1).

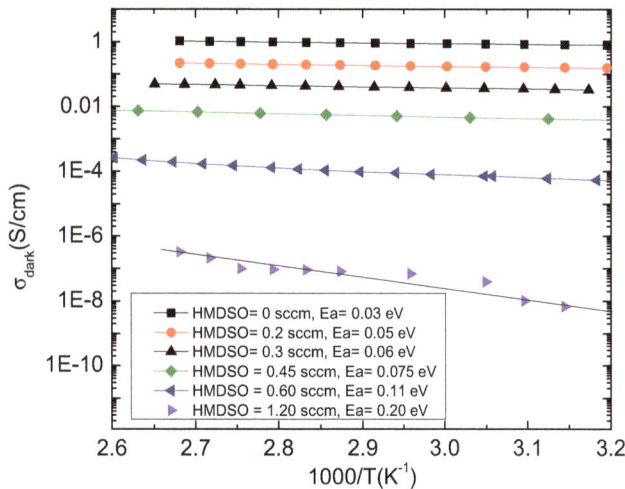

Fig. 5. Dark conductivity (σ) and activation energy (E_a) of layers deposited with different values of HMDSO flow rate.

3.2 Dark conductivity and activation energy

To check the electrical properties of the layers, the in-plane conductivity of the p-type layers was measured as a function of temperature, and the results are shown in Figure 5. Also indicated in the legend are the activation energies (E_a) of the layers (obtained from a linear fit of the data on a log(σ) vs. $1000/T$ plot), which for a p-type layer, provide a measure of the energy difference between the valence band edge and the Fermi level.

Figure 5 shows that the in-plane dark conductivity of the layers at room temperature decreases from 1 S/cm to 10^{-8} S/cm as the HMDSO flow rate increases from 0 to 1.2 sccm. This is expected because the increasing oxygen and carbon content makes the layers amorphous [25] and electrical conductivity decreases. Also the activation energy increases from 0.03 eV to 0.20 eV, possibly due to the decrease in doping efficiency as the oxygen and carbon content in the layer increases and as the layers get more amorphous. Curiously, the sharp transition in X_c between HMDSO flow rate of 0.3 sccm to 0.45 sccm is not reflected in σ or E_a.

However, we must note that we measure here only the in-plane conductivity of the layers. For solar cells, the out-of-plane conductivity is more important. It has been demonstrated by series resistance measurements in solar cells that a layer with an in-plane conductivity below 10^{-10} S/cm may have out-of-plane conductivity above 10^{-5} S/cm [9]. To check if this is true for our samples, we performed conductive probe AFM (CP-AFM) measurement on one of the samples (HMDSO = 0.45 sccm). The sample was prepared by depositing such a p-type μc-SiO$_x$C$_y$:H layer on a conductive ZnO:Al substrate, such that the current could flow vertically from the surface to the bottom contact. This configuration simulates the actual operation of the layer in a solar cell. Figure 6 shows the CP-AFM scan obtained for a layer with an HMDSO flow rate of 0.45 sccm.

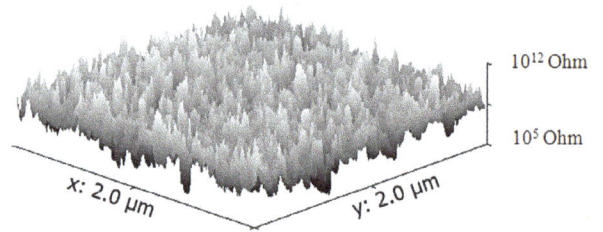

Fig. 6. Conductive probe AFM image of the layer with HMDSO flow rate of 0.45 sccm. Note that the resistance is indicated on a logarithmic scale.

Figure 6 shows that the resistance of the layer can vary by several orders of magnitude. There are some regions with a resistance of the order of 10^{12} Ω while there are other regions with resistance of the order of 10^5 Ω. We expect that the regions with low resistance values ($\sim 10^5$ Ω) correspond to microcrystalline regions in the layer [26]. It is difficult to calculate absolute values of conductivity with this resistance data (due to tip size effects and contact resistance) but the seven orders of magnitude difference in resistance is consistent with the μc-SiO$_x$:H model of microcrystalline silicon filaments in an amorphous silicon oxide (oxycarbide) matrix [9]. The low dopant levels (suitable for the more ordered microcrystalline regions) and amorphous nature of the matrix combine to give very low carrier concentrations in these regions.

3.3 Fourier Transform Infrared Spectroscopy

To study the gas dissociation and deposition pathways, we analyze the deposition process both by measuring optical absorption in the gas phase (including the plasma), as well as in the solid phase (the thin films), both through the FT-IR technique. The gas phase FT-IR system is installed downstream of the dry process pump. Figure 7 shows the FT-IR absorption spectrum of pure HMDSO in the gas phase [27], as well as the absorption due to the gases and species once the plasma (H$_2$+SiH$_4$+HMDSO+B$_2$H$_6$/Ar) is turned on. Also shown is the FT-IR absorption of an HMDSO p-μc-SiO$_x$C$_y$:H layer (p-type layer) deposited on an intrinsic FZ c-Si wafer. It should be noted that for the downstream gas-phase FT-IR system, some of the species may stick to the walls of the connecting lines and pump. Furthermore, as the range of this FT-IR is 1100–4000 cm^{-1}, it was not possible to detect the Si-O-Si peak ($\sim$1070 cm^{-1}) in the gas phase.

Figure 7 shows that HMDSO has two strong absorption peaks: Si-CH$_3$ (1250 cm^{-1}) and CH$_3$ (2950 cm^{-1}). These bond configurations are also seen when the plasma (H$_2$+SiH$_4$+HMDSO+B$_2$H$_6$/Ar) is turned on. When the FT-IR spectrum of the plasma is compared to the p-type layer (HMDSO = 0.2 sccm), there are no species of Si-CH$_3$ (1250 cm^{-1}) and CH$_3$ (2950 cm^{-1}) in the layer, whereas Si-O-Si bending (830 cm^{-1}) and Si-O-Si asymmetric stretching (1070 cm^{-1}) are visible. This confirms that HMDSO dissociates in the plasma, and species such as Si-O-Si originating from HMDSO contribute to the

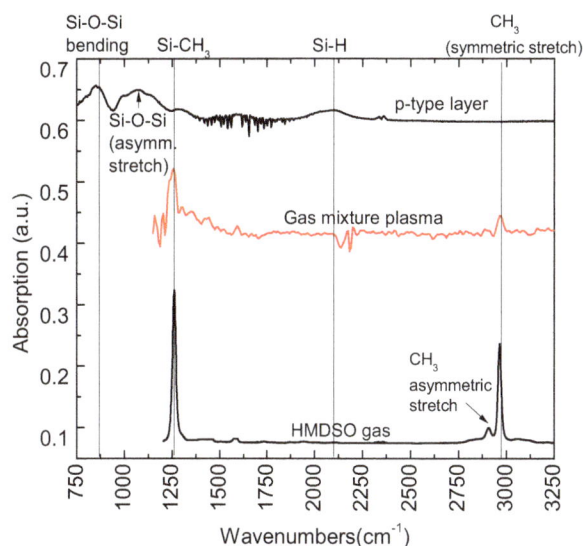

Fig. 7. FT-IR spectra of pure HMDSO in gas phase, deposition gas mixture with plasma ignited, and p-type layer (HMDSO = 0.2 sccm) deposited on silicon wafer.

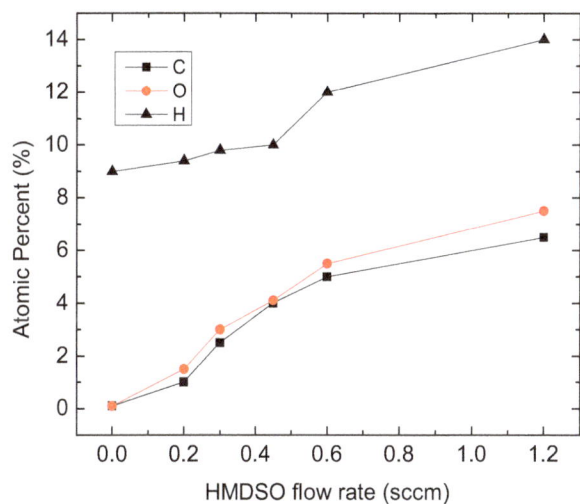

Fig. 8. Elemental composition of layers as a function of HMDSO flow rate.

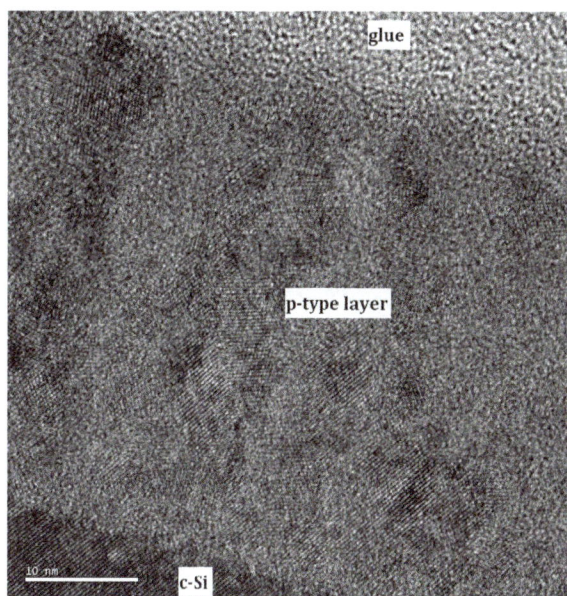

Fig. 9. HRTEM image of p-type layer at HMDSO flow rate of 0.45 sccm.

deposition of the layers, while CH_3 is not as efficiently incorporated into the films.

3.4 Secondary ion mass spectrometry

To accurately compare the elemental composition of p-μc-SiO_xC_y:H films deposited at different HMDSO flow rates, two stacks of three HMDSO p-μc-SiO_xC_y:H layers each were deposited on FZ wafers for SIMS measurements. Assuming the atomic density of the layers to be that of amorphous silicon ($\sim 5 \times 10^{22}$ at/cm^3), the atomic concentration of elements (C,O,H) in the layers was determined and is shown in Figure 8.

Figure 8 shows that as HMDSO flow rate increases, the concentrations of all three elements (C,O,H) are increasing, as would be expected. In a n-μc-SiO_x:H layer de-

posited with CO_2 as the O source, the atomic percentage of C was estimated around 2% [25]. In our case, HMDSO is an additional source of C and the percentage goes up to 6.5% for 1.2 sccm of HMDSO. We expect that this extra C in the layers is promoting the transformation to an amorphous (and eventually dielectric) layer, in addition to the impact of O. The hydrogen content in the layers is in the range of 10–15% which is also similar to the value reported in reference [25]. Also, we note that the percentage of O and C in the layers is comparable, which is striking given that the ratio of C to O in the HMDSO molecule is 6 to 1, and therefore indicating that O is incorporated six times more efficiently than C. The similar ratio of O and C in the layers suggests that p-type microcrystalline silicon oxycarbide (p-μc-SiO_xC_y:H) is an appropriate description of the material. Furthermore, the lack of CH_x peaks in the FT-IR spectra indicate that the carbon incorporated is mostly bonded with silicon.

3.5 Transmission electron microscopy

To examine the structure of the deposited layers by transmission electron microscopy (TEM), the p-type layers were deposited on crystalline silicon (c-Si) wafers. We note that the growth mechanism of the layers is different on the glass substrates and c-Si wafers [28], but to develop a first understanding of the material, we nevertheless present such samples. Cross-section samples for TEM observation were cut into slices and glued face to face in a sandwich configuration, before being thinned, first mechanically then by ion milling, until they became transparent for the electrons.

Figure 9 shows a high resolution TEM (HRTEM) image of the layer deposited with 0.45 sccm of HMDSO flow rate, showing the grains of microcrystalline silicon. Parts

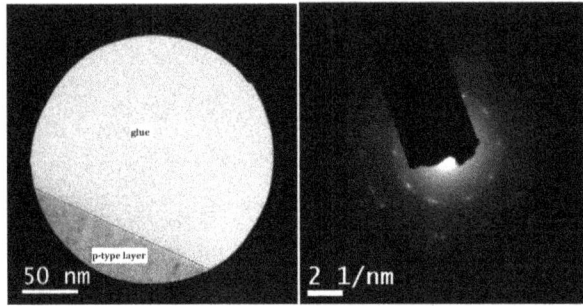

Fig. 10. Left: selected zone for diffraction pattern measurement. Right: diffraction pattern of p-type layer and glue.

of the image exhibiting no crystalline structure stem from the presence of amorphous regions in the film. This image is very similar to the one reported in reference [14] with clearly identifiable regions of microcrystalline silicon.

In order to see the crystalline quality of this sample, we applied the smallest aperture for area selection on a zone which includes the deposited layer and the glue as indicated in Figure 10 (left). The recorded diffraction pattern shown in Figure 10 (right) reveals three rings that are the signatures of {111}, {220} and {311} crystallographic planes of silicon. On verifying the inter-planar distances of family of planes of silicon carbide [29] ({111} for cubic SiC or {100}, {002} and {101} for hexagonal SiC), we note that there is no contribution from crystalline SiC in the diffraction pattern. The diffused part is mainly caused by the glue, but there is also a small contribution of the amorphous phase of the layer.

3.6 Single junction p-i-n solar cells

To test the p-type layers in devices, single junction p-i-n solar cells were fabricated. Figure 11 shows the resulting solar cell device parameters, displayed as a function of HMDSO flow rate. As indicated in Section 2, there were vacuum breaks at the p/i and i/n interfaces. To examine if the p/i interface break has a greater impact on layers using HMDSO (relative to ones using no oxygen source or CO_2 as the oxygen source), as well as to allow comparison of the performance of our p-μc-SiO$_x$C$_y$:H layers relative to more standard layers, reference p-i-n solar cells were also fabricated with the same vacuum breaks. The p-type layers for these reference cells are p-a-Si:H ($\sigma \sim 10^{-5}$ S/cm, $X_c \sim 0\%$) deposited with no oxygen source and p-μc-SiO$_x$:H ($\sigma \sim 10^{-2}$ S/cm, $X_c \sim 32\%$) deposited with CO_2 as oxygen precursor. These cells are indicated on Figure 11 at an HMDSO flow rate of 1 sccm ($X_c \sim 0\%$) and 0.30 sccm ($X_c \sim 35\%$), respectively.

Figure 11 shows that as the HMDSO flow rate increases, the device parameters (J_{SC}, V_{OC}, FF, Efficiency) increase, up to 0.45 sccm of HMDSO, and then decrease for higher flows. The maximum device efficiency (5.5%) is obtained in the transition region from amorphous silicon to microcrystalline silicon at an HMDSO flow rate of 0.45 sccm, in agreement with previous reports [13]. We also note that at an HMDSO flow rate of 0.45 sccm,

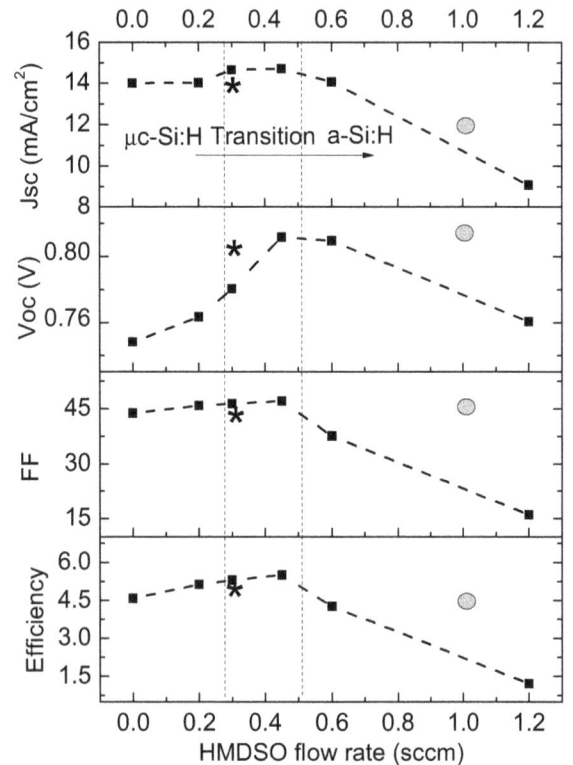

Fig. 11. p-i-n device parameters as a function of HMDSO flow rate. The parameters of a reference solar cell deposited with p-a-Si:H is shown by ○ and another reference solar cell with p-μc-SiO$_x$:H (CO_2 as the O source) is shown by *.

$V_{OC} = 0.81$ V and $J_{SC} = 14.7$ mA/cm^2. The reference cell (p-a-Si:H) has $V_{OC} = 0.82$ V and $J_{SC} = 12$ mA/cm^2 and reference cell (p-μc-SiO$_x$:H using CO_2) has $V_{OC} = 0.80$ V and $J_{SC} = 14$ mA/cm^2 and. Although poorer quality material is expected at the p/i and i/n interfaces because of the exposure of the layers to air (vacuum break) between subsequent depositions, this does not seem to impact the HMDSO layers any more than the reference p-a-Si:H and p-μc-SiO$_x$:H layer. The desired effect of optical transparency of p-type layers is reflected by higher J_{SC} in solar cells for p-μc-SiO$_x$C$_y$:H and p-μc-SiO$_x$:H layers relative to the p-a-Si:H layer.

The most deficient cell parameter for all the p-type layers is the FF, which is at most 45% rather than the expected >70% for high quality devices. To help explore the cause of this (i.e. if it is due to series resistance, shunting effects, or recombination), Figure 12 shows the J-V curves under light (measured between –1 V to 1.5 V) of p-i-n solar cells for different HMDSO flow rates.

Figure 12 shows that as the HMDSO flow rate increases, the J-V curves get more and more "S-shaped". Even for the lowest HMDSO flow (0.2 sccm), a kink can be observed in the J-V curve. The presence of such S-shapes indicates the presence of a counter-diode in the current path (as opposed to a straighter line at V_{OC} or J_{SC} which would indicate a true series resistance or "shunt resistance", respectively). As the V_{OC} of the devices remain reasonable, it is most likely that the junction between the

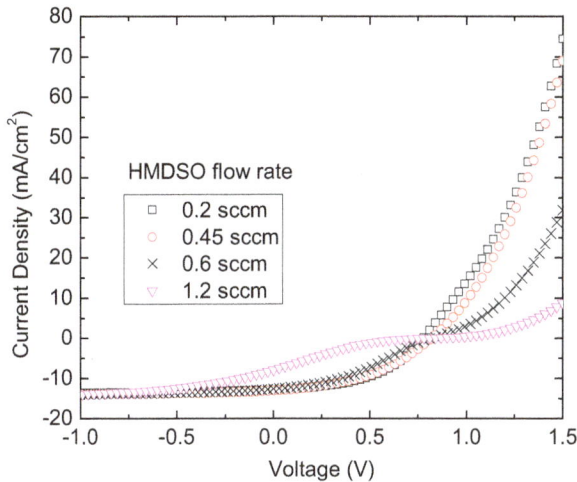

Fig. 12. *J-V* curves of single junction p-i-n solar cells, for HMDSO flow rates from 0.2 to 1.2 sccm measured between –1 V to 1.5 V under light.

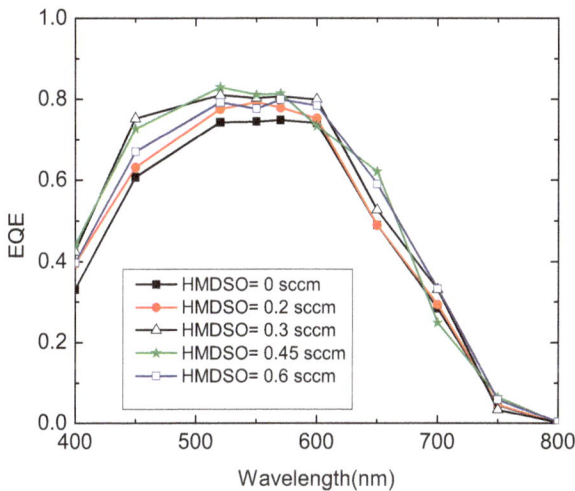

Fig. 13. EQE of the single junction p-i-n solar cells with increasing HMDSO flow rate.

ZnO:Al and p-type layers is not ohmic, and this is limiting cell performance. This effect would be in addition to the negative effects of the vacuum breaks and any other process limitations (i-layer quality, ZnO:Al) causing relatively poor solar cell performance.

Finally, to check the absorption in the solar cells at different wavelengths, the external quantum efficiency (EQE) was measured under zero voltage bias and is presented in Figure 13.

Figure 13 shows that as the HMDSO flow rate increases, so does the EQE in the low wavelength region (400–500 nm) up to 0.45 sccm of HMDSO flow rate. This confirms the desired transparency effect of higher O and C content in the layers. However, the EQE at low wavelengths decreases with subsequently increasing HMDSO flow. This is most likely due to the impact of the S-curve shown in Figure 12. The reason for the impact on the EQE in the higher wavelength region (~650 nm) is not as clear –

it may be due to poorer collection, but more likely variability in the texture and thickness of the ZnO:Al substrates.

4 Conclusion

In this work, we demonstrate the use of an easy-to-handle oxygen and carbon precursor, hexamethyldisiloxane (HMDSO) to deposit p-μc-SiO$_x$C$_y$:H layers. It is seen that with an increase in HMDSO flow rate, the microcrystalline fraction in the layers decreases and the layers become more transparent because of the incorporation of O and C. In comparison to literature results for p-μc-SiO$_x$:H, these layers have higher refractive indices for similar crystalline fractions. The additional incorporation of C in the films is effective in lowering the absorption (by forming a-SiC:H regions), but not as effective as incorporating the equivalent amount of O. It is possible to control the refractive index in the range from 2.5 to 3.85 and dark conductivity in the range from 1 to 10^{-8} S/cm. Fourier transform infrared spectroscopy (FT-IR) studies reveal that HMDSO dissociates in the plasma and Si-O-Si bonds are incorporated in the deposited layers. Secondary ion mass spectrometry (SIMS) analysis indicates that the atomic fractions of O and that of C in the layers are similar, indicating that the incorporation of O is six times more efficient than C when HMDSO is utilized as a precursor. A high resolution transmission electron microscopy (HR-TEM) image identifies regions of μc-Si:H in the deposited layer, and a selective area diffraction pattern (SAD) confirms the presence of {111}, {220} and {311} silicon families of planes in the p-type layer. The layers were tested in single junction p-i-n solar cells and the best performance of the solar cells was observed in the transition region from μc-Si:H to a-Si:H. Presently, the non-ohmic junction between ZnO:Al and p-type layer seems to be the limiting factor for these solar cells.

The work was carried out under a CIFRE Contract (2012/0892) between Air Liquide and LPICM, facilitated by the "*Association Recherche National Technologie*" (ANRT) in France. The authors would like to thank Olivier Schneegans at the LGEP for the conductive AFM image.

References

1. M. Berginski et al., Sol. Energy Mater. Sol. Cells **92**, 1037 (2008)
2. K. Ding et al., Sol. Energy Mater. Sol. Cells **95**, 3318 (2011)
3. F. Demichelis, C.F. Pirri, E. Tresso, Philos. Mag. **66**, 135 (1992)
4. P. Sichanugrist, T. Sasaki, A. Asano, Y. Ichikawa, H. Sakai, Sol. Energy Mater. Sol. Cells **34**, 415 (1994)
5. P. Cuony et al., Appl. Phys. Lett. **97**, 213502 (2010)
6. S. Inthisang, T. Krajangsang, I.A. Yunaz, A. Yamada, M. Konagai, C.R. Wronski, Phys. Status Solidi C **8**, 2990 (2011)
7. A. Sarker, A.K. Barua, Jpn J. Appl. Phys. **41**, 765 (2002)

8. S.N. Abolmasov et al., EPJ Photovolt. **5**, 55206 (2014)

9. P. Cuony et al., Adv. Mater. **24**, 1182 (2012)

10. P.D. Veneri, L.V. Mercaldo, I. Usatii, Prog. Photovol.: Res. Appl. **21**, 148 (2013)

11. A. Lambertz et al., Sol. Energy Mater. Sol. Cells **119**, 134 (2013)

12. P. Buehlmann et al., Appl. Phys. Lett. **91**, 143505 (2007)

13. K. Sriprapha, N. Sitthiphol, P. Sangkhawong, V. Sangsuwan, A. Limmanee, J. Sritharathikhun, Curr. Appl. Phys. **11**, S47 (2011)

14. C.-N. Li et al., Int. J. Photoenergy. **2012**, 1 (2012)

15. A. Lambertz, F. Finger, R.E.I. Schropp, U. Rau, V. Smirnov, Prog. Photovol.: Res. Appl. **23**, 939 (2015)

16. V. Smirnov, A. Lambertz, S. Tillmanns, F. Finger, Can. J. Phys. **92**, 932 (2014)

17. K. Haga, K. Yamamoto, M. Kumano, H. Watanabe, Jpn J. Appl. Phys. **25**, L39 (1986)

18. A. Hartstein, D.R. Young, Appl. Phys. Lett. **38**, 631 (1981)

19. V.A. Gritsenko et al., Microelectron. Reliab. **43**, 665 (2003)

20. D.S. Kim, Y.H. Lee, N.-H. Park, Appl. Phys. Lett. **69**, 2776 (1996)

21. D. Trunec et al., J. Phys. D **43**, 225403 (2010)

22. A. Sonnenfeld et al., Plasmas Polym. **6**, 237 (2001)

23. J. Muller, B. Rech, J. Springer, M. Vanecek, Sol. Energy **77**, 917 (2004)

24. G.E. Jellison, B.C. Sales, Appl. Opt. **30**, 4310 (1991)

25. P. Cuony, Optical Layers for Thin-film Silicon Solar Cells, Ph.D. Thesis, EPFL, 2011

26. G. Yue, L. Sivec, J.M. Owens, B. Yan, J. Yang, S. Guha, Appl. Phys. Lett. **95**, 263501 (2009)

27. D. Magni, C.H. Deschenaux, C. Courteille, A.A. Howling, C.H. Hollenstein, P. Fayet, MRS Proc. **544**, 65 (1998)

28. P. Roca i Cabarrocas, N. Layadi, T. Heitz, B. Drèvillon, I. Solomon, Appl. Phys. Lett. **66**, 3609 (1995)

29. S.R. Nutt, F.E. Wawner, J. Mater. Sci. **20**, 1953 (1985)

Plasma monitoring and PECVD process control in thin film silicon-based solar cell manufacturing

Onno Gabriel[1,a], Simon Kirner[1], Michael Klick[2], Bernd Stannowski[1], and Rutger Schlatmann[1]

[1] PVcomB, Helmholtz-Zentrum Berlin für Materialien und Energie GmbH, Schwarzschildstr. 3, 12489 Berlin, Germany
[2] Plasmetrex GmbH, Schwarzschildstr. 3, 12489 Berlin, Germany

Abstract A key process in thin film silicon-based solar cell manufacturing is plasma enhanced chemical vapor deposition (PECVD) of the active layers. The deposition process can be monitored in situ by plasma diagnostics. Three types of complementary diagnostics, namely optical emission spectroscopy, mass spectrometry and non-linear extended electron dynamics are applied to an industrial-type PECVD reactor. We investigated the influence of substrate and chamber wall temperature and chamber history on the PECVD process. The impact of chamber wall conditioning on the solar cell performance is demonstrated.

1 Introduction

Plasma-enhanced chemical vapor deposition (PECVD) of thin film silicon is a key process in various industrial applications. Thin film silicon material is used in flat panel displays [1], as passivation layers in crystalline silicon and hetero junction solar cells [2,3], and as absorber layers in thin film silicon-based solar cells and modules [4–7]. The material can be deposited by means of plasma-enhanced chemical vapor deposition (PECVD) in two different allotropes: hydrogenated amorphous (a-Si:H) and microcrystalline (μc-Si:H) silicon. Both types of materials are generally deposited using process gas mixtures of silane (SiH_4) and hydrogen (H_2), but with differences in the process conditions such as the gas flow composition, chamber pressure and RF power density [4,8]. Both types are applied in silicon-based tandem device solar cells [5,7], where the intrinsic (i) layer of the top cell is made from a-Si:H, while the i-layer of the bottom cell is made from μc-Si:H material. Another photovoltaic application is the hetero junction solar cell, where thin layers of a-Si:H are deposited on crystalline silicon wafers as passivation layers [2]. Doped p- and n-layers can be deposited by small admixtures of doping gases such as $B(CH_3)_3$ or PH_3 to the SiH_4/H_2 process gas mixture. Moreover, the refractive index and the electronic band gap and, thus, the optical absorption of the material can be varied by alloying of the material with carbon and oxygen, e.g. by the admixture of CH_4 or CO_2 to the process gas [9–13]. Due to the wide spread and large variety of applications of these films the growth mechanism of thin film silicon is the topic of ongoing research.

The plasma parameters determining the thin film growth remain mostly unknown. Insights into the plasma chemistry and the plasma-surface interaction, including thin film growth can generally be obtained by means of in situ plasma diagnostics. These diagnostics directly reveal the plasma parameters such as gas phase composition, electron and ion densities, energy distributions and temperatures. All these plasma parameters influence the growth process directly and, thereby, the properties of the resulting thin films as well. Each property is usually detected by a special diagnostic technique, each with its advantages and limitations. Among the plasma diagnostics proven to be useful for investigations of thin film silicon growth are optical emission spectroscopy (OES) [14–17], laser-induced fluorescence (LIF) [14,18], Fourier-transform infra-red spectroscopy (FTIR) [16,19], tunable diode laser absorption spectroscopy (TDLAS) [20,21], quadrupole mass spectrometry (QMS) [15,22,23], or cavity ring down spectroscopy (CRDS) [24]. Measurements of the electron density and the dynamics of electrons in the plasma gas phase, however, remain challenging under PECVD conditions.

An often neglected but relevant impact on the plasma chemistry is the influence of previously deposited films on substrate, electrode and chamber walls, which results in released atoms and molecules into the plasma by surface etching and surface association processes [25,26]. Although chamber wall conditioning by the deposition of

---
[a] e-mail: `onno.gabriel@helmholtz-berlin.de`

a defined layer prior to the actual thin film deposition is a well-established method to prevent an unpredictable influence of the reactor wall condition on the film quality, the actual influence of the chamber wall condition on the plasma chemistry is an often unknown parameter. Atoms, radicals and molecules are not only lost to surfaces surrounding the active plasma zone, but are also released from these surfaces even under deposition conditions. They can have a strong impact on the plasma chemistry. For example, in hydrogen rich a-Si:H or μc-Si:H deposition regimes, the flow rate of SiH_4 molecules and SiH_x radicals originating from recombination and etch processes from the surfaces back into the plasma gas phase can be a substantial fraction of the total SiH_4 process gas flow into the reactor [17]. The properties and structure of a chamber wall conditioning layer influences the deposition process, as we will show for the μc-Si:H deposition process. Moreover, pre-treatments of conditioning layers or of previously deposited layers on the substrate itself can be beneficial for the layer quality, e.g. by H_2 or CO_2 plasma treatment to improve the start of the μc-Si:H growth process [27, 28]. Additionally, plasma treatments are used for chamber cleaning [29] or to improve the interface properties between two different layers [30].

In this paper we present results of our research regarding plasma-surface interactions during PECVD processes used for thin film silicon deposition. We focus on three different in situ plasma diagnostics, which can readily be applied to an industrial-type PECVD process chamber, namely optical emission spectroscopy (OES), residual gas analysis/mass spectrometry (MS) and a novel technique called non-linear enhanced electron dynamics (NEED). The applied diagnostics deliver detailed and complementary information on the plasma chemistry. We will give examples – measured in experiments and also during solar cell production – how chamber walls and the properties of the surface material influences plasma chemistry during PECVD and how these changes in plasma chemistry influence the material properties and, thus, the resulting solar cell performance.

2 Experiment

All thin film silicon layer depositions were performed in an Applied Materials AKT 1600A cluster tool consisting of three PECVD process chambers connected to a central transfer chamber (Fig. 1) [13]. Up to six glass substrates of 30×30 cm^2 area were loaded via a load lock to a storage chamber. These substrates were automatically transferred by a central substrate mover from the storage via the transfer chamber to the PECVD process chambers, where the p-i-n/p-i-n layer structure of a-Si:H/μc-Si:H tandem solar cells were deposited. Software, process control and PECVD plasma conditions are very similar to those of the larger AKT 60k cluster tools used in industry to manufacture solar modules and flat panel displays of up to 220×260 cm^2 area (Gen8.5).

Within the PECVD chambers a shower head electrode ("diffuser") is mounted in the chamber top cover and

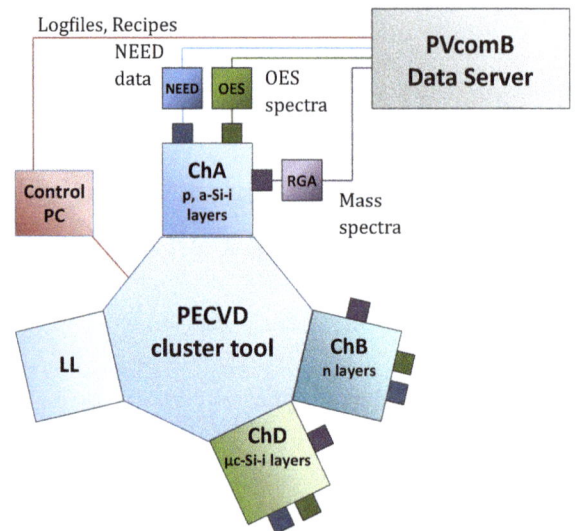

Fig. 1. Scheme of the AKT 1600A PECVD cluster, plasma diagnostics tools and the data management structure at PVcomB.

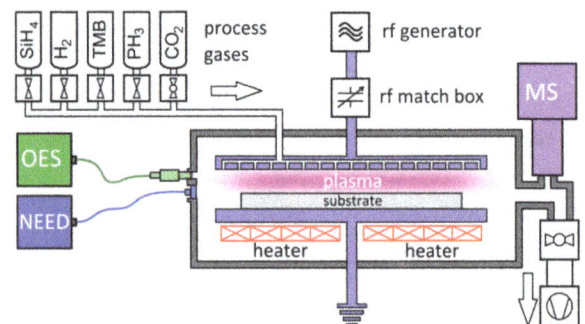

Fig. 2. Scheme of the process chamber with the parallel plate PECVD reactor and the three plasma diagnostics OES, MS and NEED.

powered by a 13.56 MHz radio frequency (RF) generator with RF powers between some 10 W and 1000 W (Fig. 2). Process gases flow into the chamber through the shower head electrode. The glass substrate is located on the grounded electrode ("susceptor"), which is heated up to typically 200 °C. All chamber walls are heated to 80 °C. The susceptor is movable in vertical direction and allows a variable electrode spacing from 5 to 75 mm. PECVD processes were performed typically at pressures in the range 200–1300 Pa with total gas flows rates between some 100 sccm up to 10 000 sccm. The chambers are usually cleaned after each layer deposition using a NF_3/Ar etch plasma followed by the deposition of a thin silicon layer ("conditioning layer") to ensure stable and reproducible process start conditions and reduce contamination. Amorphous and microcrystalline layers were deposited using process gas mixtures of SiH_4 and H_2. Doping of both materials is achieved by the admixture of trimethylboron, $B(CH_3)_3$ or TMB, to the SiH_4/H_2 process gas mixture (resulting in p-layers) or phosphine, PH_3 (resulting to n-layers). Silicon alloys were deposited using an admixture

Table 1. Wavelengths of line and band transitions of electronically excited atoms, radicals and molecules detected by OES during a-Si:H/μc-Si:H deposition and Ar/NF$_3$ plasma chamber cleaning.

Species	Transition	Wavelength
Si	$3s^2 3p^2\ {}^1\mathrm{D}\ 2 - s^2 3p4s\ {}^1\mathrm{P}^\circ\ 1$	288.3 nm
SiH	$\mathrm{X}^2 \Pi - \mathrm{A}^2 \Delta$ band	409$-$422 nm
H	$n = 3 - n = 2$ ("Balmer H$_\alpha$")	656.3 nm
H	$n = 4 - n = 2$ ("Balmer H$_\beta$")	486.1 nm
H$_2$	$2s^3 \Sigma_g^+ - 3p^3 \Pi_u^-$ band ("Fulcher-α")	570$-$640 nm
F	$2s^2 2p^4({}^3\mathrm{P})3s\ {}^2\mathrm{P}\ {}^3/_2 - 2s^2 2p^4({}^3\mathrm{P})3p\ {}^2\mathrm{P}^\circ\ {}^3/_2$	703.7 nm
Ar	$3s^2 3p^4({}^3\mathrm{P})4s\ {}^2\mathrm{P}\ {}^1/_2 - 3s^2 3p^4({}^3\mathrm{P})4p^2\ \mathrm{D}^\circ\ {}^3/_2$	496.5 nm

of CH$_4$, resulting in SiC$_x$:H films, or of CO$_2$, resulting in SiO$_x$:H films.

The deposition process conditions were monitored by means of optical emission spectroscopy (OES), mass spectrometry (MS) and by an electrodynamics- and model-based technique called non-linear electron dynamics (NEED) (see Fig. 2). OES and MS do not depend on any assumption about the plasma. The missing link to the plasma physics can be provided by NEED (see Sect. 2.3), which is based on a discharge model and provides a series resonance frequency, in first order depending on the electron density in the plasma, and the RF resistivity of the plasma, which is proportional to the quotient of collision rate for momentum transfer and electron density. Both parameters are provided at an absolute scale. The time resolution of the plasma diagnostics is about 1 s, while the data for chamber pressure, gas flow rates, RF power, temperatures, etc. are recorded with 0.25 s time resolution. All data from plasma diagnostics and AKT process data are stored in a central data base allowing user access to all recorded data, including data visualization and analysis.

2.1 Optical emission spectroscopy (OES)

One of the easiest applied in situ plasma diagnostics is OES. Only one transparent window port is needed to measure the emission spectrum generated by excited species in the plasma [4, 17]. On the other hand, the interpretation of measured spectra is more difficult and results often in indirect and relative properties of the plasma [15, 16]. The measured intensity $I'_{''}$ of an emission line ("peak") in a spectrum depends not only on the transition probability $A'_{''}$ between the upper (') and lower ('') electronic state of the transition, but also on the density of atoms or molecules in the upper electronic state $N_{'}$:

$$I'_{''} \sim A'_{''} N_{'} .$$

The upper state is generally populated by excitation processes due to collisions of electrons with atoms and molecules, which are nearly all in the electronic ground state. Therefore, $N_{'}$ depends not only on the ground state density of an atom or molecule, but also on the electron density and the electron energy distribution function (EEDF). In more complicated cases, the upper state

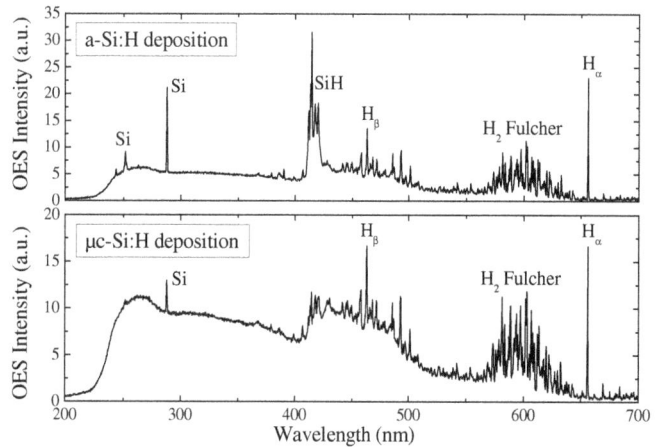

Fig. 3. Optical emission spectra measured during the deposition of amorphous (top) and microcrystalline thin film silicon (bottom). Some prominent optical emission peaks and band structures are labelled.

is populated additionally by other processes such as de-excitation from even higher states or gas phase reactions such as dissociative attachment. A full interpretation of measured spectra can only be given by detailed collisional-radiative models [15], which take all these reactions and transitions into account. Such a model is beyond the scope of this contribution.

From an experimental point of view, problems arise from the spatial distribution of the light emission from the plasma, which is usually higher in the plasma bulk and lower at the surroundings, e.g. near the substrate or electrodes (plasma sheaths). Moreover, the thickness of the plasma sheaths depend strongly on process conditions such as the chamber pressure. Therefore, a measured spectrum always depends on the optics and view direction into the chamber, i.e. the integral over the full line of sight through the plasma zone. We measured optical emission spectra using a fiber optics coupled directly to a quartz window. The same fiber window mount is installed on all three process chambers allowing exactly the same integrated views into the plasma. The light is coupled via a fiber into a two channel AvaSpec-2048 spectrometer by Avantes. The most important measured peaks and bands appearing during our PECVD processes are listed in Table 1. Example spectra during a-Si:H and μc-Si:H deposition are shown in Figure 3. Figure 4 shows measured

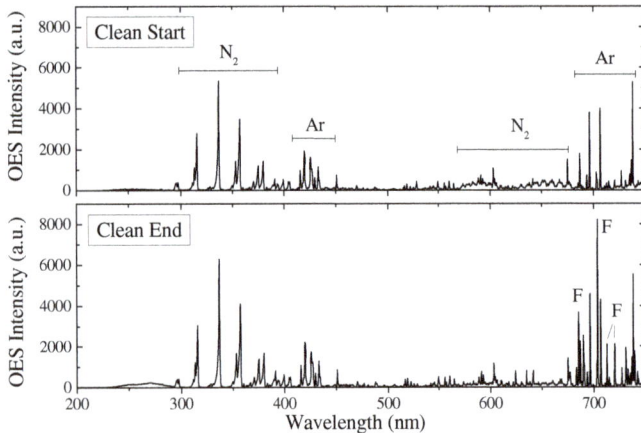

Fig. 4. Optical emission spectra measured during chamber cleaning in an Ar/NF_3 etch plasma. The clean/etch start is shown on top and the end on the bottom of the figure. F line intensities increase after the end point of the etching process.

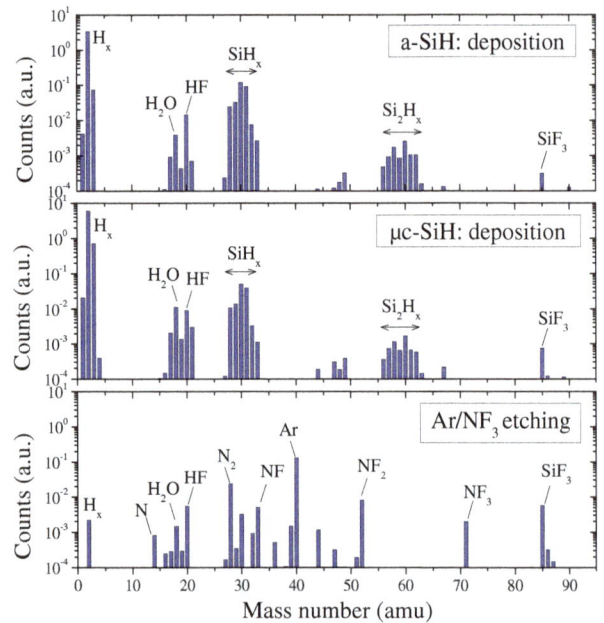

Fig. 5. Mass spectra measured during typical a-Si:H and μc-Si:H deposition (top and middle) and during Ar/NF_3 plasma chamber clean (bottom).

spectra during the start and end of an Ar/NF_3 plasma etch for chamber cleaning. The intensities of emission peaks from F atoms increase strongly after the end point of the etching process, because F is not consumed anymore by the etching process, which lead to the formation of SiF_4. Therefore, the F lines can be used for etch stop detection.

2.2 Mass spectrometry (MS)

A residual gas analyzer/mass spectrometer (MKS Vision2000c) was applied near the process chamber exhaust exit [13]. The position of the MS orifice is about 10 cm away from the outer edge of the RF electrode. The RGA consists of a quadrupole mass spectrometer and can detect masses up to 300 amu. A special orifice system is mounted in front of the MS enabling measurements at pressures up to 1333 Pa (10 Torr) during deposition as well as during chamber cleaning (NF_3 plasma etching). We use only a range up to 90 amu to increase the time resolution during measurements (about 0.5−3 s depending on accuracy). Most of the characteristic fingerprints of interesting atoms and molecules can be found in this range. MS provides only relative densities and the measured peaks are often ambiguous, i.e. they can originate from more than one detected species. Figure 5 shows mass spectra measured during a-Si:H and μc-Si:H deposition. Both are very similar apart from the difference in the H_2/SiH_4 ratio, which originates from the process gas ratio. In the bottom of Figure 5 a mass spectrum is shown measured during Ar/NF_3 plasma etch with characteristic fingerprints of etch radicals (F) and molecules (NF_x) and etch products (HF, SiF_x). The SiF_3 signal detected during deposition is a residue in the RGA originating from the NF_3 plasma etch, which is performed directly before each deposition process. RGA can detect a very broad range of stable atoms and molecules and is very useful for measurements of the gas composition. A disadvantage is the

impact of measurement history, e.g. a long residence of fluorine in the spectrometer, and the lower time resolution compared to OES.

2.3 Non-linear extended electron dynamics (NEED)

The RF plasma properties were measured by means of NEED, a plasma sensor provided by Plasmetrex GmbH, which is based on self-excited electron resonance spectroscopy (SEERS) [31]. With this technique resonance effects in capacitively coupled RF plasmas are detected, which occur due to an exchange between the kinetic energy of the electrons in the plasma bulk and the electric field energy in the plasma sheaths. Strong non-linear current-voltage characteristics within the sheaths (see equivalent circuit in Fig. 6) lead to the generation of higher harmonics, which can be measured in the displacement current at the chamber wall (see measured raw data example in Fig. 6). The sensor head can be implemented as a simple isolated metal plate at the chamber wall. Because the RF displacement current is not influenced by any thin film coating on the sensor head, the sensor operates under PECVD conditions as well as during NF_3 plasma chamber cleaning without drift in time.

NEED is based on a 3D fluid model for the isothermal plasma electrons, describing the RF current and potential in the plasma including a non-linear model of the boundary sheath, which result in two parameters. First, the normalized plasma resistivity χ, which is equal to $\nu_{\text{eff}} \times \omega_{gen}/ \omega_e^2$, with the effective electron collision rate ν_{eff}, the RF generator frequency ω_{gen} and the electron plasma frequency ω_e. The latter is generally defined as $\omega_e^2 = n_e \times e^2/ (m_e \times \varepsilon_0)$, with the electron density n_e, the elementary

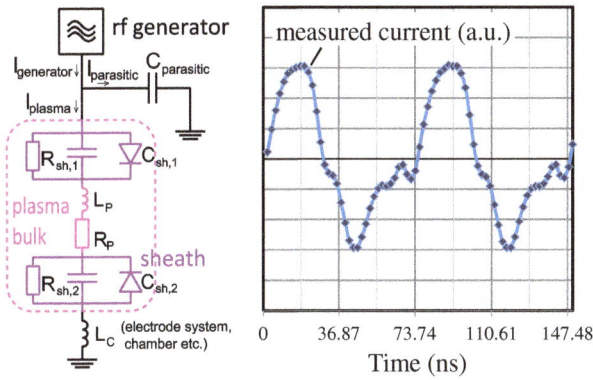

Fig. 6. Left: Scheme of the RF equivalent circuit used in the NEED model. The plasma part is marked by a dotted line. Right: Example of an electric current distorted by higher harmonics measured by NEED during one 13.56 MHz RF cycle.

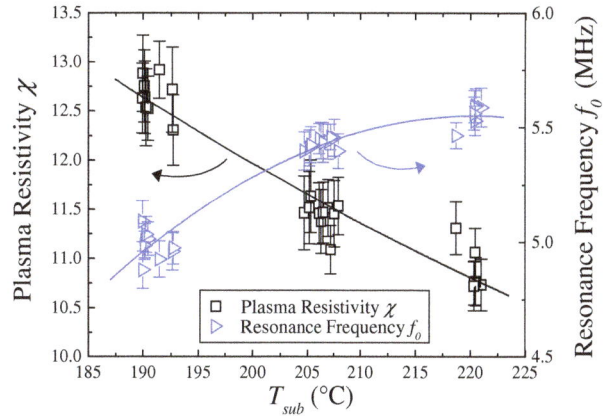

Fig. 7. Plasma resistivity and resonance frequency measured by NEED during 25 a-Si:H depositions at substrate temperatures 190, 205 and 220 °C.

charge e, the electron mass m_e and the vacuum permittivity ε_0. Therefore, χ is directly proportional to ν_{eff}/n_e. The second parameter is the resonance frequency f_0, which comprises the inertia effect of the chamber (mainly the inductance of the movable substrate holder) and the inertia effect of the plasma electrons, both at the RF scale, (compare Fig. 6). f_0 is equal to $(2\pi\sqrt{(L_P + L_C)C_S})^{-1}$ with the plasma inductance L_P, the sheath capacitance C_S and the inductance L_C of the chamber/substrate holder. L_P is equal to $m_e l_{pl}(e^2 n_e A)^{-1}$ with the plasma bulk thickness l_{pl} and the electrode area A. For our reactor geometry L_p has a value in the order of 100 nH, assuming an electron density of 10^{14} m^{-3}. The sheath capacitance $C_S = \frac{\varepsilon_0 A}{s_{tot}}$ is determined by the area of the electrodes A and the total sheath thickness s_{tot}. The value for L_C depends also on the position of the moveable susceptor. The exact value is unknown for our reactor, but has values in the range of $0.1-1$ μH and is, therefore, in the same order or even larger than L_p. Thus, only in case of sufficient grounding of the substrate holder also the plasma density can be provided directly, while in case of larger PECVD reactors f_0 is strongly influenced by L_C. Nevertheless, χ and f_0 are two basic plasma properties valuable for plasma process monitoring.

3 Results and discussion

In Sections 3.1 and 3.2 will be shown how the substrate temperature and the substrate surface material influence the plasma physics and chemistry. Even small changes can be detected by sensitive in situ diagnostics. In Section 3.3 we will show how a plasma itself can be used for diagnostics by a technique called etch product detection in a H$_2$ etch plasma, and how this technique is applied to the detection of the phase transition in a-Si:H/μc-Si:H deposition. The last Section 3.4 deals with the influence of the chamber wall condition on the material properties of deposited doped μc-SiO$_x$:H layers and how it affects fill factor and open circuit voltage of a-Si:H single junction cells.

3.1 Influence of the substrate temperature

Figure 7 shows χ and f_0 measured during the deposition of 25 intrinsic a-Si:H layers at substrate temperatures between 190 °C and 220 °C, while all other process parameters are constant (40 W, 330 Pa). The data show strong correlations with the substrate temperature T_{sub}, which has been varied between 190 °C and 220 °C. The substrate temperature is directly influencing the gas temperature of the plasma. With the gas temperature changes the gas density and, thus, the electron collision frequency. In this process window, χ is directly proportional to $1/T_{sub}$ according to the ideal gas law. On the other hand, the change in the gas density is changing the bulk property of the plasma, which also leads to the change in the resonance frequency via a change of the electron energy distribution and finally the electron density. NEED proves to be very sensitive: a change in the thermodynamic temperature of only about 6% leads to a 18% change in χ and a 9% change in f_0.

Here we can also validate our initial estimation of the electron density. The collision rate ν (for momentum transfer) can be estimated to be 4×10^9 / (100 Pa s) for both, Ar and H$_2$ [32]. For a typical process pressure of 330 Pa, one can estimate ν to be about 1.3×10^{10} s^{-1}. With $\chi = \nu_{\text{eff}} \times \omega_{gen}/\omega_e^2 = 10$ as shown in Figure 7, the electron density can be estimated to be 4×10^{13} m^{-3}. This is a typical value for an electronegative plasma and in agreement to the assumption in Section 2.3.

3.2 Influence of surface material

The condition of the surfaces surrounding the plasma has a strong influence on the plasma chemistry. Molecules and atoms are released from these surfaces, e.g. RF electrode and chamber walls, back into the gas phase by surface etching and by the association of ions and radicals at the surface forming new gas phase molecules [25, 26]. These new molecules and atoms are a source in addition

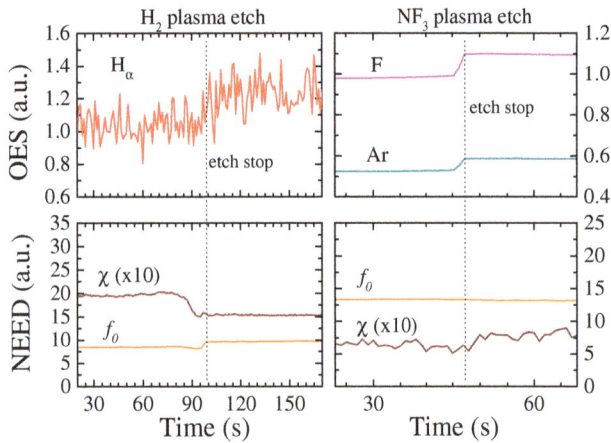

Fig. 8. Optical emission (top) and NEED (bottom) data measured during the etching of a 20 nm a-Si:H layer in a H$_2$ plasma (left) and of a 50 nm a-Si:H layer in an Ar/NF$_3$ plasma (right).

to the process gas and, thereby, change the plasma chemistry. The impact of surfaces to the plasma chemistry can be demonstrated in a simple experiment. A glass substrate covered with a 20 nm a-Si:H layer was exposed to a H$_2$ plasma in a previously cleaned reactor. The OES and NEED signals measured in this plasma are shown in the left part of Figure 8. The H$_\alpha$ emission, as well as χ and f_0 measured by NEED are constant during the H$_2$ plasma, but only until the a-Si:H layer is fully etched away by the H$_2$ plasma after about 100 s. The etch stop is clearly visible as a step in the measured signals. The etch product are SiH$_4$ molecules. This etching results in a SiH$_4$ concentration in the reactor similar to a case where the SiH$_4$ flow rate is set to 3.8 sccm, a value calculated using the etch rate (0.2 nm/s), layer density and area. This additional flow of SiH$_4$ into the plasma gas phase is strongly changing the plasma chemistry. The change is most pronounced in the plasma resistivity χ measured by NEED, as it will be explained in more detail in Section 3.3.

Another example is the NF$_3$/Ar plasma etch for chamber cleaning. The right part of Figure 8 shows the F and Ar emission (OES) and χ and f_0 measured by NEED during a NF$_3$/Ar plasma etching a 50 nm a-Si:H layer. The etch rate is much faster (1.1 nm/s) and the equivalent gas flow rate of the etch product is 20 sccm (of SiF$_4$ in this case). Again, the etch stop is clearly visible, but here only in the OES data. The NEED values χ and f_0, measures for the electron density and collision rate, remain unaffected. This points out the different plasma chemistries in the two types of etching plasmas. More insight would be given by applying a collisional-radiative model, which is beyond the scope of our work.

3.3 Plasma-surface interaction during a-Si:H/μc-Si:H deposition

PECVD of μc-Si:H thin films using 13.56 MHz RF discharges is usually performed with SiH$_4$ strongly diluted in H$_2$ gas under high-pressure-high-power (hphP)

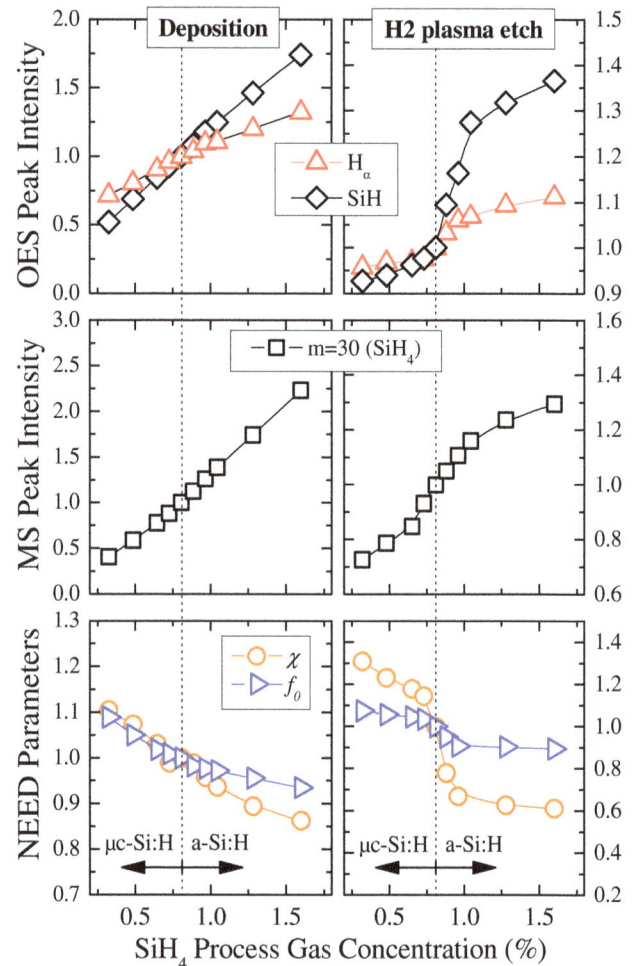

Fig. 9. OES, MS and NEED data measured during μc-Si:H deposition (left) and subsequent H$_2$ plasma etching (right). The μc-Si:H $\leftrightarrow$ a-Si:H phase transition is indicated with a dotted line at 0.81% and all curves are normalized to their values at 0.81%.

conditions [4, 5, 7, 16, 33], i.e. the deposition under higher chamber pressures and RF powers resulting in a high SiH$_4$ process gas depletion. In our PECVD system for sufficient layer quality and uniformity over large electrode areas the hphP process regime requires high total gas flows and a high hydrogen dilution of the process gas resulting in very low SiH$_4$ concentrations in the order of 1% [8]. Lower SiH$_4$ concentrations lead to μc-Si:H material growth, while higher SiH$_4$ concentrations lead to amorphous growth. Best solar cells are achieved at μc-Si:H growth very close to the phase transition to a-Si:H growth [16, 17]. The process window is very small, requiring stable and reproducible process conditions and/or active process monitoring and control [8, 17, 34]. The impact of the SiH$_4$ concentration on the plasma properties during the deposition process is shown in the left part of Figure 9. The concentration was varied between 0.32% and 1.6%, where the phase transition is at 0.81% (with 6160 sccm H$_2$ flow and 1200 Pa pressure). The optical emission of the H$_\alpha$ line and the SiH band (top part of Fig. 9) is proportional to the

Table 2. Influence of chamber condition on the Raman crystallinity and refractive index at 632 nm of deposited μc-SiO$_x$:H p- and n-layers.

Chamber conditioning	μc-SiO$_x$:H p-layer		μc-SiO$_x$:H n-layer	
layer material	Raman F_c (%)	n @ 632 nm	Raman F_c (%)	n @ 632 nm
a-Si:H	55.3	3.30	28.1	2.51
μc-SiO$_x$:H	51.4	3.26	26.7	2.42

SiH$_4$ concentration in the plasma, with the SiH emission increasing slightly faster than the H$_\alpha$ emission. The density of SiH$_4$ molecules in the chamber was measured by MS at mass number 30 (middle part of Fig. 9). The increase of SiH radicals is due to the increase of the SiH$_4$ density in the plasma, because SiH is a direct dissociation product of the SiH$_4$ process gas. The simultaneous increase of the H$_\alpha$ emission indicates an increase in the electron density n_e, which leads directly to an increase in the production of excited H atoms by electron collisions with ground state H. This increase in n_e results from different ionization energy thresholds $E_{ion,th}$ of process gas molecules and their direct products: while $E_{ion,th}$ is equal to 15.4 eV and 13.6 eV for H$_2$ and H [35], respectively, $E_{ion,th}$ is lower for SiH$_4$ (11.0 eV) and even lower for its radicals SiH$_3$, SiH$_2$, SiH and Si (8.0, 8.2, 7.9, and 8.2 eV) [36]. Therefore, the increasing SiH$_4$ concentration leads to an increase in the electron density in the plasma by enhanced ionization processes. This is confirmed by the measured plasma resistivity χ, which decreases reciprocally with the electron density (bottom part of Fig. 9). Even small changes in the SiH$_4$ gas flow strongly influence the measured plasma properties. However, the measured values change only linearly and do not show any indication at which SiH$_4$ concentration the μc-Si:H $\leftrightarrow$ a-Si:H phase transition occurs.

The right part of Figure 9 shows data measured during H$_2$ plasma etch steps (each of 45 s duration, 500 W, 1200 Pa, 6000 sccm H$_2$), which were performed directly after each of the deposition shown in the left part of Figure 9. The data is plotted over the SiH$_4$ concentration of the preceding deposition step. This technique called etch product detection was proposed by Dingemans et al. [17] and is used to detect the phase transition in thin film silicon deposited in a SiH$_4$ gas flow variation series. The measured data show a clear non-linear behavior: the H$_2$ plasma etch leads to an increased amount of etch products (primarily in the form of SiH$_4$ molecules) due to an enhanced etching of the previously deposited film on substrate, RF electrode, and chamber walls, because a-Si:H is etched more easily by hydrogen radicals than μc-Si:H material [17]. Additionally to previously measured SiH emissions [17] we also found the optical emission of other lines to be dependent on the crystallinity of the etched material, shown for the H$_\alpha$ line in Figure 9 (top right). As indicated by the SiH signal and here substantiated by mass spectrometry, the amount of etched silicon material is responsible for the s-like shape of the curves with SiH$_4$ as the main etch product. Clearly visible is the trend and the s-shape also in the NEED data curves, where χ and f_0 decrease with increasing electron density resulting from increasing SiH$_4$ concentrations in the gas phase.

3.4 Impact of chamber wall condition

In previous sections it was shown how surfaces (be it reactor walls or substrate) can influence the plasma chemistry. Vice versa, any change in the plasma has an impact on the thin film growths process. Therefore, the material properties of a deposited thin film layer depend on the pre-conditioning of the chamber wall. Since the start of a PECVD process usually determines the interface region of two sequential layers, the chamber conditioning has also a large impact on the performance of the resulting solar cell device. We found this dependence to be strongest for the deposition of microcrystalline material, strongly influenced by the equilibrium of deposition and etching processes as described in Sections 3.2 and 3.3. The development of proper chamber conditioning layers is an important part of the whole optimization of PECVD processes for solar cell manufacturing.

Any effect on the start of a PECVD process is strongest in case of the deposition of the thin doped p- and n-layers, because the total deposition time is in the order of only one minute, which is much shorter than the deposition time for i-layers. The p- and n-layers in our a-Si:H/μc-Si:H tandem solar cells are made from doped μc-SiO$_x$:H material. As compared to μc-Si:H such layers are deposited with extremely low SiH$_4$ concentrations below 1% [13]. To reveal the impact of the conditioning layer on the material quality, we deposited 100 nm p- and n-type μc-SiO$_x$:H layers on glass substrates, each in two different runs: one in a chamber conditioned with an a-Si:H layer and one conditioned with a μc-SiO$_x$:H layer. The crystallinity F_c measured by Raman spectroscopy and the refractive index n at 632 nm measured by a photo-spectrometer [13] for the p- and n-layers are listed in Table 2. The resulting values depend on the type of the conditioning layer: both, the p- and the n-layer, have a lower crystallinity if deposited in a chamber conditioned with a μc-SiO$_x$:H layer compared to the chamber with a-Si:H conditioning. With F_c also the refractive index n at 632 nm decreases slightly. This seems to be in contrast to the findings in Section 3.3, where an a-Si:H layer is etched faster than a μc-Si:H layer and the additional SiH$_4$ originating from the etch process should result in the deposition of more amorphous material. However, the additional SiH$_4$ also results in a change in the SiH$_4$/CO$_2$ process gas ratio, which determines F_c and n [10, 11, 13]. Thus, for μc-SiO$_x$:H deposition the decrease in F_c and n can be explained by a decrease in SiH$_4$/CO$_2$ ratio in the plasma gas phase at PECVD start by the change from an a-Si:H to a μc-SiO$_x$:H conditioning layers.

The last example is about the influence of the conditioning layer material on the p-layer of p-i-n solar cells

Table 3. Influence of chamber condition on the (initial) I-V characteristics of a-Si:H p-i-n single junction cells deposited on ZnO:Al front TCO. Data is averaged over 60 1-cm^2 solar cells.

Chamber conditioning layer material	J_{sc} (mA/cm^2)	V_{oc} (mV)	FF (%)	η (%)
p-type a-Si:H	14.5 ± 0.1	912 ± 7	68 ± 2	9.0 ± 0.3
p-type μc-Si:H	14.5 ± 0.1	925 ± 6	69 ± 2	9.3 ± 0.3

and, thereby, on the performance of the resulting solar cell device. The p-i-n layer stack was deposited on textured ZnO:Al front TCO and finished by a ZnO:Al/Ag back reflector/contact. The p-i-n layers are the same as used for top cells in our standard tandem cell device on ZnO:Al [37], i.e. the p-layer is made from μc-Si:O$_x$:H material. Prior to the p-layer deposition the PECVD chamber was conditioned in two ways: the deposition of p-type a-Si:H and p-type μc-Si:H material. The chamber conditioning influences the performance of the resulting solar cell device. The resulting I-V parameters measured under AM 1.5 illumination are listed in Table 3. While the short circuit current density J_{sc} remains unaffected, the open circuit voltage V_{oc} and the fill factor FF are increased by the application of the μc-Si:H conditioning layer, resulting in an increase in solar cell efficiency about 0.3% absolute.

We applied the in situ diagnostics to the PECVD of the p-layers for these two different chamber conditionings. The PECVD recipe starts with a 30 s Ar plasma treatment to clean the ZnO:Al surface e.g. from H$_2$O molecules. Then follows the deposition of a very thin p-type μc-Si:H seed layer and then the p-type μc-SiO$_x$:H layer. The in situ data measured during Ar treatment and seed layer deposition is shown in Figure 10. The impact of the chamber conditioning material on the plasma chemistry is clearly visible. During the Ar plasma the Si and SiH signals measured by OES are about a factor two lower in case of μc-Si:H conditioning. Both signals arise from etch products from the conditioning layer in form of SiH$_x$ molecules. The SiH$_4$ signal measured at $m = 30$ by mass spectrometry is below the MS detection limit, indicating that the OES signals do not originate from SiH$_4$ but from SiH$_x$ ($x = 0$–3) radicals. Again, NEED proves to be very sensitive, because the plasma resistivity χ and resonance frequency f_0 are strongly affected by the additional SiH$_x$ in the plasma. This is even visible during the μc-Si:H seed layer deposition, which is performed after pumping and re-fill of the chamber with process gas. Even then, the plasma resistivity is slightly lower in case of a μc-Si:H conditioning layer, while OES and MS signals show no influence on the material of the conditioning layer. Therefore, the Ar plasma leads to a change on the ZnO:Al surface, probably by re-deposition of Si material etched from the chamber wall. A similar effect was previously reported by Wanka et al. [30]: the re-deposition of SiO$_x$ material on a TCO coated substrate during a pure CO$_2$ plasma treatment in a process chamber conditioned by an a-Si:H layer. In our case, the re-deposited material on the ZnO influences the start of the seed layer deposition. This chamber history effect is covered by the deposition of the seed layer

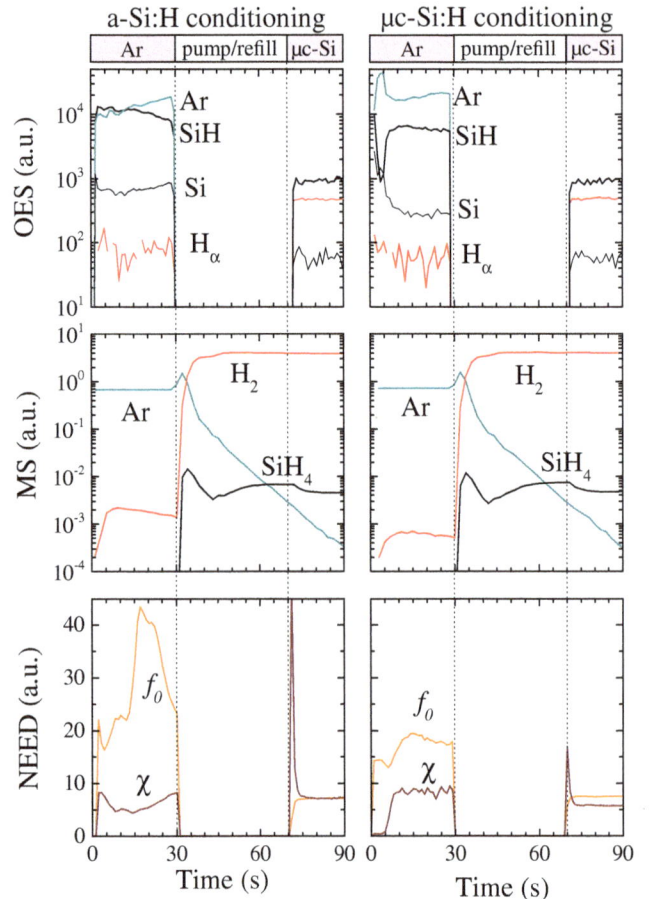

Fig. 10. OES (top), MS (middle) and NEED (bottom) data measured in situ during Ar plasma treatment and μc-Si-p-layer deposition. Left: a-Si:H chamber conditioning, right: μc-Si:H chamber conditioning.

on substrate, electrode and chamber walls. The following thicker p-type μc-SiO$_x$:H layer is not influenced by the chamber wall conditioning anymore. This explains also the equal J_{sc} values in Table 3. The influence of the conditioning layer material is only on the TCO/p interface region, which affects only FF and V_{oc}.

4 Summary

In summary, the growth of a-Si:H, μc-Si:H and μc-SiO$_x$:H thin films by PECVD were monitored by means of three complementary plasma diagnostics: OES, MS and NEED. Next to the process parameters such as chamber pressure, RF power and process gas composition, the substrate surface temperature and surface material properties influence the plasma chemistry as well and, thus, the PECVD process and properties of the resulting thin films. We showed how in situ plasma diagnostics adds valuable data for fast and effective process development on the PECVD process. Especially NEED is a very sensitive technique, which can detect even very small changes in the plasma gas phase and substrate/chamber wall temperature. Finally, we demonstrated the impact of the chamber wall conditioning on the properties of μc-SiO$_x$:H material and how the chamber conditioning influence the p-layer for p-i-n solar cells and, ultimately, their electrical performance.

The authors thank M. Zelt for technical assistance. This work was supported by the Federal Ministry of Education and Research (BMBF), by the Federal Ministry of Environment (BMU) and the state government of Berlin (SENBWF) in the framework of the program "Spitzenforschung und Innovation in den Neuen Ländern" (Grant No. 03IS2151) and the "SiliziumDS12plus" project (Grant No. 0325317C).

References

1. J. Perrin, J. Schmitt, C. Hollenstein, A. Howling, L. Sansonnens, Plasma Phys. Control. Fusion **42**, 353 (2000)
2. A. Descoeudres, L. Barraud, R. Bartlome, G. Choong, S.D. Wolf, F. Zicarelli, C. Ballif, Appl. Phys. Lett. **97**, 183505 (2010)
3. T. Mishima, M. Taguchi, H. Sakata, E. Maruyama, Sol. Energy Mater. Sol. Cells **95**, 18 (2011)
4. M. Kondo, M. Fukawa, L. Guo, A. Matsuda, J. Non-Cryst. Solids **266**, 84 (2000)
5. A.V. Shah, H. Schade, M. Vanecek, J. Meier, E. Vallat-Sauvain, N. Wyrsch, U. Kroll, C. Droz, J. Bailat, Prog. Photovolt.: Res. Appl. **12**, 113 (2004)
6. K. Yamamoto, A. Nakajima, M. Yoshimi, T. Sawada, S. Fukuda, T. Suezaki, M. Ichikawa, Y. Koi, M. Goto, T. Meguro, T. Matsuda, M. Kondo, T. Sasaki, Y. Tawada, Sol. Energy **77**, 939 (2004)
7. B. Rech, T. Repmann, M.N. van den Donker, M. Berginski, T. Kilper, J. Hüpkes, S. Calnan, H. Stiebig, S. Wieder, Thin Solid Films **511**, 548 (2006)
8. M.N. van den Donker, B. Rech, R. Schmitz, J. Klomfass, G. Dingemans, F. Finger, L. Houben, W.M.M. Kessels, M.C.M. van de Sanden, J. Mater. Res. **22**, 1767 (2007)
9. S. Inthisang, K. Sriprapha, S. Miyajima, A. Yamada, M. Konagai, Jpn J. Appl. Phys. **48**, 2402 (2009)
10. P. Cuony, D.T.L. Alexander, I. Perez-Wurfl, M. Despeisse, G. Bugnon, M. Boccard, T. Söderström, A. Hessler-Wyser, C. Hébert, C. Ballif, Adv. Mater. **24**, 1182 (2012)
11. A. Lambertz, F. Finger, B. Holländer, J. Rath, R. Schropp, J. Non-Cryst. Solids **358**, 1962 (2012)
12. K. Schwanitz, S. Klein, T. Stolley, M. Rohde, D. Severin, R. Trassl, Sol. Energy Mater. Sol. Cells **105**, 187 (2012)
13. S. Kirner, O. Gabriel, B. Stannowski, B. Rech, R. Schlatmann, Appl. Phys. Lett. **102**, 051906 (2013)
14. D. Mataras, S. Cavadias, D. Rapakoulias, J. Appl. Phys. **66**, 119 (1989)
15. U. Fantz, Plasma Phys. Control. Fusion **40**, 1035 (1998)
16. B. Strahm, A.A. Howling, L. Sansonnens, C. Hollenstein, Plasma Sources Sci. Technol. **16**, 80 (2007)
17. G. Dingemans, M.N. van den Donker, A. Gordijn, W.M.M. Kessels, M.C.M. van de Sanden, Appl. Phys. Lett. **91**, 161902 (2007)
18. H. Nomura, K. Akimoto, A. Kono, T. Goto, J. Phys. D **28**, 1977 (1995)
19. A. Howling, B. Strahm, C. Hollenstein, Thin Solid Films **517**, 6218 (2009)
20. N. Itabashi, K. Kato, N. Nishiwaki, T. Goto, C. Yamada, E. Hirota, Jpn J. Appl. Phys. **27**, L1565 (1988)
21. R. Bartlome, A. Feltrin, C. Ballif, Appl. Phys. Lett. **94**, 201501 (2009)
22. V. Lisovskiy, J.-P. Booth, K. Landry, D. Douai, V. Cassagne, V. Yegorenkov, Appl. Phys. **40**, 6631 (2007)
23. S. Nunomura, I. Yoshida, M. Kondo, Appl. Phys. Lett. **94**, 071502 (2009)
24. J.P.M. Hoefnagels, Y. Barrell, W.M.M. Kessels, M.C.M. van de Sanden, J. Appl. Phys. **96**, 4094 (2004)
25. M. Cacciatore, M. Rutigliano, Plasma Sources Sci. Technol. **18**, 023002 (2009)
26. D.C. Schram, Plasma Sources Sci. Technol. **18**, 014003 (2009)
27. P. Pernet, M. Goetz, H. Keppner, A. Shah, Mater. Res. Soc. Symp. Proc. **452**, 889 (1997)
28. A. Gordijn, J. Rath, R. Schropp, J. Appl. Phys. **95**, 8290 (2004)
29. T. Merdzhanova, J. Woerdenweber, T. Zimmermann, U. Zastrow, A. Flikweert, H. Stiebig, W. Beyer, A. Gordijn, Sol. Energy Mater. Sol. Cells **98**, 146 (2012)
30. H. Wanka, G. Bilger, M. Schubert, Appl. Surf. Sci. **93**, 339 (1996)
31. M. Klick, J. Appl. Phys. **79**, 3445 (1996)
32. Y.P. Raizer, V.I. Kisin, J.E. Allen, *Gas discharge physics* (Springer, Berlin, 1991)
33. Y. Mai, S. Klein, R. Carius, J. Wolff, A. Lambertz, F. Finger, X. Geng, J. Appl. Phys. **97**, 114913 (2005)
34. T. Kilper, M.N. van den Donker, R. Carius, B. Rech, G. Bräuer, T. Repmann, Thin Solid Films **516**, 4633 (2008)
35. R.K. Janev, D. Reiter, U. Samm, *Collision Processes in Low-Temperature Hydrogen Plasmas*, Berichte des Forschungszentrums Jülich 4038 (Forschungszentrum Jülich, Jülich, 2003)
36. R.K. Janev, D. Reiter, Contrib. Plasma Phys. **43**, 401 (2003)
37. B. Stannowski, O. Gabriel, S. Calnan, T. Frijnts, A. Heidelberg, S. Neubert, S. Kirner, S. Ring, M. Zelt, B. Rau, J.-H. Zollondz, H. Bloeß, R. Schlatmann, B. Rech, Sol. Energy Mater. Sol. Cells **119**, 196 (2013)

27

Plasma immersion ion implantation of boron for ribbon silicon solar cells

K. Derbouz[1], T. Michel[2,4,a], F. De Moro[3], Y. Spiegel[2], F. Torregrosa[2], C. Belouet[3], and A. Slaoui[1]

[1] InESS-CNRS-Univ. Strasbourg, France
[2] IBS, Peynier, France
[3] Solarforce, Bourgoin-Jallieu, France
[4] Université d'Aix-Marseille, Institut Fresnel, Marseille, France

Abstract In this work, we report for the first time on the solar cell fabrication on n-type silicon RST (for Ribbon on Sacrificial Template) using plasma immersion ion implantation. The experiments were also carried out on FZ silicon as a reference. Boron was implanted at energies from 10 to 15 kV and doses from 10^{15} to 10^{16} cm^{-2}, then activated by a thermal annealing in a conventional furnace at 900 and 950 °C for 30 min. The n$^+$ region acting as a back surface field was achieved by phosphorus spin-coating. The frontside boron emitter was passivated either by applying a 10 nm deposited SiO$_X$ plasma-enhanced chemical vapor deposition (PECVD) or with a 10 nm grown thermal oxide. The anti-reflection coating layer formed a 60 nm thick SiN$_X$ layer. We show that energies less than 15 kV and doses around 5×10^{15} cm^{-2} are appropriate to achieve open circuit voltage higher than 590 mV and efficiency around 16.7% on FZ-Si. The photovoltaic performances on ribbon silicon are so far limited by the bulk quality of the material and by the quality of the junction through the presence of silicon carbide precipitates at the surface. Nevertheless, we demonstrate that plasma immersion ion implantation is very promising for solar cell fabrication on ultrathin silicon wafers such as ribbons.

1 Introduction

Even with the new rising solar cell generations, crystalline silicon modules still dominate the market with about 80% of the global production capacity [1]. While improving the cell efficiency is the key to producing more watts per cell, manufacturing cost reduction and factory output continue to drive down the cost of photovoltaics (PV). In this way reducing the substrate cost is of great importance. One recent innovation is the use of thin silicon ribbons. Thin Si ribbons made by Solarforce using RST technology [2], allows a decrease in the consumption of silicon down to 1 to 2 g/Wp achieving costs below 0.2 €/Wp, and cells as thin as 60 μm.

Figure 1 shows the schematic of the ribbon growth method which is characterized by a vertical growth direction combined with the use of a substrate [3]. This results in a growth speed of about 3–8 cm/min; much faster than edge-defined film-fed growth (EFG) and string ribbon (SR) growth. Thus so far RST ribbon up to 8 cm in width can be grown with a thickness down to 80 μm. Another feature of the RST material is the low concentration of interstitial oxygen ($<10^{17}$ cm^{-3}) and substitutional car-

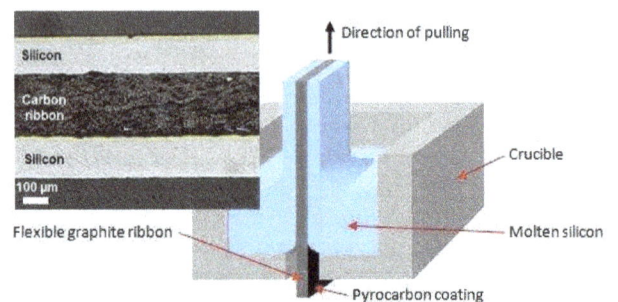

Fig. 1. Schematic views of the RST ribbon growth method and of a cross section of the resulting structure.

bon ($\approx 5 \times 10^{17}$ cm^{-3}). It is also possible to obtain either p-type or n-type silicon by simply changing the doping type of the Si precursor.

As for solar cells fabrication on these thin Si ribbons, the emitter can be processed from a gas or liquid source followed by a drive-in step. A high temperature and long duration process is often needed, especially for boron diffusion, which results in a highly surface concentrated doping profile. On the other hand, ion implantation can offer a better control of the concentration and depth of the doping distribution. While assuring great process uniformity

[a] e-mail: `thomas.michel@ion-beam-services.fr`

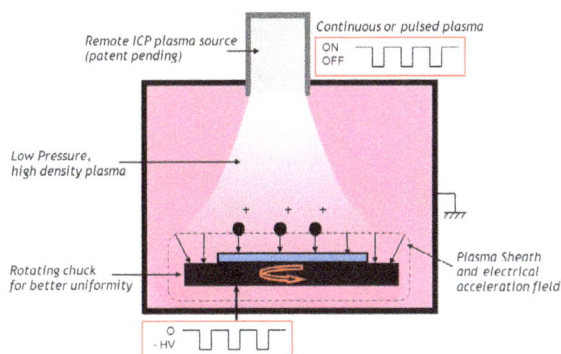

Fig. 2. PULSION® doping tool design.

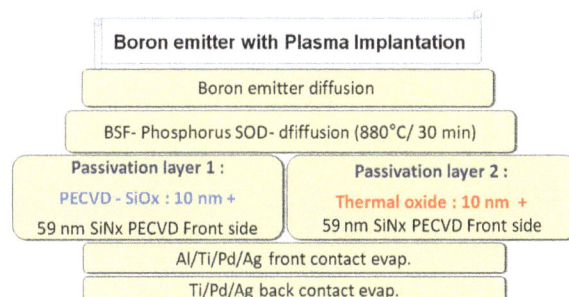

Fig. 3. Flow chart of n-type RST ribbon and FZ.

and reproducibility, ion implantation also leads to cell binning improvements [4]. Whereas classical beamline implanters do not meet the future PV industry requirements in terms of cost and productivity, new types of equipment such as plasma immersion ion implantation (PIII) reactors promise higher throughput.

Figure 2 presents the PULSION® plasma immersion tool designed by IBS. PULSION's unique features allow large-area implantations of high doses with a very high throughput, but also enable conformal doping of textured or grooved structures. Providing a better accuracy and control of concentration and depth without dead layer, PULSION® doping allows a decrease of the emitter saturation current, an enhancement of the blue response, and thus the increase of cell efficiency. After implantation, the dopants require an activation step. Furnace annealing offers the capability to grow a high quality thermal oxide for the emitter passivation. Furthermore, this emitter engineering sequence reduces manufacturing costs by eliminating the edge isolation step and the phosphosilicate glass removal.

We present for the first time results on n-type RST ribbon silicon cell process based on Boron plasma implanted emitter using PULSION®. Many results presented here concern n-type FZ silicon which acts as a reference to set the experimental parameters and to validate each step of the fabrication process flow. PULSION® doping is well suited for emitter doping, especially boron diffusion which is known to be more complicated than phosphorus. Boron was implanted through BF_3 gas precursor at energies from 10 to 15 kV and doses from 10^{15} to 10^{16} cm^{-2}. We first investigated the dependence of the sheet resistance of the p emitter on implant (energy E and dose D) and annealing conditions (temperature T and time t).

2 Experimental

The as-grown n-type RST silicon wafers, with a resistivity of 1.5–4.0 Ω cm and 80–140 μm thick, were etched in a solution (HF:HNO$_3$:CH$_3$COOH – 1:7.5:2.5 vol.) and ultrasonic bath to remove the SiC backside layer, which was in contact with the carbon ribbon during the silicon ribbon growth. The n-type FZ silicon wafers with a resistivity of 2 Ω cm and 260 μm thick were RCA-cleaned before implantation and annealing.

Figure 3 shows the flow charts for the fabrication of solar cells on RST-Si and FZ-Si samples: after chemical cleaning, the boron emitter was formed by plasma implantation. The boron was implanted at energies from 10 to 15 kV and doses from 10^{15} to 10^{16} cm^{-2}. The activation of Boron was carried out by thermal annealing in a conventional furnace at 900 and 950 °C for 30 min. As a second step, a Phosphorus doped solution (P509 from Filmtronics) was spun at the back of the RST sheets and then annealed at 880 °C for 30 min to form the n$^+$ doped region, which acts as a back surface field (BSF). After a HF-clean to remove the silicate glass layers, a plasma hydrogenation was performed at 400 °C for 2 h on the back side (through the BSF region). This process aims to passivate some dangling bonds and defects within the bulk such as grain boundaries.

The frontside boron emitter was passivated either by a 10 nm SiO$_X$ deposited by PECVD or a 10 nm thermal oxide. The anti-reflection coating (ARC) layer was formed by deposition of a 60 nm SiN$_X$ layer. The metallic contacts are patterned by photolithography. The front contacts were formed onto the silicon emitter surface with a stack of a 100 nm thick Al layer, a 50 nm thick Ti layer, a 50 nm thick Pd layer and 1 μm thick Ag layer. The backside contacts were achieved by stacking 50 nm of Ti, 50 nm of Pd and 1 μm of Ag. All cells were annealed at 350 °C for 30 min under forming gas (90%Ar and 10%H$_2$) ambient in a conventional furnace. The samples were then cut into small cells (about 1.5 cm^2) using a saw disk.

The cells were then analysed using I-V illumination set-ups (ORIEL spectrum and Sun-Voc Sinton) under AM1.5G radiation, as well as by spectral response measurements.

3 Results and discussion

The sheet resistance R_{sq} was measured for the p$^+$ front emitter made on FZ silicon as a function of implantation and annealing conditions. The results are shown in Figure 4. Annealing the plasma implanted samples at 900 °C results in a sheet resistance of about 120–140 Ω/sq. Higher energy E or higher dose D results in lower sheet resistances. On the other hand, an annealing at 950 °C for 30 min allows sheet resistance values as low as 70 Ω/sq.

As expected, a more efficient boron activation is obtained with high temperature annealing. Such low R_{sq} values are appropriate for solar cell fabrication.

Table 1. Extraction of V_{OC}, J_{SC}, FF and η from I-V measurement using Oriel and SunVoc (SINTON) setups.

Silicon	PIII Boron Emitter (IBS)	Passivation layer	Setup	V_{OC} (mV)	J_{SC} mA/cm^2	FF (%)	η (%)
Fz-Si-1	BF3:	PECVD	Oriel	593.1	38.8	70.1	16.1
200 μm	$E:8$ kV						
2 Ω cm	$D:5 \times 10^{13}$ cm^{-2}	$SiO_X + SiN_X$	Sun-Voc	582.0		75,2	16.9
Fz-Si-2	BF3:	PECVD	Oriel	596.9	38.7	69.3	16.7
200 μm	$E = 15$ kV						
2 Ω cm	$D = 5 \times 10^{15}$ cm^{-2}	$SiO_X + SiN_X$	Sun-Voc	586.0		75.0	17.7
Fz-Si-2	BF3:	Thermal	Oriel	559:1	36.2	63.4	12.9
200 μm	$E = 15$ kV						
2 Ω cm	$D = 1 \times 10^{16}$ cm^{-2}	$SiO_2 + SiN_X$	Sun-Voc	555.0		73.3	14.7
n-RST	BF3:	Thermal	Oriel	534.0	27.4	55.7	82
100 μm	$E = 15$ kV						
2 Ω cm	$D = 5 \times 10^{15}$ cm^{-2}	$SiO_2 + SiN_X$	Sun-Voc	532.0		68.6	10.0

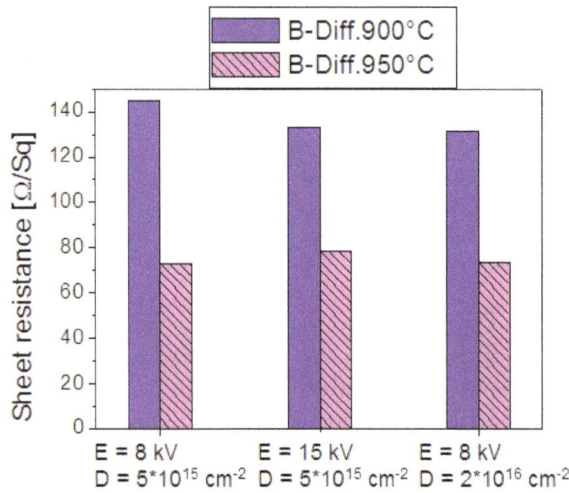

Fig. 4. Sheet resistance of boron implanted emitters as a function of implantation and annealing conditions.

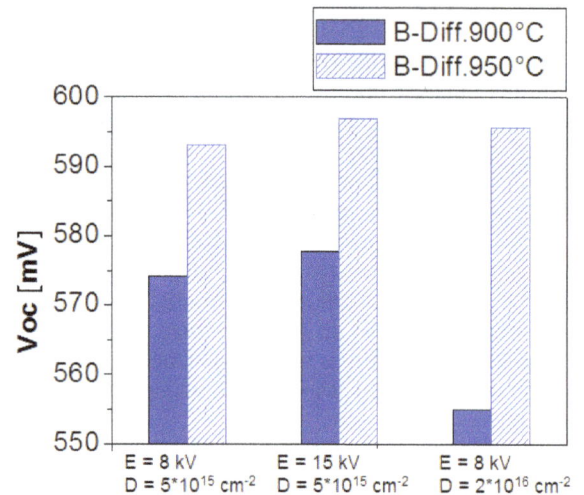

Fig. 5. Extraction of open circuit voltage of PULSION® boron emitter solar cells on FZ wafers.

Figures 5 and 6 plot open circuit voltages (V_{OC}) and conversion efficiencies (η) values as a function of implant conditions and thermal annealing for solar cells made on n-type FZ silicon wafers.

These data show that the best conditions to reach the highest values are an energy $E = 15$ kV and dose $D = 5 \times 10^{15}$ cm^{-2} combined with a thermal activation at 950 °C for 30 min in order to optimize the activation of the dopants. The best V_{oc} is approaching 600 mV and the efficiency is close to 17%. It has to be noticed that neither plasma immersion implantation nor annealing were optimised. Therefore much better solar cell performances were expected with these optimisations and the implementation of a surface texturization.

Table 1 gives the values of open circuit voltage V_{SC}, short circuit current density J_{SC}, fill factor FF, and efficiency η, extracted from the I-V characteristics under AM1.5G illumination.

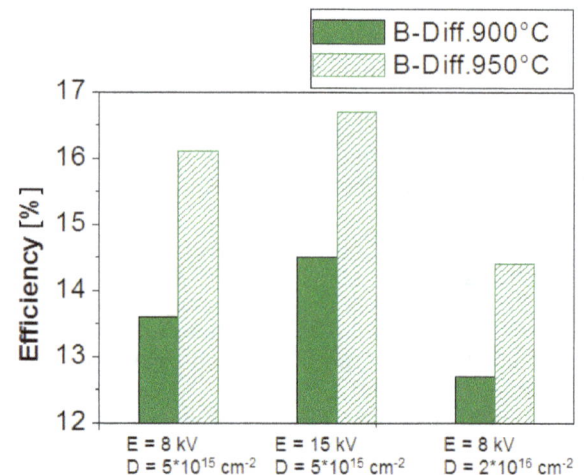

Fig. 6. Efficiency η of the PULSION® boron emitter solar cells on FZ wafers.

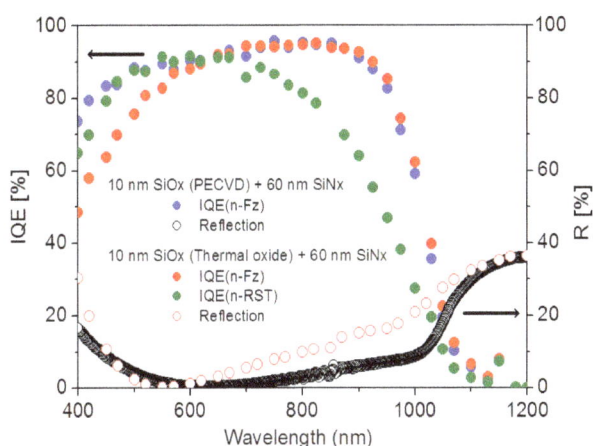

Fig. 7. Spectral responses (IQE and reflectance) on n-FZ and n-RST silicon.

The solar cells passivated with the SiO_Y/SiN_X stack performed by PECVD show a higher conversion efficiency compared to those passivated by the SiO_2(thermal)/SiN_X stack. These low performances are mainly caused by the implementation of an oxidizing annealing aiming to grow a thin passivating oxide. We have observed a degradation of the material lifetime due to aggressive cooling ramps of the annealing. We assume that, with such high impurity concentration, boron diffusion into the oxide during annealing should also be detrimental for the emitter passivation through the probable surface depletion of the boron emitter [5]. Further experiments are expected to improve the boron emitter passivation.

The RST-Si based cells suffer from a low FF which might be due to some shunts in the bulk of the material as well as due to the presence of residual SiC precipitates that partially shunt the emitter. As a result, the cell efficiency of the n-type ribbon based cell made by this method is about 8.2%.

To get a better insight on the limiting factors of the quality of the cells, spectral responses were measured. Figure 7 plots the internal quantum efficiency (IQE) and the reflectance of the n-type ribbon and FZ silicon solar cells. The strong drop of the IQE in the short wavelength range (below 600 nm) for both passivation conditions means that the passivation of the boron emitter is not good enough. The emitter region can obviously be improved by lowering the doping level at the surface while maintaining a good sheet resistance and by optimising the surface passivation of the emitter by depositing an Al_2O_3 layer.

In the near-infrared region, the response of the n-RST solar cell is quite poor compared to that of FZ silicon, which is probably due to a low minority carrier lifetime. The effective minority carrier diffusion length value (L_{eff}) deduced from the spectral response of the RST cell is of about 68 ± 5 μm, which is lower than the RST ribbon thickness (about 90 μm). This extracted diffusion length value corresponds to an effective minority carrier lifetime between 3 and 5 μs if we consider a diffusivity of about 10 cm^2/s. It was noticed that the back surface field insured by the phosphorus doped region should be likely to improve.

4 Summary

This work presents the perspective of forming boron emitter in n-type silicon by the promising technique of plasma immersion ion implantation. Required sheet resistance values can be obtained by choosing appropriate implant conditions and thermal budget for damage removal and electrical activation of boron. Solar cells with efficiencies approaching 17% on FZ silicon have already been achieved without specific optimisation of the operational parameters. The cells still suffer from low quality passivation. The first attempts to make solar cells on n-type ribbon silicon resulted in efficiencies a bit above 8%. We demonstrate that performances of such cells are limited by recombinations at the surface and in the bulk. The presented preliminary results using the PULSION® tool show that further improvement in ribbon material quality and cell processing, such as bulk hydrogenation, will allow cell efficiency to reach above 15% for a 100 μm thick ribbon using a very simple process.

Finally, the capability of PULSION® to implant Boron as well as Phosphorus could offer the perspective to reduce manufacturing costs by easily performing all doping steps followed by a single common annealing.

This work was partially funded by the *Agence de l'Environnement et de la Maîtrise de l'Énergie* (ADEME), France, under project DEMOS; the coordinator of the project is Solarforce.

References

1. European Photovoltaic Industry Association (EPIA), Solar Generation **6**, 20 (2011)
2. Solarforce S.A., Ribbon Technology, http://www.solarforce.fr/en/technology.html
3. C. Belouet, J. Cryst. Growth **82**, 110 (1987)
4. A. Gupta et al., in *Proc. 25th EU PVSEC, Valencia, Spain, 2010*, pp. 1158–1162
5. J. Benick et al., in *Proc. 33rd IEEE PVSC, San Diego, CA, USA, 2008*, pp. 1–5

28

Cell interconnection without glueing or soldering for crystalline Si photovoltaic modules

Johann Summhammer[a] and Zahra Halavani

TU Wien, Institute of Atomic and Subatomic Physics, Solar Cells Group, Stadionallee 2, 1020 Vienna, Austria

Abstract In order to maximize the power output of polycrystalline silicon PV-modules, in previous work we have already tested rectangular cells of 39×156 mm which are overlapped along the long sides. The low current density at the cell overlap allows interconnections which need neither soldering nor glueing, but use metallic strips inserted between the cells in the overlap region. The contact is established by the pressure applied to the module during lamination and is retained by the slightly bent cells in the solidified encapsulant. Here we report on the long term stability of different contact materials and contact cross sections applied in eight modules of the 240 W class monitored for up to 24 months of outdoor operation and in a variety of small 5-cell modules exposed to rapid ageing tests with up to 1000 thermal cycles. Cells with three different electrode designs were tested and the contact materials were Cu, Ag, SnPbAg and Sn. Focussing especially on series resistance, fill factor and peak power, it is found that Ag-coated contact strips perform equally well and have practically the same stability as soldered cell interconnections. Due to 70–90% savings in copper and simpler manufacturing the cost of PV-modules may thus be reduced further.

1 Introduction

Despite their relative maturity, solar cells and photovoltaic modules based on mono- or polycrystalline silicon still have potential for improvement, in terms of higher conversion efficiency as well as in terms of cheaper manufacturing [1]. Here we present experimental results on an interconnection method of solar cells, which permits higher power output per module area and may also lower production costs. Some preliminary results have been published before [2,3]. The basic idea contains two technological elements:

– First, the format of the solar cells is changed from quadratic to rectangular, the short side being much smaller than in commercial quadratic cells, and the cells are interconnected at the long sides. This leads to short electrical paths and reduces series resistance. It also reduces the linear current density to be transmitted to the next cell. This relaxes the constraints on contact resistance, which in turn allows to overlap the cells and place the interconnection in the overlap region, thereby eliminating shading from metallization and area loss from cell spacing. The advantages of the

rectangular format are already exploited in commercially available modules with half-size cells which show a notable decrease of series resistance [4], and in sliver cells [5, 6], which go to an extremely small ratio of width to length.

– And second, neighbouring cells are actually overlapped, but the interconnection is not made by soldering or the innovative and more gentle methods of glueing with conductive adhesives [7–9] or any other rigid method, but relies on pure pressure between the front and rear contact areas in the overlap region of two neighbouring cells. This is similar to the technology used in NICE-modules, although there it is applied to solar cells of the usual quadratic format without overlap [10, 11].

The cell format used here is 39×156 mm [12], which is one quarter of today's commercial standard of 156×156 mm. It is a compromise between the possible gains and the number of additional manufacturing steps like cell cleaving by laser and more individual cell handling. Only polycrystalline silicon solar cells were used, because of the expense and limited availability of full square 156×156 mm monocrystalline wafers.

The optimum benefit of cells of 39×156 mm can be obtained when their metallization pattern on the front and back side is adapted to the overlap interconnection

---
[a] e-mail: `summhammer@ati.ac.at`

method. Such special cells, which we called QuarterCells to distinguish them from quartered standard commercial cells, have been made for this investigation, but we have also employed quartered standard cells.

The reported experiments tested different kinds of overlap interconnections between the cells, all of which relied on pressure only. For this purpose so called contact strips were placed in the overlap region between two neighbouring cells. These contact strips had different cross sections and different conductive coatings. Although interconnection of metallic conductors by pressure only is a method used universally in power transmission (e.g., the cabling of buildings and machinery, sockets, switches and the multitude of plug connectors), one may doubt the reliability of such a form of interconnection for solar cells in a photovoltaic module. For instance, the contact area between the touching surfaces might decrease over time due to a relaxation of the contact force or a creep of the materials, or there might be slow oxidation or other detrimental chemical reactions of these surfaces. These effects could lead to higher series resistance and lower output power and perhaps even to an interruption of an interconnection. Therefore, the main purpose of this study was to find out how large this increase of series resistance and loss of power will be for different kinds of metallic surfaces and different shapes and thicknesses of the contact strips between the cells.

In order to obtain credible results in a reasonable time, two different sorts of modules were produced and investigated in parallel: small modules with only 5 cells each were subjected to rapid ageing in a climate chamber. And large modules with a nominal power around 240 W, which consisted of either 246, 252 or 276 cells connected in series, were mounted outdoors and their IV-curves monitored permanently for up to 28 months, at present (the experiments are still continued). These modules had operating voltages around 120 V and currents in the range of 2 A or more. Such high voltages of a single module have advantages for use of single-module inverters or for system-layout in building integration, as we have argued before [2, 3, 11], but the results to be presented here will focus on the physical and technical aspects of the cell interconnection only. In the following sections we describe the technology of the modules, present the data on rapid ageing of the small modules and on the outdoor ageing of the large modules, and analyze the results in relation to the properties of the contact materials between the cells and to other module parameters. We begin with a discussion of the characteristics of the specially made QuarterCells.

2 Concept of QuarterCells

We introduced a first version of the concept of Quarter-Cells for crystalline silicon photovoltaic modules already in reference [12] and the version for overlapping of cells in reference [3], both with the aim of increasing the output power of the modules without changing the basic cell technology. The idea was initiated by the realization that the

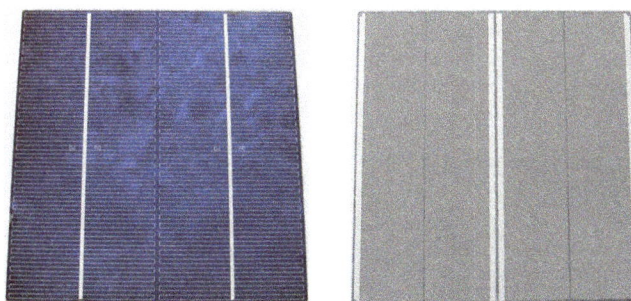

Fig. 1. Front and back side of a 156×156 mm polycrystalline silicon solar cell to be cut into four QuarterCells. The cuts by laser are done in the three non-metallized stripes at the back, and do not go through the wafers. The final separation is done by breaking along the laser lines. This also breaks the screen printed front bus bars in two (several types of metallization have been tried, the one shown here is called QuarterCell "P").

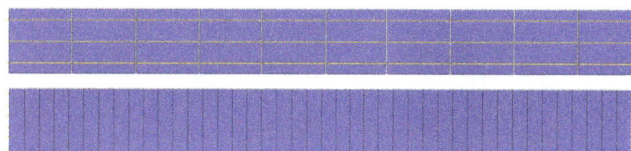

Fig. 2. Standard string of 10 cells with three busbars and string of similar length with 42 QuarterCells (overlap of cells: 2 mm).

industrial standard cells of 156×156 mm have high current and small voltage, which – even after optimizing the trade-off between minimum shading and minimum electrical series resistance – leads to appreciable optical and ohmic power losses. By slightly adapting the metallization pattern and cutting such a cell into several equal slices, and then connecting these slices in series along the *long* sides, it is possible to reduce these losses significantly. Especially, if these cells have a thin front bus bar and a similar rear bus bar at opposite long sides of the cell [13]. For our studies we have chosen to make four rectangular cells of 39×156 mm out of a quadratic cell and have called them QuarterCells. A quadratic cell from which the four pieces are cut is shown in Figure 1. The QuarterCells can be connected in an *overlap fashion*, such that the rear bus bar of a cell lies directly on top of the front bus bar of the previous cell. A comparison of a string of ten standard cells with three busbars to a string of similar length made of QuarterCells (Fig. 2) shows that the latter can yield about 6.3% more power under standard testing conditions (STC), when normalized to the same string length. The various contributions to this gain in power and the underlying assumptions are listed in Table 1. But the interconnection of the QuarterCells must not be rigid when such strings are laminated into a module, because solar glass and silicon have different thermal expansion coefficients. In order to solve this problem, one can think of different ways of soldering or glueing to interconnect the overlapped cells. The method we investigate here is pure pressure applied to the contact areas of the cells.

Table 1. Power losses of standard string and string with QuarterCells, relative to power theoretically obtainable from the string area if there were no metallization shading and no ohmic losses in the metallization. Normalized to same string length. Assumptions for standard string: Cell spacing 2 mm, cross section of Cu solder ribbons: 0.3×1.5 mm^2.

	String with 10 standard cells of 156×156 mm with three busbars	String with 42 QuarterCells of 156×39 mm, overlap 2 mm
Area loss by bus bars	2.9%	0.0%
Area loss by cell spacing	1.3%	0.0%
Ohmic loss in bus ribbons at STC	2.3%	~0.0%
Ohmic and area loss because of additional metal fingers	n.a.	0.2%
TOTAL	6.5%	0.2%

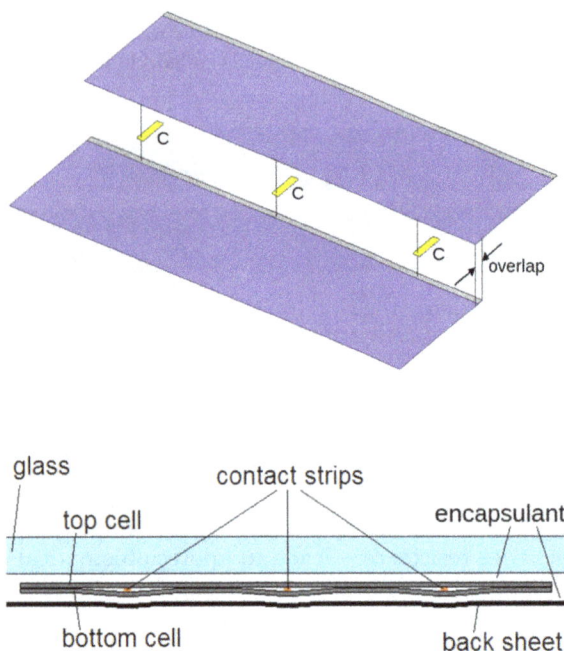

Fig. 3. Top: scheme of contacting two QuarterCells by overlap and with contact strips. Bottom: cross section through finished module at the overlap of two cells. In the overlap region the cells bend slightly around the contact strips.

3 Tests in this study

The technology investigated here is a method of interconnecting such overlapped cells by pressure only. The essential elements are small metallic contact strips which, in the overlap region, are placed on the front bus bar of the lower cell and which touch the rear bus bar of the next cell when it is placed on top (Fig. 3, top). The force for good contact is built up during lamination into a module. It is important that the module is of the type glass – encapsulant – cells – encapsulant – back sheet. The pressure exerted on the elastic back sheet during lamination bends the cells slightly around the contact strips, and these bends become permanent, if the lamination pressure is kept on during the cooling phase of the lamination until the encapsulant has sufficiently hardened. The permanent bends of the cells create sufficient contact force to ensure

good electrical connection between the cells (Fig. 3, bottom). Since there is no rigid connection, the cells can still slide against each other, and in this manner the string will follow the thermal expansion or contraction of the solar glass during changes in temperature, or when the module is exposed to mechanical loads, with little or no mechanical stress on the solar cells themselves.

The concept of connecting cells in this manner is not restricted to the type of QuarterCells shown in Figures 1 and 3, but can also be applied to normal cells, which are cut into four (or any other number) equal slices perpendicular to the bus bars and then connected in an overlap fashion. This has also been investigated in this work.

In order to assess the feasibility of this type of interconnection, a number of small and large modules were produced. Since this was an explorative study, and as no similar study seems to have been done before, only general information on the functioning of electrical contacts could be relied on. For instance, a mechanical electrical connection is usually established through many small holes in the oxide layers between the two contacting surfaces and the number of these holes increases with contact force [14]. This suggested to use malleable metals and to create a rather strong curvature of the cells. Therefore the various small and large modules differed by the materials used in the contact strips, their width and thickness, as well as by the number of contact strips between two cells, and by the type of cells used. In several cases only one module with a specific material, width and thickness of the contact strips was made. This was based on the reasoning that, since there are many identical contact strips in a module, already a single module's behaviour should provide valuable information on that type of contact strip.

For all modules, except one small module, the encapsulant was EVA. The small modules consisted of 5 Quarter-Cells each, or of 5 slices of 39×156 mm cut from standard cells perpendicular to their busbars. The large modules were of the 240 W class, and consisted each of six strings, a string containing between 41 and 46 QuarterCells (or quarters of cells cut from standard cells). In each of the large modules, all strings were connected in series, which resulted in operating voltages around 120 V. The small modules were subjected to rapid ageing tests by temperature cycling. After about every 50 cycles their IV-curves

Table 2. Cell types and contact strips used in the small modules. The first column shows pictures of the front and back side of the cells used. Symbols: d, thickness of the coating of the contact strips. T, W, thickness and width of the contact strips, respectively. N, how many contact strips were used between two neighbouring cells. $R_{s,in}$, initial series resistance, determined from a fit of the theoretical IV-curve of the one-diode model to the data of the IV curves measured right after production of the modules. *, "tinned iron", a core of Fe with a thin layer of Sn, whose thickness was below 1 μm.

| Cell type | Module name | Contact strips | | | | $R_{s,in}$ [mΩ] | Number of temperature cycles |
		Coating d [μm]		$T \times W$ [mm]	N		
Standard	S1	SnPbAg	15	0.15×2.5	2	193	760
	S2	SnPbAg	15	0.20×2.0	2	80	760
	S3	Cu		0.20×2.5	2	165	796
	S4	Ag	$\sim$1	0.30×2.2	2	95	546
QuarterCell K	K1	Ag	6.5	0.19×5.0	4	114	1006
	K2	Ag	6.5	0.19×5.0	4	92	1006
	K3	Ag	6.5	0.25×5.0	3	113	493
	K4	Ag	$\sim$1	0.30×2.2	3	296	546
	K5	Cu		0.20×5.0	3	226	796
QuarterCell P	P1	SnPbAg	6.5	0.05×3.5	7	65	685
	P2	SnPbAg	6.5	0.05×3.5	7	66	685
	P3	Ag	6.5	0.05×5.0	7	44	685
	P4	Ag	6.5	0.05×5.0	7	54	685
	P5	Ag	6.5	0.19×2.5	3	36	796
	P6	Sn on Fe*	$<$ 1	0.20×2.0	3	129	350

were recorded under STC. Although rapid ageing in the laboratory does not necessarily show the same degradation mechanisms as real outdoor wheathering, they are usually a fast indicator of possible failure modes [15]. The large modules were put into outdoor operation and IV-curves were recorded every two minutes.

4 Rapid ageing tests of small modules

An overview of the investigated small modules is shown in Table 2. As can be seen there, three different kinds of cells were used. One type were quartered standard cells with two bus bars. The other two types were two variations of QuarterCells. One was the QuarterCell of type P, already shown in Figure 1. The other one was called QuarterCell of type K. Its front metallization was the same as that of type P, but at the back side the bus bars consisted of 6 pads of 2×17 mm^2 and started only 3 mm from a long edge (see the pictures in Tab. 2). All cells were from p-doped polycrystalline silicon of a thickness between 220 and 240 μm with an n-doped emitter and the metallization was made by screen printing. They were produced by the company Falconcell (operations ceased in 2013). The contact strips were made from different foils, which usually consisted of a core of copper plus a coating (except module P6, see Tab. 2). Since the coating is the relevant material, its thickness is listed separately in Table 2. The only type of contact strips which had no coating were the pure copper contact strips. More details of the contact strips will be given in the presentation of the individual modules. The front glass of the modules was normal 2 mm float glass. As encapsulant two sheets of 0.5 mm thick EVA were used, one on each side of the cell string, except for module S4, where the silicone-based Tectosil by Wacker

was used. Further tests with Tectosil showed that, since it becomes very liquid during lamination, either the pressure is not fully conveyed to the cells or the contact strips obtain a thin insulating film, and no reliable cell interconnections could be produced. Therefore no additional modules with Tectosil were made. The back sheet of all modules was 0.17 mm thick Tedlar and usually black, in some modules it was transparent. The lamination pressure was 820 mbar, except for modules P1 and P3, where it was only 620 mbar.

4.1 Small modules with quartered standard cells

With this type of cell four small modules were investigated (S1–S4 in Tab. 2). The overlap of the cells was 2 mm. The contact strips were placed along the bus bars and were allowed to stick out on the front side of the cells by 2 to 3 mm, as can be seen in Figure 4b. The length of the contact strips was between 12 and 20 mm (depending on the material and form used) so that there was usually much more contact area with the rear bus bars than with the front bus bars. The connections at the ends of the modules were done by soldering SnPbAg-coated solar ribbon with copper core to the first and the last cell. Since there was no mechanical connection between the cells, as well as between the cells and the contact strips, they were held together by high temperature adhesive tape (PPI 1040 W, thickness 50 μm) as is visible in Figure 4c. Module S1 had contact strips made from ordinary solar soldering ribbon with copper core and SnPbAg-coating, but they were not soldered to the cells. Module S2 also used contact strips from such ribbon but of different cross section. Module S3 used pure copper strips and module S4 used copper ribbon with a thin galvanic layer of silver (supplied by the company Ulbrich of Austria).

(a)

(b)

(c)

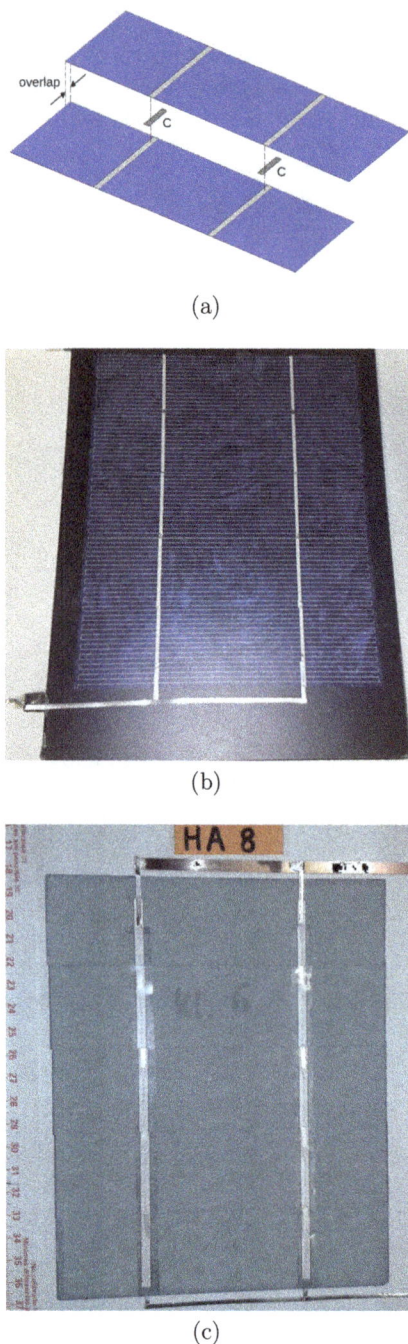

Fig. 4. (a) Interconnection scheme used in all small modules with quartered standard solar cells (S1–S4). (b) Front side of module S2. (c) Back side of module S4 before lamination.

4.2 Small modules with QuarterCells K

Here, five modules were investigated (K1–K5). Modules K1 and K2 were made in identical manner. Since the QuarterCells K had their back side bus pads relatively far from the edge (see photograph in Tab. 2) the cell overlap in these two modules was 5 mm, so that the front bus bar of one cell was directly under the back side bus bar of the next cell (Fig. 5a). The contact strips were made

from silver coated copper ribbon with a cross section of 5.00×0.05 mm^2, the thickness already including 6.5 μm of silver on each side (supplied by Schlenk Metallfolien). Pieces of about 30 mm length were folded once to obtain strips of 15 mm length. These were then folded again at 2 mm from the bent end to obtain an end section of 0.19 mm thickness (after pressing), a length of 2 mm and a width of 5 mm. These served as the actual contact pieces between the cells. The rest of the length of these strips was needed to attach these pieces to the back side of the cells with adhesive tape. Figure 5a shows the connection scheme and Figure 5b the back side of one of these modules before lamination.

Modules K3, K4 and K5 used a cell overlap of only 2 mm. Therefore front and rear bus bars of neighbouring cells were not on top of each other. In the finished module this would lead to pressure of the contact strips onto the front bus bars, but not on the rear bus pads. In order to get contact pressure there, too, the contact strips received an embossing at a position which lay exactly on the rear bus pads as shown in Figure 5c. Since, at the back side of the cells, the contact strips were covered with adhesive tape, after lamination this tape together with the EVA encapsulant and the back sheet had a small deformation rising above the smooth surface of the module's back side and this exerted a force on the contact strip and kept it pressed onto the bus pads. This method is a deviation from the original idea of forming contact by placing front and rear bus bars of two neighbouring cells right on top of each other with a contact strip in between. However, it was tried out, because this option offered the possibility to interconnect the QuarterCells K with only 2 mm overlap instead of 5 mm (the QuarterCells of type K were originally developed for soldered cell interconnections similar to the US-patents [13], in a project with company Kioto Photovoltaics).

The contact strips of K3 were made from the same silver coated copper ribbon as in modules K1 and K2, but the ribbon was folded to obtain five-fold thickness, such that the dimensions of the contact strips were $0.25 \times 5 \times 12$ mm $(T \times W \times L)$ after pressing, and then the embossing was made. Module K4 used copper contact strips with a thin galvanic layer of silver (the same silver coated copper ribbon as in module S4 was used). Module K5 used contact strips made from pure copper ribbon.

4.3 Small modules with QuarterCells P

The back side bus bar of these cells was right at the edge of a long side and was uninterrupted (see photograph in Tab. 2). This permitted the special kind of interconnection shown in Figure 6a. It was applied in the modules P1, P2, P3 and P4. The overlap of the cells was 2 mm.

The special connector consisted of two strips of adhesive tape, which were laid on top of each other with the non-sticky sides facing each other. Then V-shaped pieces of contact ribbon were wrapped around them and stuck to the tapes. This connector was then first pressed onto the front bus bar of one cell and thus stuck to it. Then the

(a)

(b)

(c)

(d)

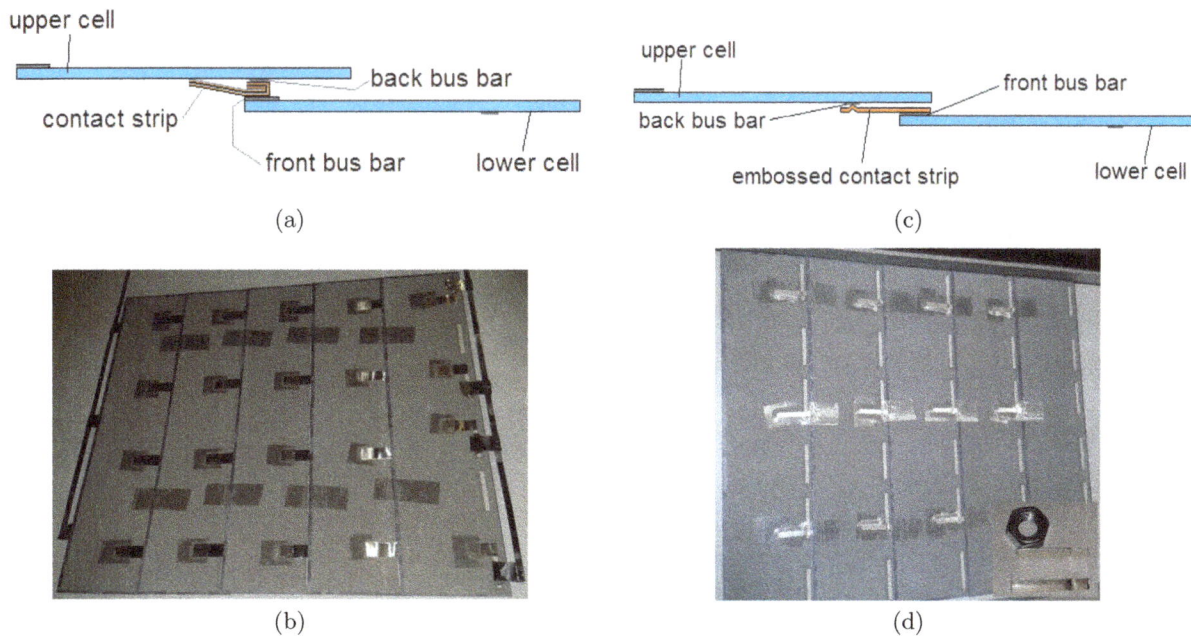

Fig. 5. (a) Side view of connection scheme of two cells as used for modules K1 and K2 (not to scale). (b) Back side of module K1 before lamination. At the ends of the module solar ribbon was soldered to the cells. (c) Side view of connection scheme of two cells as used for modules K3, K4 and K5 (not to scale). (d) Back side of module K4 before lamination. The insert in the lower right corner shows contact strips with embossing.

second cell was laid on top, and also stuck to the connector. In this manner, the cells were already electrically connected and held in place, but they could still slide against each other a little bit, because the bends in the V-shaped contact strips would allow a little motion. Once the cells were embedded in the module, the cells were slightly bent around the contact strips, which caused additional contact force. As will be seen, this type of connection showed the lowest series resistance and suffered virtually no deterioration in hundreds of temperature cycles. For modules P1 and P2 the contact strips were made from solar ribbon of 50 μm thickness, with a 37 μm copper core and 6.5 μm SnPbAg coating on each side. The width of the ribbon was 3.5 mm. (Ribbon supplied by Schlenk Metallfolien.) For modules P3 and P4 the same Ag-coated copper ribbon was used as in other modules before (e.g. K1, K2), but it was not folded, and its thickness was thus only 50 μm.

The modules P5 and P6 used straight contact strips as shown in Figure 6c. Those of P5 were Ag-coated. Those of P6 were made from tinned iron, the thickness of the Sn coating was not determined exactly but was below 1 μm.

4.4 Results of thermal cycling of small modules

The rapid ageing tests were done by repeated thermal cycling. The available climate chamber permitted heating to about 90 °C and cooling down to –26 °C. The first 400 cycles were done between 85 °C and –26 °C. Then it was changed to the range 87 °C to –24 °C. The temperature was monitored in four different positions in the chamber. With a fan it was held homogeneous to within +– 0.7 °C. One cycle lasted a little over 8 h. The rise and fall

of temperature were exponential, the operational modes of the climate chamber being either constant heating or constant cooling. The modules were open circuited during the cycles. After about every 50 cycles the IV-curves of the modules were recorded under STC. Since the modules were added to the climate chamber as soon as they were produced, not all modules have undergone the same number of cycles to date. The module with the smallest number is P6 with 350 cycles.

There are several ageing effects to be expected. Aside from deficiencies of the embedding material, like diffusion of water vapor [16] or yellowing of the EVA [17], one could expect an increase in series resistance of the cells due to abrasion or slow oxidation or other chemical reactions of the contact strips. This might go as far as complete loss of electrical contact between cells. At the bends of the cells around the contact strips microcracks might develop, which could lead to smaller parallel resistance of the cells and module [18]. Cells might also obtain macroscopic cracks in these regions, which could reduce the short circuit current [19]. The analysis of the IV-curves revealed, however, that all effects are negligible, except for an increase in series resistance which strongly depends on the material of the contact strips. The increase of series resistance entails a decrease in fill factor and thus a decrease of power output. Figures 7 and 8 show this decrease for the three different types of cells. In the following the ageing effects will be discussed separately for each coating material of the contact strips, because this is the dominant determining factor, independent of cell type.

SnPbAg: The strongest deterioration is observed on modules with SnPbAg-coated contact strips which were

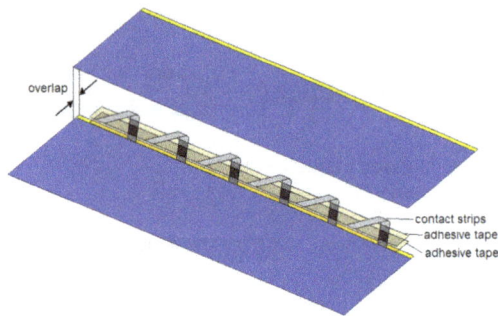

(a)

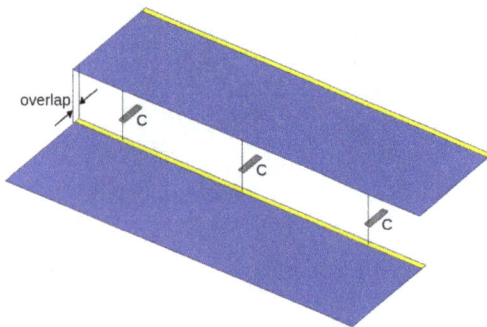

(b)

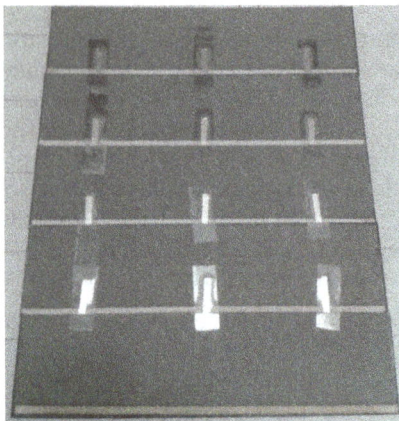

(c)

Fig. 6. (a) Connection scheme for modules P1, P2, P3, P4. (b) Connection scheme for modules P5 and P6. (c) Back side of module P5 before lamination.

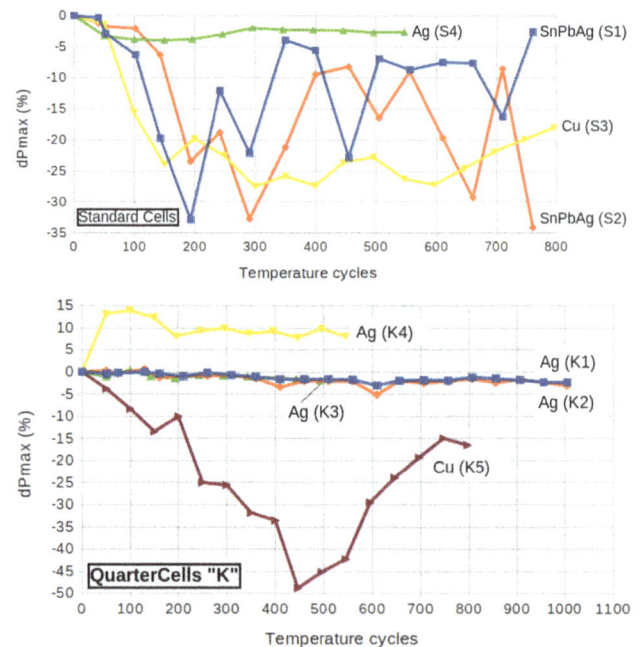

Fig. 7. Relative decrease of power output (dPmax) under STC for small modules as a function of the number of temperature cycles. Top: modules with quartered standard cells. Bottom: modules with QuarterCells K. The labels indicate the coating of the contact strips and the name of the module.

made from ordinary solar ribbon. This kind of contact strips was used for two modules with quartered standard cells (S1, S2) and for two modules with QuarterCells P (P1, P2). In all four modules the power loss reaches 5% or more after only around 100 cycles, and from then on it deteriorates fast and in an erratic manner with occasional recovery. This is probably caused by abrasions of the relatively soft SnPbAg (Brinell hardness 16–17 MPa) on the harder surfaces of the screen printed bus bars when the cells slide against each other (the busbars are mainly of sintered silver particles, Brinell hardness 206–250 MPa). The repeated contractions and expansions of the front

glass of the module are enforced on the cells, because the glass is much thicker than the cells, and each cell can be considered attached to the glass at its midpoint due to the encapsulant. This model has been found to agree reasonably well with observed cell displacements during thermal cycling [20, 21]. To verify the existence of scratches in the SnPbAg-layer, after the thermal cycles module S1 has been opened at two different positions to take out the contact strips and inspect them with a microscope (Fig. 9). In Figures 9b and 9c one clearly observes scratches parallel to the length of the strips where the edge of the top cell pressed onto them. Further to the right, in the 2 mm long overlap zone, one also finds scratches in the same direction. Since they are not visible in the whole overlap zone, one may infer that the contact strips were not pressed flat completely during the lamination process and that, in particular, the approximately 0.15 mm thick copper core of these strips was too hard. With a thermal expansion coefficient of the used plate glass of 9×10^{-6} [22] and of 2.6×10^{-6} for silicon [23] a theoretical length of the scratches of about 25 μm could be expected, caused by the temperature range –24 °C to +87 °C. However, the scratches tend to be up to 2 or 3 times as long. We attribute this to the much larger expansion coefficient of the Tedlar back sheet (28×10^{-6} [24]) and a similarly large expansion coefficient, but also a material transition below about 0 °C of EVA [25], which caused actual bending and re-straightening of these small modules during the thermal cycles.

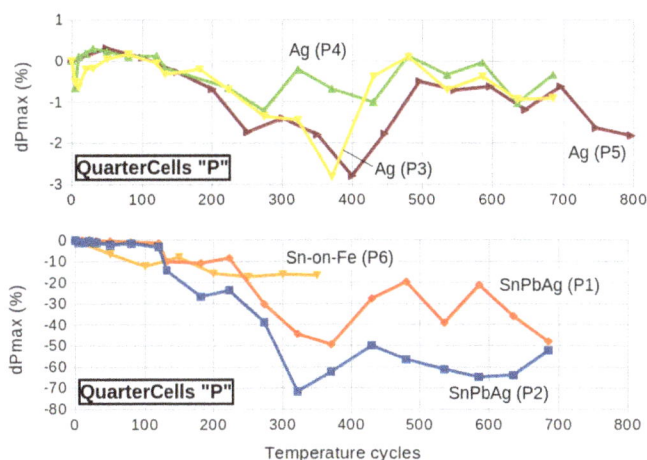

Fig. 8. Relative decrease of power output (dPmax) under STC for small modules made from QuarterCells P as a function of the number of temperature cycles. Top: modules with Ag-coated contact strips. Bottom: modules with other contact strips. The labels indicate the coating of the contact strips and the name of the module. Note that the top graph has a much narrower range of dPmax.

In principle, the contact resistance between the contact strips and the contacted areas of the cells should decrease with pressure [14]. But a relationship of the power loss with stronger contact force is not apparent, as module S2 tended to deteriorate faster than module S1, although it had thicker contact strips and therefore more strongly bent cells resulting in higher contact force. Similarly, module P2 performed worse than the identical module P1, although the former was laminated with 820 mbar and the latter with only 620 mbar. Here it must be mentioned, however, that the pressure during lamination may not be so important, as long as it is sufficiently high to bend the cells around the contact strips. Of more relevance may be the high pressure which the cells should experience from the contraction and hardening of EVA in the cool periods of the thermal cycles [26, 27]. As evidenced by the scratches, the rough surface of the bus bars causes permanent indentations on the SnPbAg coating, and these could become so deep, that in the relaxed state of EVA at STC-temperatures and above the SnPgAg coating formed only loose contact to the cells and actually *decreased* the contacted area.

Copper: Pure copper contact strips were used in module S3 with quartered standard cells and in module K5 with QuarterCells K. Since copper has a higher hardness than silver or SnPbAg (Brinell hardness 235–878 MPa) one could expect that these contact strips adapt less to the rough surface of the sintered bus bars during lamination, resulting in smaller contact area and higher initial series resistance. This is true for K5, but for S3 only in tendency, compared to the similar modules with SnPbAg, S1 and S2. Then, during thermal cycling, the two modules show a fast decrease of power, but it is less erratic than with modules with SnPbAg-coated contact strips. One cause may be permanent indentations, in this case into the softer silver

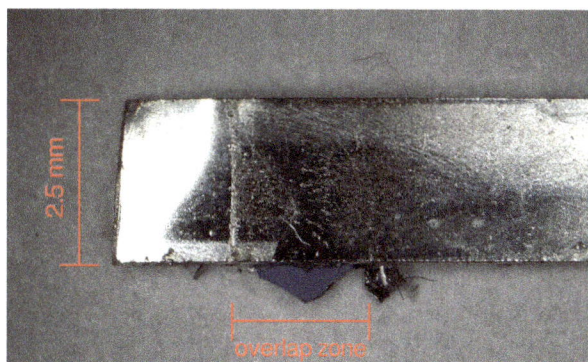

(a)

(b)

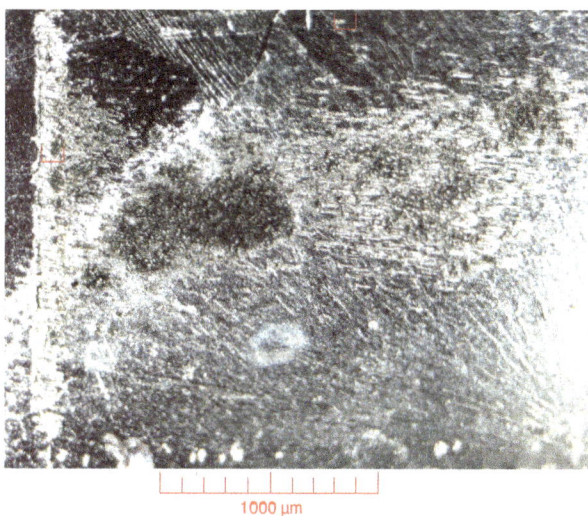

(c)

Fig. 9. (a) Top view of a contact strip from module S1 consisting of 0.15 mm Cu-core with 15 μm SnPbAg solder coating. The region where it was pressed on by the back side of the cell lying on it starts about 2 mm from the left side. Scratches are visible. (b) Enlarged view of the contact region with scratches. (c) Enlarged top view of another contact strip, with visible scratches.

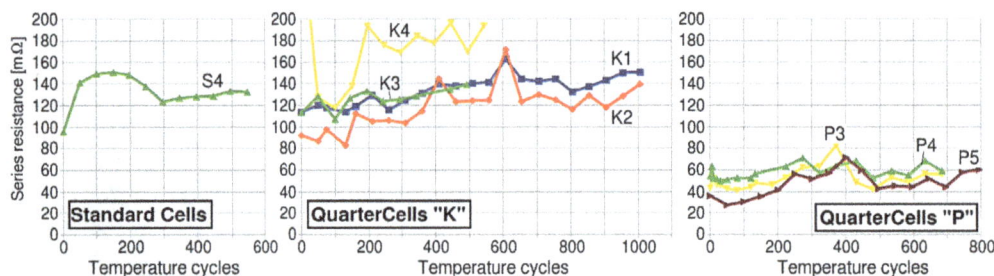

Fig. 10. Series resistance of all small modules with Ag-coated contact strips as a function of the number of temperature cycles. The contact strips of modules S4 and K4 were thicker and had thinner Ag-coating than all the others (see Tab. 2). Modules P3 and P4 had the connector with the largest contact area, shown in Figure 6a.

particles of the bus bars, during the cold stages of the thermal cycles, which then resulted in areas of no contact at ambient temperatures. Another may be oxidation, as Cu may react with acidic by-products in EVA [28]. However, current flow through oxide layers is unlikely. Both CuO, which forms first, and Cu_2O, which makes up the bulk of the oxyde layer, are semiconductors [29], and the conductivity of Cu_2O is found in the range of 120 [30] to several thousand [31] Ω cm. Assuming 1000 Ω cm, an oxide layer as thin as 10 nm, and a contact cross section as large as 4 mm^2 would result in a resistance of 250 mΩ at one interface. This is too high to account for the observed increases in series resistance, so that direct contact of the metals through asperities in the oxide layer must be the path of conductance [14]. With module K5 the very strong decrease of almost 50% in output power and subsequent steady increase to 16% after 800 cycles may have to do with the fact that this was one of the modules where the contact force on the back side of the cells only originated in the bends in the adhesive tape and the back sheet and the probably inhomogeneous distribution of EVA. The repeated softening and hardening of these materials may have led to additional rubbing of the Cu contact strips on the bus pads, which must have cured the contact properties by creating more asperities in the oxide layers. A small effect of curing seems to be present in S3, too. So, in both cases a change of the elastic properties of EVA may have played a role [32, 33].

Sn on Fe: There was only one module with contact strips made from iron coated with Sn (P6). The choice of this material was based on long term corrosion resistance of Sn [34]. Although data of only 350 cycles are available until now, one observes a slower deterioration than with pure Cu contact strips or with SnPbAg-coated contacts strips. Also, the deterioration is less erratic than with these materials. Nevertheless, it is strong enough to exclude this type of contact strips from further considerations.

Silver: The eight modules which used Ag-coated contact strips show the smallest loss of power, and it is very similar for all three kinds of cells. Since these were the best performing small modules, the evolution of their series resistance values is also shown in Figure 10. Interestingly, in some of these modules (P3, P4, P5, K1, K4) the power actually increases during the first 100 thermal

cycles and from then on decreases only slightly. The series resistance exhibits a corresponding inverse tendency. This behavior may be due to the adaptation of the Ag-coating of the contact strips to the morphology of the screen printed bus bars and may have led to increased direct metal-metal contact area during the first cycles. The effect is particularly strong for module K4, which used QuarterCells K and was one of those, where the contact pressure on the rear bus bars was only caused by the elasticity of the adhesive tape, the EVA, and the back sheet (see Fig. 5c). The thermal cycles might have lead to additional shrinkage of the EVA [27, 33], thereby increasing contact force. Aside from these mechanical aspects, the chemical properties of silver also favor a good performance of modules with Ag-coated contact strips. Silver does not react directly with O_2, but with atomic oxygen, thus slowing down the growth of a semiconducting Ag_2O layer to only some 1.2 nm in 4 to 5 hours in ambient air, while during the same time Cu would develop a layer of Cu_2O four times as thick [35]. Moreover, the electrical resistivity of Ag_2O is only in the range of 4.2×10^{-4} to 5.2×10^{-3} Ω cm [36]. This is at least a factor of 10^{-5} less than the value for Cu_2O mentioned before. In air, silver also reacts with sulfur-containing molecules to a tarnish of semiconducting Ag_2S [37], which has a room temperature resistivity around 10 Ω cm, and which decreases with increasing film thickness [38]. This resistivity is still a factor of 0.1 to 0.01 less than that of Cu_2O. Although one can assume that, here too, the electrical conduction from contact strips to busses proceeds mainly through small holes in the oxide and sulfide layers where metal-metal contact is established [14], the good conductivity of Ag_2O and Ag_2S is very advantageous.

When asking which type of cells and which kind of interconnection performed best, the answer clearly is QuarterCells P with the special interconnector shown in Figure 6a and with contact strips with Ag-coating. The initial values of series resistance of modules with this type of connector are distinctly among the lowest of all modules (P3, P4, see Fig. 10). As can be seen in Table 2, this was also the case for the two modules with this type of connector whose contact ribbon was coated by SnPbAg (P1, P2). The most likely explanation is that this type of connector offers the largest contact area. Incidentally, it also leads to the smallest bending of the cells, and this might reduce the

Fig. 11. Modules mounted for outdoor monitoring at south west facade of Institute of Atomic and Subatomic Physics at TU Wien. The outer dimensions of modules M1-M7 are 1629×982 mm, those of M8 are 1564×982 mm.

contact force. But with this connector the contact force to the busbars is to some degree also generated by the adhesive tape, which meanders up and down, and its tendency to straighten itself creates additional contact force. There is one module with Ag-coated contacts (P5), which had a still lower initial series resistance and did not have this special connector, although the difference is in the range of the reproducibility of the used method of extracting the series resistance from the IV-curve. This module used only three contact strips (Fig. 6c). What may certainly have helped to reduce the series resistance of this module is that, in the regions between the contact strips, the front and rear busbars of any two overlapped cells touched each other over several centimeters and this yielded a large additional contact area between the sintered silver particles. This can be visualized with the help of the lower graph of Figure 3 and with Figures 6b and 6c.

5 Outdoor tests of large modules

Eight large modules with cell type of either quartered standard cells or QuarterCells K and different kinds of contact strips were produced and mounted with an inclination of about $10°$ from vertical at a south west facade of the institute (Fig. 11). Unfortunately, QuarterCells P, to which the special connector (Fig. 6a) could be applied and which showed the best results with the small modules, were no longer available. As with the small modules, the cells for the large modules were produced by Falconcell, except those of module M6, which used cells with 3 busbars cut from quadratic cells from Gintech. One module did not use contact strips, but its cells were connected by soldering with SnPbAg-coated solder ribbon (M4). This module could serve as a reference, since it could be expected that the interconnetions would show little or no degradation. Cutting of cells and manufacturing of strings was done by the authors. Lamination and framing with aluminium profiles was done by the company PVT-Austria. Just as with almost all small modules the encapsulant was EVA. However the front glass was 4 mm thick hardened solar glass with random structured surface for reduced reflection and the back sheet was from a softer PE-based material and not the Tedlar of the small modules. The colour of the back sheets was black, except for the reference module M4, where it was white. As soon as a module was finished it was mounted on the stand. This is why they have different start dates and different periods of exposure until now. The operational cycle of the modules consisted in alternating open-circuit periods with short-circuit periods. Each of these periods lasted for several days to weeks. During the open-circuit periods current-voltage curves of each module were recorded every two minutes from sunrise to sunset. Table 3 gives an overview of the technical parameters of the modules.

The typical degradation of the current voltage curves over periods from 12 to 18 months are shown in Figure 12 for four different modules. The curves have been selected with the same short circuit current (1.2 A). Since for a given module the temperature of the initial curve and of the second curve from the later date were not the same, the open circuit voltages of the two curves to be compared were also different. Fortunately, this has only a small influence on the values of the fill factor, which are quoted in the figures. A recent review of theoretical temperature dependences is given in [39]. The expected temperature coefficient of the fill factor for crystalline silicon solar cells has been evaluated between -0.06 and $-0.09\%/K$ (percentage relative to the value of the fill factor) [40]. In practice, a temperature coefficient of $-0.19\%/K$ has been found for a small module of two silicon solar cells connected in series [41]. This value has been adopted for the calculation of the temperature-corrected fill factor of the later date for each module, which is quoted in brackets in Figure 12. For these corrections the temperature coefficient of the open circuit voltage was also needed, and the field value of $-0.5\%/K$ obtained in [42] was used.

The essential information in Figure 12 is the change of slope of the IV-curve near the open circuit voltage. This section is more or less a straight line. The inverse of its slope has the dimension of resistance. Although this is not directly the series resistance, one can easily verify with the one- or two-diode model that a change of the series resistance of the module causes a numerically almost identical change of the inverse of this slope. Keeping this in mind, the changes in Figure 12 between initial and final curves show an increase in series resistance of all four modules and associated change in fill factor, although

Table 3. Technical parameters of the eight large modules for outdoor measurements. "2BB": made from standard cells with 2 busbars. "3BB": made from standard cells with 3 busbars.

Module	Number of cells	Cell type	Contact strips coating, $T \times W$ [mm], number	Most similar small module
M1	252	Standard, 2BB	SnPbAg, 0.20×2.0, 2	S2
M2	252	Standard, 2BB	Ag, 0.19×2.5, 2	–
M3	252	Standard, 2BB	Ag, 0.24×2.5, 2	–
M4	252	Standard, 2BB	Soldered reference 0.10×2.0, 2	–
M5	276	QuarterCell K	Ag, 0.19×5.0, 4	K1, K2
M6	252	Standard, 3BB	Ag, 0.24×2.5, 3	–
M7	252	Standard, 2BB	Cu, 0.20×2.5, 2	S3
M8	246	QuarterCell K	Ag, 0.25×5.0, 3	K3

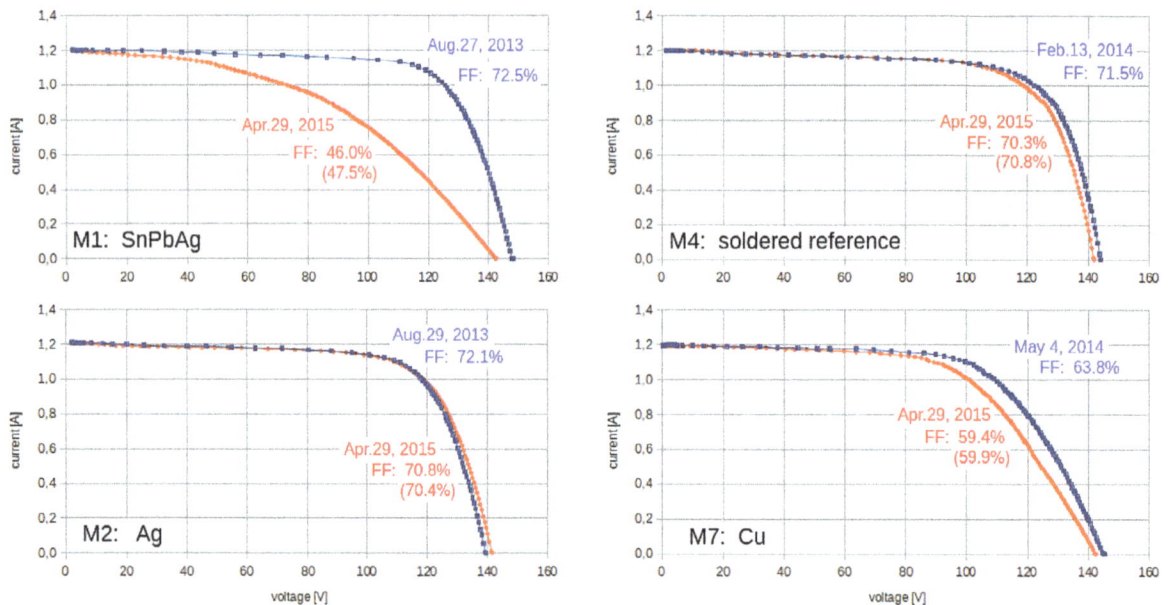

Fig. 12. IV-curves with short circuit current of 1.2 A for modules M1, M2, M4 and M7 at their first day in operation outdoors (blue) and after operation between 12 and 18 months (red). The different open circuit voltages of first and respective last day are almost solely due to different temperatures at the two days. For the later dates the measured fill factor (FF) and the fill factor corrected for the different temperature and different open circuit voltage are given (the latter one in brackets).

to quite a different degree, depending on the kind of cell interconnection. As could be expected from the results of the small modules, the module with SnPbAg-coated contact strips (M1) showed the strongest deterioration of series resistance. Also the module with contact strips from pure Cu (M7) showed quite a considerable increase in series resistance. On the other hand, the module with Ag-coated contact strips (M2) suffered little deterioration. And so did the reference module (M4), for which hardly any deterioration was expected, because it had soldered cell interconnections.

Figure 13 gives an overview of the *changes* of the monthly averages of series resistance relative to the first month of operation of each module. The selection criteria of the IV-curves used for this analysis were a short circuit current between 1.4 and 1.5 A (but 1.3 and 1.4 A for module M5, because of the smaller illuminated area of each cell), a voltage at the maximum power point larger than 85 V, and a fill factor larger than 40% (to elimi-

nate data from partially shaded modules). These graphs confirm the relatively strong increase of series resistance for the module with SnPgAg-coated contact strips (M1) and the more erratic behavior of the module with Cu-contact strips (M7), compared to the stable performance of the modules with Ag-coated contact strips. For these latter modules it can also be noted that there is a twelve-month period of the changes of the series resistance. A closer inspection revealed a higher series resistance during the warm summer months. This agrees with the known change of series resistance of silicon solar cells [43] and the positive temperature coefficient of electrical resistance of metals.

As the outdoor modules virtually never work under conditions of STC, one has no direct information on how the changes of series resistance affect the nominal output power of the modules. Therefore the performance at STC was estimated in the following way: For each module, an IV-curve from the first few days with a short

Table 4. Relative change of maximum power at STC normalized to one year (ΔP/year) estimated from outdoor IV data. For modules M1 to M7 data until Apr. 29, 2015 are included. For module M8 until Dec. 31, 2015, in order to cover a minimum of one year.

Module	First day	Contact strip coating	Maximum power measured outdoor [W]	ΔP/year [%]	Decline of FF [% abs./year]
M1	Aug. 23, 2013	SnPbAg	183.7	−29.6	−14.38
M2	Aug. 29, 2013	Ag	226.5	−2.4	−1.28
M3	Feb. 13, 2014	Ag	217.4	−2.4	−0.59
M4	Feb. 13, 2014	soldered reference	215.7	−2.5	−1.26
M5	Feb. 13, 2014	Ag	209.4	−2.5	−1.27
M6	May 2, 2014	Ag	222.4	+0.2	−0.76
M7	May 2, 2014	Cu	168.0	−4.5	−2.74
M8	Sep. 4, 2014	Ag	204.4	−2.5	−1.99

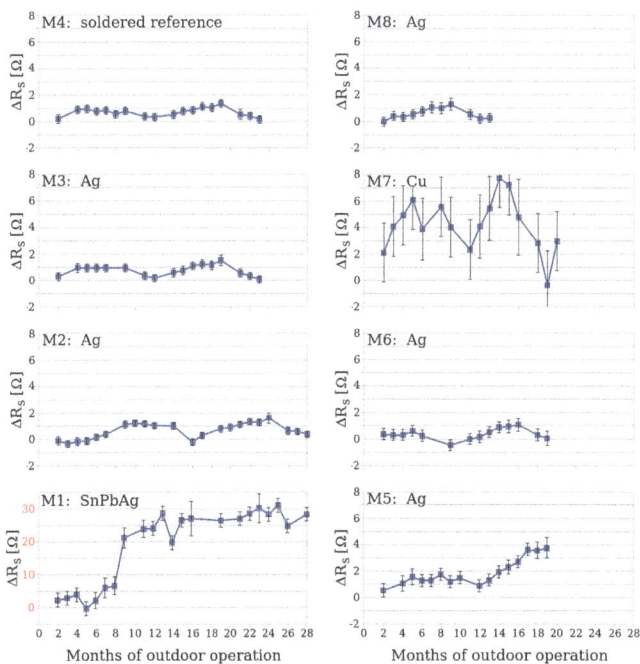

Fig. 13. Monthly averages of the *change* of series resistance (ΔR_s) relative to the first month of outdoor operation of each module. Extracted from IV-curves with short circuit current between 1.4 and 1.5 A (1.3 and 1.4 A for M5). Note the different scale for module M1.

circuit current around 1.2 A is selected and extrapolated to an IV-curve of same short circuit current, but with observed decline of the fill factor some 12 months up to about 18 months later (depending on module). This was done by fitting the IV-curve to the two diode model and then increasing the series resistance until the fill factor of the later point in time was reached. Then both curves were extrapolated to illumination of 1000 W/m^2, by changing the short circuit current to that known to apply at STC for that particular module. Then the maximum power of these curves could be compared, and the relative decrease in output power, normalized to the period of one year, could be calculated. The results are shown in Table 4. This table also lists the maximum power which each module ever delivered outdoors. This was ususaly around the

end of March, because with the given orientation and inclination, this is the time of year in which these modules can generate the highest power. The final column of Table 4 shows the decrease of the fill factor of each module, normalized to one year. The values are in absolute percent, not relative to the initial value. For this calculation the same selection criteria of IV-curves were used as for the data shown in Figure 13, especially because higher current was of interest (short circuit current between 1.4 and 1.5 A, (1.3–1.4 A for M5), because an increase of series resistance affects the fill factor more at higher currents, open circuit voltage larger 85 V, fill factor larger 40% to eliminate partial shading of modules, which was possible for the lower modules M1, M2 and M5, M6 due to some bushes, especially in winter in the hours before sunset).

For the modules M1, M4, M6 and M7 the evolution of this fill factor is shown on a day-by-day basis in Figure 14. One notes the strong deterioration of module M1 due to its contact strips made from SnPbAg. An actual jump of deterioration occurred during the short circuited period of March–April 2014. It is conceivable that during this period some contact strips became hot enough to melt and re-solidify their SnPbAg coating, which would first form good contact to the affected busbars of the cells, but upon cooling in the following night this would be broken again, leading to greater surface roughness and hence higher resistance (soldering would be unlikely, since there was no solder flux). Module M7, with its copper contact strips, also showed quite some deterioration, and it already started with a lower fill factor than other modules. This, and the relative magnitudes of fill factor decline of these two modules are in line with those found in the rapid ageing tests of the small modules with analogous contact strips (S2 and S3, respectively).

Interestingly, module M6, with Ag-coated contact strips shows a smaller decrease in fill factor than the soldered reference module M4. Two factors come into play here: on the one hand, there is some statistical error in the estimates of the decline of the fill factor, which should amount to less than +−0.5% abs./year. On the other hand, module M6 was produced with cells from another company than all the other modules, and it already had the newer technology of three bus bars, which may

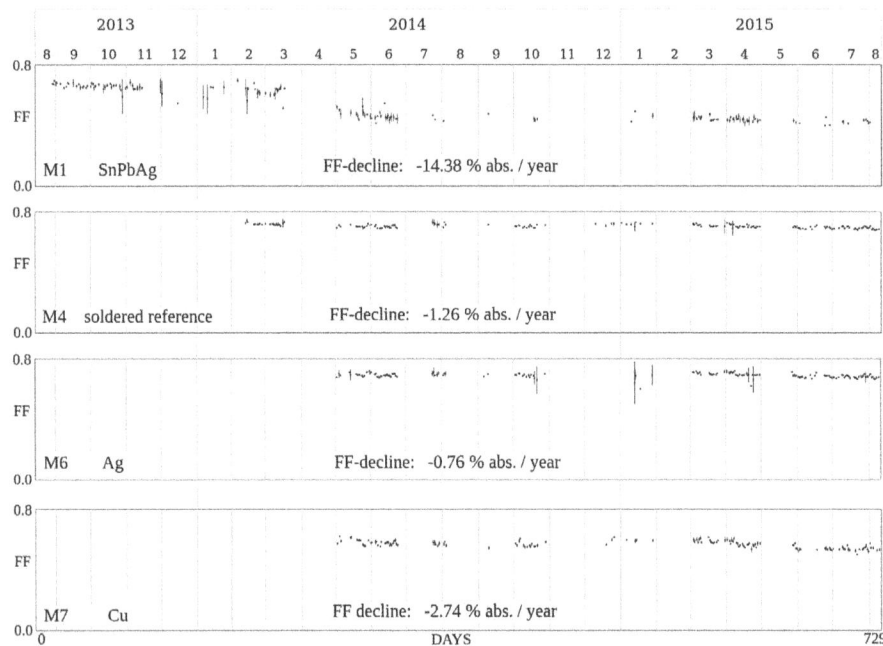

Fig. 14. Evolution of fill factor. Each data point represents the average of the fill factor of one day of the IV-curves with short circuit current between 1.4 and 1.5 A, a voltage at the maximum power point larger than 85 V, and a fill factor larger than 40%. The vertical bars indicate the standard deviation.

be an indication, that the silicon used in these cells or the screen printed metallization was of higher quality, too. This opens the possibility that the decline of fill factor of the reference module M4 may not be due to an increase of series resistance of the soldered cell interconnections, but to other ageing effects of the cells [44–46].

In Table 4 it should also be noted that module M8 showed a tendency towards a larger decline of fill factor, although its contact strips were silver coated. This module used the same interconnection scheme as the small module K3 (see Figs. 5c and 5d), which means that on the back sides of the cells the contact strips were pressed onto the cells only by the adhesive tape, the encapsulant, and the back sheet of the module. Most likely this pressure is quite a bit lower than that exerted on a contact strip between two overlapped cells, and it may get smaller with repeated warming and cooling of the module, due to a change of viscoelastic properties of EVA [27, 32, 33].

6 Summary and conclusion

Comparing the results of rapid ageing of the small modules by thermal cycling and the slower ageing by outdoor exposure of the large modules, one notes that with respect to the different coating materials of the contact strips they show the same trends: Modules with SnPbAg coated contact strips quickly accumulate a high series resistance in both cases, modules with pure Cu contact strips show less degradation, but it is still unacceptably high. Also the one small module with Sn-coated contact strips showed a high loss in power after only a few hundred cycles, although it performed better than the modules with pure Cu contact strips. Only the modules with

Ag-coated contact strips exhibit high reliability and appear to degrade no faster than modules with soldered cell interconnections. One large module with Ag-coated contact strips showed an even slower decline of fill factor than the soldered reference module, but possibly this must be attributed to its cell quality. On the other hand it underlines the quality of its contact strips.

Of the three different types of polycrystalline cells employed in this investigation – quartered standard cells, and the special QuarterCells of type K and P, respectively – the type QuarterCell P performed best. The reason is that its metallization pattern on front and back side was optimally laid out for cell overlap of 2 mm, and maximum contact area between any two cells was achieved.

An interesting observation on small modules with Ag-coated contact strips was that after the initial thermal cycles their series resistance became smaller before it started the expected slow increase. This feature is as yet unexplained but a combination of changing of the properties of EVA in a kind of post processing during thermal cycling [33], and a better adaptation of the silver coating of the contact strips to the grainy surface of the screen printed bus bars on the cells seem plausible.

Another open question is how frequently micro cracks will occur in the bent regions of the cells. This will depend on the curvature imposed on the cell, which in turn will depend on the thickness of the contact strips and their spacing. Micro cracks need not have a noticeable effect on the power output of the module [18], but in time they may develop into real cracks which separate a part of a cell electrically from the rest. A first check has been done with an electroluminescence scan on one of the worst performing small modules (S2), and it did indeed reveal a small crack at a position of a contact strip [47]. However, the performance of the large modules, with as many as 276 cells connected in series, and especially that of the small modules with cell type QuarterCell P and special connector, indicates that micro cracks may be a minor issue.

A final remark concerns the potential savings of PV modules with crystalline silicon solar cells made with cell interconnections solely based on contact pressure. As we have shown, the overlap concept alone permits more than 6% higher power density of a module without changing the cell technology. And if pressure based contact strips are used instead of the traditional soldering or glueing of metallic ribbons, only very little metal is needed for the cell interconnections. We estimate that between 70% and 90% of the copper currently needed in a typical module with 60 quadratic cells of 156×156 mm can be saved in a module of the same power but made with the kind of cells and interconnections presented here. Therefore, although the overlap concept requires a few percent more silicon, the overall material savings and the higher power density should make it an economically more attractive option.

We would like to thank J.L. Lang for development and mechanical measurements on the interconnections of the small modules during the starting phase of the project. We are also grateful to Prof. Marko Topic and Dr. Kristijan Brecl from the University of Lubljana in Slovenia for a first electroluminescence scan of a small module.

References

1. J. John, V. Prajapati, B. Vermang, A. Lorenz, C. Allebe, A. Rothschild, L. Tous, A. Uruena, K. Baert, J. Poortmans, EPJ Photovolt. **3**, 35005 (2012)
2. Z. Halavani, J.L. Lang, J. Summhammer, Results of Pressure-Only Cell Interconnections in High Voltage PV-Modules, in *29th EUPVSEC, Amsterdam, 2014*, p. 64
3. J. Summhammer, Z. Halavani, High-Voltage PV- Modules with Crystalline Silicon Solar Cells, in *28th EUPVSEC, Paris, 2013*, p. 3119
4. J. Schneider, S. Schönfelder, S. Dietrich, M. Turek, Solar Module with Half Size Solar Cells, in *29th EUPVSEC, Amsterdam, 2014*, p. 185
5. E. Franklin, A. Blakers, K. Weber, V. Everet, Sliver Solar Cell Technology: Pushing the Material Boundaries, in *MRS Proceedings, 2011*, p. 1323
6. E. Franklin, V. Everett, A. Blakers, K. Weber, Adv. Optoelectron. **2007**, 35383 (2007)
7. R. Ebner, M. Schwark, G. Újvári, W. Mühleisen, C. Hirschl, L. Neumaier, M. Pedevilla, J. Scheurer, A. Plösch, A. Kogler, W. Krumlacher, H. Muckenhuber, Increased Power Output Of Crystalline Silicon Solar Modules By Application Of New Module Concepts, in *29th EUPVSEC, Amsterdam, 2014*, p. 171
8. A. Schneider, R. Harney, S. Aulehla, S. Hummel, E. Lemp, S. Koch, K. Schröder, Conductive Gluing as Interconnection Technique towards Solar Cells without Front Busbars and Rear Pads, in *27th EUPVSEC, 2012*, p. 335
9. G. Beaucarne, I. Kuzma-Filipek, F. Campeol, X. Young, J. Wei, Y. Yu, R. Russell, F. Duerinckx, Energy Procedia **67**, 185 (2015)
10. J. Dupuis, E. Saint-Sernin, P. Lefillastre, A. Vachez, E. Pilat, D. Bussery, R. Einhaus, Cell-ribbons contacts interface study in NICE modules, in *26th EUPVSEC, Hamburg, 2011*, p. 3117
11. O. Nichiporuk, J. Dupuis, E. Saint-Sernin, J. Degoulange, P. Lefillastre, R. Einhaus, Secured Intrinsic Under-Pressure in the NICE Modules – The Oxygen Gettering Approach, in *27th EUPVSEC, 2012*, p. 3449
12. J. Summhammer, H. Rothen, Rectangular silicon solar cell with more power and higher voltage modules, in *24th EUPVSEC, Hamburg, 2009*, p. 2221
13. US Patent in 1966, http://www.google.com/patents/US3459597 and in 1972, http://www.google.com/patents/US3769091
14. *Electrical contacts: principles and applications*, edited by P.G. Slade (Marcel Dekker Inc. New York, 1999). Especially Chap. 1.2 and 1.3
15. N. Bogdanski, W. Herrmann, Weighting of Climatic Impacts on PV-Module Degradation – Comparison of Outdoor Weathering Data and Indoor Weathering Data, in *26th EUPVSEC, Hamburg, 2011*, p. 3093
16. P. Hülsmann, M. Heck, M. Köhl, J. Mater. **2013**, 102691 (2013)
17. K.R. McIntosh, J.N. Cotsell, J.S. Cumpston, A.W. Norris, N.E Powell, B.M. Ketola, The effect of accelerated aging tests on the optical properties of silicone and EVA encapsulants, in *24th EUPVSEC, Hamburg, 2009*, p. 3475
18. T.M. Pletzer, J.I. Van Mölken, S. Rißland, O. Breitenstein, J. Knoch, Prog. Photovolt.: Res. Appl. **23**, 428 (2015)
19. Review of Failures of Photovoltaic Modules, Report IEA-PVPS T13-01:2014. Available at: http://iea-pvps.org/index.php?id=275
20. U. Eitner, M. Koentges, R. Brendel, Measuring Thermodynamical Displacements of Solar Cells in Laminates Using Digital Image Correlation, in *Proceedings of 34th IEEE PVSEC, Philadelphia, 2009*, pp. 1280–1284
21. U. Eitner, M. Köntges, R. Brendel, Sol. Energy Mater. Sol. Cells **94**, 1346 (2010)
22. www.engineeringtoolbox.com/linear-expansion-coefficients-d_95.htm
23. C.A. Swenson, J. Phys. Chem. Ref. Data **12**, 179 (1983)
24. http://www.dupont.com/content/dam/dupont/products-and-services/membranes-and-films/pvf-films/documents/DEC_Tedlar_GeneralProperties.pdf
25. M. Paggi, S. Kajari-Schröder, U. Eitner, J. Strain Anal. Eng. Design **46**, 772 (2011)
26. M. Schmidt, P. Guttmann, K. Berger, Y. Voronko, M. Knausz, G. Oreski, G. Eder, T. Koch, G. Pinter, Polymer Testing **44**, 160 (2015)
27. U. Eitner, M. Pander, S. Kajari-Schröder, M. Köntges, H. Altenbach, Thermomechanics of PV Modules Including the Viscoelasticity of EVA, in *26th EUPVSEC, Hamburg, 2011*, p. 3267
28. M.D. Kempe, G.J. Jorgensen, K.M. Terwilliger, T.J. McMahon, C.E. Kennedy, T.T. Borek, Ethylene-Vinyl Acetate Potential Problems for Photovoltaic Packaging, in *Conference Record of the 2006 IEEE 4th World Conference on Photovoltaic Energy Conversion, 2006*, Vol. 2, pp. 2160–2163

29. A.A. Ogwu, T.H. Darma, E. Bouquerel, J. Achiev. Mater. Manuf. Eng. **24/1**, 172 (2007)

30. T. Suehiro, T. Sasaki, Y. Hiratate, Thin Solid Films **383**, 318 (2001)

31. L. De Los Santos Valladares, D. Hurtado Salinas, A. Bustamante Dominguez, D. Acosta Najarro, S.I. Khondaker, T. Mitrelias, C.H.W. Barnes, J. Albino Aguiar, Y. Majima, Thin Solid Films **520**, 6368 (2012)

32. D. Wu, J. Zhu, D. Montiel Chicharro, T.R. Betts, R. Gottschalg, Influences of Different Lamination Conditions on the Reliability of Encapsulation of PV Modules, in *29th EUPVSEC Amsterdam, 2014*, p. 3415

33. C. Hirschl, L. Neumaier, W. Mühleisen, M. DeBiasio, G. Oreski, A. Rauschenbach, G.C. Eder, B.S. Chernev, M. Kraft, Post-Crosslinking in Photovoltaic Modules under Different Conditions, in *29th EUPVSEC Amsterdam, 2014*, p. 3133

34. J. Song, L. Wang, A. Zibart, C. Koch, Metals **2**, 450 (2012)

35. A. De Rooij, ESA J. **13**, 363 (1989)

36. P. Narayana Reddy, M. Hari Prasad Reddy, J.F. Pierson, S. Uthanna, ISRN Optics **2014**, 684317 (2014)

37. http://www.chemicool.com/elements/silver.html

38. M.M. El-Nahass, A.A.M. Farag, E.M. Ibrahim, S. Abd-El-Rahman, Vacuum **72**, 453 (2004)

39. O. Dupré, R. Vaillon, M.A. Green, Sol. Energy Mater. Sol. Cells **140**, 92 (2015)

40. P. Singh, N.M. Ravindra, Sol. Energy Mater. Sol. Cells **101**, 36 (2012)

41. S. Chander, A. Purohit, A. Sharma, S.P. Nehra, M.S. Dhaka, Energy Rep. **1**, 175 (2015)

42. P. Kamkird, N. Ketjoy, W. Rakwichian, S. Sukchai, Procedia Eng. **32**, 376 (2012)

43. S. Bensalem, M. Chegaar, Revue des Energies Renouvelables **16**, 171 (2013)

44. C.R. Osterwald, J. Pruett, T. Moriarty, Crystalline silicon short circuit current degradation study: initial results, in *Conference Record of the 31st IEEE Photovoltaic Specialists Conference, Lake Buena Vista, FL, 2005*, pp. 1335–1338

45. C. Peike, S. Hoffmann, P. Hülsmann, B. Thaidijsmann, K.A. Weiß, M. Koehl, P. Bentz, Sol. Energy Mater. Sol. Cells **116**, 49 (2013)

46. N.C. Park, W.W. Oh, D.H. Kim, Int. J. Photoenergy **2013**, 925280 (2013)

47. M. Topic, University of Lubljana, customer measurement report, June 2015 (unpublished)

29

Substrate and p-layer effects on polymorphous silicon solar cells

S.N. Abolmasov[1,a], H. Woo[1], R. Planques[1], J. Holovský[2], E.V. Johnson[1], A. Purkrt[2], and P. Roca i Cabarrocas[1]

[1] LPICM, CNRS, École Polytechnique, 91128 Palaiseau, France
[2] Institute of Physics, ASCR v.v.i. Cukrovarnická, Prague, Czech Republic

Abstract The influence of textured transparent conducting oxide (TCO) substrate and p-layer on the performance of single-junction hydrogenated polymorphous silicon (pm-Si:H) solar cells has been addressed. Comparative studies were performed using p-i-n devices with identical i/n-layers and back reflectors fabricated on textured Asahi U-type fluorine-doped SnO_2, low-pressure chemical vapor deposited (LPCVD) boron-doped ZnO and sputtered/etched aluminum-doped ZnO substrates. The p-layers were hydrogenated amorphous silicon carbon and microcrystalline silicon oxide. As expected, the type of TCO and p-layer both have a great influence on the initial conversion efficiency of the solar cells. However they have no effect on the defect density of the pm-Si:H absorber layer.

1 Introduction

General trends in thin film silicon (Si) photovoltaics are the reduction of the production costs and improvement of the conversion efficiency. The state of the art in thin film silicon solar cells utilizes tandem structures which pair a hydrogenated amorphous Si (a-Si:H) top cell with a hydrogenated microcrystalline Si (μc-Si:H) bottom cell. The efficiency and cost of these tandem cells are mainly limited by the light-induced degradation of a-Si:H and the low absorption coefficient/deposition rate of μc-Si:H, respectively [1]. Therefore, there has been considerable interest recently in wide band gap nanostructured Si materials which may offer better performance/stability to the top cell and allow reducing the thickness of the bottom cell using enhanced light trapping techniques. These nanostructured Si materials mainly include hydrogenated amorphous and microcrystalline silicon oxide (a- and μc-SiO:H) [2–8], protocrystalline (pc-Si:H) [9] and polymorphous (pm-Si:H) silicon [10] that could be used as alternatives to standard doped and absorber layers, respectively. On the other hand, the light trapping is mainly realized by texturing of transparent conductive oxide (TCO) layers on glass substrates [11]. However, this approach tends to deteriorate the electrical parameters of the cells with both a- and μc-Si:H absorber layers [12,13]. The goal of this study is to evaluate the performance of single-junction (p-i-n) pm-Si:H solar cells on different textured TCO substrates which also incorporate p-type silicon oxide films as the window layer. To examine the substrate and p-layer effects on pm-Si:H solar cell performance, the cells were fabricated with the same i- and n-layers.

2 Experimental

Silicon films were deposited in a 13.56 MHz radiofrequency plasma enhanced chemical vapor deposition (RF-PECVD) system, consisting of a single vacuum vessel containing three independent plasma chambers [14]. Mixtures of $H_2/SiH_4/CO_2/B(CH_3)_3$, H_2/SiH_4 and $H_2/SiH_4/PH_3$ were used as source gases for the growth of p-type a-SiO:H, intrinsic pm-Si:H and n-type a-Si:H or μc-Si:H, respectively. The p-i-n devices were fabricated on textured Asahi U-type fluorine-doped tin oxide (SnO_2:F), low-pressure chemical vapor deposited (LPCVD) boron-doped (ZnO:B) ZnO (type A), and RF sputtered and etched (sp-e) aluminum-doped zinc oxide (ZnO:Al) coated glass substrates (1 mm thick). Details on TCO deposition processes along with TCO characteristics can be found in references [15–17]. Aluminum back contacts were deposited by thermal evaporation. Sputtered ZnO/Ag stacks were also used as back reflectors to enhance the light trapping from the back side of the device. The cells areas were defined by using a shadow mask (0.126 cm^2). Fourier transform photocurrent spectroscopy (FTPS) [18] was used to determine the effect of the TCO substrate on the absorption coefficient (α) and defect density of pm-Si:H in the cells. Dark conductivity (σ_d) of p-type silicon oxide on Corning 7059 glass was measured using a coplanar Al electrode configuration. The optical properties and thickness of the films were measured

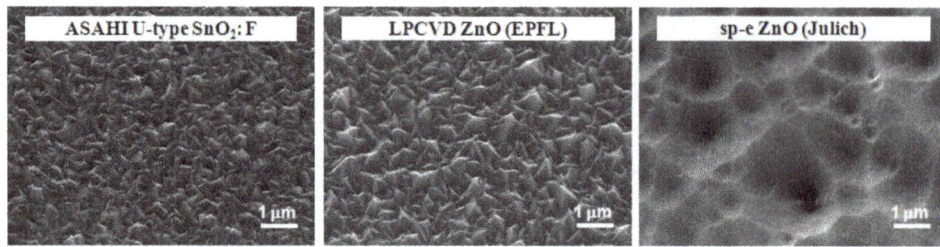

Fig. 1. Scanning electron microscope (SEM) images of textured TCO substrates used in this study (top view under an angle of 30°).

with UV-visible spectroscopic ellipsometry (SE). The J-V-characteristics and external quantum efficiency (EQE) of the solar cells were measured under 1 sun illumination at 25 °C. A 1-mm-diameter orifice was used for the EQE and reflectivity measurements. Most results were obtained on cells in the annealed state (40 min at 150 °C). Accelerated light-induced degradation (LID) tests were performed at approximately 2 suns in a home-made LID system using Oriel Apex illuminator (Model 71223) with a quartz tungsten halogen lamp. The cell temperature was kept at 50 °C. In this case, the J-V-characteristics were measured in situ using the same lamp and a neutral density filter, resulting in a light exposure equivalent to 1 sun.

3 Results and discussion

The surface morphologies of the textured TCO substrates used in this study are shown in Figure 1. It is obvious that the front TCO layer has to exhibit high transparency in the useful spectrum range and a sufficient conductivity to limit the series resistance of the cell. However, it must also have sufficient surface roughness to optimize the light confinement [11]. It can be seen in Figure 1 that there is a significant difference in the surface roughness of the three TCO layers. Both SnO_2 and LPCVD ZnO have pyramidal features with sizes below 500 nm that are well suited for effective light trapping in a-Si:H solar cells. On the other hand, sp-e ZnO has much smoother surface with lateral features that are more than 1 μm in size and therefore is more suited for the use in μc-Si:H solar cells. Because the properties of TCO may also vary significantly, appropriate window p-layers should be developed. For example, it is well known that a textured SnO_2 front contact is susceptible to reduction by hydrogen [19, 20]. Therefore, ZnO has recently been investigated as an alternative that is both resistant to hydrogen plasma induced darkening and has higher transmission. However, fabrication of p-i-n devices on ZnO results in a lower fill factor (FF), which is believed to be due to a rectifying contact that is formed between the p-type a-Si:H and the n-type ZnO layer [21]. There have been numerous attempts to overcome this problem [22–25], but there is still lack of a complete understanding of the phenomenon. The most efficient way to overcome the problem is to use an ultra-thin μc-Si:H or silicon oxide (μc-SiO:H) p-layer, which is more conducting than a hydrogenated amorphous silicon carbide (a-SiC:H) p-layer [2, 4, 5].

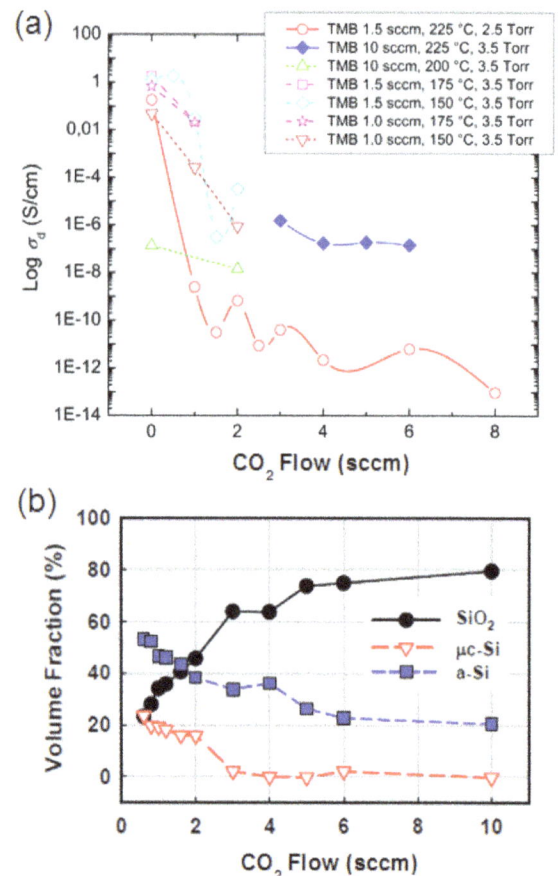

Fig. 2. (a) Dark conductivity (at room temperature) of p-type silicon oxide thin films as a function of CO_2 flow rate (the films were grown on Corning glass at H_2 (500 sccm)/SiH_4 (10 sccm) =50 and with 2% $B(CH_3)_3$ in H_2) and (b) evolution of SiO_2, a-Si and μc-Si volume fractions of optimized μc-SiO:H p-layer with increasing CO_2 flow rate, as deduced from the BEMA modeling.

Like the TCO layer, the window layer should also exhibit high transparency and sufficient conductivity. For the silicon oxide p-layer deposition, a wide range of deposition parameters such as RF power, inter-electrode distance, substrate temperature, gas composition, pressure and flow were optimized. It turns out that lowering the CO_2 and $B(CH_3)_3$ concentration favors the growth of conductive silicon oxide p-layers, as shown in Figure 2a. The results of SE measurements/modeling for optimized p-layers grown at $p = 3.5$ Torr, $T = 175$ °C and a TMB

Table 1. Initial photovoltaic parameters of pm-Si:H p-i-n devices fabricated on three different TCO substrates along with the absorption coefficient and the Urbach energy of their intrinsic pm-Si:H layer. All the cells incorporate the same i-pm-Si:H (300 nm) and n-a-Si:H layers and evaporated Al back contacts.

Sample #	TCO/p-layer description	α (1.2 eV) (cm^{-1})	E_U (meV)	FF (%)	V_{OC} (V)	J_{SC} (mA/cm^2)	Eff. (%)
1207271	ASAHI SnO$_2$ (ref) a-SiC p-layer	0.15 ± 0.03	39.9 ± 2	71.77	0.868	12.65	7.88
1208071I	LPCVD ZnO μc-SiO$_x$ p-layer #1	0.18 ± 0.02	39.6 ± 0.8	69.71	0.853	11.19	6.65
1208071J	sp-e ZnO μc-SiO$_x$ p-layer #1	0.13 ± 0.05	37.8 ± 1.5	66.23	0.881	10.08	5.88

Table 2. Initial photovoltaic parameters of pm-Si:H p-i-n devices fabricated on three different TCO substrates. All the cells incorporate the same p-μc-SiO:H, i-pm-Si:H (250 nm) and n-μc-Si:H layers and sputtered ZnO-Ag back contacts.

Sample #	TCO/p-layer description	FF (%)	V_{OC} (V)	J_{SC} (mA/cm^2)	Eff. (%)
1301245I	LPCVD ZnO μc-SiO$_x$ p-layer #2	68.54	0.903	14.83	9.2
1301245J	sp-e ZnO μc-SiO$_x$ p-layer #2	61.09	0.918	13.54	7.6
1301245A	ASAHI SnO$_2$ μc-SiO$_x$ p-layer #2	55.27	0.793	13.69	6.0

flow rate of 1.5 sccm indicate that the optical band-gap energy (E_g) of these thin films can be increased from 1.95 to 2.8 eV by increasing the CO$_2$ flow rate from 1 to 10 sccm. The increase in E_g is mainly due to an increase in the oxygen concentration, whereas the conductivity of these films is controlled by their μc-Si fraction as determined from the modeling of the spectroscopic ellipsometry data. The measured spectra were modeled using the Bruggeman's effective medium approximation (BEMA) [26], in which the silicon oxide material was considered as being a physical combination of three distinct phases formed by SiO$_2$, a-Si and μc-Si. Figure 2b shows that the μc-Si fraction in our optimized silicon oxide p-layer, according to the BEMA model, does not exceed 20% and sharply decreases with increasing the CO$_2$ mass flow rate above 2 sccm [27]. In our case the silicon oxide p-layer was grown at a CO$_2$ mass flow rate of 1 sccm. Such a layer has a 19% μc-Si fraction and thereby referred to as μc-SiO:H in the paper.

In the first series of experiments, the TCO/p-layer and p/i interface regions of the cells deposited on the three types of TCO substrates were systematically improved by introducing thin buffer layers. By optimizing the doping profile (by adding one or more layers grown with different gas composition) and thickness of the window p-layer, the cell current was improved while maintaining relatively high FF and open circuit voltage (V_{OC}). Typically the p-layer was a 3-layer-stack consisting of an ultra-thin μc-Si:H, a heavily doped μc-SiO:H and a lightly doped μc-SiO:H layer. Such multilayer structure provides a wider process window, enabling the use of much thinner window p-layer. Consequently, short-circuit current J_{SC} and V_{OC} were enhanced with only a small drop in FF. Since the growth of both μc-SiO:H and pm-Si:H requires a hydrogen rich plasma, an a-SiC:H p-layer was used in the case of

SnO$_2$ substrate and served as a reference. Moreover, FTPS was used to check the quality of the pm-Si:H in the cells grown on the three TCO substrates. Interestingly enough, the FTPS results, obtained for the cells with a 300 nm thick intrinsic layer and an a-Si:H n-layer, shown in Figure 3 indicate very similar pm-Si:H material properties in the cells with a defect density below 5×10^{15} cm^{-3} and Urbach energy about 40 meV. This contrasts with quite different values for the solar cell parameters summarized in Table 1, which did show a strong dependence on the TCO and p-layer type, thus suggesting that the surface roughness as well as the TCO/p-layer and p/i interface regions play very important roles in determining the overall cell performance. Similar results have been reported for μc-Si:H p-i-n devices and it was also demonstrated that the FTPS method is sensitive to bulk material properties but not to local shunts induced by TCO morphology [28].

Further optimization of the μc-SiO:H p-layer/i interface and introduction of ZnO/Ag back contacts along with a μc-Si:H n-layer, however, resulted in great enhancement of J_{SC} and V_{OC} in the cells on ZnO substrates, as can be seen in Table 2. Among the three different TCO substrates, sp-e ZnO showed the highest $V_{OC} \sim 0.92$ V, whereas LPCVD ZnO led to the highest $J_{SC} \sim 14.8$ mA/cm^2 for the same p-i-n/metal contact configuration. These two facts could be related directly to the TCO surface morphology: a smooth surface has smaller number of defects than a rough one, simply because of the difference in the surface area, resulting in higher V_{OC}. On the contrary, a rough surface results in light scattering and, in turn, in better light trapping. The use of the same μc-SiO:H p-layer on SnO$_2$ substrates, however, resulted in a deterioration of all solar cell parameters due to hydrogen plasma induced reduction of SnO$_2$, as mentioned above. Figure 4 shows a comparison

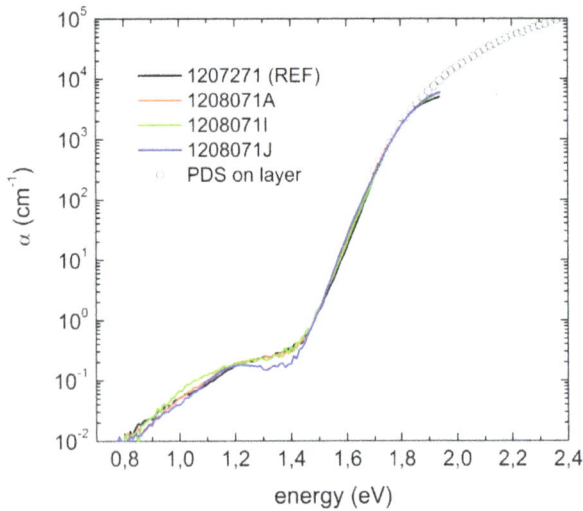

Fig. 3. Absorption coefficient of pm-Si:H absorber layer in p-i-n devices fabricated on Asahi SnO$_2$, LPCVD ZnO and sp-e ZnO substrates versus photon energy measured by FTPS. The absolute scaling is made relative to i-layer on Corning glass measured by photothermal deflection spectroscopy (PDS).

of EQE and reflectivity curves of the three p-i-n devices with a 250 nm thick pm-Si:H layer, indicating excellent spectral response in the blue part of the spectrum for the p-i-n device on LPCVD ZnO, thanks to our newly developed μc-SiO:H p-layer. One can see (Fig. 4b) that the p-i-n on SnO$_2$ has higher reflectivity due to SnO$_2$ reduction by hydrogen, as mentioned above. Note that the highest J_{SC} value for LPCVD ZnO was also confirmed independently by EQE measurements at EPFL and PTB. Nevertheless, this value is still below that (16.75 mA/cm^2) obtained in the champion p-i-n device with an a-Si:H absorber layer of the same thickness (250 nm) and Oerlikon LPCVD ZnO (front and back contact) [29]. This is partly due to the fact that pm-Si:H has larger optical band gap and lower optical absorption compared to a-Si:H. For example, E_{04}, defined as the energy at which the absorption coefficient is equal to 10^4 cm^{-1}, lies approximately in the range between 1.7–1.8 eV and 1.8–1.9 eV [27,30] (see also Fig. 3) for a-Si:H and pm-Si:H, respectively. It should also be mentioned that the champion a-Si:H solar cell has an anti-reflection coating and a LPCVD ZnO/white paint back contact.

Last but not least, it has been shown recently that the regions around the window p-layer could also have an influence on the stability of single-junction pm-Si:H solar cells. In particular, localized delamination of the TCO/p-layer interface in pm-Si:H solar cells fabricated on textured SnO$_2$ substrates was observed during the initial stage of light-soaking [31]. Note that pm-Si:H was produced in a hydrogen rich plasma and hence contains more hydrogen than standard a-Si:H. It was proposed in reference [31] that the light-soaking of pm-Si:H introduces structural changes related to the diffusion of molecular hydrogen in the material with subsequent accumulation at the TCO/p-layer interface which finally causes localized delamination of the interface, affecting the solar cell

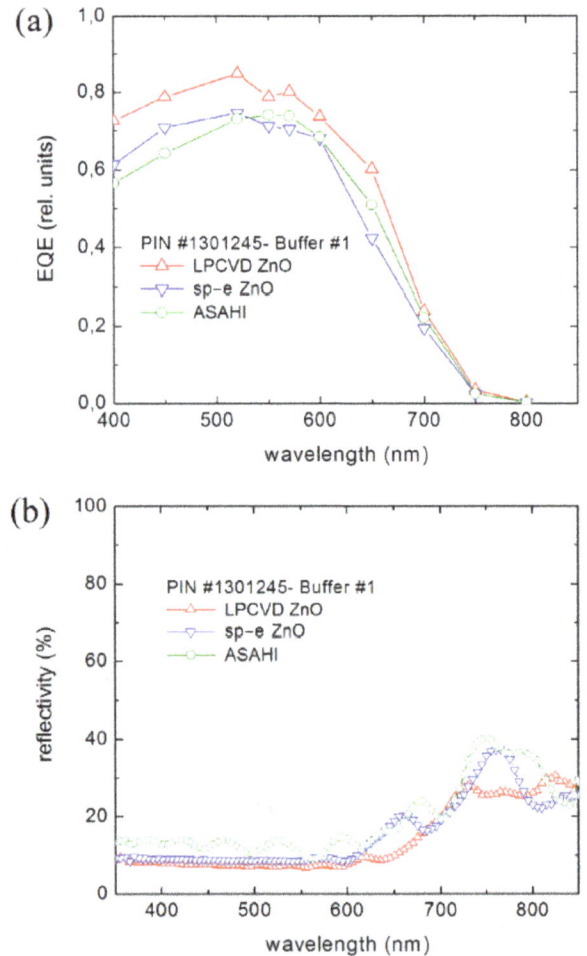

Fig. 4. External quantum efficiency of pm-Si:H solar cells with the same p-i-n/metal contact configuration fabricated on three different TCO substrates. A thin a-Si:H buffer layer is used at the p/i interface. The cells are measured in their initial state. Photovoltaic parameters of these devices are summarized in Table 2.

parameters in an irreversible way (no recovery with thermal annealing). It is reasonable therefore to assume that the use of different TCO and/or p-layers may affect localized delamination of the interface. Note, however, that in reference [31] a xenon lamp was used to perform the light-soaking. To examine the effect further we used a home-made LID system in which a halogen lamp was employed instead of Xe lamp. Using this LID system it was found that the type of TCO and TCO/p-layer interface have no effect on the initial light-induced degradation dynamics of pm-Si:H solar cells and no irreversible behavior was observed in our solar cells. This discrepancy is most likely due to fact that the two lamps mentioned above have very different spectral compositions: quartz tungsten halogen lamps have the spectrum with a broad maximum around 800 nm and a sharp decrease between 500 and 350 nm, whereas Oriel Apex Xe lamps have an almost flat maximum in the range 350–800 nm [32]. Note that the maximum of solar radiation spectrum (AM1.5) lies in the

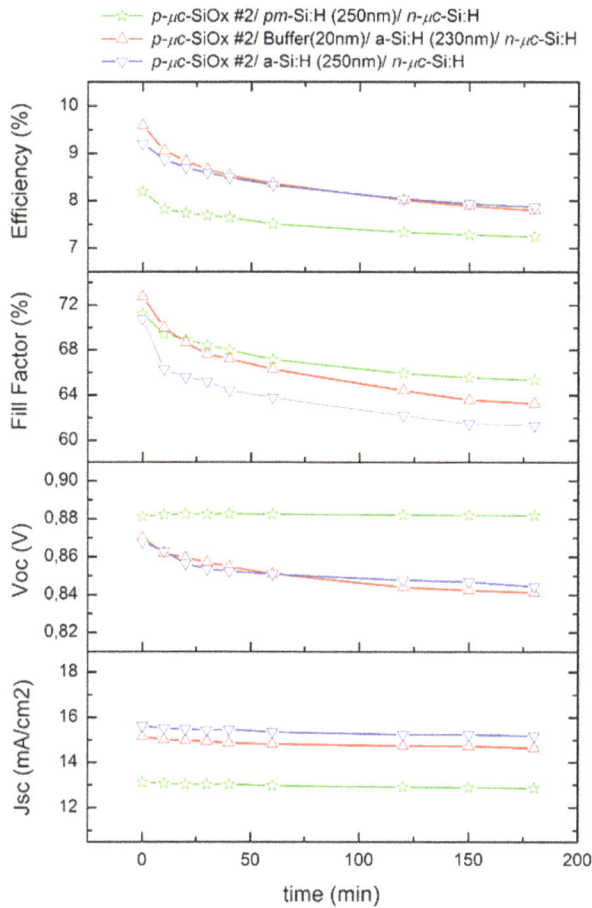

Fig. 5. Evolution of photovoltaic parameters of three different p-i-n solar cells during the light-soaking under a halogen lamp illumination at 2 suns. All three cells are fabricated on LPCVD ZnO using sputtered ZnO-Ag back contacts; their p-i-n structures are shown at the top of graph. Note that the J-V-characteristics are measured using the same halogen lamp at 1 sun and 50 °C.

range of 500–600 nm. Taking into account the fact that SnO$_2$ substrates are quite transparent to the blue light it is likely that a long exposure of p-i-n devices on their basis using a Xe lamp can result in severe damage of the region near the p/i interface.

Because the state-of-the-art tandem silicon solar cells incorporate a-Si:H top cells it is interesting to make a direct comparison of the LID dynamics in a-Si:H and pm-Si:H cells with the same thickness of intrinsic layers. Figure 5 shows such a comparison including an a-Si:H solar cell with a thin (20 nm) pm-Si:H buffer layer. It is seen that degradation of J_{SC} is negligibly small for the three cells with the initial J_{SC} values controlled by i-layer band gap while most of degradation is caused by a substantial decrease in FF – a common drawback among a-Si:H p-i-n devices [33, 34]. The decrease in FF is however more severe for a-Si:H cell than for pm-Si:H cell. Even though the insertion of a thin pm-Si:H buffer layer can improve the initial FF and conversion efficiency of a-Si:H cell, it has no effect on the conversion efficiency after 1 h of light-soaking:

the efficiency of a-Si:H cell with a pm-Si:H buffer layer becomes equal to that of pure a-Si:H cell. Still the pm-Si:H cell is the one which maintains the highest FF value on the long run, as recently reported for light-soaking times up to 1000 h [35]. Another distinguished feature of solar cells with absorber layers grown at high hydrogen dilution is that V_{OC} of these cells usually increases with light-soaking time [31, 36], as can be seen in Figure 5 for pm-Si:H p-i-n device. However, despite of their better stability the efficiency of pm-Si:H solar cells is mainly limited by lower optical absorption, as shown above. Thus, realization of effective light-trapping in single-junction pm-Si:H solar cells on textured TCO substrates is a challenging task and so far LPCVD ZnO shows the highest J_{SC}. Nevertheless, higher values of stabilized FF and V_{OC} for pm-Si:H material make it attractive for the use in multi-junction devices, e.g. in triple junction solar cells that do not require high J_{SC}.

4 Conclusions

The effects of textured TCO substrate and p-layer on the performance of single-junction (p-i-n) pm-Si:H solar cells have been evaluated. It turns out that the type of TCO and p-layer both have a great influence on the initial conversion efficiency of the solar cells. A μc-SiO:H p-layer was successfully implemented in the cells based on ZnO substrates, resulting in great enhancement of their performance. In particular, the cells grown on LPCVD ZnO had the highest J_{SC} and initial η, whereas those on spe ZnO had the highest V_{OC}. Unlike the strong influence on the initial efficiency, the TCO type and TCO/p-layer interface had no effect on the defect density of pm-Si:H absorber layer. The obtained results allow to conclude that pm-Si:H is more suitable for the use in multi-junction devices such as in 3J solar cells rather than in single-junction devices, mainly due to its improved stability compared to a-Si:H.

This work was carried out in the framework of the FP7 project "Fast Track", funded by the EC under grant agreement No. 283501. The authors thank the Forschungszentrum Jülich and IMT's PV-lab (EPFL) for preparation of textured ZnO substrates. The help of Dr. Ingo Kröger (PTB, Brunswick) and Dr. Martin Foldyna (LPICM) in validation of EQE measurements is also gratefully acknowledged.

References

1. C.A. Wolden et al., J. Vac. Sci. Technol. A **29**, 030801 (2010)
2. K. Schwanitz, S. Klein, T. Stolley, M. Rohde, D. Severin, R. Trassl, Sol. Energy Mater. Sol. Cells **105**, 187 (2012)
3. A. Lambertz, F. Finger, B. Hollaender, J.K. Rath, R.E.I. Schropp, J. Non-Cryst. Solids **358**, 1962 (2012)
4. S. Inthisang, T. Krajangsang, I.A. Yunaz, A. Yamada, M. Konagai, C.R. Wronski, Phys. Stat. Sol. C **8**, 2990 (2011)

5. P. Cuony, M. Marending, D.T.L. Alexander, M. Boccard, G. Bugnon, M. Despeisse, C. Ballif, Appl. Phys. Lett. **97**, 213502 (2010)

6. A. Sarker, A.K. Barua, Jpn J. Appl. Phys. **41**, 765 (2002)

7. P. Sichanurgist, T. Yoshida, Y. Ichikawa, H. Sakai, J. Non-Cryst. Solids **164–166**, 1081 (1993)

8. H. Watanabe, K. Haga, T. Lohner, J. Non-Cryst. Solids **164–166**, 1085 (1993)

9. C.R. Wronski, J.M. Pearce, R.J. Koval, A.S. Ferlauto, R.W. Collins, in *World Climate and Energy Event Proceedings, Rio de Janeiro, Brazil, 2002*, p. 67

10. P. Roca i Cabarrocas, Y. Djeridane, Th. Nguyen-Tran, E.V. Johnson, A. Abramov, Q. Zhang, Plasma Phys. Control. Fusion **50**, 124037 (2008)

11. M. Boccard et al., IEEE J. Photovolt. **2**, 229 (2012)

12. M. Despeisse, G. Bugnon, A. Feltrin, M. Stueckelberger, P. Cuony, F. Meillaud, A. Billet, C. Ballif, Appl. Phys. Lett. **96**, 73507 (2010)

13. M. Python, O. Madani, D. Dominé, F. Meillaud, E. Vallat-Sauvain, C. Ballif, Sol. Energy Mater. Sol. Cells **93**, 1714 (2009)

14. P. Roca i Cabarrocas, J.B. Chevrier, J. Huc, A. Lloret, J.Y. Parey, J.P.M. Schmitt, J. Vac. Sci. Technol. A **9**, 2331 (1991)

15. K. Sato, Y. Gotoh, Y. Wakayama, Y. Hayashi, K. Adachi, H. Nishimura, Reports Res. Lab. Asahi Glass Co. Ltd. **42**, 129 (1999)

16. S. Faÿ, J. Steihauser, N. Oliveira, E. Vallat-Sauvain, C. Ballif, Thin Solid Films **515**, 8558 (2007)

17. J. Müller, B. Rech, J. Springer, M. Vanecek, Sol. Energy **77**, 917 (2004)

18. J. Holovsky, in *Fourier Transforms – New Analytical Approaches and FTIR Strategies*, edited by G. Nikolic (InTech, 2011), p. 257

19. H. Schade, Z.E. Smith, J.H. Thomas, A. Catalano, Thin Solid Films **117**, 149 (1984)

20. S. Kumar, B. Drevillon, J. Appl. Phys. **65**, 3023 (1989)

21. E. Bohmer, F. Siebke, B. Rech, C. Beneking, H. Wagner, Mater. Res. Soc. Symp. Proc. **420**, 519 (1996)

22. M. Kubon, E. Boehmer, F. Siebke, B. Rech, C. Beneking, H. Wagner, Sol. Energy Mater. Sol. Cells **14–42**, 485 (1996)

23. K. Minegishi, Y. Koiwai, Y. Kikuchi, K. Yano, M. Kasuga, A. Shimizu, Jpn J. Appl. Phys. Part 2 **36**, L1453 (1997)

24. G. Ganguly, D.E. Carlson, S.S. Hegedus, D. Ryan, R.G. Gordon, D. Pang, R.C. Reedy, Appl. Phys. Lett. **85**, 479 (2004)

25. S. Baek, J.C. Lee, Y.J. Lee, S.M. Iftiquar, Y. Kim, J. Park, J. Yi, Nanoscale Res. Lett. **7**, 81 (2012)

26. G.E. Gellison, in *Ellipsometry Handbook*, edited by H.G. Tompkins, E.A. Irene (William Andrew, Highland Mills, 2005)

27. K.H. Kim, Ph.D. thesis, École Polytechnique, 2012

28. G. Bugnon, G. Parascandolo, T. Söderström, P. Cuony, M. Despeisse, S. Hänni, J. Holovský, F. Meillaud, C. Ballif, Adv. Funct. Mater. **22**, 3665 (2012)

29. S. Benagli, D. Borrello, E. Vallat-Sauvain, J. Meier, U. Kroll, J. Hötzel, J. Spitznagel, J. Steinhauser, L. Marmello, G. Monteduro, L. Castens, in *Proceedings of the 24th European Photovoltaic Solar Energy Conference, Hamburg, Germany, 2009*, p. 2293

30. P. Roca i Cabarrocas, A. Fontcuberta i Morral, Y. Poissant, Thin Solid Films **403-404**, 39 (2002)

31. K.H. Kim, E.V. Johnson, P. Roca i Cabarrocas, Sol. Energy Mater. Sol. Cells **105**, 208 (2012)

32. Newport Spectral Irradiance Data, www.newport.com

33. H. Sakai, T. Yoshida, S. Fujikake, T. Hama, Y. Ichikawa, J. Appl. Phys. **67**, 3494 (1990)

34. A. Kolodziej, Opto-Electron. Rev. **12**, 21 (2004)

35. K.H. Kim, S. Kasouit, E.V. Johnson, P. Roca i Cabarrocas, Sol. Energy Mater. Sol. Cells **119**, 124 (2013)

36. P. Siamchai, M. Konagai, Appl. Phys. Lett. **67**, 3468 (1995)

30

Silicon-Light: a European project aiming at high efficiency thin film silicon solar cells on foil

W. Soppe[1,a], J. Krc[2], K. Leitner[3], F.-J. Haug[4], M. Duchamp[5], G. Sanchez Plaza[6], and Q.-K. Wang[7]

[1] Energy Research Centre of the Netherlands (ECN), 5656 AE Eindhoven, The Netherlands
[2] University of Ljubjana, 12 1000 Ljubjana, Slovenia
[3] Umicore Thin Film Products AG, 9496 Balzers, Liechtenstein
[4] Ecole Polytechnique Federale de Lausanne, 1015 Lausanne, Switzerland
[5] Forschungszentrum Jülich, 52428 Jülich, Germany
[6] Universidad Politechnica de Valencia, 46022 Valencia, Spain
[7] Shanghai Jiaotong University, Shanghai, P.R. China

Abstract In the European project Silicon-Light we developed concepts and technologies to increase conversion efficiencies of thin film silicon solar cells on foil. Main focus was put on improved light management, using NIL for creating light scattering textures, improved TCOs using sputtering, and improved silicon absorber material made by PECVD. On foil we achieved initial cell efficiencies of 11% and on rigid substrates stable efficiencies of 11.6% were achieved. Finally, the project demonstrated the industrial scale feasibility of the developed technologies and materials. Cost of ownership calculations showed that implementation of these technologies on large scale would enable the production of these high efficiency solar modules at manufacturing cost of 0.65 €/Wp with encapsulation costs (0.20 €/Wp) being the dominant costs. Life cycle analysis showed that large scale production of modules based on the technologies developed in Silicon-Light would have an energy payback time of 0.85 years in Central European countries.

1 Introduction

In order to be competitive with other PV technologies, thin film silicon PV needs higher efficiencies and lower production cost levels than the presently available technologies. In the EU project Silicon-Light we addressed both issues. By improving the light management, the silicon absorber layers and the front-side TCO, the project aimed at stable cell efficiencies of more than 11%. In the project these developments were up-scaled to large area roll-to-roll fabrication methods, potentially leading to significant reduction in production costs in comparison with conventional batch production on glass panels.

The project Silicon-Light was carried out by a consortium of 9 partners which were strongly complementary in the fields which are dealt with. The activities were separated into 5 different work packages: 1) light management, 2) silicon absorber layers and interfaces; 3) TCOs; 4) high efficiency cells; 5) integration and upscaling. In the first half of the project, the focus was on the development of the individual elements of the cells (WP 1–3). The best and most promising results then were used in the second half of the project to make high efficiency cells (WP 4), followed by up-scaling and implementation in pilot scale production in WP 5.

2 Results

2.1 Light management

Through a careful interplay between opto-electrical modeling and device fabrication we have identified ideal 2D periodic textures for the back reflector in nip/nip a-Si/μc-Si tandem cells (a-Si cell on top).

The typical device structure used in the modeling is shown in Figure 1. Note that this device structure contains two light management layers: one synthetic structure made by nano-imprint lithography (NIL) on the substrate foil, and one natural structure on the 2000 nm ZnO:B layer, grown by LPCVD.

In the modeling, the non-conformal growth of the ZnO/Ag back contact and of the silicon layers was taking into account [1]. This was done by introducing an isotropic growth factor g as explained in Figure 2. Using this empirical growth model, in which the factor g was determined

---

[a] e-mail: soppe@ecn.nl

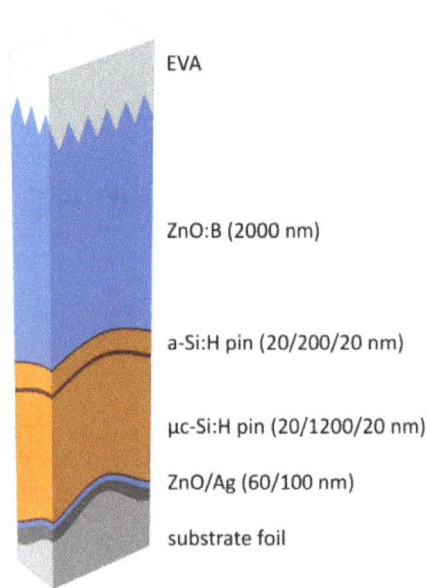

Fig. 1. Structure of micromorph a-Si:H/μc-Si:H solar cell as used in the opto-electrical simulations.

by analyzing cross sections of ZnO/Ag/nip-Si layer stacks on various substrate morphologies we were able to achieve a very realistic description of the stack geometry as input for the optical modeling.

The optical modeling was used to determine the optimum 2D periodic texture of the rear side reflector/contact, taking into consideration that the structures should be experimentally feasible to make and that textures with high aspect ratios commonly lead to formation of micro-cracks in the silicon layers [2]. We found that to avoid formation of micro-cracks, the opening angle φ of the structures at the growth surface (see Fig. 3) should be larger than 130°.

We selected 2D-sinusoidal structures as experimentally most feasible structures, and then determined the J_{sc} that can be achieved for a-Si/μc-Si tandem devices with a bottom layer thickness of 1200 nm, for various periods and heights of the structures [3]. The results of J_{sc} of the bottom cell are presented in Figure 4, and indicate that currents of more than 14 mA/cm^2 could be achieved for 2D-sinusoidal structures with a period of about 1100 nm and a height of about 1000 nm. These optically ideal structures however do not fulfill the $\varphi > 130°$ criterion and will lead to substantial crack formation in the micro-crystalline silicon layers. To fulfill the $\varphi > 130°$ criterion, the height of the sinusoidal structures must be decreased to about 300 nm and for these structures the model predicts maximum currents of about 12.5 mA/cm^2. Further improvements can be done by including low-absorbing doped SiO$_x$ layers and anti-reflecting coatings in the structure.

These 2D periodic textures were successfully made as small masters and subsequently up-scaled to larger shims for roll-to-roll Nano-Imprint-Lithography (NIL). Finally the industrial feasibility of the concept was demonstrated by imprinting these textures on foil of a width of 30 cm (see Fig. 5).

2.2 Silicon layers and interfaces

Detailed analysis of interfaces by advanced micro-copy techniques showed the diffusion of conductive TCO material in voids along the grain boundaries in microcrystalline silicon [4]. In interplay with the texturization activities, we identified growth regimes of microcrystalline silicon where crack formation on textured substrates can be reduced largely, reducing shunting of the solar cells.

We also applied electron energy-loss spectroscopy (EELS) to determine the parasitic absorption losses in the Ag back reflector. EELS can be used to measure the energy-loss that electrons suffer when travelling through a sample in a TEM. The physical processes involved in energy-loss events can be (i) band gap transitions, (ii) plasmon collective oscillations and (iii) interatomic transitions. Therefore, EEL spectra can be used to provide information about (a) the band gap energy, (b) the carrier density (through the plasmon energy) and (c) the local atomic composition with nanometer spatial resolution [5]. With recent improvements in TEM technology, energy losses as low as 1 eV are accessible, allowing the measurement of surface and bulk plasmon resonances in Ag with high spatial resolution [6]. Previously, plasmon absorptions were measured in back-contact Ag layers using visible light leading to a limited spatial resolution. Figure 6a shows a Scanning TEM (STEM) image of a Ag back reflector grown on a ZnO back contact. The Ag grain sizes are between 50 and 200 nm which is much smaller than the diameter of the optical beam previously used to probe plasmon losses. The TEM beam was positioned at the surface and at the centre of a grain, as indicated in Figure 6a, to measure plasmon absorption. The two spectra are shown in Figure 6b. A bulk plasmon peak at 3.8 eV is visible when measuring at the center of the Ag crystal. More importantly, when measuring at the Ag grain edge, where the reflection of light in solar cells happens, a strong absorption peak, centered at ∼2.8 eV and extended down to 2 eV, is present. As the absorption range of a thin-film Si solar cell is between ∼1.5 and 3 eV, this peak is indicative of parasitic absorption and can decrease the performance of the solar cell in case of thin absorber layers [7].

One way to reduce the absorption due to the Ag/Si surface plasmon peak is to apply a dielectric buffer layer between the Ag and Si layers [8]. This approach could be one of the positive side-effects of including a ZnO layer between the Ag back reflector and the nip solar cell. It has also been shown that annealing the Ag layer before the Si PECVD process improves the reflectivity below ∼1000 nm (above ∼1.2 eV) [9]. We have not yet assessed whether this effect is caused by a reduction in the Ag plasmon absorption or by reduction of other parasitic absorptions.

EELS was also used to determine the boron (B) concentration profile in the p- and adjacent layers. The p-layer in a nip Si solar cell should be both highly conductive and transparent in the absorption range of the Si layer. The conductivity can be adjusted by varying the B concentration, while the transparency window can be extended by adding either carbon (C) or oxygen. The drawback of

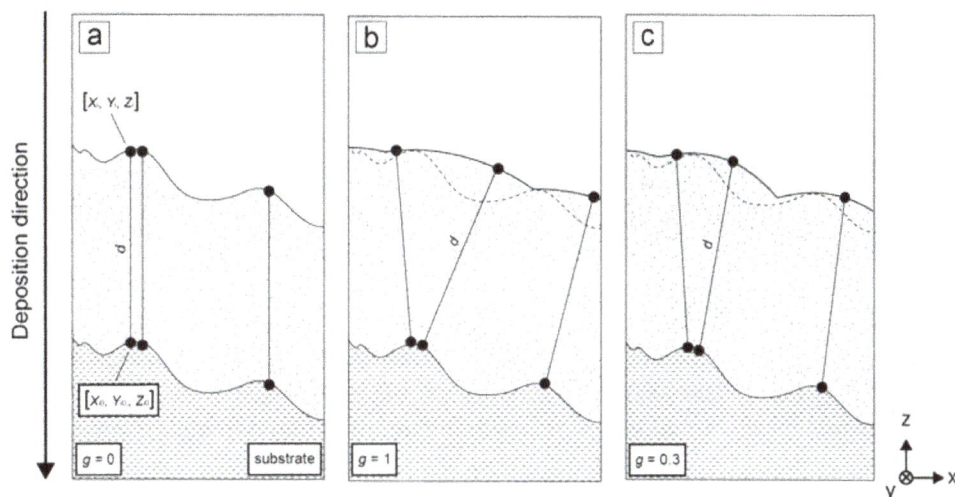

Fig. 2. Vertical cross-sections of a thin-film layer grown on a substrate as calculated by the developed 3-D growth model by considering: (a) fully conformal ($g = 0$), (b) fully isotropic ($g = 1$) and (c) combined growth type ($g = 0.3$). The dashed lines on the top surfaces in (b) and (c) indicate the reference texture obtained by conformal growth type.

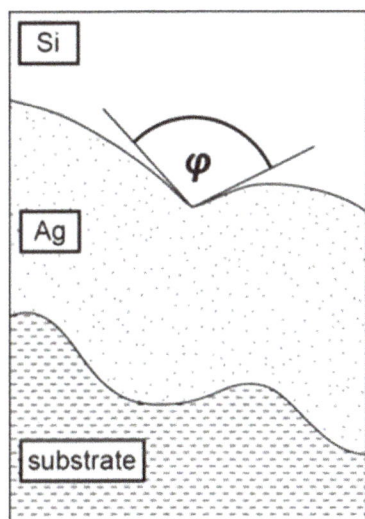

Fig. 3. Opening angle φ.

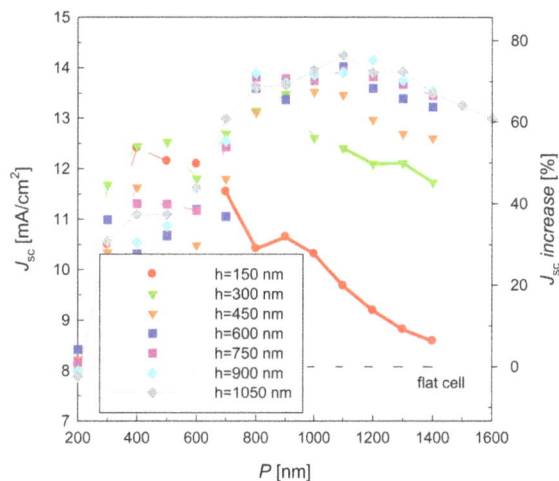

Fig. 4. J_{sc} calculated for the bottom cell of a-Si/μc-Si tandem devices with a bottom layer thickness of 1200 nm, for various periods and heights of 2D sinusoidal structures of the rear side reflector. Only the bold part of the curves fulfill the criterion $\varphi > 130°$.

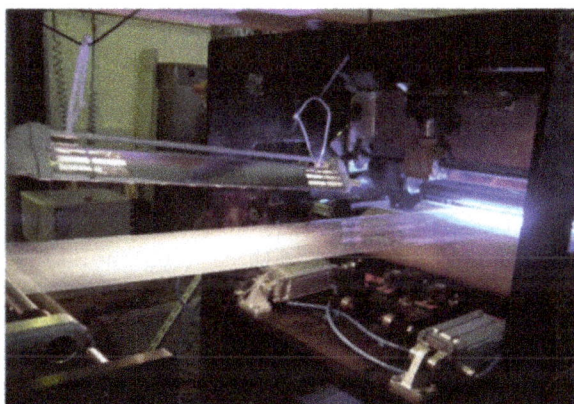

Fig. 5. Nanoptics R-2-R-pilotmachine demonstrating NIL on PEN film.

increasing the C content is a decrease in the conductivity. Therefore, an optimal C/B ratio has to be found to fulfill the requirements for an a-Si solar cell. Figure 7a shows calculations of the expected cell efficiency by varying the optical band gap E_g and the activation energy E_{act}. The best cell efficiency is expected for high E_g and low E_{act} of the p-a-SiC layer. By varying the experimental growth conditions within a roll-to-roll compatible process line, we are striving to reach this high efficiency region. The values of the experimentally accessible region are shown by the black dashed line in Figure 7a and show cells with an efficiency of close to 8% [10].

In practice, this layer has to remain as thin as possible (i.e. a few nm) to limit the optical absorption and the carrier recombination in the layer. A standard characterization tool to directly access to doping concentration is secondary ion mass spectrometry (SIMS) however SIMS

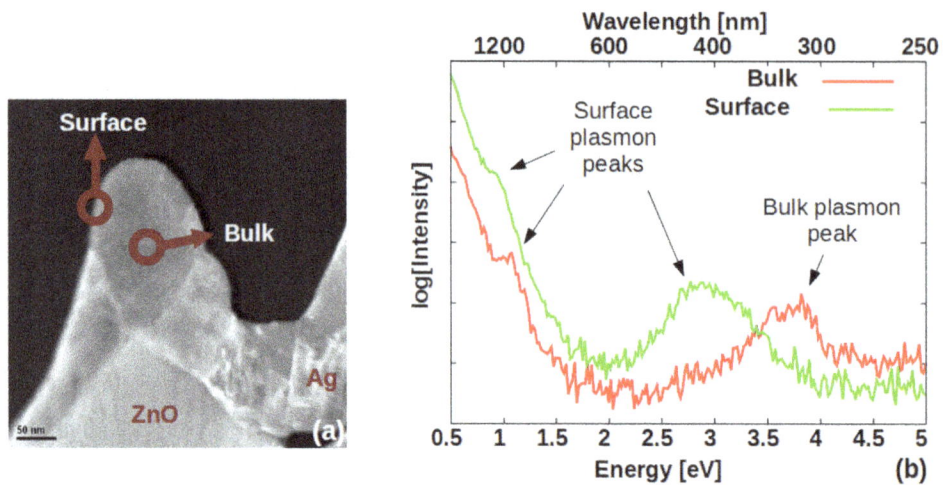

Fig. 6. (a) STEM image of an as-grown Ag back reflector on top of a 2 μm thick ZnO layer. The "bulk" and "surface" locations refer to positions where EEL spectra were recorded. (b) EEL spectra taken at the "bulk" and "surface" regions.

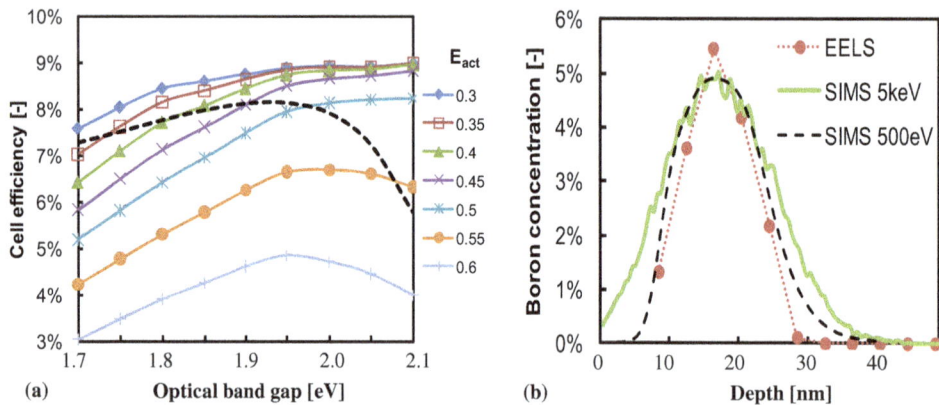

Fig. 7. (a) Calculated efficiency of a-Si solar cells as a function of the optical band gap of the p-a-SiC layer for seven values of the activation energy (E_{act}). The efficiencies for grown solar cells are given by the dashed line. (b) Relative boron concentration profiles determined by EELS (markers) and SIMS (lines).

has limited depth resolution and averages the doping concentration over distances of several micrometers, which is not suitable for devices on rough substrates. To overcome this problem, we have measured core loss signals using EELS [11]. Recorded graphs are shown in Figure 7b. The relative B concentration measured by EELS is higher when compared to SIMS values at the center of the layer. The thickness of the p-layer is determined to be about 15 nm.

Although core-level EELS measures the concentration of B atoms and ions, it is not sensitive to the active dopant concentration. In contrast, low-loss EELS shows an absorption peak related to the plasmon absorption, which is located at 17.2 eV for Si. We examined a test sample which consisted of a stack of three ~200 nm SiC layers with no (i-SiC), standard (p+-SiC) and double (p++-SiC) amounts of B doping sandwiched between standard back (ZnO/Ag) and front (ITO) contact layers. In Figure 8, the plasmon energy (open symbols) is plotted as a function of depth. For comparison, a SIMS profile of the B concentration is shown as a solid line. In order to estimate the relative B concentration obtained by SIMS, we assume that

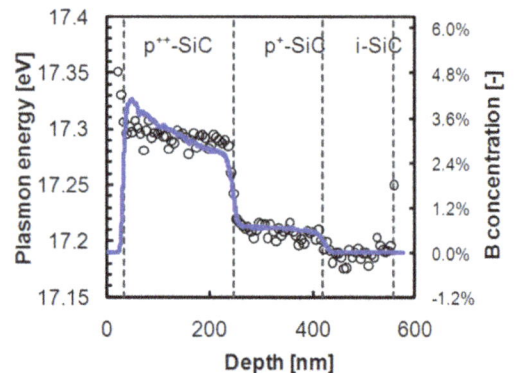

Fig. 8. Plasmon energy determined by EELS (data points) and boron profile from SIMS (line). The dashed lines indicate the interfaces in the sample.

the total density of atoms is 5×10^{22} cm^{-3}; the typical error when determining B in Si is ~10 at.% [12]. We observe a strong relationship between the B concentration and the

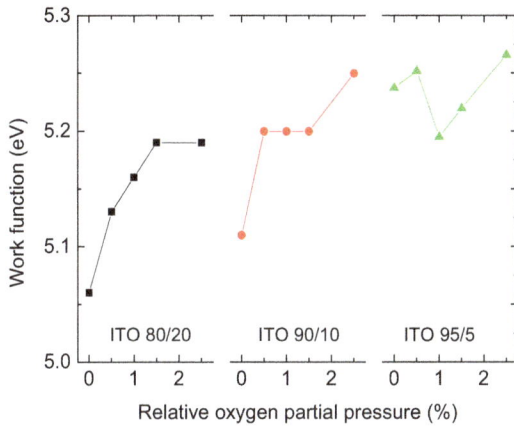

Fig. 9. Dependence of the work function on target composition and oxygen partial pressure during deposition.

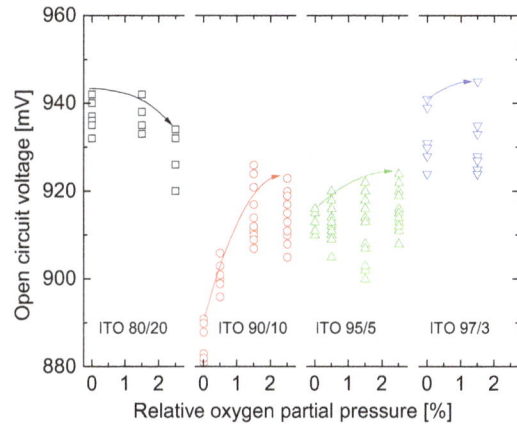

Fig. 10. Variation of the open circuit voltage V_{oc} with the oxygen content in the sputtering plasma of the ITO front electrode.

plasmon energy. By assuming that the density of atoms and the carrier mobility are independent of the B-doping level, we can deduce that the increase in B concentration increases the valence electron density. Furthermore, the TCO front and back contacts and the Ag layer can be identified by the measurements of their plasmon energies at the expected depths (not shown in the graph).

2.3 TCO development

TCO development in the project Silicon-Light was focused on development of new TCOs by sputtering. The reference material in these investigations was ITO and one of the topics was to eliminate a possible collection barrier between TCO and p-layer. Band diagram analysis shows that the interface between commonly used TCOs like AZO and ITO and the p-layer provides such a theoretical collection barrier. This barrier can be reduced if the work function of the TCO can be increased. Exact determination of the work function of sputtered TCOs is not straightforward but we established a method to determine work functions of TCOs accurately and reproducible by Kelvin Probe measurements. Applying reactive sputtering, with oxygen as reactant we were successful in increasing the work function of ITO [13]. Figure 9 shows the results of these experiments which indicate that highest work functions are obtained for sputter targets with the highest indium content, and with highest oxygen concentrations during the sputter process. However, the V_{oc}s of the a-Si cells that we made with these layers (see Fig. 10) do not follow straightforwardly the trend of the work function, but the correlation between V_{oc} and oxygen pressure seems to be statistically significant.

2.4 High efficiency cells

By combining best results from the light management activities, silicon-layer development and TCO development we were able to improve cell efficiencies significantly.

The aim was to achieve a-Si/μc-Si tandem nip/nip devices with stabilized efficiencies of 11%. For this purpose we had to improve both the μc-Si bottom cells and the a-Si top cells. Applying NIL to fabricate textured back contacts, we were able to achieve J_{sc} of 24 mA/cm^2 for single junction nip μc-Si cells with a μc-Si absorber layer of only 1000 nm. These currents were obtained for both a random texture (replica of Asahi-U glass) and for 2D sinusoidal structures. Due to shunting through microcracks, however, the fill factors of the cells on the random texture were significantly lower than on the sinusoidal structures. Cross sectional TEM images (see Fig. 11) show the successful suppression of crack formation of μc-Si cells grown on 2D sinusoidal structures [14].

To increase the current in the a-Si top cells, we applied p-doped SiO$_x$:B as window layer. This is achieved by adding CO$_2$ to the precursor gas mix during deposition [15]. Figure 12 shows the effect on the EQE in the blue wavelength regime. Best solar cells were achieved for a CO$_2$/SiH$_4$ ratio of 0.4, with a V_{oc} of 920 mV, J_{sc} = 14.6 mA/cm^2 and a FF of 68% leading to an efficiency of 9.0%.

Finally we made nip a-Si/μc-Si tandem cells, both on glass and on steel substrates.

2.4.1 Tandem cells on glass substrates

The stack configuration of these cells was: glass/ZnO:B/Ag/ZnO/nip μc-Si/ZnO/nip a-Si/ZnO.

The glass was covered with a 5 micron thick layer of ZnO grown by LPCVD. The natural texture of this ZnO layer serves as the light scattering texture of the back reflector, which was completed by sputtering a Ag layer plus a thin ZnO buffer layer on top of it.

Best nip tandem cells on glass had an initial efficiency of 13.2% and a stable efficiency of 11.6%. This device contains a μc-Si bottom cell with a thickness of 1.7 micron and a intermediate reflector layer of ZnO with a thickness of 1.1 micron. The spectral response of this cell is shown in Figure 13.

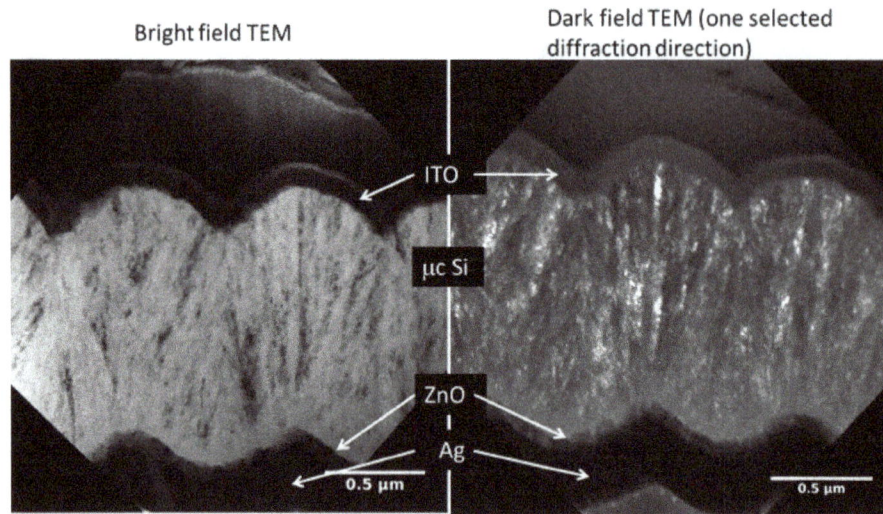

Fig. 11. Cross sectional TEM images made of microcrystalline silicon solar cells made on steel foil with a 2D sinusoidal texture.

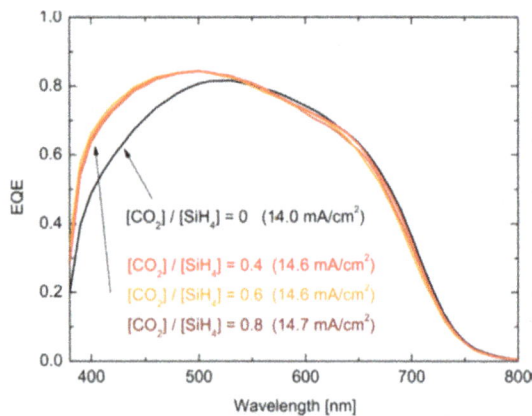

Fig. 12. EQEs of a-Si nip solar cells with p-layers made from SiO_x with increasing oxygen content.

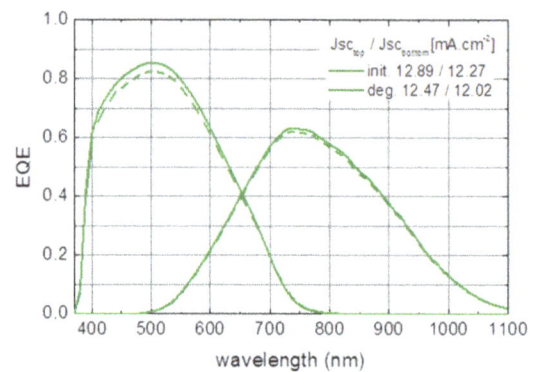

Fig. 13. EQEs of the best stable device in initial (full line) and stable state (dashed line).

2.4.2 Tandem cells on steel foil

The stack configuration of the tandem cells on steel foil is sketched in Figure 14. On the foil we apply a UV curing lacquer for NIL, which also serves as an insulating layer to enable monolithic series interconnection. The metal grid is applied to enable current collection from the front TCO: a 80 nm thin ITO layer.

As for the single junction μc-Si cells on foil, we applied both a random master (replica of Asahi-U texture) and a 2D periodic (sinusoidal) master to make imprints in the UV lacquer. The fill factor of the tandem cells appear to be less vulnerable for microcracks in the μc-Si bottom cell and so we obtained very similar results for both types of textures. For tandem cells with a bottom cell thickness of 1000 nm and a top cell thickness of 300 nm, we obtained (initial) efficiencies of 11.0% (V_{oc} = 1381 mV; J_{sc} top-cell = 11.84 mA/cm^2; J_{sc} bottom-cell = 12.31 mA/cm^2, FF = 67%). The spectral response curve of these cells is shown in Figure 15. Note that these cells do not contain an intermediate reflector yet.

2.5 Implementation in pilot production and calculations of cost and environmental impact

Implementation of project results into pilot production was hindered by financial problems at the end user (VHF Technologies). Nevertheless they were able to increase the stabilized efficiency for a-Si/a-Si tandem cells in their production line towards 8.0% by implementing light management technologies developed by the project.

VHF Technologies applies ITO as top TCO and cost reductions can be realized if the indium consumption could be reduced. For this purpose we investigated co-sputtering of ITO and AZO (Al doped ITZO). In order to simulate co-sputtering of ITO and AZO, a chess-pattern sputtering target containing both ITO and AZO tiles was prepared. An industrial scale target was prepared by Umicore for VHF Technologies PVD machines. Small ITO and AZO segments were bonded onto a molybdenum backing plate in a chess pattern design (see Fig. 16). Using this target, single junction a-Si cells and a-Si/a-Si tandem cells were made in the production line and compared with reference cells with ITO as top TCO.

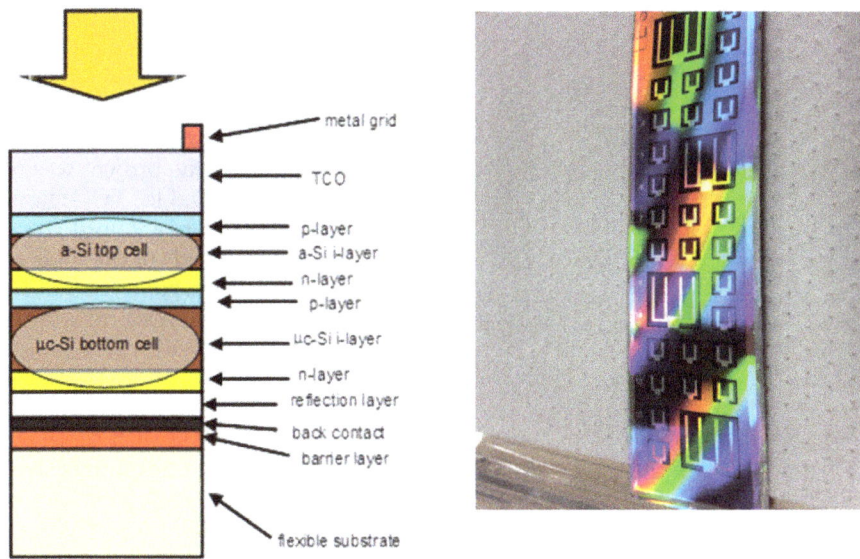

Fig. 14. Left: Stack configuration of tandem cells on (steel) foil. The UV curing lacquer for NIL is indicated as "barrier layer"; right: a photograph showing 4×4 mm^2 and 10×10 mm^2 cells. The rainbow effect is due to the 2D structure of the back contact.

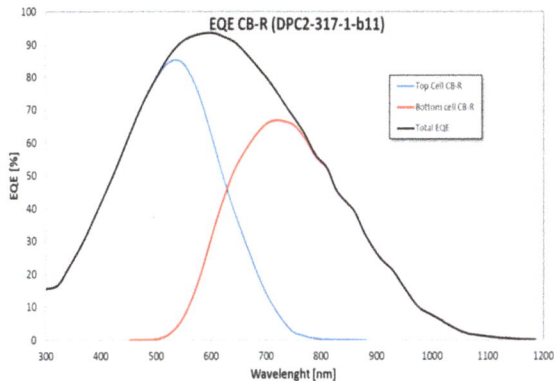

Fig. 15. EQE spectrum of a 11% efficiency tandem cell on steel foil.

Table 1. IV characteristics of a-Si/a-Si tandem cells with top TCO made with ITO/AZO mosaic target with various amounts of oxygen added to the sputter gas argon.

Target	Oxygen addition	V_{oc} (V)	J_{sc} (mA/cm^2)	FF (%)
ITO	ITO std	1.739	5.54	72.4
ITO/AZO	3% oxygen	1.772	5.46	70.2
	2% oxygen	1.711	5.49	70.1
	1% oxygen	1.824	5.51	72.5

Fig. 16. ITO/AZO mosaic target (chess pattern) for sputtering tests under industrial conditions at VHF Technologies. Dimension 750×125 mm, 48 segments.

The solar cells made with the ITO/AZO top TCO were as good as the reference cells. In Table 1 an overview is given of an a-Si/a-Si tandem cells showing that high V_{oc}s and good fill factors can be obtained with these TCOs.

Using data available in the public literature – e.g. [16,17] – we also attempted a cost and an environmental analysis of the production process of thin film silicon solar cells on foil, in which the technologies developed in the project silicon light would be implemented. So we assumed a production line of a-Si/μc-Si tandem cells on steel foil, with a capacity of 120 MWp/yr and a module efficiency of 10%. In such a line, modules could be produced at a cost of 650 €/kWp, which can be divided in 159 €/kWp Fixed Costs, 156 €/kWp Operational Costs and 334 €/kWp Material Costs. So the largest costs in the production are Materials Costs and if we look at a breakdown of the costs per process step (Fig. 17) we see that these costs are dominated by the material costs involved in the "final assembly", namely the cost for lamination

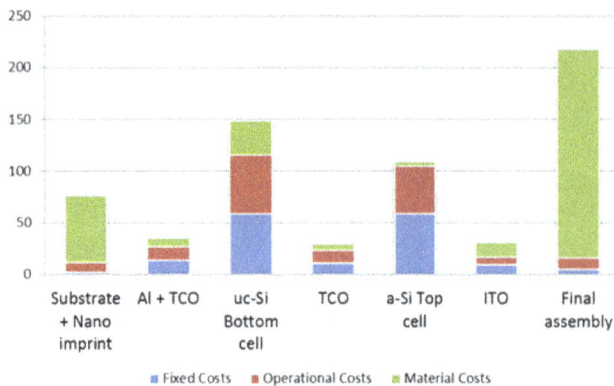

Fig. 17. Cost break down per major process step. Series interconnection, lamination and assembly of junction box are comprised in the "final assembly".

(ETFE) and the junction box. Most significant cost reductions could therefore be achieved if cheaper encapsulants could be applied.

A life cycle assessment was carried out in which three kinds of flexible thin-film silicon PV modules were compared. A comparison was made between an a-Si module (benchmark-1), an a-Si/μc-Si micromorph module (benchmark-2) and an a-Si/μc-Si micromorph module containing the improvements developed in the Silicon-Light project (Silicon-Light). The three modules differed from each other in the configurations of the photoactive layer as well as in the texture of the back-reflector.

The analyses show that the Silicon-Light PV module, i.e, the device implementing the technology achievements of the Silicon-Light project, shows the best environmental profile of the three devices. This is because of the advanced back-reflector texture fabricated by nanoimprint lithography. The advanced texture allows the application of a thinner photoactive layer while maintaining a high module efficiency (10%). This photoactive layer of the Silicon-Light device consists of a 180 nm a-Si and a 1000 nm μc-Si layer, whereas without NIL a (flat) bottom cell of more than 2000 nm would be required to achieve the same J_{sc}.

The environmental profile of the Silicon-Light PV module is characterized by an embedded energy of 617.5 MJ/m^2, an energy payback time of 0.85 years and a carbon footprint of 20.1 g CO$_2$-eq./kWh.

3 Conclusions

Silicon-Light is a European FP7 project that aimed at the development of high efficiency thin film silicon solar cells on foil. The project focused on (a) improved light management through implementation of nano-imprint lithography; (b) improved silicon material and (c) novel TCO materials made by sputtering. In Silicon-Light we investigated methods to create light-scattering textures at the rear side of the cell. For the fabrication of these textures, with structures on nanometer scale, methods from

the semiconductor industry like e-beam lithography were applied. To demonstrate that these textures can be manufactured on large scale, these methods were combined with large scale production methods for nano-imprint lithography (NIL) which are used in the holographic industry.

Another aim of the project was to develop new Transparent Conductive Oxide (TCO) layers for thin film silicon solar cells. TCO layers are needed to collect the generated current at the front side of the solar cell. Indium Tin Oxide (ITO) is technically a good candidate but the scarceness of indium requires to investigate alternative materials. Zinc-oxide (ZnO) is a possible alternative but has certain disadvantages related to its stability in humid environments. In Silicon-Light new TCOs were developed that combine the advantages of ITO with those of ZnO.

Integration of the novel light management techniques and new TCOs into high efficiency solar cells was one of main objectives of the project. We achieved thin film silicon solar cells with initial efficiencies of 13.2%.

Finally, the project demonstrated the industrial scale feasibility of the developed technologies and materials. Cost of ownership calculations showed that implementation of these technologies on large scale would enable the production of these high efficiency solar modules at manufacturing cost of 65 Eurocents per Wp. Life cycle analysis showed that large scale production of modules based on the technologies developed in Silicon-Light would have an Energy Payback Time of 0.85 years in Central European countries.

This work was funded by the European FP7 project Silicon-Light (GA No. 241277).

References

1. M. Sever, B. Lipovšek, J. Krč, A. Čampa, G. Sánchez Plaza, F.-J. Haug, M. Duchamp, W. Soppe, M. Topič, Sol. Energy Mater. Sol. Cells **119**, 59 (2013)
2. M. Python, O. Madani, D. Domine, F. Meillaud, E. Vallat-Sauvain, C. Ballif, Sol. Energy Mater. **93**, 1714 (2009)
3. M. Sever et al., in *Proceedings of the 27th EU PVSEC, 24–28 September 2012*, pp. 2129–2131
4. M. Duchamp, M. Lachmann, C.B. Boothroyd, A. Kovacs, F.-J. Haug, C. Ballif, R.E. Dunin-Borkowski, Appl. Phys. Lett. **102**, 133902 (2013)
5. R.W. Kelsall et al., *Nanoscale Science and Technology* (John Wiley & Sons, Ltd, Chichester, 2005)
6. M. Duchamp et al., in *Proceedings of the 26th EU PVSEC, 2011*, p. 2554
7. F.-J. Haug et al., MRS Symposium Proceedings **1321**, 63 (2011)
8. F.-J. Haug, T. Söderström, O. Cubero, V. Terrazzoni-Daudrix, C. Ballif, J. Appl. Phys. **104**, 064509 (2008)
9. K. Söderström, F.-J. Haug, J. Escarré, C. Pahud, R. Biron, C. Ballif, Sol. Energy Mater. Solar Cells **95**, 3585 (2011)
10. B. Van Aken, M. Duchamp, C. Boothroyd, R. Dunin-Borkowski, W. Soppe, J. Non-Cryst. Solids **358**, 2179 (2012)

11. M. Duchamp, C. Boothroyd, M. Moreno, B. van Aken, W. Soppe, R. Dunin-Borkowski, J. Appl. Phys. **113**, 093513 (2013)
12. G. Stingeder, Anal. Chim. Acta **297**, 231 (1994)
13. F.-J. Haug, R. Biron, G. Kratzer, F. Leresche, J. Besuchet, Ch. Ballif, M. Dissel, S. Kretschmer, W. Soppe, P. Lippens, K. Leitner, Prog. Photovolt.: Res. Appl. **20**, 727 (2012)
14. W. Soppe, M. Dorenkamper, J.-B. Notta, P. Pex, W. Schipper, R. Wilde, Phys. Stat. Sol. A **210**, 707 (2013)
15. R. Biron, C. Pahud, F.-J. Haug, J. Escarré, K. Söderström, C. Ballif, J. Appl. Phys. **110**, 124511 (2011)
16. D. Richard, in *Photon International, March 2010*, p. 118
17. D. Richard, in *Photon International, November 2010*, p. 142

PERMISSIONS

LIST OF CONTRIBUTORS

F. Dadouche
Institut d'Électronique du Solide et des Systémes (InESS), CNRS, 23 rue du Loess, BP 20 CR, 67037 Strasbourg Cedex 2, France

O. Béthoux, M.E. Gueunier-Farret, C. Marchand and and J.P. Kleider
Laboratoire de Génie Électrique de Paris, CNRS UMR 8507, SUPELEC, Université Paris-Sud, UPMC Univ Paris VI, 11 rue Joliot-Curie, Plateau de Moulon, 91192 Gif-sur-Yvette Cedex, France

E.V. Johnson and P. Roca i Cabarrocas
Laboratoire de Physique des Interfaces et Couches Minces, École polytechnique, CNRS, 91128 Palaiseau, France

Antoine Salomon
Total ´Energies Nouvelles, La Défense, France

Guillaume Courtois
Total Énergies Nouvelles, La Défense, France
LPICM, CNRS – École Polytechnique, Palaiseau, France

Pere Roca i Cabarrocas
LPICM, CNRS – École Polytechnique, Palaiseau, France

Parsathi Chatterjee
Energy Research Unit, Indian Association for the Cultivation of Science, Kolkata, India
LPICM, CNRS – École Polytechnique, Palaiseau, France

Veinardi Suendo
Inorganic and Physical Chemistry Research Division, Institut Teknologi Bandung, Indonesia Abstract

H. Koshino, Z. Tang, S. Sato and H. Shirai
Graduate School of Science and Engineering, Saitama University, 255 Shimo-Okubo, Sakura, 338-8570 Saitama, Japan

H. Shimizu
Saitama Industrial Technology Centre (SAITEC) 3-12-28 Kami-Aoki, Kawaguchi, 333-0844 Saitama, Japan

Y. Fujii3, T. Hanajiri
Bio-Nano Electronics Research Centre, Toyo University, 2100 Kujirai, Kawagoe, 350-8585 Saitama, Japan

J. John
IMEC, Kapeldreef 75, Leuven, Belgium

V. Prajapati, B. Vermang, A. Lorenz and C. Allebe
IMEC, Kapeldreef 75, Leuven, Belgium
Katholieke Universiteit Leuven, Leuven, Belgium

A. Rothschild, L. Tous, A. Uruena, K. Baert and and J. Poortmans
Katholieke Universiteit Leuven, Leuven, Belgium

S. Chakraborty
Indian Association for the Cultivation of Science, 700032 Kolkata, WB, India

P. Chatterjee
Indian Association for the Cultivation of Science, 700032 Kolkata, WB, India
Laboratoire de Physique des Interfaces et des Couches Minces, École Polytechnique, 91128 Palaiseau, France

R. Cariou and P. Roca i Cabarrocas
Laboratoire de Physique des Interfaces et des Couches Minces, École Polytechnique, 91128 Palaiseau, France

M. Labrune
Laboratoire de Physique des Interfaces et des Couches Minces, École Polytechnique, 91128 Palaiseau, France
Total S.A., Gas & Power – R&D Division, 92400 Courbevoie, France

E. Bunte, H. Zhu, J. Hüpkesa and J. Owen
Institut für Energie- und Klimaforschung, IEK5-Photovoltaik, Forschungszentrum Jülich GmbH, 52425 Jülich, Germany

S. Chakraborty
Energy Research Unit, Indian Association for the Cultivation of Science, Jadavpur, 700032 Kolkata, India

A. Datta
Energy Research Unit, Indian Association for the Cultivation of Science, Jadavpur, 700032 Kolkata, India
Haltu High School for Girls (H.S.), Neli Nagar, Haltu, 700 078 Kolkata, India

P. Chatterjee
Energy Research Unit, Indian Association for the Cultivation of Science, Jadavpur, 700032 Kolkata, India
Laboratoire de Physique des Interfaces et Couches Minces, CNRS, École Polytechnique, 91128 Palaiseau, France

M. Labrune
Laboratoire de Physique des Interfaces et Couches
Minces, CNRS, ´Ecole Polytechnique, 91128 Palaiseau,
France
Total S.A., Gas & Power – R&D Division, 92400
Courbevoie, France

P. Roca i Cabarrocas
Laboratoire de Physique des Interfaces et Couches
Minces, CNRS, École Polytechnique, 91128 Palaiseau,
France

**A. Grimm, D. Kieven, I. Lauermann, M.Ch. Lux-
Steiner and R. Klenk**
Helmholtz-Zentrum Berlin f¨ur Materialien und
Energie, Hahn-Meitner-Platz 1, 14109 Berlin, Germany

F. Hergert and R. Schwieger
Bosch Solar CISTech, M¨unstersche Str. 24, 14772
Brandenburg an der Havel, Germany

**R. Kotipallia, R. Delamare, O. Poncelet, X. Tang, L.A.
Francis and D. Flandre**
ICTEAM, Université catholique de Louvain, Place du
Levant 3, 1348 Louvain-la-Neuve, Belgium

V. Tvarozek, I. Novotny, M. Vojs and S. Flickyngerova
Institute of Electronics and Photonics, Slovak University
of Technology, Ilkovicova 381219 Bratislava, Slovakia

K.S. Shtereva
Institute of Electronics and Photonics, Slovak University
of Technology, Ilkovicova 381219 Bratislava, Slovakia
Department of Electronics, University of Rousse,
Studentska 8, 7017 Ruse, Bulgaria

P. Sutta
New technologies-Research Center, University of West
Bohemia, Plzen Czech Republic

M. Milosavljevic
VINČA Institute of Nuclear Sciences, Laboratory for
Atomic Physics, Belgrade, Serbia

A. Vincze
International Laser Centre, Bratislava, Slovakia

**M. Schmid, R. Caballero, R. Klenk, T. Rissom and
M.Ch. Lux-Steiner**
Helmholtz Zentrum Berlin für Materialien und
Energie, Hahn-Meitner-Platz 1, 14109 Berlin, Germany

J. Krč and M. Topič
University of Ljubljana, Faculty of Electrical
Engineering, Tržaška 25, 1000 Ljubljana, Slovenia

**Ana-Maria Teodoreanu, Felice Friedrich, Rainer
Leihkauf and Christian Boit**
Technische Universität Berlin, Semiconductor Devices
Division, PVcomB, Einsteinufer 19, Sekr. E2, 10587
Berlin, Germany

Caspar Leendertz and and Lars Korte
Helmholtz-Zentrum Berlin, Institute for Silicon
Photovoltaics, Kekuléstrasse 5, 12489 Berlin, Germany
Abstract

K.H. Kim
TOTAL S.A., Gas & Power – R&D Division, Courbevoie,
France
Laboratoire de Physique des Interfaces et des Couches
Minces (UMR 7647 CNRS), École Polytechnique, 91128
Palaiseau, France

E.V. Johnson, A. Abramov and P. Roca i Cabarrocas
Laboratoire de Physique des Interfaces et des Couches
Minces (UMR 7647 CNRS), École Polytechnique, 91128
Palaiseau, France

A. Jäger-Waldaua
European Commission, Joint Research Centre;
Renewable Energy Unit, via E. Fermi 2749, TP 450,
21027 Ispra (VA), Italy

R. Evans, D. Ong, O. Kunz, U. Schubert and R. Egan
Suntech R&D Australia, Pty., Ltd. 82-86 Bay St.,
Botany, NSW 2019, Australia

J. Dore, B. Eggleston and K.H. Kim
Suntech R&D Australia, Pty., Ltd. 82-86 Bay St.,
Botany, NSW 2019, Australia
University of NSW Sydney, NSW 2052, Australia

S. Varlamov, J. Huang and M. Green
University of NSW Sydney, NSW 2052, Australia

T. Zhang, I. Perez-Wurfl and G. Conibeer
The University of New South Wales, UNSW, Sydney,
NSW 2052, Australia

B. Berghoff and S. Suckow
Institute of Semiconductor Electronic, RWTH Aachen
University, Aachen, Germany

F. Delachat, F. Antoni, P. Prathap and A. Slaoui
CNRS-UdS InESS, Strasbourg, France

C. Cayron and C. Ducros
CEA-Liten, Grenoble, France

**Marcus Rennhofer, Angelika Dangel and Bogdan
Duman**
Photovoltaics System, Austrian Institute of Technology,
1220 Vienna, Austria

Ankit Mittal
Photovoltaics System, Austrian Institute of Technology, 1220 Vienna, Austria
Department of Physics, University of Vienna, 1010 Vienna, Austria

Victor Schlosser
Department of Physics, University of Vienna, 1010 Vienna, Austria

Jae Sung Yun, Cha Ho Ahn, Miga Jung, Jialiang Huang, Sergey Varlamov and Martin A. Green
University of New South Wales, NSW, 2033, Kensington, Australia

Kyung Hun Kim
University of New South Wales, NSW, 2033, Kensington, Australia
Suntech R&D Australia, Pty., Ltd. 82-86 Bay St., BSW 209, Botany, Australia

Martin Theuringa, Stefan Geissend¨orfer, Martin Vehse, Karsten von Maydell and Carsten Agert
NEXT ENERGY – EWE Research Centre for Energy Technology at Carl von Ossietzky University, Carl-von-Ossietzky-Straße 15, 26129 Oldenburg, Germany

Per Ingemar Widenborg
Solar Energy Research Institute of Singapore, National University of Singapore, 7 Engineering Drive 1, 117574 Singapore, Singapore

Armin Gerhard Aberle
Solar Energy Research Institute of Singapore, National University of Singapore, 7 Engineering Drive 1, 117574 Singapore, Singapore
Department of Electrical and Computer Engineering, National University of Singapore, 117583 Singapore, Singapore

Avishek Kumar
Solar Energy Research Institute of Singapore, National University of Singapore, 7 Engineering Drive 1, 117574 Singapore, Singapore
Department of Electrical and Computer Engineering, National University of Singapore, 117583 Singapore, Singapore
Institute of Materials Research and Engineering, A*STAR (Agency for Science, Technology and Research), 3 Research Link, 117602 Singapore, Singapore

Goutam Kumar Dalapati, Gomathy Sandhya Subramanian
Institute of Materials Research and Engineering, A*STAR (Agency for Science, Technology and Research), 3 Research Link, 117602 Singapore, Singapore

T. Schutz-Kuchly and A. Slaoui
Laboratoire des sciences de l'Ingénieur, de l'Informatique et de l'Imagerie (ICUBE) UMR 7357, UdS/CNRS, 23 rue du Loess, BP 20 CR, 67037 Strasbourg Cedex 2, France

J. Zelgowski and A. Bahouka
IREPA-Laser Pole API – Parc d'innovation, 67400 Strasbourg, France

M. Pawlik and J.-P. Vilcot
Institut d'Électronique, de Microélectronique et de Nanotechnologie (IEMN) UMR 8520, Université Lille Sciences et Technologies, CS 60069, 59652 Villeneuve d'Ascq, France

E. Delbos
KMG Group, 45 avenue des États-Unis, 78035 Versailles, France

M. Bouttemy
Institut Lavoisier de Versailles UMR 8180, Université de Versailles-St-Quentin en Yvelines, 45 avenue des États-Unis, 78000 Versailles, France

R. Cabal
CEA-INES, 50 avenue du Lac Léman, 73375 Le Bourget du Lac, France

Nies Reininghausa, Martin Kellermann, Karsten von Maydell and Carsten Agert
NEXT ENERGY, EWE Research Centre for Energy Technology at the University of Oldenburg, Carl-von-Ossietzky-Str. 15, 26129 Oldenburg, Germany

Thomas Mambrini
Université Paris-Sud 11, UMR 8507, LGEP, Bâtiment 301, 91405 Orsay Cedex, France
SUPELEC, LGEP, UMR 8507, 3 rue Joliot-Curie, Plateau de Moulon, 91192 Gif-sur-Yvette Cedex, France
CNRS, LGEP, UMR 8507, 11 rue Joliot-Curie, Plateau de Moulon, 91192 Gif-sur-Yvette Cedex, France

Anne Migan Dubois
SUPELEC, LGEP, UMR 8507, 3 rue Joliot-Curie, Plateau de Moulon, 91192 Gif-sur-Yvette Cedex, France
CNRS, LGEP, UMR 8507, 11 rue Joliot-Curie, Plateau de Moulon, 91192 Gif-sur-Yvette Cedex, France
Sorbonne Universités, UPMC Univ. Paris 06, UMR 8507, LGEP, 5 Place Jussieu, 75005 Paris Cedex, France

Christophe Longeaud
SUPELEC, LGEP, UMR 8507, 3 rue Joliot-Curie, Plateau de Moulon, 91192 Gif-sur-Yvette Cedex, France
CNRS, LGEP, UMR 8507, 11 rue Joliot-Curie, Plateau de Moulon, 91192 Gif-sur-Yvette Cedex, France

Jordi Badosa and Martial Haeffelin
LMD, Institut Pierre-Simon Laplace, CNRS, Ecole Polytechnique, 91128 Palaiseau Cedex, France

Laurent Prieur and Vincent Radivoniuk
SOLEÏS Technologie, 4 allée Jean-Paul Sartre, 77186 Noisiel, France

Prabal Goyal
Air Liquide, Centre de Recherche Paris Saclay, 78354 Jouy-en-Josas, France
Laboratoire de Physique des Interfaces et des Couches Minces, CNRS, Ecole Polytechnique, 91128 Palaiseau, France
LPICM- LPICM, CNRS, Ecole polytechnique, Université Paris-Saclay, 91128 Palaiseau, France

Junegie Hong
Air Liquide, Centre de Recherche Paris Saclay, 78354 Jouy-en-Josas, France
LPICM- LPICM, CNRS, Ecole polytechnique, Université Paris-Saclay, 91128 Palaiseau, France

Farah Haddad, Jean-Luc Maurice, Pere Roca i Cabarrocas and Erik Johnson
Laboratoire de Physique des Interfaces et des Couches Minces, CNRS, Ecole Polytechnique, 91128 Palaiseau, France
LPICM- LPICM, CNRS, Ecole polytechnique, Universit´e Paris-Saclay, 91128 Palaiseau, France

Onno Gabriel, Simon Kirner, Bernd Stannowski and Rutger Schlatmann
PVcomB, Helmholtz-Zentrum Berlin für Materialien und Energie GmbH, Schwarzschildstr. 3, 12489 Berlin, Germany

Michael Klick
Plasmetrex GmbH, Schwarzschildstr. 3, 12489 Berlin, Germany

K. Derbouz and A. Slaoui
InESS-CNRS-Univ. Strasbourg, France

Y. Spiegel and F. Torregrosa
IBS, Peynier, France

T. Michel
IBS, Peynier, France
Université d'Aix-Marseille, Institut Fresnel, Marseille, France

C. Belouet and F. De Moro
Solarforce, Bourgoin-Jallieu, France

Johann Summhammera and Zahra Halavani
TU Wien, Institute of Atomic and Subatomic Physics, Solar Cells Group, Stadionallee 2, 1020 Vienna, Austria

S.N. Abolmasov, H.Woo, R. Planques, P. Roca i Cabarrocas and E.V. Johnson
LPICM, CNRS, École Polytechnique, 91128 Palaiseau, France

J. Holovský and A. Purkrt
Institute of Physics, ASCR v.v.i. Cukrovarnická, Prague, Czech Republic

W. Soppe
Energy Research Centre of the Netherlands (ECN), 5656 AE Eindhoven, The Netherlands

J. Krc
University of Ljubjana, 12 1000 Ljubjana, Slovenia

K. Leitner
Umicore Thin Film Products AG, 9496 Balzers, Liechtenstein

F.-J. Haug
Ecole Polytechnique Federale de Lausanne, 1015 Lausanne, Switzerland

M. Duchamp
Forschungszentrum J¨ulich, 52428 J¨ulich, Germany

G. Sanchez Plaza
Universidad Politechnica de Valencia, 46022 Valencia, Spain

Q.-K.Wang
Shanghai Jiaotong University, Shanghai, P.R. China

Index

www.ingramcontent.com/pod-product-compliance
Lightning Source LLC
Chambersburg PA
CBHW080630200326
41458CB00013B/4581